Sacramento-San Joaquin Delta Historical Ecology Investigation:
EXPLORING PATTERN AND PROCESS

PREPARED FOR THE CALIFORNIA DEPARTMENT OF FISH AND GAME
AND ECOSYSTEM RESTORATION PROGRAM
AUGUST 2012

SFEI
AOSOC

Prepared by:

San Francisco Estuary Institute-Aquatic Science Center

Alison Whipple

Robin Grossinger

Daniel Rankin[1]

Bronwen Stanford

Ruth Askevold

Additional contributions by: Carie Battistone[1], Erin Beller, Elise Brewster[2], Bronwyn Hogan[1], Gena Lasko[1], Amy Lyons[1], Maika Nicholson, Jenny Rempel, Melissa Runsten, Micha Salomon, Ciprian Simon[1], Chuck Striplen

[1] California Department of Fish and Game, Sacramento
[2] Brewster Design Arts

In Cooperation with and Funded by:
California Department of Fish and Game
Ecosystem Restoration Program

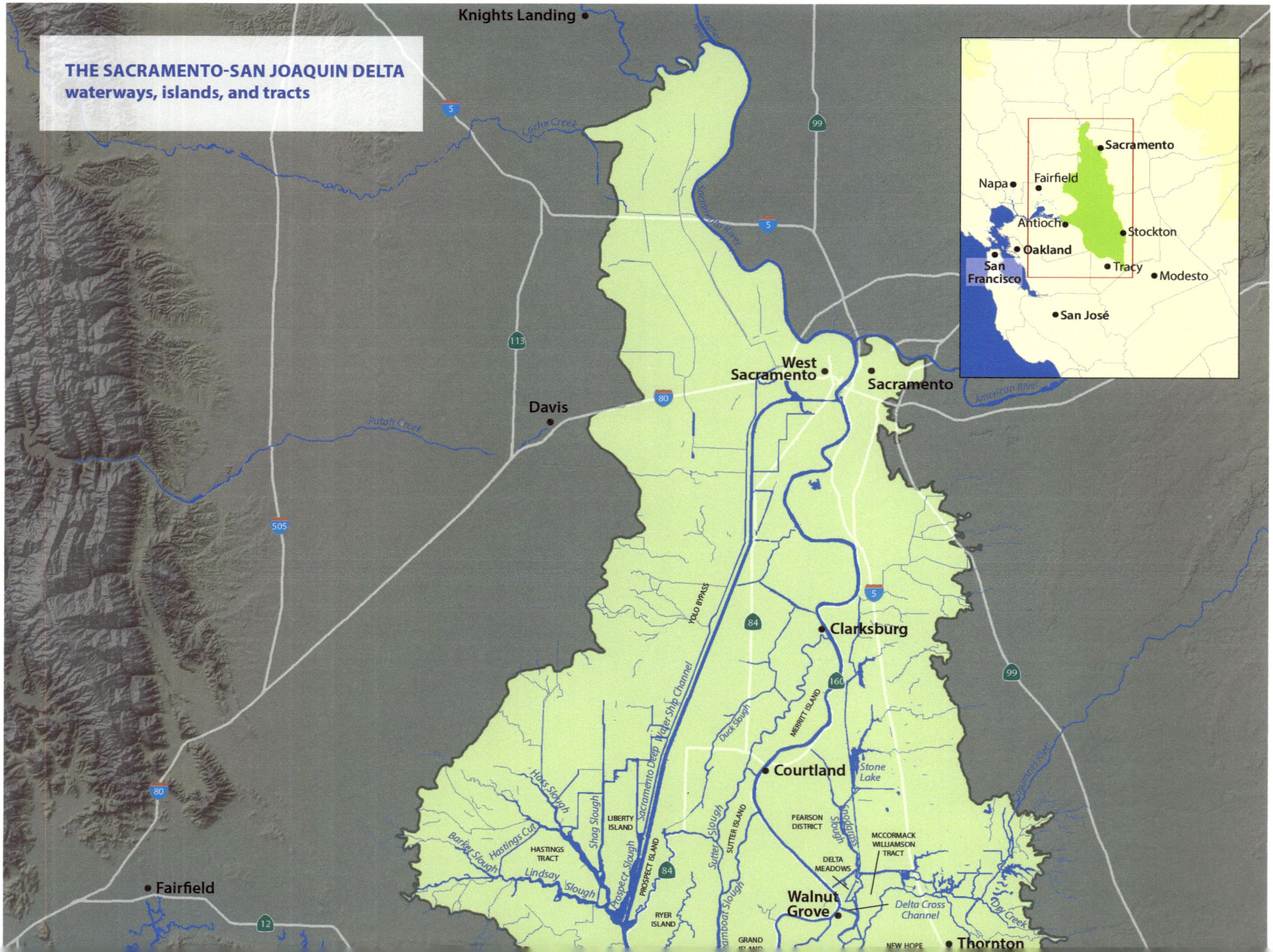

THE SACRAMENTO-SAN JOAQUIN DELTA
waterways, islands, and tracts

Sacramento

Stockton

Modesto

Tracy

San José

Fairfield

Antioch

Oakland

Napa

San Francisco

American River

99

Sacramento

Clarksburg

5

160

Courtland

Walnut Grove

Thornton

Dog Creek

Delta Cross Channel

NEW HOPE

McCORMACK WILLIAMSON TRACT

Stone Lake

Snodgrass Slough

DELTA MEADOWS

PEARSON DISTRICT

MERRITT ISLAND

99

West Sacramento

Sacramento River

5

84

Duck Slough

SUTTER ISLAND

Sutter Slough

Steamboat Slough

GRAND ISLAND

YOLO BYPASS

Sacramento Deep Water Ship Channel

PROSPECT ISLAND

84

RYER ISLAND

Prospect Slough

LIBERTY ISLAND

Shag Slough

Lindsey Slough

Putah Creek

80

Davis

HASTINGS TRACT

Hass Slough

Barker Slough

Hastings Cut

113

Knights Landing

Cache Creek

Putah Creek

5

505

80

12

Fairfield

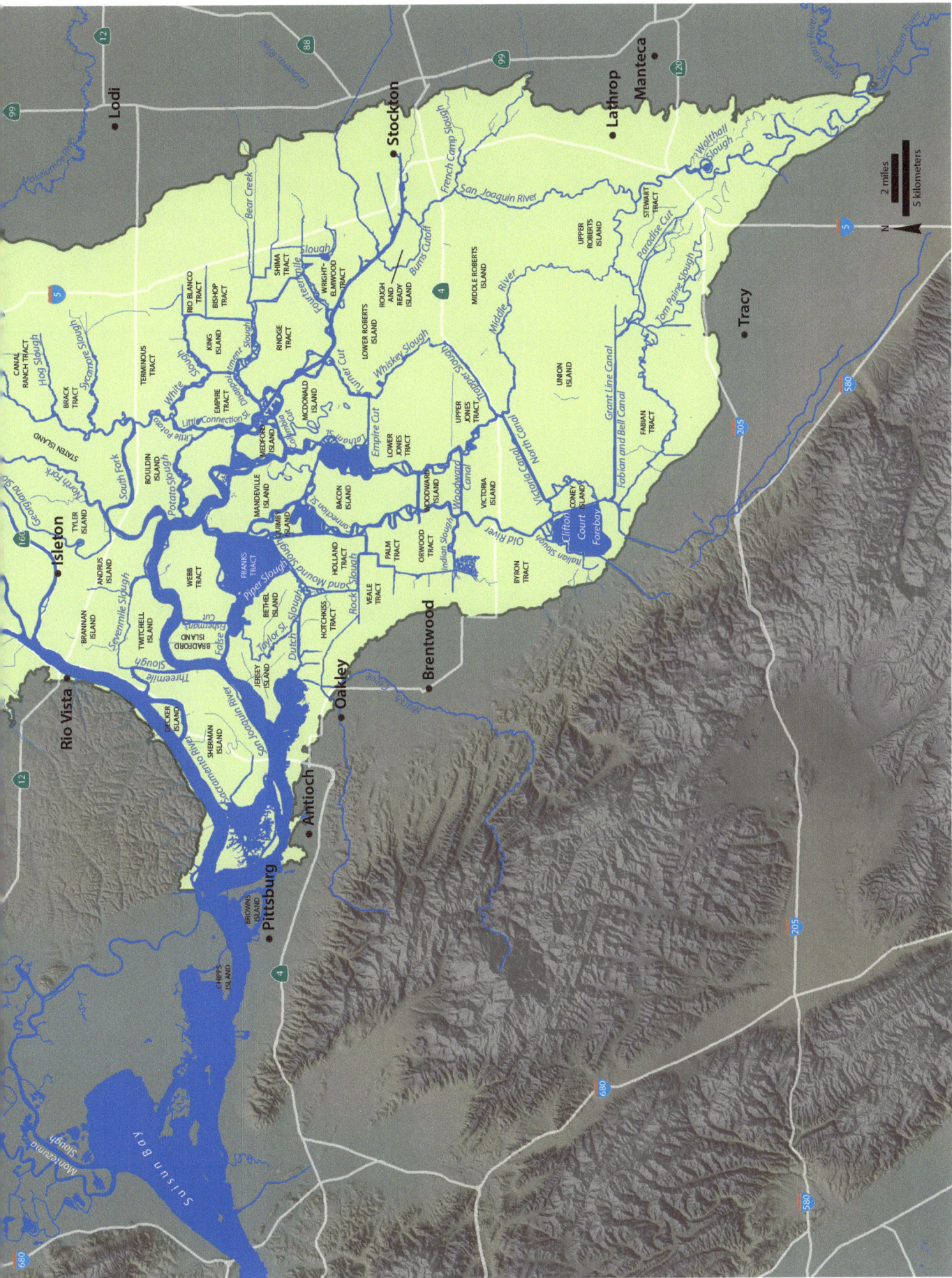

The Sacramento–San Joaquin Delta

Cities: Lodi, Stockton, Lathrop, Manteca, Tracy, Brentwood, Oakley, Antioch, Pittsburg, Rio Vista, Isleton

Islands and Tracts: Canal Ranch Tract, Brack Tract, Staten Island, Tyler Island, Andrus Island, Brannan Island, Twitchell Island, Bradford Island, Jersey Island, Bethel Island, Webb Tract, Franks Tract, Decker Island, Sherman Island, Browns Island, Chipps Island, Rio Blanco Tract, Bishop Tract, King Island, Terminus Tract, Rindge Tract, Empire Tract, Shima Tract, Wright-Elmwood Tract, McDonald Island, Medford Island, Mandeville Island, Venice Island, Quimby Island, Bouldin Island, Rough and Ready Island, Lower Roberts Island, Middle Roberts Island, Upper Roberts Island, Stewart Tract, Columbia, Bacon Island, Lower Jones Tract, Upper Jones Tract, Woodward Island, Victoria Island, Union Island, Fabian Tract, Byron Tract, Orwood Tract, Holland Tract, Palm Tract, Veale Tract, Hotchkiss Tract, Coney Island

Waterways: Mokelumne River, Cosumnes River, Calaveras River, Stanislaus River, San Joaquin River, Sacramento River, Bear Creek, French Camp Slough, Walthall Slough, Fourteenmile Slough, Disappointment Slough, White Slough, Sycamore Slough, Hog Slough, Potato Slough, Little Potato Slough, Connection Slough, Georgiana Slough, North Fork, South Fork, Sevenmile Slough, Threemile Slough, Whiskey Slough, Burns Cutoff, Empire Cut, Turner Cut, Trapper Slough, Middle River, Old River, Grant Line Canal, Fabian and Bell Canal, Victoria Canal, North Canal, Paradise Cut, Tom Paine Slough, Italian Slough, Clifton Court Forebay, Woodward Canal, Indian Slough, Rock Slough, Dutch Slough, Taylor Sl., False River, Fishermans Cut, Piper Slough, Mound Slough, Marsh Creek, Montezuma Slough, Suisun Bay, Stone Slough

Scale: 2 miles / 5 kilometers

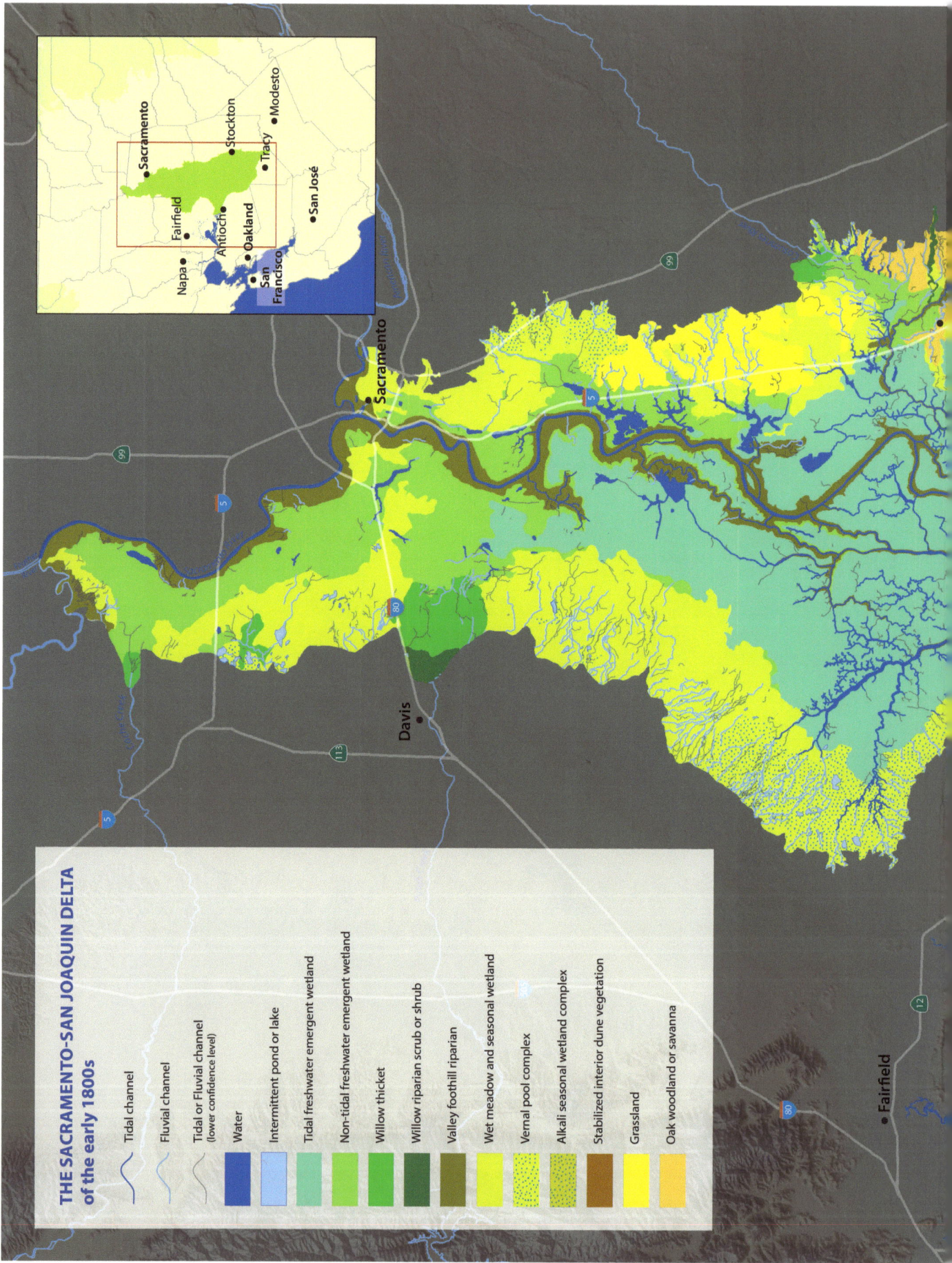

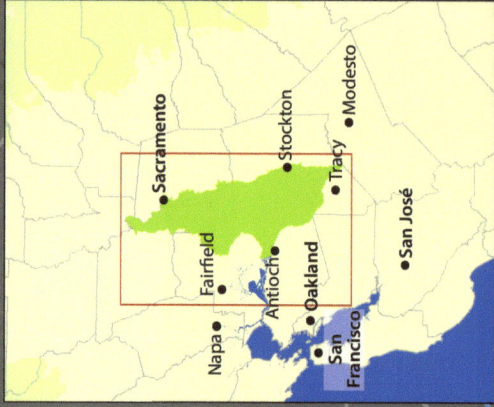

THE SACRAMENTO-SAN JOAQUIN DELTA
of the early 1800s

Tidal channel

Fluvial channel

Tidal or Fluvial channel
(lower confidence level)

Water

Intermittent pond or lake

Tidal freshwater emergent wetland

Non-tidal freshwater emergent wetland

Willow thicket

Willow riparian scrub or shrub

Valley foothill riparian

Wet meadow and seasonal wetland

Vernal pool complex

Alkali seasonal wetland complex

Stabilized interior dune vegetation

Grassland

Oak woodland or savanna

Sacramento

Stockton

Tracy

Modesto

Fairfield

Antioch

Oakland

Napa

San
Francisco

San José

Davis

Fairfield

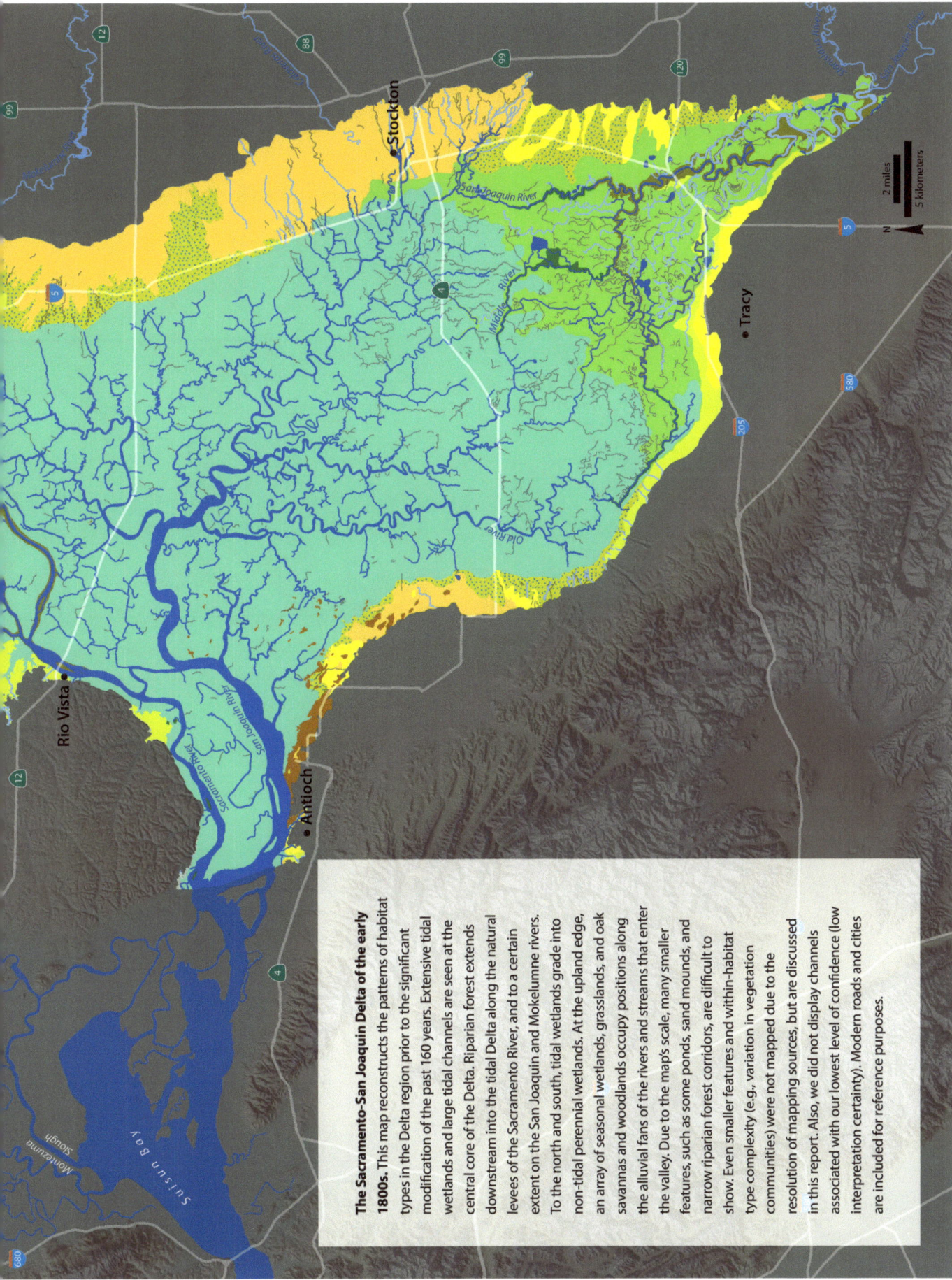

The Sacramento-San Joaquin Delta of the early 1800s. This map reconstructs the patterns of habitat types in the Delta region prior to the significant modification of the past 160 years. Extensive tidal wetlands and large tidal channels are seen at the central core of the Delta. Riparian forest extends downstream into the tidal Delta along the natural levees of the Sacramento River, and to a certain extent on the San Joaquin and Mokelumne rivers. To the north and south, tidal wetlands grade into non-tidal perennial wetlands. At the upland edge, an array of seasonal wetlands, grasslands, and oak savannas and woodlands occupy positions along the alluvial fans of the rivers and streams that enter the valley. Due to the map's scale, many smaller features, such as some ponds, sand mounds, and narrow riparian forest corridors, are difficult to show. Even smaller features and within-habitat type complexity (e.g., variation in vegetation communities) were not mapped due to the resolution of mapping sources, but are discussed in this report. Also, we did not display channels associated with our lowest level of confidence (low interpretation certainty). Modern roads and cities are included for reference purposes.

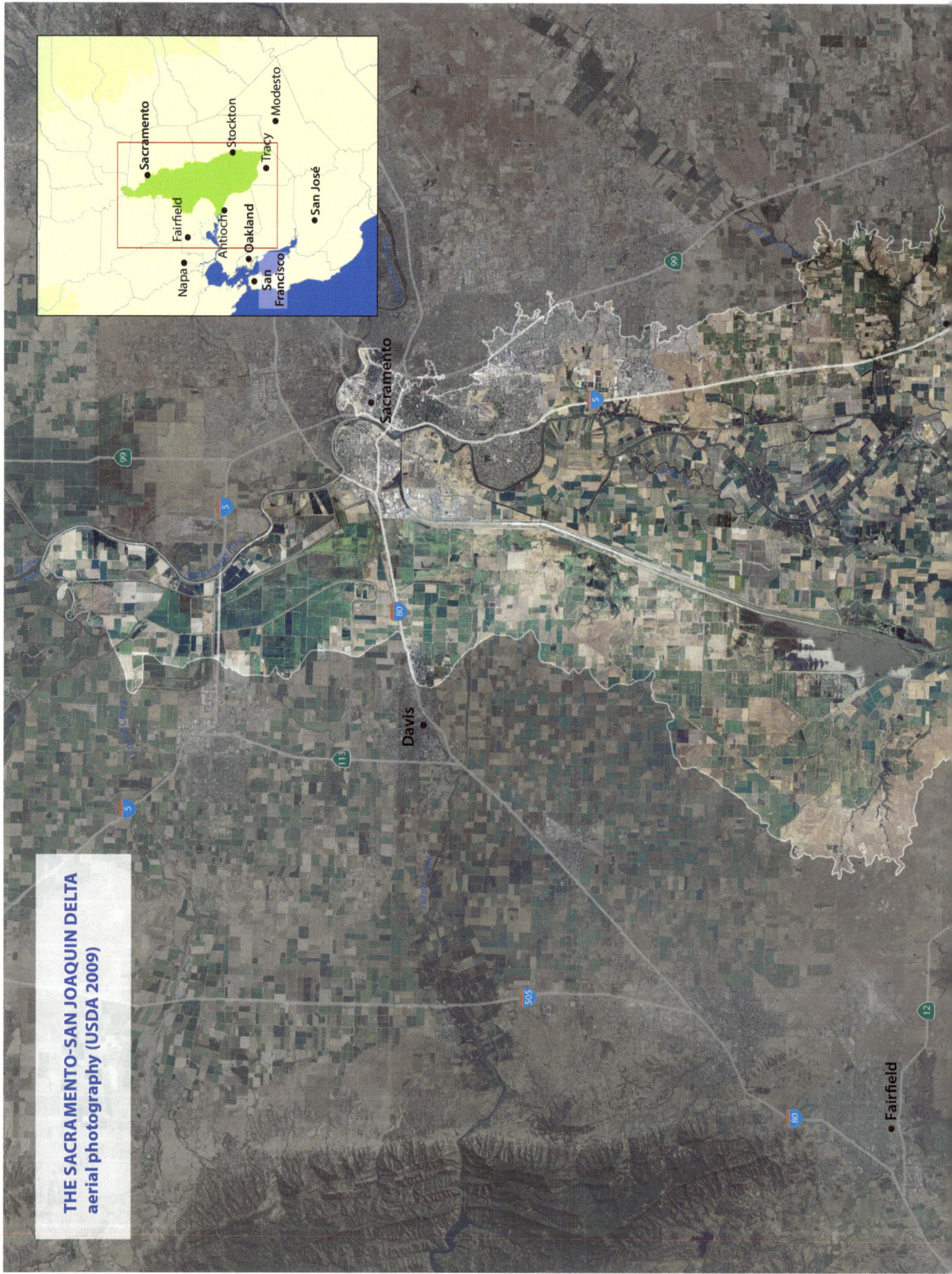

THE SACRAMENTO-SAN JOAQUIN DELTA
aerial photography (USDA 2009)

Sacramento

Stockton

Modesto

Tracy

San José

Fairfield

Antioch

Oakland

Napa

San Francisco

Sacramento

Davis

Fairfield

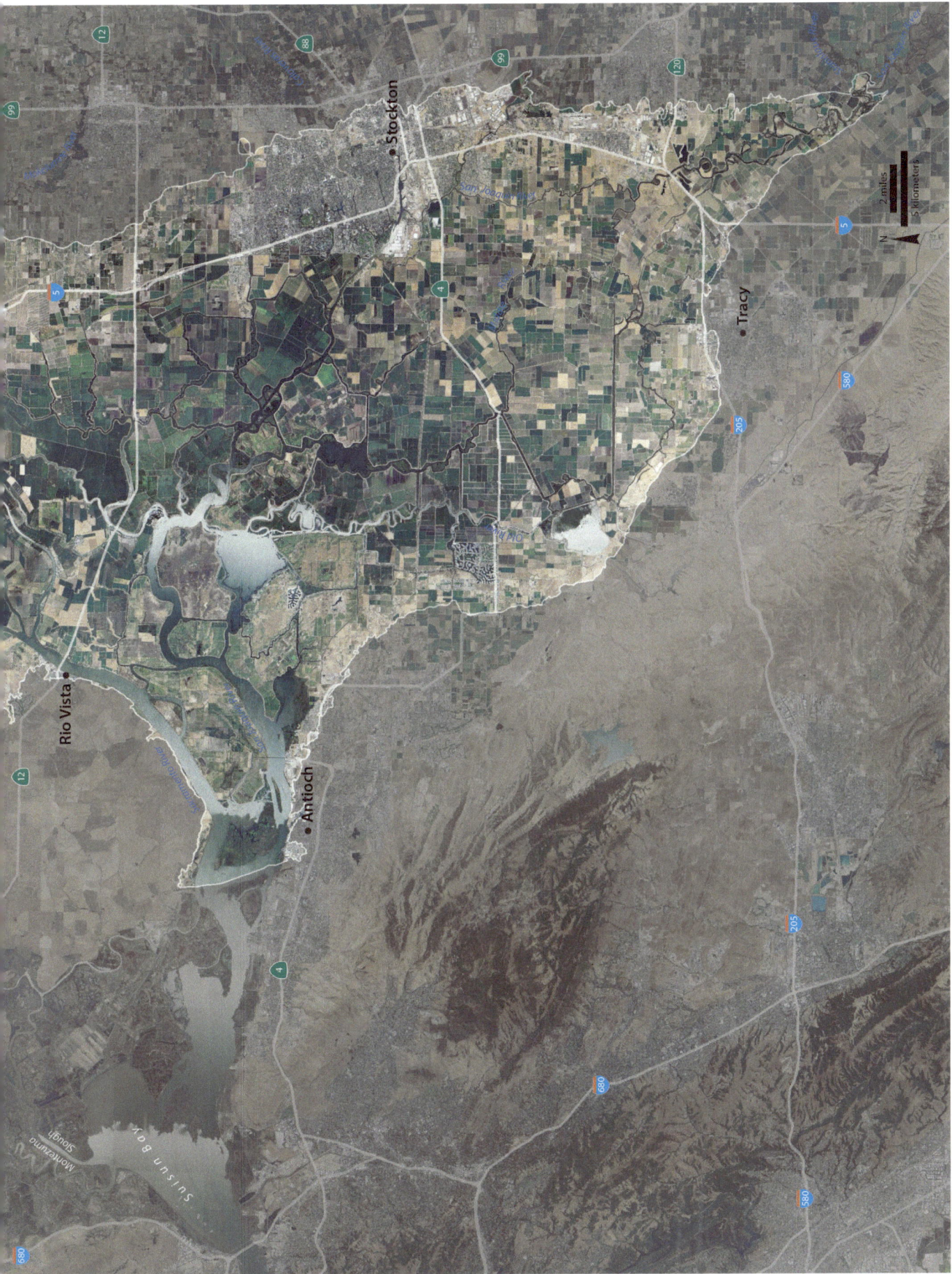

Stockton

Tracy

Rio Vista

Antioch

San Joaquin River

Suisun Bay

Montezuma Slough

2 miles

5 kilometers

N

12

88

99

120

99

5

4

4

12

4

205

580

5

580

205

680

680

Suggested Citation:

Whipple AA, Grossinger RM, Rankin D, Stanford B, Askevold RA. 2012.
Sacramento-San Joaquin Delta Historical Ecology Investigation: Exploring
Pattern and Process. Prepared for the California Department of Fish and Game
and Ecosystem Restoration Program. A Report of SFEI-ASC's Historical Ecology
Program, SFEI-ASC Publication #672, San Francisco Estuary Institute-Aquatic
Science Center, Richmond, CA.

Report and GIS layers are available on SFEI's website, at www.sfei.org/
DeltaHEStudy.

Front cover, from left to right and top to bottom: land grant map (Von Schmidt
1859, courtesy of The Bancroft Library, UC Berkeley), sailboat on waterway
(photo by Gilbert 1905, courtesy of the USGS Photographic Library), riparian
forest (ca. 1910, courtesy of the California History Room, California State
Library, Sacramento), cutting tule (Tule 1916, Holland Land Co., D-118,
courtesy of Special Collections, University of California Library, Davis).

CONTENTS

FIGURES

TABLES

BOXED TEXT

EXECUTIVE SUMMARY

The Sacramento-San Joaquin Delta has been transformed from the largest wetland system on the Pacific Coast of the United States to highly productive farmland and other uses embodying California's water struggles. The Delta comprises the upper extent of the San Francisco Estuary and connects two-thirds of California via the watersheds that feed into it. It is central to the larger California landscape and associated ecosystems, which will continue to experience substantial modification in the future due to climate change and continued land and water use changes. Yet this vital ecological and economic link for California and the world has been altered to the extent that it is no longer able to support needed ecological functions. Approximately 3% of the Delta's historical wetland extent remains wetland today; the Delta is now crisscrossed with agricultural ditches replacing the over 1,000 miles of branching tidal channels.

Imagining a healthy Delta ecosystem in the future and taking bold, concrete steps toward that future requires an understanding and vision of what a healthy ecosystem looks like. For a place as extensive, unique, and modified as the Delta, valuable knowledge can be acquired through the study of the past, investigating the Delta as it existed just prior to the substantial human modifications of the last 160 years. Though the Delta is irrevocably altered, this does not mean that the past is irrelevant. Underlying geologic and hydrologic processes still influence the landscape, and native species still ply the waters, soar through the air, and move across the land. Significant opportunities are available to strategically reconnect landscape components in ways that support ecosystem resilience to both present and future stressors.

Project objectives

The Sacramento-San Joaquin Delta Historical Ecology Study was conducted to provide foundational information needed to develop sound large-scale restoration efforts in the Delta. This research has been performed at the request of the California Department of Fish and Game (CDFG) and the Ecosystem Restoration Program (ERP). This report and accompanying Geographic Information System (GIS) document early 1800s pattern and process in the Delta. Historical habitat type extent and distribution are described, the landscape context explored, and driving hydrological and other physical processes examined. To do this, we synthesized thousands of historical cartographic, textual, photographic, and artistic materials to interpret and reconstruct the historical Delta. Information was compiled into a GIS, where sources and features could be compared across space and time. With this and other information, we mapped historical habitat types, including tidal freshwater emergent wetland, tidal channels, ponds and lakes, seasonal wetlands, and riparian forest.

This report complements the mapping with additional details, context, and analysis of the Delta's historical landscapes. The report is organized into six chapters. Chapter 1 provides an overview of the project, establishes the environmental setting, and outlines the land use history. Methods used to conduct the research and map the historical Delta are discussed in detail in Chapter 2. The regional summary in Chapter 3 is an important part of the report, in which overall results from the GIS are discussed, the past habitat types' extent and distribution is compared to the present-day Delta, the three primary landscapes of the Delta are introduced, and a summarized section on the primary findings of the study is presented. Chapters 4-6 document the historical characteristics of the central, north, and south Delta, respectively.

Delta landscapes

Central to developing landscape-scale restoration strategies is understanding not just extent, distribution, and characteristics of habitat types, but how components fit together across the physical gradients to form functional landscapes that offer ecological benefits greater than the sum of the parts. The conditions at every scale are the result of interactions among climate, geology (including hydrology), and land use. These interactions control the quantity, distribution, and quality of water, sediment, and vegetation, which in turn control form, structure, and function. Probably the most significant underlying physical gradient is the tidal to fluvial gradient expressed across the Delta. Others include salinity at the Delta mouth, temperature (including maritime influences), edaphic, geologic, hydrologic, and topographic gradients. Looking to the past illustrates the influence of underlying landforms, providing insight into how the future Delta may look and adapt along those physical gradients.

The historical reconstruction in this study revealed large-scale patterns that existed within the Delta. We describe three primary Delta landscapes: the central Delta, where a freshwater tidal wetland was interwoven with myriad tidal channels; the north Delta, with flood basins lying parallel to the riparian forests of the Sacramento River and its distributaries; and the south Delta, where branching distributary networks supported a broad floodplain that gradually merged with tidal wetlands (Fig. i).

The central Delta tidal islands landscape consisted primarily of tidal freshwater emergent wetland, supporting a matrix of tule, willows, and other species. These wetlands were tidally influenced, being

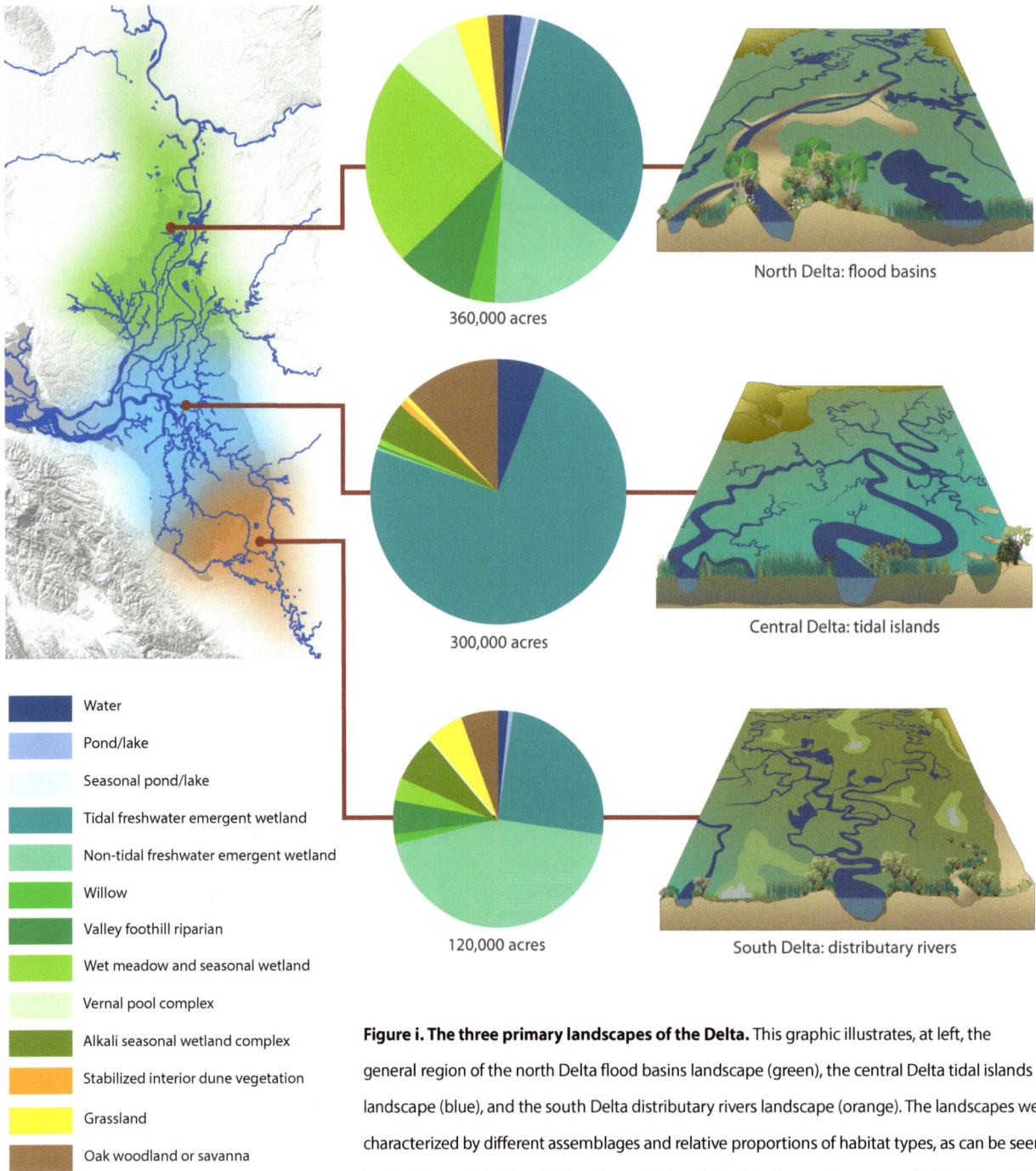

xxii

Legend:
- Water
- Pond/lake
- Seasonal pond/lake
- Tidal freshwater emergent wetland
- Non-tidal freshwater emergent wetland
- Willow
- Valley foothill riparian
- Wet meadow and seasonal wetland
- Vernal pool complex
- Alkali seasonal wetland complex
- Stabilized interior dune vegetation
- Grassland
- Oak woodland or savanna

360,000 acres

North Delta: flood basins

300,000 acres

Central Delta: tidal islands

120,000 acres

South Delta: distributary rivers

Figure i. The three primary landscapes of the Delta. This graphic illustrates, at left, the general region of the north Delta flood basins landscape (green), the central Delta tidal islands landscape (blue), and the south Delta distributary rivers landscape (orange). The landscapes were characterized by different assemblages and relative proportions of habitat types, as can be seen in the pie graphs in the middle column. Although the landscapes share many habitat types, the way they were arranged along the differing Delta landforms was distinct. Habitat characteristics also differed between landscapes. For example, channels were more sinuous in the central Delta, ponds and lakes were generally smaller and more connected to major river channels in the south Delta, and natural levees were large and hosted a wide and complex riparian forest in the north Delta. Conceptual diagrams illustrating these landscapes are shown in the third column.

wetted daily and inundated by the monthly spring tides, if not more frequently. Topographic relief was slight, with the marsh plain approximating high tide levels. During high river stages in the wet season, entire islands were often submerged with several feet of water. Large tidal sloughs with low banks intersected to form islands. Like capillaries, numerous small branching tidal channels wove through the wetlands, bringing the tides onto the wetland plain. Channel density and sinuosity in the central Delta were greater than in the less tidally dominated northern and southern parts of the Delta (but lower than the brackish and saline marshes of the estuary downstream). Distinctive to the western central Delta, sand mounds rose like islands above the wetland plain, providing dry land in an otherwise wet landscape. Alkali seasonal wetlands, grassland, oak savannas, and oak woodlands could be found at the upland transitions.

The flood basins of the north Delta lay parallel to the rivers, accommodating large-magnitude floods occurring regularly on the Sacramento River and other streams that discharged their annual flows at the basin margins. Inundation could persist for several months. The north Delta flood basins contained broad zones of non-tidal freshwater emergent wetland relatively free of channel, which graded into tidal freshwater emergent wetland. Dense stands of tules over ten feet (3 m) tall grew in these basins. Large lakes occupied the lowest and most isolated positions within the expansive wetlands, and few tidal channels penetrated far into the dense emergent vegetation. Some areas within tidal elevations may have been seasonally isolated from the tides due to supra-tidal natural levees along the rivers. The adjoining natural levees were covered by a dense multi-layered riparian forest, usually between a half a mile to a mile (0.8-1.6 km) in width. The upland margin was lined primarily by seasonal wetlands. Also at the upland margin of the north Delta, willow thickets could be found at the "sinks" (distributary networks) of larger creeks as they entered the flood basins.

The south Delta was shaped by the three distributary branches of the San Joaquin River. These branches produced numerous secondary overflow channels that serviced the floodplain, which broadened downstream and merged gradually with tidal wetlands. This complex network of distributary channels with associated levees of variable height intersected the fluvial-tidal transition zone, conveying floodwaters toward the tidal central Delta. Some parts of the main channels were prone to accumulating large woody debris, which likely obstructed flow. Ponds and lakes were generally smaller, less numerous, and more closely tied to the river than in the north Delta. A variety of habitat types were interspersed within the emergent wetland, including willow thickets, seasonal wetlands, and grasslands, as well as perennial and seasonal ponds and lakes. In comparison to the north Delta flood basin landscape, a greater portion of the natural levee riparian vegetation was composed of willows and other shrubs. Also, particularly in the most southern extent, the floodplain was occupied by willows and other trees as well as tule. Whereas wetlands and vernal pools made up a significant proportion of the upland edge at the Delta margin in the north Delta, alkali seasonal wetland complex, grassland, and oak woodland or savanna habitat types were found along the south Delta edge.

Application

To support the landscape-scale restoration currently taking shape in the Delta, the information provided in this report should be integrated with contemporary research, monitoring, and ecological theory in order to more explicitly link landscape pattern and process to ecological functions provided. Efforts are currently underway to address this through a study funded by the CDFG ERP, entitled "Management Tools for Landscape-Scale Restoration of Ecological Functions in the Delta." Tools such as conceptual models, restoration principles, and target metrics will help support the goals of current planning efforts – including the ERP, Bay Delta Conservation Plan, and the Delta Plan – to perform large-scale restoration of heterogeneous, interconnected habitats that support native species. Below, we summarize some of the findings and implications of this study for use in the next steps of adaptive management and restoration in the Delta.

Main points

- **A diverse array of habitat types was found within the historical Delta.** This included deep and broad sloughs, small dendritic tidal channels branching into the wetland plain, perennial and seasonal ponds and lakes at backwater locations, extensive freshwater emergent wetlands dominated by tule, willow-fern swamps within the tidal wetland complex, complex riparian forest with multiple vertical layers, willow thickets where upland drainages spread at the Delta's edge, a range of seasonal wetlands along the perennial wetland perimeter, stabilized interior dune vegetation including live oaks occupying the small but pronounced sand mounds of the western Delta, and grasslands, oak savannas and woodlands at the upland Delta margins.

- **The Delta consisted of multiple landscapes.** The central Delta's tidal freshwater wetlands of tule and willow, with its numerous winding channels, looked and functioned differently than the north Delta's broad flood basins, occupied by tule marsh and lakes and bordered by broad riparian forest on the natural levees of the Sacramento River and its distributaries. These landscapes, in turn, were different from the floodplain of the southerly San Joaquin River distributary branches, which was composed of tidal wetlands merging southward into a floodplain wetland interspersed with side channels, lakes and ponds, willows along channels, and patches of seasonal wetland.

- **Landscape-scale habitat patterns were a reflection of the Delta's broad physical gradients and landforms.** Patterns shifted depending on gradients, including tidal to fluvial influence (e.g., flood frequency, duration, magnitude, and extent), brackish to fresh water, low to high elevations, hot to cool temperatures, and peat to clay to loam soils. Landscape-scale patterns reflected the primary landforms of sub-tidal waterways intersecting Holocene peat deposits lying at tide elevation. Supra-tidal natural levees lined the rivers, and small sand mounds rose above the wetland plain. Peat deposits at the wetland edge overlapped the toes of alluvial fans along the Central Valley floor.

- **The historical landscapes exhibited gradual transition zones** between habitat types that allowed movement and adaption along physical gradients, in contrast to the sharp edges that exist today. The river and floodplain, as well as the north-south tidal to fluvial gradient, are largely disconnected today due to the leveeing of the main rivers, damming and filling of secondary channels, and reductions in flood flows. The loss of interconnected habitat mosaics, or increase in habitat fragmentation, limits habitat opportunities for species and the ability of the ecosystem to withstand physical and biological stressors.

- **The Delta is unique in its shape.** Characteristics such as the Delta's freshwater character, overall channel planform, and stability of features owe themselves, in part, to the fact that the channels of the Sacramento and San Joaquin rivers meet at the Delta's constricted mouth and flow into the highly enclosed San Francisco Bay, rather than directly into the Pacific Ocean.

- **Temporal variability was overlaid on a less changeable physical template.** Within the context of relatively stable landscape patterns, the Delta experienced droughts and deluge that generated great variability in environmental conditions.

- **Seasonal variation was expressed differently in different Delta landscapes.** While the influence of daily tides muted seasonal differences in flows and water availability within the central Delta, more seasonal variation was evident in the north and, particularly, south Delta.

- **A small percentage of the "natural" habitats within the Delta today is remnant of the former landscape.** The majority of the approximately 106,000 acres of natural habitat within the mutual area of the legal Delta and study area did not exist historically in their present locations. For example, seasonal wetlands are found where perennial wetlands once existed, and willow thickets on artificial levees are now present where tidal wetland edges once met water. The Delta has undergone an almost complete transformation, due to land use and water management.

- **Modern anthropogenic modifications occurred early in the Delta.** Changes due to leveeing, agriculture, ditching, clearing of riparian forests, grazing, and other impacts were evident in the 1850s. This affected how floodwaters moved through the Delta and substantially reduced the extent of perennial wetlands. Hydraulic mining debris impacted channel bed levels, among other effects. Most emergent wetlands of the central Delta were leveed and farmed by the 1880s. Habitats of native species thus were significantly altered or absent well over a century ago.

Management implications

- **Consider that native species were adapted to the patterns and processes of the past.** Developing functional landscape units reflective of historical patterns should improve chances of restoration success.

- **Recognize that restored habitats will not necessarily be the same as historical habitats, and will continue to evolve over time.** The many non-native species throughout the Delta, subsidence, climate change, and other large scale changes, will cause future habitats to have many differences from historical habitats, even if they provide function in similar ways.

- **Manage restoration to be reflective of current physical parameters and processes.** Historical habitat reconstruction does not provide a location-specific template for restoration. Instead, by better understanding how habitats reflect physical landforms and processes, more effective restoration can be created that is consistent with the physical gradients within the present-day and possible future. Consider options for managing physical processes to support more functional habitats and leverage restoration efforts by considering physical parameters.

- **Take advantage of physical gradients in the landscape and consider how these may shift in the future.** The Delta is part of the San Francisco Estuary, lying at the upper end of the estuarine continuum. With sea level rise over time, areas at the edge of tidal influence may be intertidal in the future; adequate room for estuarine transgression should be established along these gradients. Tidal wetlands and adjacent natural upland habitats can thus provide a buffer, supporting greater resilience to climate change. By designing landscapes to involve and be reflective of whole physical gradients, we are more likely to achieve a wider range of habitat characteristics that will provide opportunities for adaptation. This will support the continued evolution of plants and animals by maintaining populations at the limits of local habitat conditions.

- **Remove rigidity in the present Delta where possible.** The historical Delta was adapted to shifting conditions along broad gradients. Broad ecotones would better equip the ecosystem to handle the type of future changes expected in the Delta. With the sharp edges and discontinuities in the Delta today, there is little room for the natural adjustments that gave the historical Delta much of the resiliency that is missing in the contemporary system.

- **Recognize what large and interconnected habitats might mean.** The study of landscape patterns can help define these terms more concretely. For instance, supporting basin landscapes may only require one side of the Sacramento River, but requires adequate flood flows. Supporting San Joaquin floodplain processes at the tidal margin may involve allowing the river to meander on both sides of the channel.

- **Employ a landscape perspective and manage toward assemblages of connected habitats,** recognizing that an isolated restoration project will likely provide much less ecosystem benefit than a restoration of the same size and habitat type that is connected to multiple other habitat types. The ecological value of individual habitat types is magnified by their surrounding landscape. Given limited land and financial resources, these considerations are especially important. The landscape perspective helps target broad assemblages of ecological functions, as opposed to specific conditions required for individual species.

- **Promote habitat connection and disconnection in the appropriate places.** The ecological functions of many Delta habitats were provided through the connectivity of features (e.g., side channel habitat connected to riparian forest and backwater ponds and lakes). Improving understanding of historical conditions supports the developing consensus of the importance of floodplain habitat and its connections to riverine processes. At the same time, discontinuities were important (e.g., blind tidal channels, flood basin and river), increasing residence time and heterogeneity. Deciding where to increase and decrease connectivity must be done at a landscape scale and can be informed by conceptual models of the historical landscape.

- **Heterogeneous landscapes are less sensitive to extreme events.** The historical Delta provided a wide array of conditions; places of refuge could be found in times of flood and places with ample water could be found in the dry season.

- **Use Delta freshwater inflows to their greatest potential.** Historically, freshwater inflows encountered and influenced a much broader range of habitats than they do today. Questions about where water should go are valuable in addition to asking how much water is needed. Understanding the role of hydrology becomes more critical when addressing the current and future challenges related to climate change. Such challenges include potentially large floods unknown in recent times related to loss of Sierra Nevada snowpack and large storm events.

- **Different ecological functions can be provided by the same habitat types, depending on the position of those habitats within different landscapes.** In the historical Delta, driving physical processes and habitat connectivity meant that different functions were provided depending on a feature's location. For example, a large lake within a broad wetland flood basin served a different array of functions than a small pond along a side channel system created by woody debris in the river.

- **Recognize that every habitat or function cannot be supported everywhere.** Certain places will provide some functions better than others. Also, certain functions may not be possible, or, may be significantly limited in the contemporary or future Delta. Consider both altered physical conditions (e.g., hydrodynamics) to determine limitations and opportunities identified using the historical perspective. Think in terms of functional landscape units that provide different groups of functions.

- **Match functional targets to the appropriate scale of restoration.** Many desired Delta functions are likely scale dependent, requiring components of certain sizes. Restoration at scales smaller than landscape patterns and processes may not produce the desired characteristics. For example, restoring a functional tidal island may require a restored tidal wetland of sufficient size in order to support a blind tidal channel network. There is a risk that small restoration projects may not achieve desired characteristics. To avoid this pitfall, individual restoration projects should be embedded within a larger vision of a future functional Delta.

- **Think at the large-scale and in the long-term.** Attaining sustainable ecosystems will require reconnecting pattern and process at a landscape scale, in perhaps different places and scales than what occurred in the historical landscape. This should involve re-imagining functional landscapes in new places that leverage existing natural habitats and landforms. Long-range plans should be developed such that individual projects or transformations today can, in the future, become part of an interconnected and diverse complex of both natural and cultural elements that more successfully addresses ecological needs.

ACKNOWLEDGMENTS

This project was funded by the California Department of Fish and Game (CDFG) through the Ecosystem Restoration Program (ERP). We had the opportunity to work with extraordinary CDFG staff over the course of the project, including Carie Battistone, Daniel Burmester, Scott Cantrell, Gena Lasko, Bronwyn Hogan, Amy Lyons, Daniel Rankin, Ciprian Simon, Carl Wilcox, and Dave Zezulak. In particular, Daniel Rankin was deeply involved every step of the way and critical to the success of the project, offering his GIS expertise and witty remarks. We give special thanks to Carl Wilcox, who initiated the project and was critical to its success.

We would like to express our deep gratitude to the volunteers and staff at the regional archives and institutions we visited over the course of the project, including The Bancroft Library, the California Historical Society, California State Archives, California State Library, UC Davis Shields Library, University of the Pacific Library, and the Water Resources Center Archives. We thank Matt Fossum from the California State Lands Commission , who worked with us extensively to explore the CSLC's rich collection of historical maps. The California State Archives was a valuable resource of maps and early engineering documents and we thank Lynda for all her help viewing and acquiring these sources. We also thank the surveyors and recorders offices of Sacramento, San Joaquin, Solano, Yolo, and the Solano County Public Works. The UC Davis Shields Library, Earth Sciences & Map Library of UC Berkeley, and the Contra Costa County Public Works contributed early aerial photos. We also thank staff at the Bureau of Land Management for providing General Land Office surveys.

We are indebted to the staff and volunteers at local historical societies and archives, including the Bank of Stockton Archive, Center for Sacramento History, Contra Costa County Historical Society, Dutra Museum of Dredging, East Contra Costa Historical Society & Museum, Haggin Museum, Isleton Brannan-Andrus Historical Society, Reclamation District 999, Rio Vista Museum, Sacramento River Delta Historical Society, San Joaquin County Historical Society and Museum, Solano County Archives, West Sacramento Historical Society, and the Yolo County Archives. Without their efforts making the rich historical record available, our research would not be possible. We thank Rebecca Crowther and the Center for Sacramento History for their patience with our interest in a large volume of their material. We appreciate the extensive assistance of archivist Leigh Johnson of the San Joaquin County Historical Society and Museum.

We greatly appreciate the input on historical conditions and change from several Delta residents. Walter Hoppe provided a wealth of information with his extensive regional knowledge and archival research. Russell van Löben Sels and his mother, Pam van

Löben Sels, offered a rich history of their family's relationship to the Delta. We also spoke with Russell Parrott about his father's experience as a river boat pilot on the Sacramento. We would also like to thank Jeff Hart and his Delta Ecotours and John Herrick and Brett Baker for making their extensive collection of archival material available. Finally, thank you to the many other individuals we have spoken to along the way at talks and at local historical societies and archives. Your many questions and valuable input was critical to the success of the project.

We would like to also thank SFEI-ASC staff and colleagues. Josh Collins contributed his expertise over the course of the project and provided review on analysis and interpretation. Mike Connor offered significant support during the initiation of the project. We thank Kristen Cayce, Shira Bezalel, and Gregory Tseng for their technical assistance. We extend our thanks to Jon Christensen and Stanford's Bill Lane Center for the American West for the support of interns Melissa Runsten, Maika Nicholson, and Jenny Rempel.

The project benefited substantially from the sound advice, encouragement, and enthusiasm expressed by scientific reviewers and regional experts along the way: Brian Atwater (USGS), Matthew Booker (North Carolina State University), Brian Collins (University of Washington), Val Connor (State and Federal Contractors Water Agency), Cliff Dahm (University of New Mexico, former CALFED Lead Scientist), Joel Dudas (CDWR), Chris Enright (Delta Science Program), Bill Fleenor (UC Davis), David Hansen (U.S. Bureau of Reclamation), Jeff Hart (Hart Restoration), Mara Johnson (Tremaine and Associates, Inc.), Todd Keeler-Wolf (CDFG), Jay Lund (UC Davis), Jeff Mount (UC Davis), Peter Moyle (UC Davis), Anke Mueller-Solger (Interagency Ecological Program), Jim O'Connor (USGS), Stuart Siegel (Wetlands and Water Resources), John Thompson (University of Illinois, emeritus), and Leo Winternitz (The Nature Conservancy).

We would like to thank our technical review team for providing input to the report draft and GIS. Reviewers were Brian Atwater, Peter Baye, Daniel Burmester, Brian Collins, Cliff Dahm, Chris Enright, Todd Keeler-Wolf, Peter Moyle, Anke Mueller-Solger, Si Simenstad, John Thompson, and Peter Vorster.

We appreciate the opportunity to work with and display the artwork of Laura Cunningham. We enjoyed working with Laura, John Hart, and David Loeb to develop these images, which were published in *Bay Nature Magazine* in 2010. Also, thank you to those whose photography is used in the report: Christopher Bronny, Daniel Burmester, Carolyn Cole, Mark Hoshovsky, Bill Miller, Jean Pawek, and Oren Pollak.

1. Overview

INTRODUCTION

This report brings together a broad range of historical data to document land cover patterns, habitat characteristics, and hydrogeomorphic conditions in the Sacramento-San Joaquin Delta during the early 1800s, prior to significant modern modification. The report and the associated geodatabase offer foundational information to inform understanding of landscape pattern and function and their relationship to governing physical processes.

Relatively little is known about the Delta ecosystem as it existed historically. The Delta has undergone dramatic change over the last 160 years, rendering its early nature virtually unrecognizable. Many fundamental alterations occurred within the first few decades after the Gold Rush of 1848. Rivers were leveed, wetlands drained, tidal sloughs dammed, riparian forests cut, and flows altered. Today, the many layers of change and unintended consequences and long-lasting repercussions of actions make it challenging to comprehend the natural ecosystem form, process and function. It is broadly recognized that the modern Delta is failing as an ecosystem; undestanding how it tended to look and work in the absence of recent modifications is essential to plan a future Delta that thrives with nature's support.

Historical ecology provides an avenue to examine the characteristics of the once highly productive and complex Delta ecosystem and to facilitate understanding of the current Delta (Fig. 1.1). The Delta was flexible and resilient; it was buffered against dramatic perturbations. It was a diverse place supporting a wide range of local and migratory species. Understanding the natural characteristics of the Delta necessitates the synthesis of diverse historical information, a process of piecing together the story of what the Delta looked like and how it functioned before European contact.

Historical ecology informs decisions about what habitat types might be desired and where they might be best supported by physical controls. Historical ecology does not offer a specific template from which to recreate the past; ecosystems don't run backward and the past cannot be reached. Nor is its purpose to despair over what has been lost. Instead, study of historical landscapes provides clues for how to foster future functional landscapes that promote ecosystem health and resilience as the controls change. The future Delta must accommodate climate change, sea level rise, and changes in land and water use.

We focused on mapping and describing conditions in the early 1800s, just prior to significant anthropogenic change. This does not represent an ideal condition nor a time to return to; rather, it is the most recent period for which it is possible to study natural process and function in detail

Figure 1.1. Delta wetlands. Top, Consumnes River wetlands and riparian forest. Bottom, Sycamore Slough (south of Woodbridge Road). Both November 8, 2011. (photos by William G. Miller, Cole~Miller Photography)

and under a roughly similar climate as today. This endeavor is feasible because of the extensive historical maps, texts, and images available. This project drew upon data from many time periods, but the early- and mid-nineteenth century sources were emphasized.

The modern Delta is an intensely studied system and many have described various aspects of the Delta as it existed historically. John Thompson's (1957) dissertation offers perhaps the best glimpse into the early settlement period in the Delta and the changes wrought to the system as a result. Brian Atwater's (1982) geologic mapping of the Delta provides valuable, detailed information concerning historical hydrography, tidal wetland extent, and primary landforms. Other studies by Atwater (e.g., Atwater and Hedel 1976, Atwater et al. 1979, Atwater and Belknap 1980) provide additional foundational knowledge of the historical Delta and its geologic history. A succinct description of Delta historical ecology within the context of its broader watershed is found in The Bay Institute's *Sierra to the Sea* (TBI 1998).

The Delta Historical Ecology Study built upon these and other prior works, seeking to draw from a broader range of historical sources (e.g., early 1937 aerial photography, land grant testimony, General Land Office survey data) and to bring a landscape ecology perspective and a focus on historical habitat patterns to the fore. The study is also distinguished by its extensive mapping and documentation of habitat types (representing land cover types, rather than habitat for a particular species). The report is detailed in its presentation of material, in order to support the overall study findings. It did not involve an extensive land and water use history or a chronology of change through time, though such information aided interpretation of historical sources (e.g., Thompson 1957, Fox 1987a, Kelley 1989).

The primary objective of this research is to describe the habitat patterns and hydrogeomorphic characteristics of the early 1800s Delta. The goal is that this information will inform large-scale restoration planning that fosters the development of more functional landscapes in the future. The information presented in this report and the associated geodatabase are intended to improve knowledge of how Delta habitat components, habitat mosaics, landscapes, and physical and biological processes interact (Hobbs 1996, Bell et al. 1997, Collins et al. 2003, Simenstad et al. 2006, Beechie et al. 2010, Greiner 2010). The study seeks to address the fundamental questions of what the Delta looked like and lend insight into how the Delta functioned: how species accessed and utilized the Delta's range of habitat types, and how the system varied along major physical environmental gradients. Such information is essential to developing appropriate ecological and hydrological restoration strategies. Historical ecology can advance scientific understanding about how ecosystems work and the conditions to which native species were adapted; it can also inform how we view the Delta, how we see it change, how it might provide a broad range of functions at the landscape level in the future, and how actions might be prioritized.

To reiterate, the historical picture is not a restoration template, but information that, along with contemporary research, generates greater understanding of ecological process and function. This understanding gives rise to a landscape perspective in planning. In order to apply this landscape perspective, it must be translated into quantitative habitat metrics, based on sound conceptual models and restoration principles. Efforts are now underway to address this through a study entitled "Management Tools for Landscape-Scale Restoration of Ecological Function in the Delta," funded by the California Department of Fish and Game (CDFG) through the Ecosystem Restoration Program (ERP; Fig. 1.2).

This project was prepared for, funded by, and conducted in collaboration with CDFG and ERP. This program is developing approaches to restoring large areas of interconnected habitats and rehabilitating natural processes and ecological functions (CDFG 2010). Much of what we discuss in the report relates to forming a basis for what "large" and "interconnected" means. This project also benefits other visioning processes that are ongoing in the Delta, including the Delta Plan and the Bay Delta Conservation Plan. This report and the accompanying geodatabase provide baseline information for developing more intimate knowledge of the richness of pattern and process once expressed in the Delta. It broadens perspective and helps build the big picture vision for the future Delta, which might necessarily be very different from the past and present.

Figure 1.2. Historical ecology context. This diagram depicts the context of historical ecology within environmental planning and management. Historical data and study of the physical landforms of an area provide the information needed to understand the historical ecology of an area (blue box). This baseline information can then be used to support interpretation of ecological functions (green box) and development of landscape planning and management strategies (yellow box; e.g., conceptual models, restoration principles, and target metrics). These steps will be furthered in subsequent studies. Importantly, the larger context of this study is its informing a collective vision of the future Delta (pink box).

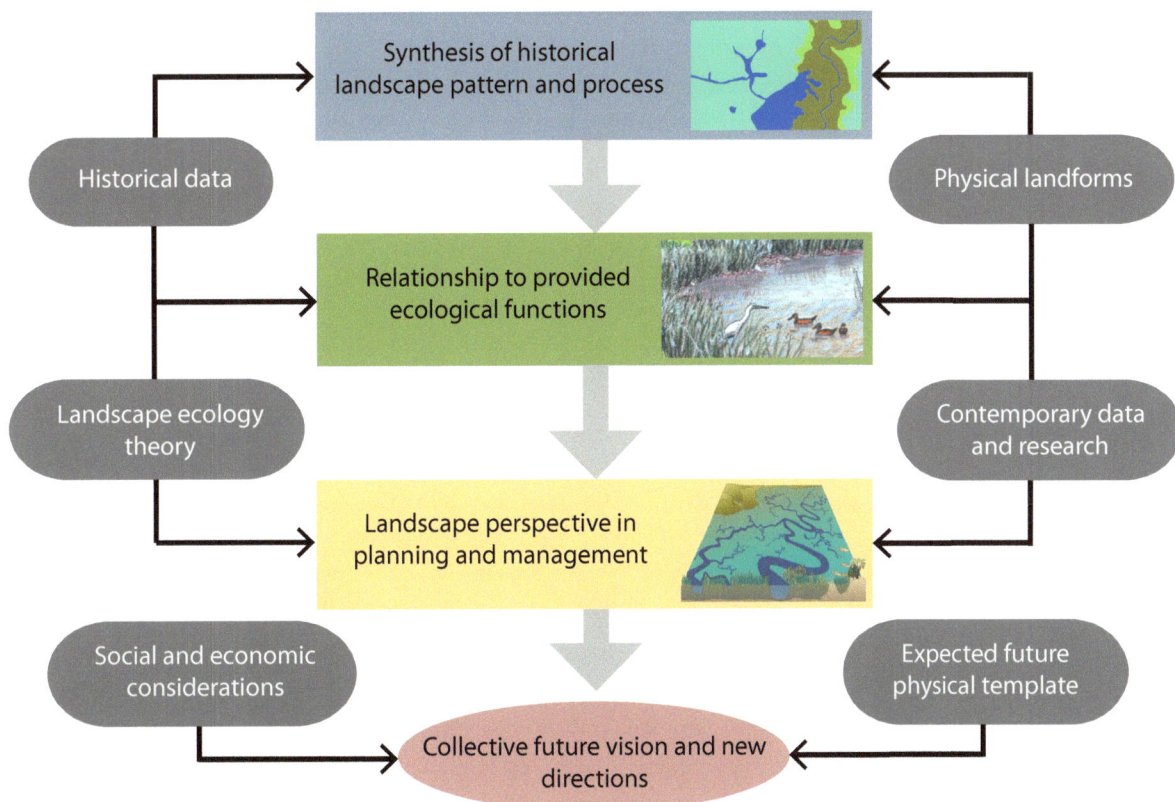

Study area

The Delta is where the downstream extents of the Sacramento and San Joaquin rivers meet the tides. It comprises the uppermost portion of the San Francisco Estuary, which is the largest estuary on the Pacific Coast of the United States. Drainage from approximately 40% of California historically flowed through the Delta, which lies at the heart of the Central Valley. It is characterized by a Mediterranean climate, with hot, dry summers and cool, wet winters. The area is hotter and drier to the south than the north. Average annual precipitation is 13-14 inches (330-356 mm) to the south near the Stanislaus River confluence and 19-20 inches (483-508 mm) to the north above the American River (Faunt 2009). The San Joaquin River flows (unimpaired) are 6.2 million acre-feet (7.6 billion m^3) annually compared to the Sacramento River flows of 21.6 million acre-feet (22.7 billion m^3).

Today, more than 400,000 people reside in the Delta region. The state's capitol, Sacramento, and other cities including Stockton, Tracy, Antioch, Rio Vista, and Davis are positioned along the Delta margins. Many communities, including Isleton, Walnut Grove, Courtland, and Clarksburg, occupy the Delta's river banks. It is an area of intensive agriculture, with over 470,000 acres (190,200 ha) of farmland. The Delta also lies at the core of the California water supply system that directs drinking and irrigation water from the north to users in Southern California, the San Joaquin Valley, and the Bay Area.

The Delta means different things to different people. For some it is the area that falls within the political legal Delta boundary (as defined by Water Code § 12220). For others it is the upper San Francisco Estuary defined by the maximum influence of tides. For others it may be a unique combination of social, political, and ecological factors that distinguish the Delta from other regions. Political boundaries aside, most agree that the Delta historically encompassed close to 400,000 acres (161,900 ha) of tidal wetlands and waterways, with the northern limit near the City of Sacramento and the southern limit just south of Old River (Gilbert 1917, Cosby 1941, Thompson 1957, Atwater et al. 1979).

In order for the study to include the extent of the Delta's historical tidal wetlands, adjacent non-tidal freshwater wetlands, plus upland transitional areas, we defined the study area as the contiguous lands lying below 25 feet (7.6 m) in elevation. This encompasses an area of about 800,000 acres, including parts of Sacramento, Yolo, Solano, Contra Costa, and San Joaquin counties (Fig. 1.3). The boundary was defined using the National Elevation Dataset (NED) 10m-Resolution (⅓-Arc-Second) Digital Elevation Model (DEM). We used GIS tools to generalize the boundary and removed upland (fluvial) channels less than 650 feet (200 m) wide. To avoid holes in the study area, we included small hillocks within the the outer boundary. We also included areas within the sinks of Putah and Cache creeks that were above the 25 foot (7.6 m) contour.

The western boundary of the study area was established at the west end of Sherman Island in order to match the historical ecology mapping previously

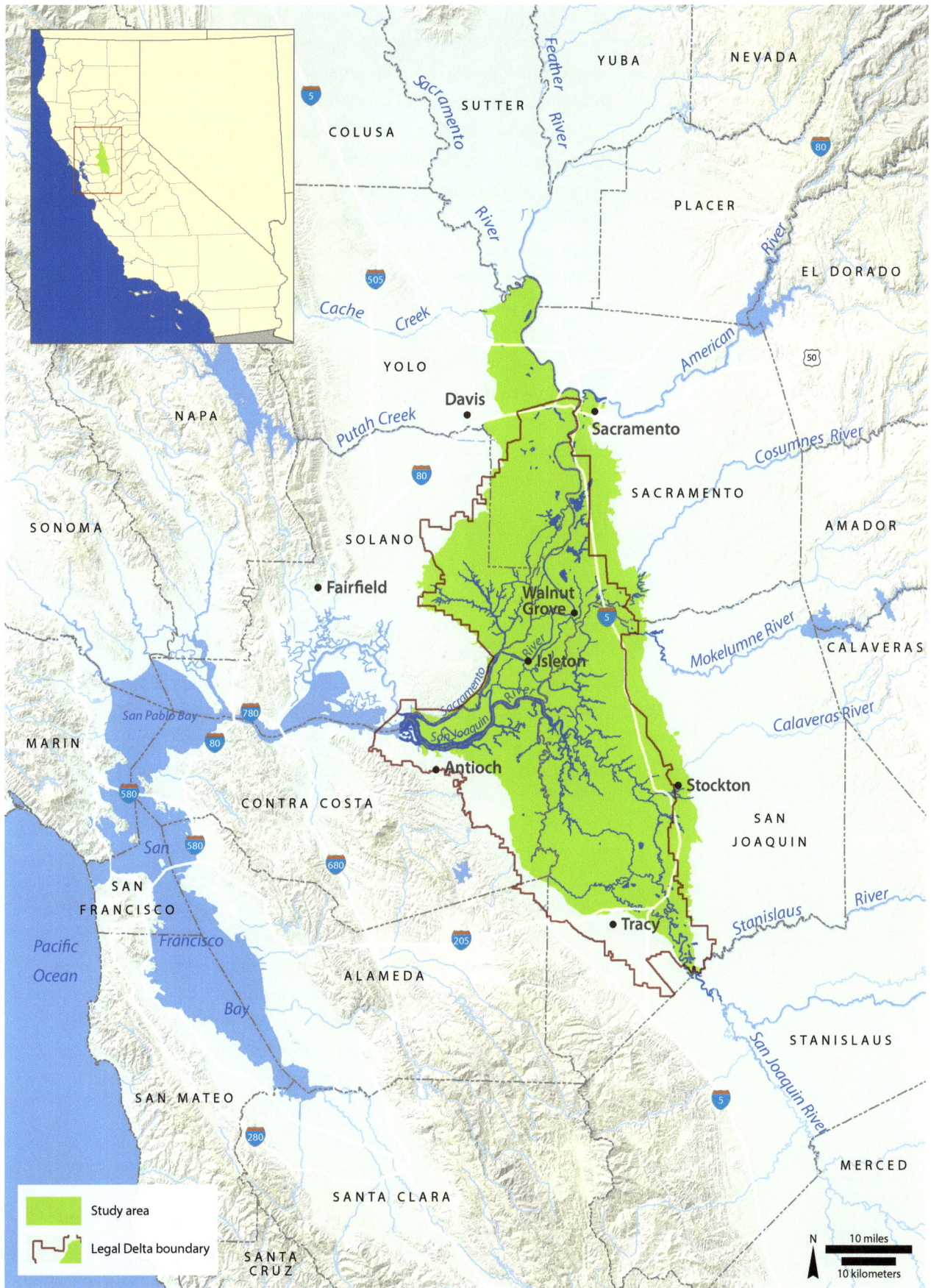

Figure 1.3. Study area and regional geographic context. The project area (green) is about 800,000 acres, including parts of Sacramento, Yolo, Solano, Contra Costa, and San Joaquin counties. The legal Delta boundary is shown in red. Historical water bodies are shown within the study area and modern outside.

completed for the Bay Area EcoAtlas and Baylands Ecosystem Habitat Goals Project (Goals Project 1999). The upstream extent of the study area falls at hydrogeomorphically logical locations. On the west side of the Sacramento River, the study area extends northward in the Yolo Basin to Knights Landing Ridge, also near where the Feather River enters the Sacramento River. Historically, this point on the river marked a dramatic shift in the character of the river (Wilkes 1849). It was the location where, as one descended the river, the Sacramento ceased to meander with tight bends and became a relatively fixed channel "consisting of a series of smooth, large bends in no way suggestive of ordinary meanders" (Bryan 1923). We did not include the American Basin on the east side of the Sacramento River between the American and Feather rivers as it was completely non-tidal and extended well above the 25 foot (7.6 m) contour. The southern extent of the study area also marks a morphologically significant location at the confluence of the San Joaquin River with the Stanislaus (Edminster 2002). Just downstream of this confluence, the San Joaquin divides into its three main distributary branches that are defining features of the Delta.

Report structure

This introduction to the study is followed by a background section, and Chapter Two describes the research methods. Chapter Three provides a brief description of the historical Delta, presents overall summaries of the historical habitat type mapping, introduces the three primary landscapes of the central, north, and south Delta by which the rest of the report is organized, and summarizes important project findings. Chapters Four through Six offer detailed descriptions of historical land cover and hydrogeomorphic conditions in the central, north, and south Delta by drawing upon the rich historical dataset and project synthesis and analysis.

BACKGROUND

The environmental setting and land use history of an area are important context for understanding the transformation of landscapes through time. They also aid interpretation of historical sources, as they are drawn from a range of time periods that represent different climatic conditions (e.g., flood or drought) and land uses. They help address the challenges of parsing the many layers of changes that have occurred at various times within the Delta, some of which affect certain habitat types more than others or have counteracting effects. This section offers this general context; first by discussing the environmental setting and then by providing a brief land and water use timeline.

Environmental setting

The Sacramento-San Joaquin Delta is part of the upstream region of the San Francisco Estuary and naturally received runoff from approximately 40% of California (before changes eliminated Tulare Basin overflows). This runoff from the watershed originates from the mountainous rivers of the

Sierra Nevada, southern Klamath, and Coast Range and passes through the main tributaries and distributaries of the Sacramento and San Joaquin rivers. Historically, much of the runoff moved through the floodplains and freshwater marshes of the Central Valley and Delta before passing through Suisun Bay, the Carquinez Straits, and into San Francisco Bay. Over thousands of years, deltaic deposits have accreted and eroded and been reshaped by natural processes through tectonic, tidal, fluvial, eolian (wind-driven) processes, and climatic fluctuations. The Delta was a complex mosaic of major and minor habitat types due to gradients in controlling factors operating at multiple sides. Gradients in tidal energy and water salinity dominated from east to west. Gradients in temperature, rainfall, and river discharge dominated from north to wouth. Unique local conditions existed at intermediate positions along these major environmental gradients.

FORMATION AND EVOLUTION OF THE DELTA The Delta's shape and position is unique among deltas throughout the world (Mount 1995). Situated in the Central Valley and confined by the parallel Coast Range to the west and Sierra Nevada to the east, the Delta did not form like traditional coastal plain deltas, which are typified by alluvial deposits that broaden toward the ocean from one river. Instead, the Delta results from a convergence of multiple streams from the Sierra Nevada – namely the Sacramento, San Joaquin, and Mokelumne rivers – that spread into numerous distributary channels before meeting at a narrow passage just east of Suisun Bay. This causes the Delta to broaden landward (Atwater and Belknap 1980, Faunt 2009). The decreasing thickness of estuarine peat soils toward the Delta margins records the estuary's landward transgression associated with sea level rise, while the seaward movement of terrigenous sediment indicates the influence of the rivers (Atwater 1982).

While the general geologic setting of the Delta was mostly in place by roughly 2 million years ago, the Delta has continued to evolve. Geologist Andrei Sarna-Wojcicki et al. (1985) concluded that the Central Valley found its present outlet at Carquinez Straight about 600,000 years ago. Before that, but after the Valley's outlet to Monterey Bay closed due to uplift after 2 million years ago, a large freshwater lake occupied the Central Valley, evidence for which is found in a layer of Corcoran Clay and in the fossil record. Sediment cores from the Bay-Delta suggest that four or more estuaries existed since the lake disappeared, fading and reemerging in response to global fluctuations in sea level (Atwater et al. 1979). Cycles of deposition and erosion occurred during these glacial and interglacial ages of the Pleistocene epoch, contributing to the formation of the Delta's underlying sedimentary features. During interglacial phases, sea water advanced into the Central Valley and estuarine sediment deposits accreted over older alluvial deposits, creating flood basins (low-lying troughs subject to overflow) and natural levees along major Delta tributaries (Atwater and Belknap 1980). During periods of glaciation, Sierra Nevada rivers carried glacial deposits to form the alluvial fans that spread across much of the valley (Atwater 1982). The lowering sea-level also exposed previously deposited fine estuarine sediments to erosion, as the vast "inland sea" that engulfed the valley receded.

Beginning approximately 15,000 years ago, a period of climatic warming at the beginning of the Holocene epoch caused glaciers to melt and sea level to rise, forming many of the modern depositional features of today's Bay-Delta watershed (Fig. 1.4).The sea rose and spread eastward; migrating from the edge of the Farallon Islands (ca. 15,000 B.P.), eastward through the Golden Gate (ca. 10,000 B.P.); through the valleys that became San Francisco Bay; and extended tidal influence through the Delta by around 6,000 B.P. (Atwater et al. 1979, TBI 1998). Sea level rise was fairly rapid (0.8 in/yr/20 mm/yr) early in the Holocene, as opposed to more recent rates over the last several thousand years of about 1-2 mm/yr (0.04-0.08 in/yr; Atwater et al. 1979, Malamud-Roam and Ingram 2004, Brown and Pasternack 2005).

By 4,000 B.P., the San Francisco Bay and Delta resembled the early 1800s extent (West 1977, Atwater et al. 1979, Malamud-Roam et al. 2007). Sedimentation rates caught up with the slowing submergence rate, resulting in thick layers of peat reaching depths of 65 feet (20 m) in the central Delta (Thompson 1957, Atwater et al. 1979). Recent research suggests that rates of peat accretion ranged between 0.03 and 0.49 cm/yr (Drexler et al. 2009a). The inland fringe of the Delta, however, was only recently influenced by tidal processes, with only a thin layer of estuarine sediments less than 3,000 years old and was still greatly influenced by fluvial processes (Brown and Pasternack 2005).

Differences in physical geography and climate resulted in distinct environments within the Delta. Intertidal wetlands, characterized by the accumulation of deep peats. A complex network of waterways wove within the intertidal wetlands, including the main riverine channels of the Sacramento and San Joaquin rivers and their distributary channels. Natural levees extended into these tidal environments (Thompson 1957, TBI 1998). They were most prevalent in the north Delta, as influenced by the flood deposits of the Sacramento River.

CLIMATIC GRADIENTS Climate and weather are important drivers of ecological patterns and change. Climate influences hydrology and therefore overall habitat distribution and abundance. Variability in climate can disrupt and alter local and region patterns of habitat conditions, as well as affect land and water use. Droughts can instigate greater reliance on groundwater, crop failure, or the use of wetlands for pasturing stock, while extreme flooding can cause levee and crop failure, redoubling of reclamation, and channelization efforts. For these and other reasons, historical landscapes should be interpreted within the context of the climate history.

During the past 2,000 years, the climate of the Delta and its watershed has shown a gradual trend towards cooling and drying, punctuated by anomalous wet and dry periods (Goman and Wells 2000, Byrne et al. 2001).

15,000 Years Ago
(End of last Ice Age -- sea level
approximately 400 feet below
present level; rivers not shown)

10,000 Years Ago
(Formation of Farallon Islands
and intrusion into the
"Golden Gate")

5,000 Years Ago
(Formation of Bay and Delta
Basins)

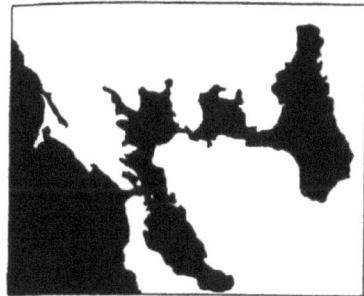

125 Years Ago
(Landward edge of undiked
tidal marsh)

Today
(Includes changes due to
hydraulic mining sediment
deposition, land reclamation,
and filling of wetland areas)

Figure 1.4. The invading estuary. Holocene transgression of the San Francisco Estuary and current extent of tidal waters as influenced by modern land use. (adapted by San Francisco Estuary Project from Atwater 1979 and Atwater et al. 1979, reprinted in TBI 1998)

Changing upper and lower tree-lines and tree-ring chronologies indicate that conditions became increasingly arid (Malamud-Roam et al. 2006). Also, isotopic compositions from estuary sediments indicate increasing salinity (above what would be expected from sea level rise alone) and marsh cores show evidence of vegetation changes supporting increasing salinities (Malamud-Roam et al. 2007). Climate reconstructions suggest a shift in the mid-1800s from a distinct cool and relatively dry several century period (the "Little Ice Age") to one of warmer and wetter conditions (Stine 1990, Stine 1996, TBI 1998). The relatively cool temperatures impacted flows by prolonging snowmelt flows, conditions likely in effect when foreigners entered the valley in the early 1800s. Since that time, climate has been relatively stable despite interdecadal variability (Dettinger et al. 1998, Malamud-Roam et al. 2007).

The present Mediterranean climate in the Delta is characterized by a cool, moist season in the winter and a dry, hotter season in the summer. Climate in the Delta is considerably affected by both interannual and decadal fluctuations, particularly by the Pacific Decadal Oscillation and the El Nino-Southern Oscillation climate patterns (TBI 1998, Malamud-Roam 2007, Stahle et al. 2011). Temperature and precipitation patterns manifest

Annual Average Precipitation

20 inches

13 inches

Average Maximum Temperatures

High 24° C

Low 17° C

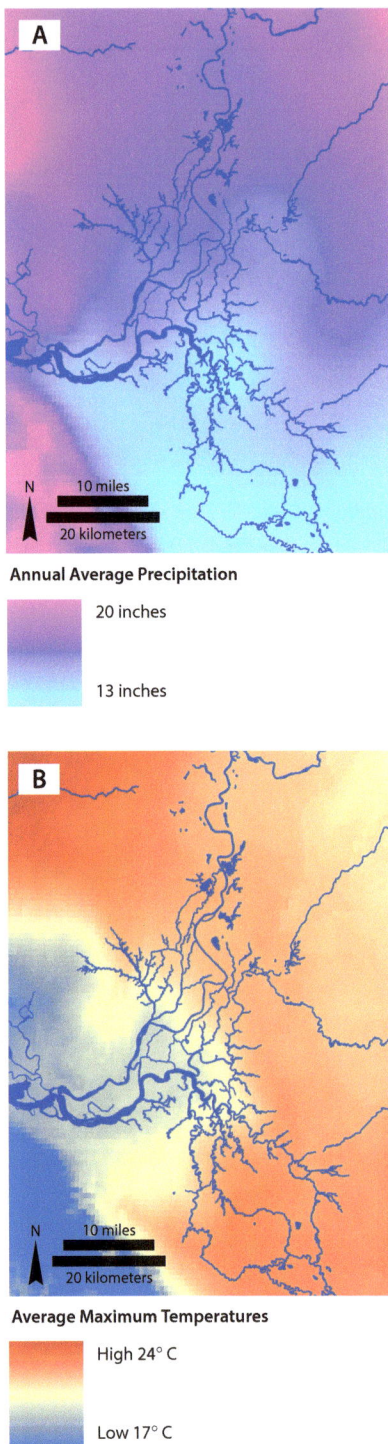

Figure 1.5. Climate gradients are seen in the distribution of average annual preciptitation (A), which ranges from approximately 13 to 20 inches (330-508 mm) south to north. Average annual maximum temperatures (B) illustrate how effective maritime influences were in reducing temperatures in the central Delta, creating an east-west gradient. This data is from the climate normal of 1971-2000. (PRISM 2006)

north-south variability, which have resulted in distinct differences in the hydrology and ecology of the Sacramento and San Joaquin systems. Average annual rainfall in the Delta is about 17.4 inches (442 mm), ranging from 13 to 20 inches (330-508 mm) south to north (Fig. 1.5a). Close to 95% of that precipitation occurs between the months of October and April. Mean temperature lies around 64°F (18°C), being about 50°F (10°C) in the winter and 70°F (21°C) in the summer (West 1977). Climate in the Delta is also affected by maritime influences and the presence of tule fog, which creates a pronounced west-east gradient and keeps the western and central Delta cooler (Fig. 1.5b).

To characterize wet and dry year conditions, we used precipitation records at various weather stations, the earliest of which is from the City of Sacramento and dates from 1850. Another important dataset indicating inter-annual climate variability is that of river inflow, as reconstructed by Meko et al. (2001) using dendrochronology. Figure 1.6 illustrates this record since Spanish explorers first viewed the Delta. Notable floods and droughts are indentified as further context for interpreting historical landscape change.

HYDROLOGY GRADIENTS Quantity, timing, and distribution of surface water flows through the Delta, as influenced by climate, shape landforms and fundamentally drive ecological patterns and processes. The natural hydrology of the Delta and its watershed is spatially and temporally variable at decadal, inter-annual, seasonal, and daily time frames.

Historically, an average of about 31.7 million acre-ft/yr (39.1 billion m³/yr) of runoff is estimated to have flowed into the Central Valley (Faunt 2009). Of the flow passing through the Delta (about 85%), most originated from the Sacramento River watershed (Malamud-Roam et al. 2006). A substantial portion of the annual flow volume was evapotranspired or went to the groundwater recharge. The natural annual flows were quite variable; sometimes the volume was less than half or more than twice as much as the average. Seasonal variation in flows was also quite substantial. About 80% of average annual flow occurs in just six months of the year, with peak flows generally occurring in the later spring. Once most of the snowpack has melted, flow drops dramatically: the lowest average monthly flow is just 3% of the highest average monthly flow on the San Joaquin River and 13% on the Sacramento River. The valley and its wetlands impacted the hydrograph: a substantial portion of the annual volume was retained and slowed by wetlands, evapotranspired, and exchanged with the groundwater, which was found within 25 feet of the surface for about 80% of the Sacramento Valley (Fig 1.7; TBI 1998).

There is substantial variability in the volume and timing of runoff between the Sacramento and San Joaquin rivers, and the Delta consequently reflects these differences. The wetter and lower maximum elevation Sacramento River watershed had annual flows that were more than three times greater than the San Joaquin River and were marked by larger peak flood events occurring earlier in the season (TBI 1998). The porous volcanic geology of the Sacramento River watershed helped sustain relatively high baseflows through the dry season (TBI 1998). Differences between the systems are clear comparing their "unimpaired" runoff (flows of the past century that would have occurred in the absence of water impoundments and diversions): the Sacramento is about 21.6 million acre-feet, while that on the San Joaquin is just 6.2 million acre-feet (Fig. 1.8; CDWR 2007). The Sacramento flows were more associated with rainfall events – resulting in higher peak events and more frequent flooding – though both hydrographs had an important snowmelt component. The maximum average monthly flow is in March on the Sacramento and May on the San Joaquin. The lag between peak rainfall and peak runoff is therefore about 2 months on the Sacramento River, but about four months on the San Joaquin. This means that the Delta was generally flooded earlier by the Sacramento and later by the San Joaquin, attenuating flood peaks. Sometimes, however, large floods affecting both systems simultaneously were produced by rain on snow events in the winter. Many differences in habitat patterns and characteristics between the north and south Delta can be linked to these differences in flow volume, timing and sediment load. This is discussed at various points in the report with regard to implications for historical conditions.

The delivery of inorganic sediments by freshwater inflows affected some differences in landscape characteristics across the Delta. Prior to the advent of hydraulic mining, estimates show that about 1.5 million cubic meters of sediment were delivered annually to San Francisco Bay from the Sacramento-San Joaquin watersheds (Atwater and Belknap 1980). The sediment load was primarily derived from the Sacramento River watershed since the granites of the San Joaquin watershed were less prone to erosion (TBI 1998, Wright and Schoellhamer 2005). The Sacramento River is still the dominant contributor of sediment today, as demonstrated by a study showing that the Sacramento contributed 85% of the sediment inflows between 1999 and 2002 (Wright and Schoellhamer 2005). Historically, much of the sediment brought down by the rivers settled in the wetlands of the Sacramento Valley without reaching the Delta (Gilbert 1917, Shoellhamer 2007). During the dry season, the rivers were described in the historical record as clear enough to "see shoals of fish sporting in it at the very bottom" (Hoag 1882).

Perhaps the dominant environmental gradient within the Delta is the upstream decrease in tidal energy. At the Delta mouth, tidal flows during times of low water are about 330,000 cfs (9,340 cms; CDWR 1993, TBI 1998). For much of the Delta, water levels in the channels (and on the marsh) rose and fell with the ebb and flow of the tide. Tidal influence diminishes upstream. At some point water stage continues to rise and fall with the tides, but flow maintains a downstream direction (today, during low water, this transition occurs at Walnut Grove on the Sacramento and

Figure 1.6. Wet and dry years of the recent past. This diagram was created from reconstructed Sacramento River runoff in million acre-feet (MAF), based on tree ring analysis by Meko et al. (2001). The bars showing the runoff magnitude are color-coded by the decile within which they fall, with the drier years represented in warm colors and the wetter in cool. The blue solid line shows unimpaired flow as reconstructed by the Department of Water Resources (2007) from the period of record (beginning in 1921). Immediately above the graph is precipitation at Sacramento

1841, August:
Captain Suter [sic] has commenced extensive operations in farming; but in the year of our visit [1841] the drought had affected him, as well as others, and ruined all of his crops.

—WILKES 1845

1850:
One of the greatest floods occurred…From the top of a high hill on the left bank of Feather River, not far from the Table Mountain, where I could command an extensive view of the valley, I estimated that one-third of the land was overflowed.

—DELANO 1857

1852:
In March, 1852, the water reached a higher point than at any time previously. During that month the rainfall was measured in Sacramento as thirteen inches.

—LEWIS PUBLISHING CO. 1890

1805:
It is said the entire Sacramento Valley was covered with water, except Marysville Buttes. This tradition was handed down by the Indians and at the time of the first white settlers in this section stories of the 'great waters' were still extant.

—TAYLOR 1913

1850:
From January 9 to 17, 1850, the entire City of Sacramento was flooded.

—THOMPSON AND WEST 1880

1852-1853:
From the Colusa hills to the Montezuma hills in Solano the west shore of the Sacramento river was under water – excepting the Indian mounds.

—GREGORY 1913

1829:
These two last seasons crops have entirely failed in that country owing to the extreme heat which accounts for the low state of the water and the several streams we found dry.

— MCLEOD AND NUNIS [1829]1968

1847:
The winter of 1846-47 was very wet and stormy…About the middle of January, 1847, the river overflowed its banks, and the whole country was under water for miles in every direction.

—LEWIS PUBLISHING CO. 1890

1854, Dec 5:
Times is very bad at present on the acount [sic] of its being so dry we have had no rain yet but a few sprinkles nothing to do any good.

—HOWLAND 1854

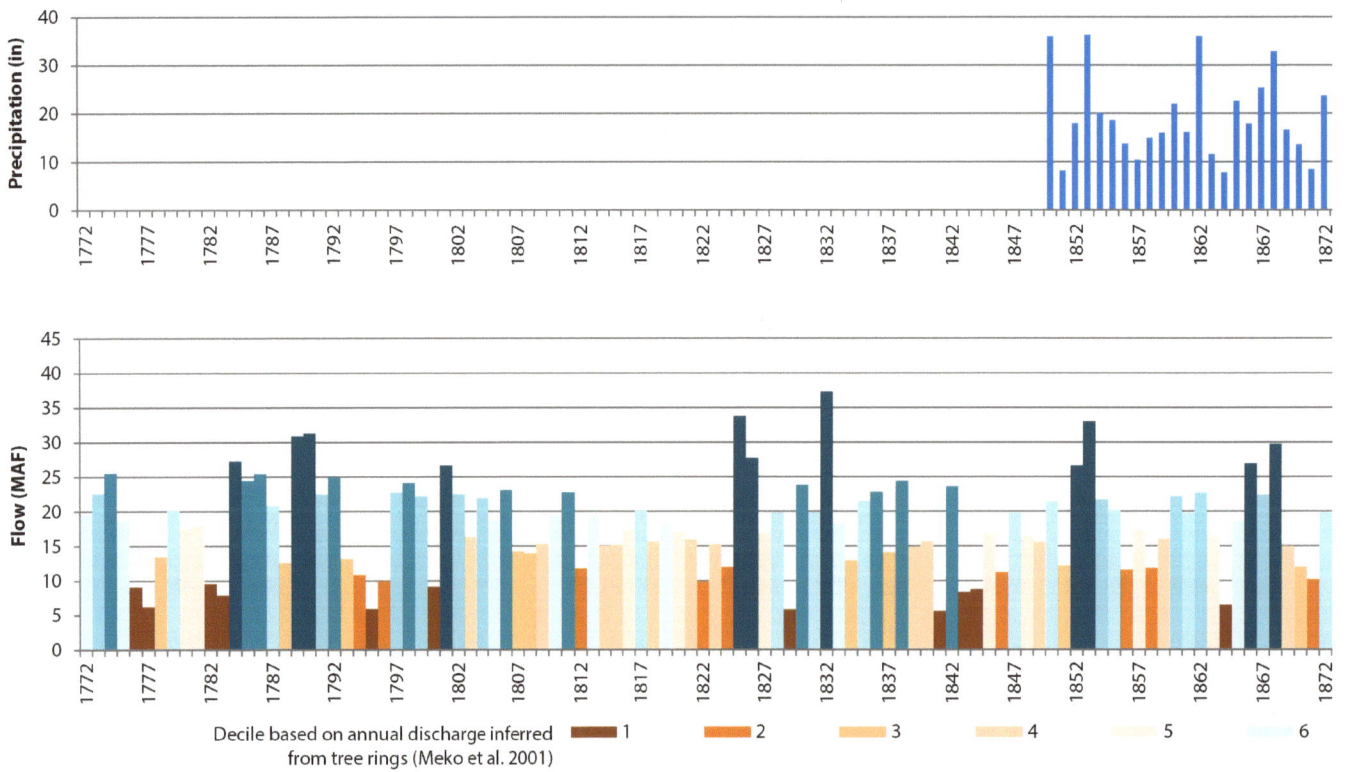

Decile based on annual discharge inferred from tree rings (Meko et al. 2001) 1 2 3 4 5 6

for the period of record (beginning in 1850). Finally, selected quotes for a number of the wet and dry years are shown above. See Fig. 1.7 for more explanation of unimpaired flows. For more detail concerning historical floods, consult Thompson's (1996) "Flood Chronologies."

1855, Dec 16:
We have had not rain of any acount [sic] yet and that makes times rather dull.
—HOWLAND 1855

1856, Dec 12:
Times are about as usual here rather dull we haven't had scarcely any rain here yet and the prospects is for a dry winter…
—HOWLAND 1856

1858, May 21:
Times is rather dull at present we have had rather a dry spring, crops I think will be rather cut short.
—HOWLAND 1858

1862:
Most singular of all, however, was the fact that the bay fishermen frequently caught fresh water fish in the bay. For from two to three months the surface portion of the entire waters of the Bay of San Francisco consisted of fresh water to a depth of from 18 to 24 inches.
—BANCROFT 1863 IN MCCLURE 1927

1865:
During the unusually dry season of the past summer…when it became evident that the hay crop in a large portion of the State must prove a failure, and consequently command a high price, many persons resorted to the tule lands at the mouths of the San Joaquin, Sacramento, and Cosumnes Rivers, in search of the desired article.
—REED 1865

1885:
The river is now, in the fore part of July, in about the same condition as during the lowest water of last year.
—PAYSON 1885

1890:
Most severe storm the people had experienced for many years.
—GREGORY 1913

1907, March:
"Extraordinary" flooding on the Sacramento River and its chief tributaries.
—FORTIER 1909

1909:
A large area of country in the neighborhoods of Bensons Ferry and Lodi has been flooded.
—TAYLOR 1913

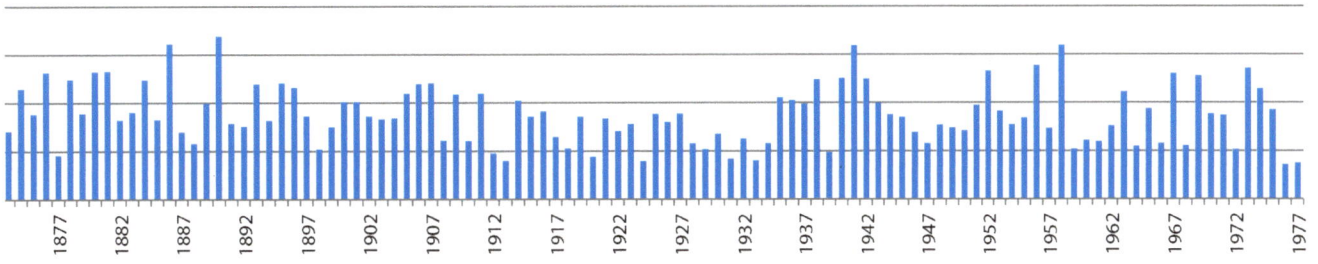

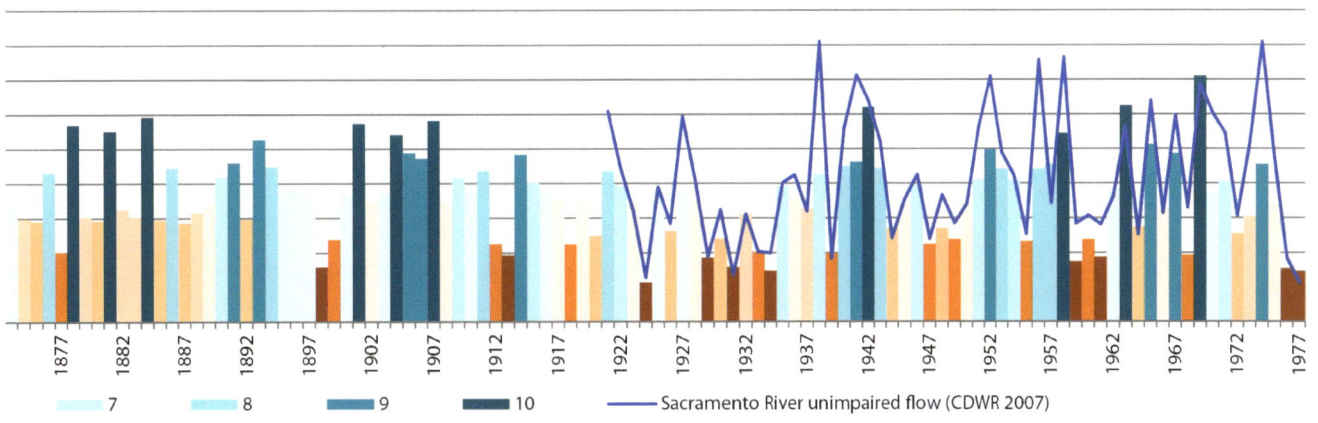

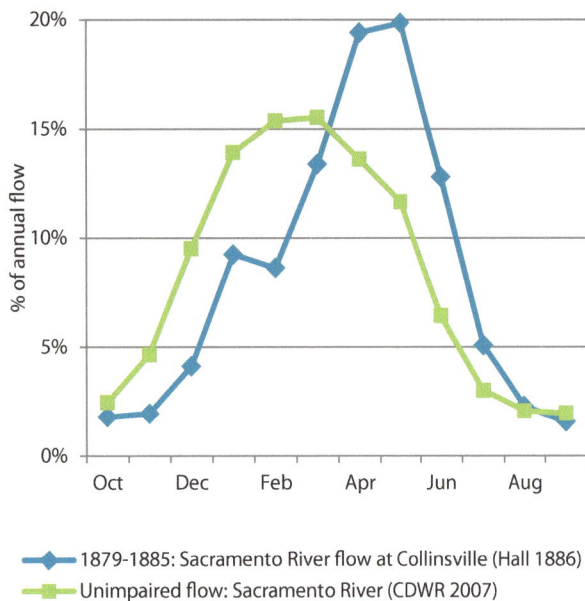

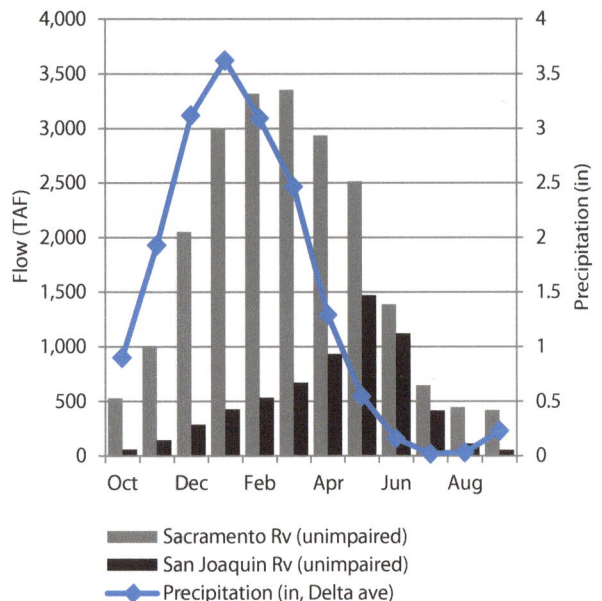

Figure 1.7. Comparison between late 1800s Sacramento River outflow and unimpaired runoff. Flows at Collinsville for the water years of 1879-1885 (Hall 1886) show later-season flows than that suggested by the unimpaired flow data for the Sacramento River (CDWR 2007), which aggregates inflow from major tributaries to the Delta. This indicates that the wetlands, flood basins, and groundwater recharge served to shift the hydrograph such that high flows were felt at the Delta mouth later in the season than it was upstream (TBI 1998). It should be noted that unimpaired flow is not actual early 1800s flows, but is runoff that would have occurred over the period of record (1921-2003) without dams and diversions, though with other modified physical conditions present. Also, the snowmelt period likely extended later into the season in the early 1800s due to the cooler conditions at the end of the "Little Ice Age" (Stine 1996).

Figure 1.8. Monthly runoff comparison between Sacramento and San Joaquin rivers. Unimpaired runoff (i.e., runoff that would have occurred without dams or diversions; CDWR 2007) is shown averaged by month for Sacramento (gray) and San Joaquin (black) in thousand acre-feet (TAF). Sacramento flows are greater due to its wetter climate. Also, because it receives a greater proportion of its water as rain instead of snow (the San Joaquin watershed has higher maximum elevation), the Sacramento's peak months of runoff occur earlier in the season than those of the San Joaquin. Also, Sacramento flows do not drop as dramatically in the late summer and early fall.

Vernalis on the San Joaquin; Enright pers. comm). During periods of low freshwater input, the upstream extent of bi-directional flow would increase. In contrast, during extreme floods the downstream extent of uni-directional flow would increase. The amount of freshwater input also affected the upstream extent of the salinity gradient. While the transition between brackish and fresh water was usually downstream of the foot of Sherman Island, during periods of low freshwater inflow, this gradient shifted upstream toward Antioch. During droughts, salinity intrusion extended even farther upstream.

Land and water use history timeline

Understanding the Delta's land use and water use history helps us comprehend the progression of changes in ecosystems that has led to current conditions and is important for interpreting historical sources. Furthermore, it is a reminder that the Delta is not a static place. Rapid human modification, beginning in the mid-1800s, came at a point along a continuum of the ecological Delta responding to climatic variation,

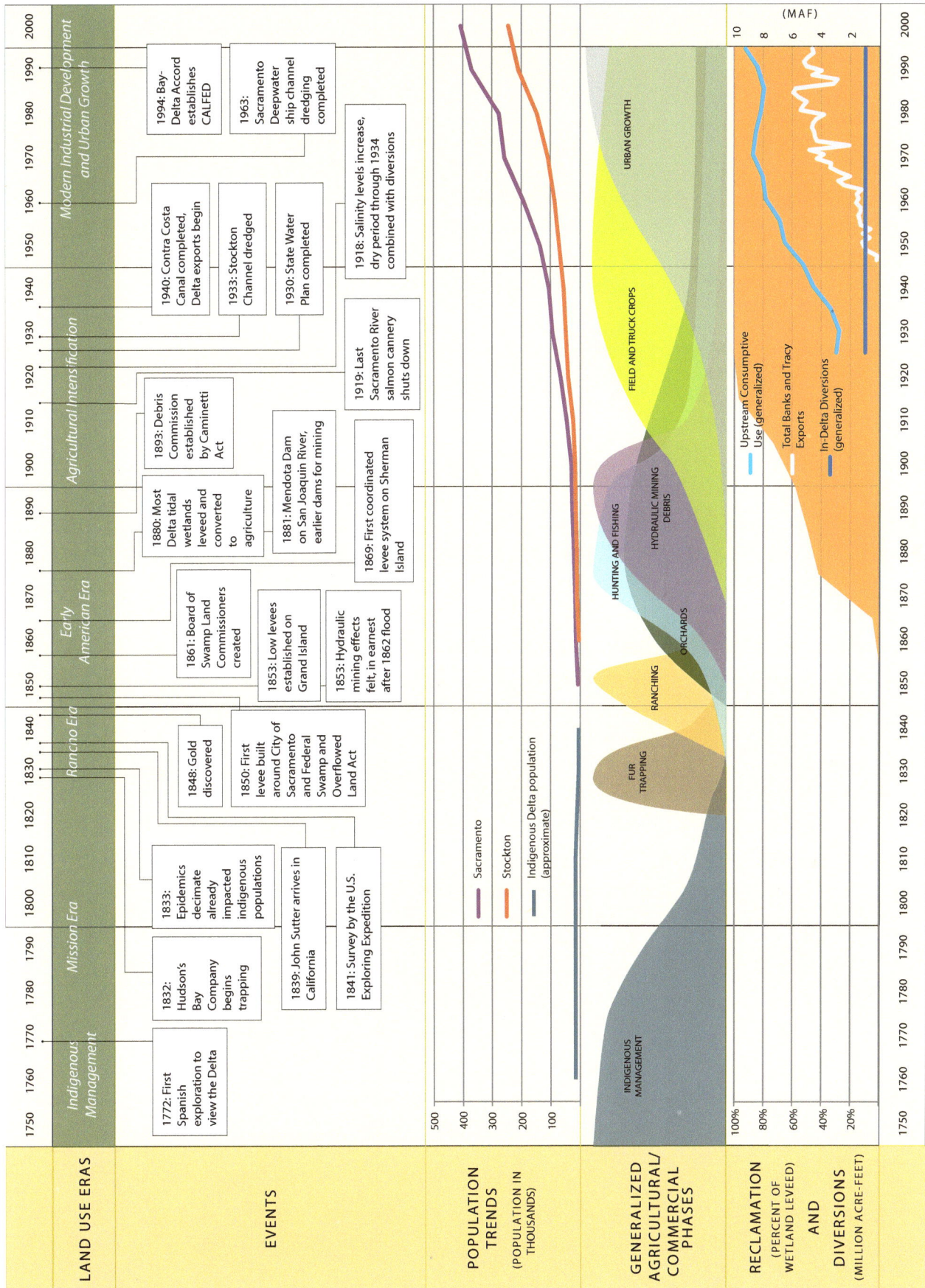

Figure 1.9. Timeline depicting land use trends in the Delta region over the past 250 years, including population growth, phases of different resource uses, reclamation sequence, major water exports (since 1930), and other significant events. (Diversion data: Governor's Delta Vision Blue Ribbon Task Force 2008)

geological trends, and catastrophic events. The Delta will continue to respond to these past events as well as future conditions.

This section presents a brief summary of the land use history in the Delta over the past several centuries. Attention is focused on the initial and rapid modifications that occurred in the nineteenth century Delta. This summary is intended to provide general context of human history. Major trends in the Delta are illustrated in Figure 1.9. For a thorough treatment of the Delta's land use history, see John Thompson's (1957) *Settlement Geography of the Sacramento-San Joaquin Delta*. Many others have followed with different emphases (e.g., Vaught 2007, Kelley 1989, Garone 2011).

PRE-1769: NATIVE LAND MANAGEMENT

Indigenous peoples have lived in the Delta since before the advent of tidal development 6,000 years ago. The rich Delta ecosystem supported a population on the order of 10,000 people of the estimated 300,000 people in California, embracing four distinct linguistic groupings and numerous smaller communities (Fig. 1.10; Cook 1955a, Thompson 1957, Blount et al. 2008). Villages, often marked by artificially constructed mounds (up to 300 feet or more in diameter) occupied the higher lands within the Delta, including natural levees and sand mounds. Groups appear to have moved around seasonally to take advantage of various resources and in response to flooding

Figure 1.10. Plano topográfico de la misión de San José. One of the earliest maps of the San Francisco Estuary, by Spanish missionary Father Narciso Durán, shows over a dozen distinct indigenous communities in the Delta. The names of these communities are written in between the lines representing the Delta waterways. The map's key shows that "I" indicates islands, the circles with crosses indicate Christian communities and the circles without crosses indicate non-Christian communities. (Durán 1824, courtesy of The Bancroft Library, UC Berkeley)

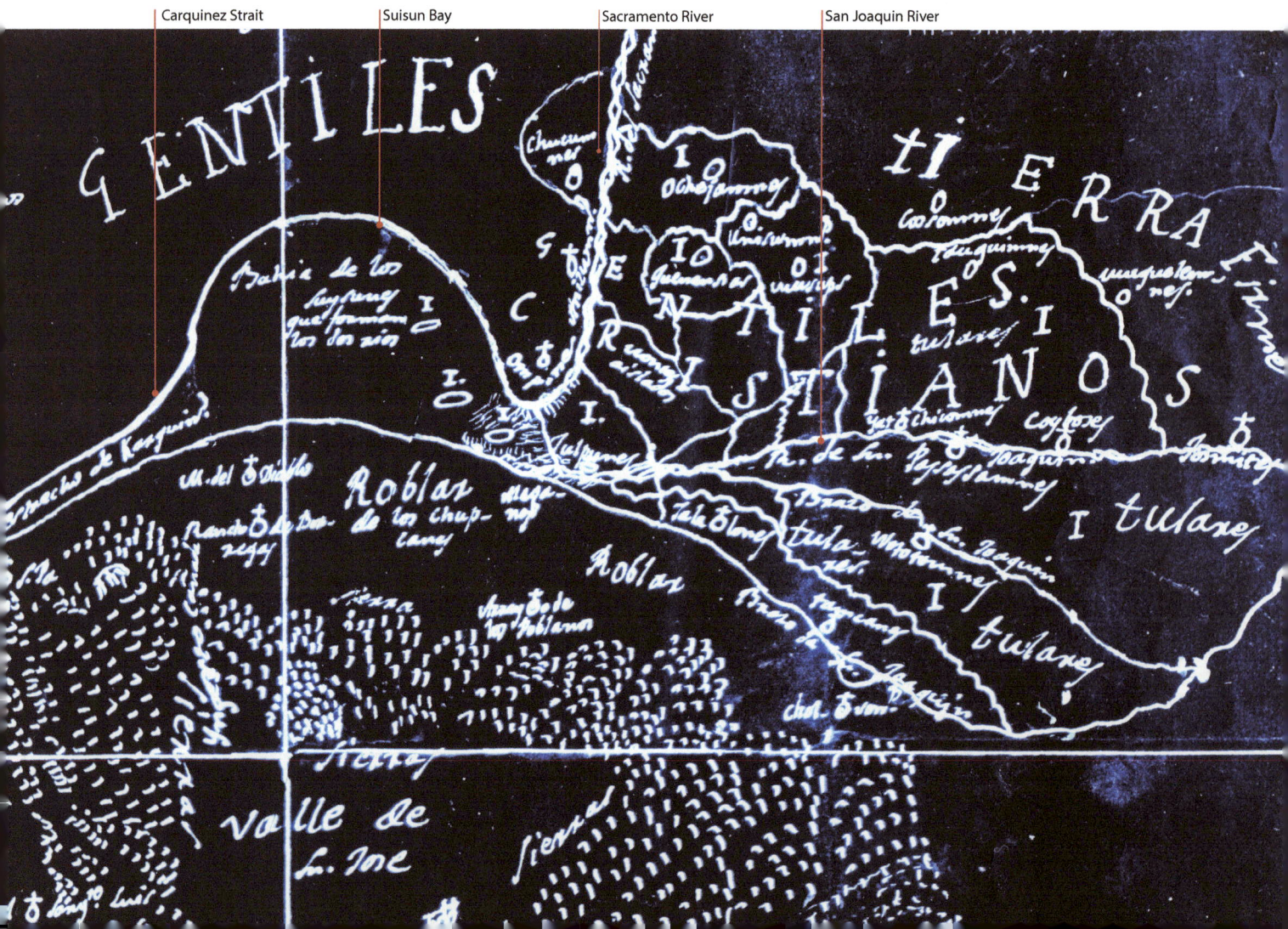

within the Delta (Belcher 1843, Robinson 1860). For thousands of years, indigenous peoples living in and around the Delta actively modified their surrounding environment through harvesting acorns, grasses, and wetland plants for food, baskets, and construction of tule balsas (rafts), huts, and mounds; hunting for large game and fishing for salmon, chub, shellfish and numerous other species; and the use of fire which altered vegetation patterns (Kroeber [1925]1976). While much has been learned about the demographics of the tribal societies, considerably less is known about the extent and character of these landscape modifications (Box 1.1). Indigenous populations declined precipitously during the early 1800s due to relocations to missions and epidemics that decimated entire villages, effectively ending native management of the Delta. The epidemic of 1833 was especially notable, reducing, by some estimates, populations by about 75%. The native population of the Central Valley was therefore significantly reduced by the time Euro-American settlement began in earnest (Cook 1955b). Accounts from the 1830s and 1840s describe past villages within the Central Valley that were large population centers, but were vacant and strewn with the bones of the villages' occupants (Brackenridge [1841]1945, Wilkes 1845).

1772: SPANISH EXPLORATION

A Spanish expedition, led by Captain Pedro Fages, viewed the Delta for the first time on March 30, 1772, from an area now known as Willow Pass in Contra Costa County marking the initiation of European contact in the region. Fray Juan Crespí wrote of the sighting: "Where these [two large rivers] united to form the estuary we saw good sized islands." (Crespí and Bolton 1927). Diaries from this and subsequent expeditions provide the first written accounts of the Delta. The subsequent establishment of missions impacted native populations, as people were forced to leave their villages. Also, livestock were introduced during this period. The era effectively ended with the secularization of missions in 1834.

1825: TRAPPING FOR FUR

The renowned American traveler and trapper, Jedediah Smith, entered California in 1826, initiating a period of intensive fur trapping. The Hudson's Bay Company trapped extensively in the Delta. It sent parties of trappers to California between 1828 and 1846 for beaver and otter, including John Work's expedition of 1832-33 (Maloney 1936, Thompson 1957, McLeod and Nunis 1968). Beaver were reportedly found to be most plentiful in the Delta (McLeod and Nunis 1968). As early as the mid-1830s, beaver were becoming scarce (Bryant 1915, Ogden 1988).

1839: RANCHING AND LAND GRANTS

As part of a system to grant land by the Mexican government for the purposes of ranching and agriculture, Jose Noriega obtained the first rancho in the region, Los Meganos, in Contra Costa County, in 1837. In 1839, John Sutter traveled up the Sacramento River and established his fort where the City of Sacramento now stands. As part of the New Helvetia Rancho, this was deeded to John Sutter in 1841. Sutter's Fort became the hub of regional settlement, with extensive cattle

BOX 1.1. ETHNOGEOGRAPHY AND NATIVE MANAGEMENT

Our understanding of the demographics of tribal societies of the Delta has improved substantially in recent years through the work of a number of diligent researchers and modern tribes. Less attention has been paid to the reconstruction and understanding of aboriginal lifeways in recent decades. Much of what is known of Delta cultures' lifeways was gathered in the early and mid-twentieth century (e.g., Barrett 1908, Gifford 1916, Schenck 1926, Kroeber 1932, Cook 1960a, and Merriam 1967) after much acculturation and community disintegration had already taken place. Even less well developed is the study of indigenous influences on the form and function of ecosystems in the Delta. Translating the impressive assemblage of demographic information, known information on lifeways and floral and faunal assemblages, information on native management practices, and memories of living tribal members into a model of indigenous human-landscape interactions has yet to be undertaken in any meaningful way. This box highlights some of the major recent demographic findings and addresses the "language group as tribe" mythology (Blount et al. 2008) that pervades much of the public consciousness, and discusses a number of areas of additional research that are essential to fully inform understanding of human-landscape interactions that may have shaped or maintained many of the Delta's ecological functions.

Historical and demographic record

The story of the indigenous people of Sacramento-San Joaquin Delta is told in large part through the record of the human experience of colonization, as detailed by the colonizers themselves. The tribes of the area had no written language prior to the arrival of Spaniards in 1769, so Spanish writings provide a first narrative glimpse of the region's first people. There are three major classes of documentation that are useful in this area: Mission records, exploration and settlement diaries, and land grant claims (Bennyhoff 1977). Much of the information in these document types, however, can be imprecise, erroneous, or otherwise unclear, so it is often necessary to "cross-walk" this information with ethnographic and archaeological data.

Synthesis of information Despite the breadth, complexity, and fragmentary nature of sources of ethnogeographic information about the Delta, there has been an extraordinarily systematic and cogent approach to adding value, interpretation, and new data to the story of the pre-colonial Delta cultures. Taylor (1861), Bancroft (1883), Schenck (1926), Cook (1955a), and Bennyhoff (1977) all used increasingly heterogeneous sources of information to build an increasingly precise distribution of Delta cultures. Over the last 40 years, Dr. Randall Milliken of Davis, CA – a student and colleague of Bennyhoff's – has assembled perhaps the most impressive synthesis of all prior work, adding new interpretations of the vast reservoir of data from the California Mission Records. This synthesis is now being incorporated into a GIS/Wiki interactive database for the Delta and much of the rest of the state (Blount et al. 2008, Milliken unpubl. data).

Linguistic groupings One of the primary challenges to making sense of historical records is synonymy of variant names. Since there were so few living speakers of the multitude of native languages in the region by the time the most rudimentary efforts to record the demographics of the region were made, scholars work to this day to assemble a defensible map of the indigenous Delta. There were at least four linguistic groups represented in the Delta and the vicinity at the time of contact: Patwin, Nisenan (Maiduan), Eastern Miwokan, and Northern Yokuts (Schenck 1926, Merriam 1967, Bennyhoff 1977, Blount et al 2008). While the tribes of the region are known colloquially by these linguistic groupings (Patwin, Miwok, Yokuts, Maidu), pre-contact tribes did not necessarily view themselves according to these academic distinctions. Table 1.1 lists the 20 known communities within these linguistic groupings in the project area. Within each community were often multiple townships, villages, or smaller family groupings primarily along permanent waterways throughout their territory.

Table 1.1. Geographic positions of linguistic grouping and associated communities within the Delta region. (Data from Blount et al. 2008)

Linguistic grouping	Community	Approximate geographic context
Patwin	Churuptoy	Woodland
	Puttoy	Davis
Nisenan (Maiduan)	Wolok	Verona
	Sakumne, Pusunlumne	Natomas
Plains/Eastern Miwokan	Gualacomne	Freeport
	Ylamne	Yolano
	Ochejamne	Courtland
	Cosomne	Wilton
	Sonolome	Clay
	Chucumne	Liberty Island
	Quenemsla	Grand Island
	Unizumne	Thornton
	Anizumne	Rio Vista
	Guaypem	Terminus
	Musupum	Andrus Island
	Muquslemne	Lodi
	Julpun	Oakley
	Ompin	Collinsville
Northern Yokuts	Tauquimne	King Island
	Yatchlcumne	Stockton

Working landscapes

What was this complex network of interrelated cultures doing in the Delta? How were they able to survive, and establish large, flourishing communities that persisted for hundreds of generations? What were the principal resources utilized, and what management activities were employed to maintain the predictability, resilience, and productivity of those resources?

Native people in California did not practice agriculture as it is typically described; however they did modify their landscape in a variety of important ways (Stewart et al. 2002, Anderson 2005, Martinez 1998). Tribal groups managed lands under their influence with practices such as seed beating, burning of scrub and grasslands, harvest of grasses, and use of digging sticks to turn the soil (c.f. Anderson 2005). Products harvested by the tribes of the Delta region included acorns, grasses and forbs, willows, a variety of wetland plants, and tule to construct rafts and innumerable other products (Kroeber [1925]1976). Archaeological research in the area has been very limited, but sites that have been studied reveal that a high diversity of shellfish, large and small mammals, birds, and small fishes were eaten by native people (Gifford 1916, Milliken 1995b, Pierce 1988). Native groups also hunted game such as deer, pronghorn antelope, and elk, which appear to have been abundant in the Delta at the time of Spanish contact (Anza 1772 in Brown 1998, Font and Bolton 1933).

Of particular interest to land managers as well as tribes today is the use of fire by local tribes to maintain and enhance the local ecosystems. Native groups used fire to control the distribution of chaparral, maintain grassland cover and forage for wildlife, control pathogens, improve access to seeds and acorns, and aid in hunting rabbits and other small game (Kroeber [1925]1976, Keeley 2002, Anderson 2005). Recent quantitative studies of fire ecology and fire behavior (Stephens and Fry 2005, Evett et al. 2007) have concluded that local tribes were the source of a large majority of all fire ignitions in many coastal and interior valley regions prior to colonization. While fire use in the Delta was not well documented, it may have been used much as if was in other, similarly populated areas. Exploration and settlement diaries record behavior by tribes that support the notion that fire was widespread in the Delta region, however these observations have yet to be systematically assembled and organized (see Box 6.1). Further investigations would likely add substantially to the array of similar observations, and would add significantly to the ability to understand the pre-settlement fire regime as well as other management techniques of the region.

ranching and the first commercial production of grain in the valley. Campo de los Franceses, within which Stockton now stands, was deeded to Guillermo Gulnac in 1844, and later obtained by Captain Weber. Additional land grants were made along the periphery of the Delta until California became a state (Thompson 1957). Also, with the increasing attention to the valley, several exploring expeditions entered the Delta, notably Captain Belcher's trip up the Sacramento in 1837 and the Captain Wilkes U.S. Exploring Expedition in 1841.

1847: TRANSPORTATION VIA STEAMBOAT

The first steamboat, the *Sitka*, traveled to Sacramento from San Francisco in 1847 (Hutchings 1859). Regular steamboat travel along the San Joaquin River began in 1849. Steamboats became the primary mode of transportation within the Delta, shipping products to market in San Francisco and later bringing miners part way to their diggings in the Sierra Nevada. The steamboat era came to an end around 1878, as hydraulic mining debris reduced channel depths and railroads became the more economical means of transportation.

1848: GOLD DISCOVERED

The discovery of gold marked the beginning of a flood of settlement in the Delta region. Settlements and small gardens and orchards quickly sprang up along the Sacramento River, and the wood cleared from the land was used to fuel steamboats (Fig. 1.11). Riparian forests were affected early. Homesteads were established along the high lands of the natural levees, while the wetter land behind was used as pasture in the summer (Van Löben Sels 1902, Hoppe pers. comm.). The Delta had relatively little human modification until this time.

1850: FEDERAL SWAMP AND OVERFLOWED LAND ACT

The Swamp and Overflowed Land Act was passed in 1849 and extended to California in 1850, when it became a state. The Act transferred "swamp and

Figure 1.11. Orchards on Grand Island, along Steamboat Slough. The natural levee lands along the river were some of the first areas to be cultivated in the Delta and were prime sites for orchard crops. (Thompson and West 1880, courtesy of SharingHistory.com)

overflowed" lands from the federal government to the state for sale to the public (see Box 2.3). This was intended to incentivize drainage of wetlands.

1850: LEVEE BUILDING BEGINS

The first levees constructed in the Delta were to protect the City of Sacramento after the 1850 flood. Levee building in the tidal wetlands began, among other places, on upper Grand Island in 1853, Andrus Island in 1855, Roberts Island in 1856, Union Island in 1857, and Brannan Island in 1859 (Tucker 1879a-f, Rose et al. 1895). These initial efforts involved the damming of many small sloughs that branched and headed within the wetlands (Fig. 1.12). Such actions modified hydrologic conditions by reducing tidal prism, tidal channel length, and tidal wetland area. Also, as levees were raised steadily higher to keep floodwaters from spilling into the wetlands, this effectively raised flood levels, such that unprotected lands were flooded to a greater extent than before (Gilbert 1917).

1851: SWAMP AND OVERFLOWED LAND FIRST SOLD

The state passed its first law to grant swamp and overflowed land to individuals. Later, the California Reclamation District Act of 1855 governed the sale of these lands for $1 per acre, with a limit of 320 acres (Thompson 1957). The limit was increased to 640 acres in 1859. The early period of reclamation was fraught with battles between private control and state and federal assistance. Some argued that reclamation should be left to private interests to guarantee that the land would be used, while others challenged that projects were too large to be coordinated and implemented effectively by private landholders.

1855: RAILROADS INTRODUCED

The Sacramento Valley Railroad was established in 1855. It wasn't until 1871, however, that the California Pacific Railroad Company, Central Pacific Railroad Company, and Western Pacific began expanding through the valley.

Figure 1.12. Damming sloughs. In the early years of reclamation, most of the smaller sloughs that brought tides to the interior of Delta islands were dammed either with tide gates or by building a levee across them. (Unknown ca. 1900, MS 229, Dyer Photograph Album, courtesy of Holt-Atherton Special Collections, University of the Pacific Library)

1856: HYDRAULIC MINING EFFECTS FELT

Hydraulic mining for gold began in the Sierra Nevada in 1853 and over the next several decades washed entire hillsides into the streams and rivers (Fig. 1.13). Sediment inflow increased more than nine times over historical levels, from about 1.5 million cubic meters (2.0 million yd^3) annually to about 14 (31.4 million yd^3; Gilbert 1917, Atwater and Belknap 1980). Suisun Bay accumulated about 115 million cubic meters (150 million yd^3) between 1867 and 1887 (Cappiella et al. 1999). Effects were felt in the Sacramento Valley as early as 1856 (Bailey [1919]1927), but were particularly apparent in the Delta after the 1862 flood. The sediment raised the beds of streams which exacerbated the effects of floods, dramatically decreased tidal range, and hampered navigation, eliminated important spawning grounds for fish, and increased river turbidity (Box 1.2; Gilbert 1917). In some cases (mostly upstream of the Delta) the sediment covered valuable agricultural land (Nesbit 1885). It was also contaminated with mercury used in the mining operations, which (along with mercury from the Coast Range) continues to affect downstream ecosystems today. This was outlawed in the 1884 *People v. Gold Run Ditch and Mining Company* case, whch ended with the Supreme Court decision, *Woodruff v. North Bloomfield et al*, in 1892. The apex of debris from the American River is estimated to have passed Sacramento in 1896, though waves from other rivers passed through later (Gilbert 1917). Over the past century, sediment supply has diminished substantially, with a 50% reduction in the Sacramento River's sediment supply between 1957 and 2001 (Wright and Schoellhamer 2004) and significant erosion in Suisun Bay (Cappiella et al. 1999).

1861: RECLAMATION DISTRICT ACT PASSED

The California Legislature created the State Board of Swamp Land Commissioners to encourage the establishment of reclamation districts to levee the Delta. These districts would join land together within natural boundaries to facilitate reclamation. The first districts to be created were District #1, the American Basin; District #2, the Sacramento Basin; and District #3, Grand Island.

1862: FLOOD

The largest flood in recorded history passed through the Central Valley in the winter of 1861-1862. Accounts describe the Sacramento Valley as almost entirely flooded. It washed away the town of Rio Vista and blocked flood tide currents at the water surface. Mining debris from the Sierra Nevada contributed to the damages caused by the flood.

1862: AMERICAN RIVER MOUTH REALIGNED

In order to relieve pressure on the levees protecting the City of Sacramento from flooding by the American River, efforts were initiated in 1862 to realign the American River at its mouth. It was diverted from its channel curving to the south and made to enter the Sacramento about a half a mile upstream.

1863: SEVERE DROUGHT

The large floods of 1862 were followed by several years of severe drought. As cattle and sheep died of starvation, ranchers sought new grazing land, including the Delta.

Figure 1.13. Hydraulic mining. The caption of this photograph, which is shown as Plate I of Grove Karl Gilbert's treatise on hydraulic mining debris (1917), reads: "A hydraulic gold mine. The water is conveyed by pipe, under a head of several hundred feet, and delivered through a nozzle that can be turned in any direction. The jet washes the auriferous earth from the cliff and thence to a sluice, seen at the left. The sluice is several hundred feet long and contains pockets of mercury by which the gold is caught. (photo by Gilbert 1908, from Gilbert 1917)

BOX 1.2. CHANNELS FILL WITH HYDRAULIC MINING DEBRIS

Hydraulic mining debris, particularly in the Sacramento River, dramatically raised the river's bed. As the debris accumulated, tidal range that had once been two feet at Sacramento in 1847 became negligible by the 1880s. By the early 1900s, the rise and fall of tides was reduced even further to be imperceptible nine miles downstream of Sacramento (Mendell 1881, Hall 1880). The increasing bed levels caused low water levels to increase close to six feet between 1849 and 1879, a year where a tide range of two inches was recorded at Sacramento (Fig. 1.14; Taylor 1913, Gilbert 1917). An 1895 report found that "the bottom of the river bed is now higher than the old low-water surface of the river (Rose et al. 1895). Tidal action was partially restored by 1913, subsequent to the 1892 *Bloomfield v. Woodruff* decision that outlawed hydraulic mining. Today, tide ranges at Sacramento are largely restored to their 1849 two foot (0.6 m) range, though the changes to channel geometry that occurred alongside the rising bed levels challenges interpretation. It is unknown what the tide range would be at Sacramento today were there still several hundred thousand acres of tidal marsh and over one thousand miles of tidal channel in the Delta.

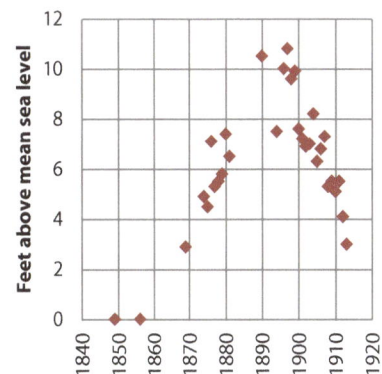

Figure 1.14. Rising bed levels through the hydraulic mining period, indicated by low-water stages of the Sacramento River at Sacramento. This graph is adapted from Table 2 (pp. 29-30) in Gilbert (1917).

1866: CONTROL OVER SWAMP AND OVERFLOWED LAND GIVEN TO COUNTIES

The Board of Swamp Land Commissioners was dissolved and control over surveying and selling the State's swamp and overflowed land was given to the counties.

1868: ACREAGE LIMITS FOR SWAMP AND OVERFLOWED LAND REMOVED

The Green Act removed acreage limitations for individuals to purchase swamp and overflowed land. Other incentives were also given for large-scale reclamation efforts. The 640-acre limit was restored in 1874.

1869: COORDINATED LEVEE BUILDING AND RECLAMATION

With the reclamation of Sherman Island in 1869 by the Tide Land Reclamation Company, a new era began. At one time, the Tide Land Reclamation Company held 120,000 acres of land in the Delta (Paterson et al. 1978). The majority of the work in the initial decades was performed by hand by Chinese laborers. In the 1870s, steam-powered dredges were used to rebuild and raise levees (Fig. 1.15). During this period, many channel meanders were cut off, particularly on the San Joaquin River, to lessen levee miles (reduce cost) and shorten travel distance. It was understood that reclamation was generally easier within the central Delta, because these peat lands could be easily cleared with fire (the land along the Sacramento River flood basins did not burn deeply because they were not deep peat soils) and levees did not have to be built to withstand the flood heights of the Sacramento River (Tide Land Reclamation Company 1872, Naglee 1879). In general, the levees within the central Delta were built to prevent tidal inundation of the land, whereas the levees to the north and south were largely to prevent overflow by major river floods. By 1900 more than half of the Delta had been leveed and an even greater area had experienced reclamation attempts (Fig. 1.16; Thompson 1957). The Delta was almost entirely reclaimed by 1930, although levees continue to break, requiring significant expenses in drainage and repair.

Figure 1.15. Clam-shell dredge building levees. The clam-shell dredge, Vulcan, is shown building the height of an artificial levee along an unknown waterway. After initial reclamation efforts, where levees were built primarily by Chinese laborers, dredges were used to build higher levees to keep out the tides and floods. (McCurry Foto Co. ca. 1910, courtesy of the California History Room, California Sate Library, Sacramento)

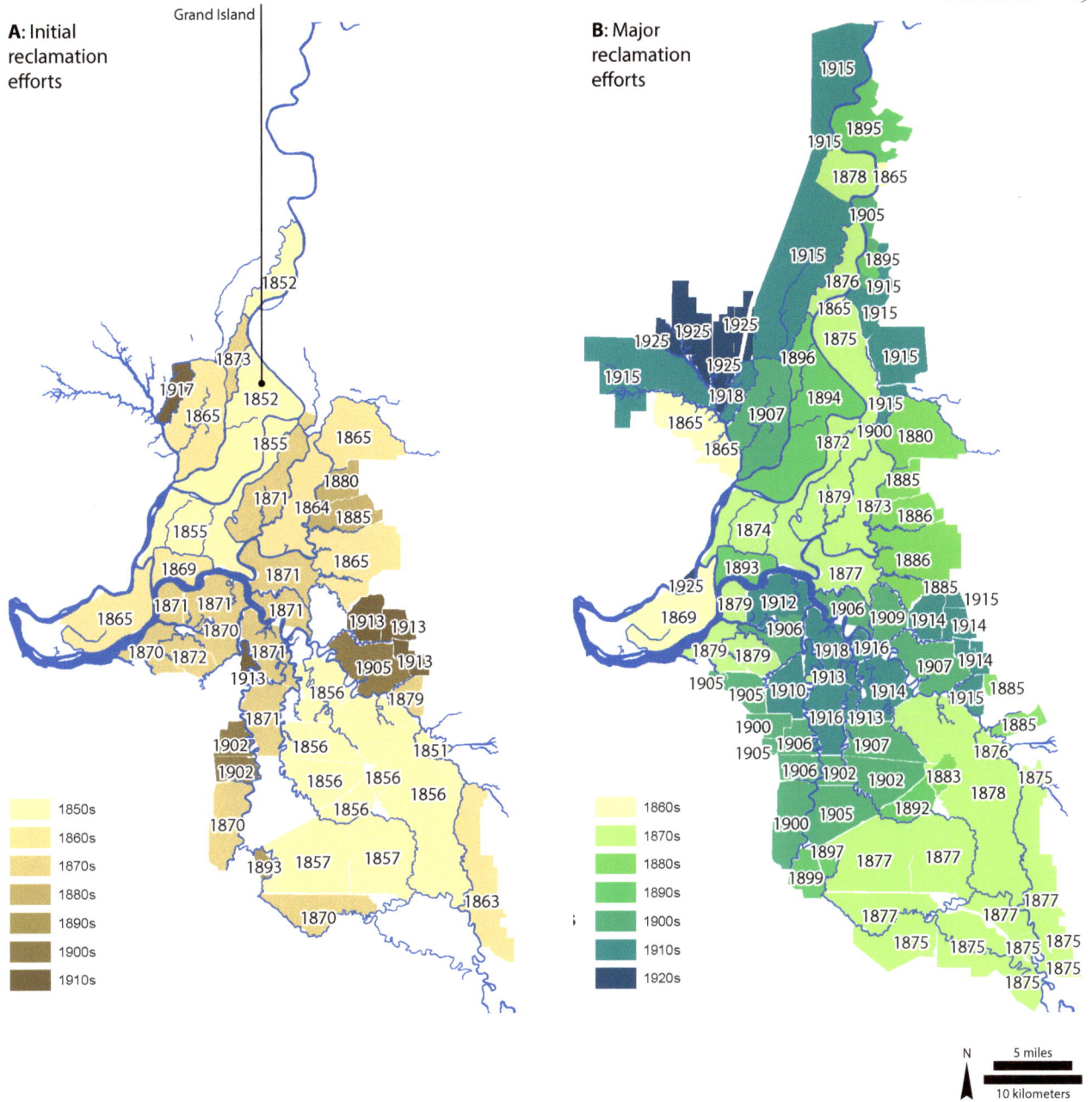

A: Initial reclamation efforts

Grand Island

1852

1873

1917

1865

1852

1855

1865

1880

1871

1864

1885

1855

1865

1869

1865

1871

1865

1871 1871

1871

1913 1913

1870

1871

1905 1913

1872

1913

1856

1879

1871

1856

1851

1902

1856 1856

1902

1856

1856

1870

1857 1857

1893

1863

1870

	1850s
	1860s
	1870s
	1880s
	1890s
	1900s
	1910s

B: Major reclamation efforts

1915

1895

1915

1878 1865

1905

1915 1895

1876 1915

1865 1915

1925 1925 1925 1875

1925 1896 1915

1915 1925 1894 1915

1918

1865 1907 1900 1880

1865 1872

1885

1879 1873 1886

1874 1886

1893 1877

1925 1885

1869 1879 1912 1906 1909 1914 1915

1906 1914

1879 1879 1918 1916 1907 1914

1905 1913 1914 1915 1885

1910 1916 1913 1885

1900 1906 1907

1905 1906 1902 1902 1883 1878 1876

1906 1892 1875

1900 1905

1897 1877 1877

1899

1877 1877

1877

1875 1875 1875 1875

1875

1875

	1860s
	1870s
	1880s
	1890s
	1900s
	1910s
	1920s

N

5 miles

10 kilometers

Figure 1.16. Reclamation sequence in the Delta. In A, the earliest date of known reclamation efforts is shown if this is different from the main date of reclamation given in historical records (B). For instance, small levees and damming of sloughs began as early as 1852 on Grand Island, but the island was not officially reclaimed until 1894. The dates given for the initial reclamation efforts do not indicate that the entire islands were reclaimed, simply that work had begun on that island or tract of land. Also, dates for major reclamation are those given for levee completion and do not necessarily indicate that the entire area was under agricultural production by that time. Information was compiled from numerous sources, including reclamation notes by the State Engineering Department (Tucker 1879a-f), John Thompson's (1957) reconstructions of reclamation sequence, those of the California State Lands Commission, and the Department of Water Resources.

1860: LARGE MAMMAL POPULATIONS ON THE DECLINE

Most large mammal populations significantly reduced or eliminated. Beaver and otter had become scarce several decades earlier due to intensive trapping.

1879: STRIPED BASS INTRODUCED INTO THE DELTA

This marked a period of other introductions in the rivers and streams of California. It also was a period when salmon hatcheries were established, in an attempt to recover already declining populations. This has led to the situation today, where the majority of salmon migrating through the Central Valley are of hatchery origin, without the unique endemism to individual rivers and streams that is found in wild populations.

1880: SALMON FISHING AND WATERFOWL HUNTING INTENSIFIES

Commercial fishing began in the 1850s and canneries were established in the 1860s (the first in 1864). A recorded 10.8 million pounds of salmon were caught in 1880, with the peak catch in 1909 at 12 million pounds (Yoshiyama et al. 1998). The last cannery on the Sacramento River went out of business in 1919. Hunting for ducks, geese, and other waterfowl for market increased in the decades after the Gold Rush. It became a significant business in the late 1800s, with markets in Sacramento and San Francisco for both the meat and eggs.

1886: RIPARIAN RIGHTS UPHELD

In *Lux v. Miller* and *Lux v. Haggin,* riparian water rights were upheld over appropriative rights.

1887: THE WRIGHT ACT PASSED

The Wright Act authorized the formation of irrigation districts, allowing and fostering the irrigation of lands not lying adjacent to rivers or streams.

1890: FISHERY AND WATER DIVERSION REGULATIONS

The State Board of Fish Commissioners began regulating fisheries and water diversions (Jacobs 1993).

1893: DEBRIS COMMISSION CREATED

With the Caminetti Act, the California Debris Commission was tasked with solving the mining debris problem. The first state appropriation for river improvement was in 1897. Detailed surveys and maps were made of the major rivers in the valley. The Dabney Commission report of 1904 presented flood control policy. The Debris Commission posed a number of solutions, including the "Minor Project" of 1907 and the "Major Project" of 1910, which involved dredging operations and included modifications such as the cutoff of Horseshoe Bend on the lower Sacramento River in 1914. This significantly increased the width and depth of the lower Sacramento River.

1902: FEDERAL RECLAMATION ACT

The passage of the Federal Reclamation Act created the U.S. Bureau of Reclamation which subsequently granted $500,000 to address hydraulic mining debris in California.

1914: WATER COMMISSION ACT

The current system of surface water rights was established with the passage of the Water Commission Act.

1920: SALINITY INTRUSION STUDIED

Salinity intrusion became significant in the 1920s, promting monitoring and a lawsuit. The intrusion caused the cities of Antioch and Pittsburg to seek new water sources in 1926. An investigation was initiated, and a report produced in 1931 on the control of salinity in the Delta, which proposed several options (CDPW 1931).

1931: STATE WATER PLAN PROPOSED

The State Water Plan, by State Engineer Edward Hyatt, laid out the system that was to transfer water from the north to the southern part of the state, though it had been proposed by others in various forms earlier. It was authorized in 1933 and was later undertaken by the Federal Government as a public works project, named the Central Valley Project. Today, it is one of the largest water infrastructure projects in the world and forms the core of the network of reservoirs, canals, and pumps that store and transport water throughout California.

1933: STOCKTON CHANNEL DREDGED

The Stockton Deep Water Ship Channel was dredged, deepening the channel substantially. This reach had been modified substantially by earlier leveeing and dredging efforts.

1938: O'SHAUGHNESSY DAM HERALDS AN ERA OF DAM BUILDING

Major changes in the hydrographs of rivers feeding into the Central Valley are brought about by an era of dam building that resulted in the damming of every major river in the Sierra Nevada, except for the Cosumnes River. One of the first major dams, the Mendota Dam on the San Joaquin, was constructed in 1881.

1940: DELTA WATER EXPORTS BEGIN

The completed Contra Costa Canal initiated water exports from the Delta. The Tracy (C.W. "Bill" Jones) Pumping Plant was completed in 1951, which brings water the Delta-Mendota Canal, all part of the Central Valley Project. Exports have continued to grow. Part of this increase is attributable to the completion of the California Aqueduct in 1973. Exports to the Bay Area and Southern California now exceed 6 million acre-feet.

1944: DELTA CROSS CHANNEL CONSTRUCTED

This channel, which connects the Sacramento River to the Mokelumne just upstream of Walnut Grove, facilitates water transfers through the Delta to the Tracy Pumping Plants. It has contributed to altering the magnitude and direction of flows in the Delta.

1955: SACRAMENTO DEEP WATER SHIPPING CANAL

Construction for the Sacramento Deep Water Shipping Channal began in 1955. The canal extends from West Sacramento southward through the Yolo Bypass to the outlet at Cache Slough. It was completed in 1963.

1966: THE FEDERAL ENDANGERED SPECIES PRESERVATION ACT

The Federal Endangered Species Preservation Act in 1966 established the protection of endangered species.

1969: ASSESSMENT OF ENVIRONMENTAL IMPACTS REQUIRED

The National Environmental Policy Act (NEPA) was passed in 1969, along with the state's Porter-Cologne Water Quality Act, which granted authority to the State Water Resources Control Board (SWRCB) regarding water quality. In 1970, the California Environmental Quality Act (CEQA) was passed, similar to NEPA.

1982: PERIPHERAL CANAL PROPOSAL REJECTED

A proposal to build a peripheral canal to divert water exports around the Delta to southern California was rejected by voters in 1982. It had been recommended by the Delta Environmental Advisory Committee in 1973.

1987: BAY-DELTA HEARINGS BEGIN

The Bay-Delta hearings were in response to the SWRCB failing to establish adequate water quality standards.

1993: LISTING OF ENDANGERED SPECIES

The Delta smelt (*Hypomesus transpacificus*), endemic to California, was listed as a threatened species in 1993 and it is now endangered. In 1999, spring-run Chinook salmon (*Oncorhynchus tshawytscha*) and Sacramento splittail (*Pogonichthys macrolepidotus*) also were added to the list of endangered species. The southern green sturgeon was listed as threatened in 2006.

1994: AGREEMENT ON WATER QUALITY STANDARDS

California and the EPA signed the Bay-Delta Accord in 1994, which formed CALFED. The California Bay-Delta Authority was created in 2002 to provide oversight for CALFED. Its authority and the Delta Science Program were later transferred to the Delta Stewardship Council.

2002: PELAGIC ORGANISM DECLINE

By 2004, researchers recognized long-term declines in abundance of important Delta fish species, including delta smelt, longfin smelt, threadfin shad, and juvenile striped bass, despite relatively good water years. The Interagency Ecological Program (IEP) has been monitoring and studying these declines since 2005 (Baxter et al. 2010).

SUMMARY

Understanding the historical landscape and how it has changed over time can help address many of the challenges associated with planning for the future. To develop the early 1800s picture of the Delta, historical ecology methods were employed, involving iterative analysis, synthesis, and interpretation of an extensive and diverse range of data sources.

Data collection and compilation (page 32) • To reconstruct the historical landscape, numerous disparate historical materials were acquired from a broad range of institutions. Journals, diaries, and newspaper articles described historical conditions. Early maps, surveys, and aerial photography provided the locations of historical features. Other important sources included landscape photography, sketches, and paintings. Collected data were then organized, read, transcribed, and georeferenced, depending on data type (page 34). Sources were drawn together along the themes of historical vegetation types, physical process, subregion, and land use. We georeferenced early maps, aerial photography, surveys, and narrative descriptions in a Geographic Information System (GIS) for purposes of mapping and spatial comparison.

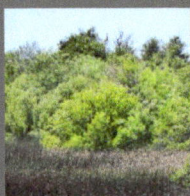

Classification (page 34) • The mapping classification consists of seventeen habitat types. Channels were classified as low order or mainstem and as tidal or fluvial. The other habitat types include freshwater pond or lake, freshwater intermittent pond or lake, tidal freshwater emergent wetland, non-tidal freshwater emergent wetland, willow thicket, willow riparian scrub or shrub, valley foothill riparian forest, wet meadow or seasonal wetland, vernal pool complex, alkali seasonal wetland complex, stabilized interior dune vegetation, grassland, and oak woodland or savanna. The classes were distinguished primarily by physical characteristics, such as landscape position and hydrology, and by general plant communities, such as oaks and sycamores on natural levees. This classification balanced the need to be consistent with the level of detail offered by the historical record and to provide classes relevant to contemporary classification. It reflects the goal of representing the extent and distribution of habitat types at the landscape scale.

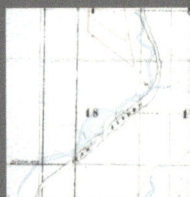

Data interpretation (page 46) • Careful interpretation of historical sources was required as diverse material spanning many decades was used to reconstruct the early 1800s landscape. The utility of individual sources was examined while considering the era, year, and season created; the methods used for production; the original purpose for creation; and the motivation and background of the creator. The climate, seasonal context, and land use changes present at the time the sources were created was also considered.

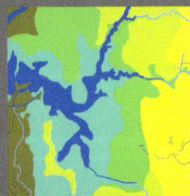

Mapping methodology (page 48) • We used GIS to synthesize data and represent historical landscape characteristics. Each mapped feature was attributed individually with supporting sources and relative certainty level (high, medium, or low) for its interpretation, size, and location (page 49). We used a minimum mapping unit of five acres (2 ha) and/or 50 feet (15 m) for the primary habitat (polygon) layer (page 51). A second geospatial data layer (lines) of rivers, streams and tidal channels consisted of the center lines of polygon channels as well as smaller channels (page 56). Different sources were key to mapping certain features. Early 1900s USGS topographic mapping and California Debris Commission survey maps often were the main sources used to map the larger channels in the Delta. Mapping smaller channels was aided by historical aerial photography, in which channels' tonal signatures are visible in the agricultural landscape. To map the perennial wetland boundary, we linked the edge of tule to appropriate elevations, using survey and other spatially accurate data (page 62). Riparian forest could often be mapped based on topographic contours that matched maps and survey data (page 68). Soil survey maps and descriptions aided identification and delineation of seasonal wetland types (page 73). Overall, mapping was supported through recognition of relationships between habitat types, landforms, and physical gradients in factors such as topography, soils, salinity, and hydrology. The resulting habitat type map does not represent exact conditions at any single point in time, but rather represents overall landscape patterns as they existed in the early 1800s Delta.

INTRODUCTION

Understanding the landscape of the Sacramento-San Joaquin Delta in the early 1800s depends upon the application of sound research methods to compile numerous unique datasets (Fig. 2.1). Historical ecology research is a process of integrating these diverse datasets to produce reliable, authoritative scientific products. Methods must address the uncertainties associated with using data that originate from many sources, often created for non-ecological purposes. The many possible interpretations that can be derived from each source must be carefully scrutinized and evaluated relative to the limitations of the data (Foster and Motzkin 2003). Furthermore, to appropriately interpret and describe the historical landscape, the process of synthesizing information must draw upon a wide range of disciplines, including ecology, geography, geology, hydrology, history, and anthropology. The research presented here has benefited from methods developed for other projects of the Historical Ecology group at the San Francisco Estuary Institute-Aquatic Science Center (SFEI-ASC; e.g., Grossinger et al. 2007), as well as from research performed by others in the field (e.g., Swetnam et al. 1999, Manies and Mladenoff 2000, Egan and Howell 2001, Foster 2002, Collins et al. 2003, Sanderson 2009).

Historical landscape reconstructions in the United States usually focus on the period just prior to significant Euro-American modification (Egan and Howell 2001). Over the span of many thousands of years, geologic and climatic conditions in the Delta have varied substantially, such that the reconstruction of habitat types of the early 1800s represents only a short period of time in the natural history of the Delta (see pages 7-8). However, while earlier conditions and processes provide context, the early 1800s is a relevant period for understanding how climatic and geologic processes formed and maintained habitats within a large-scale geomorphic and climate context that is relatively similar to today. The early 1800s picture – prior to substantial changes resulting from the fur trade, cattle, missionization of tribes, as well as the Gold Rush – provides opportunity for greater study of the physical and biological processes that were interacting to produce the arrays of habitat types to which species were adapted. Data drawn from the historical record can be used to reconstruct a picture of the relative influence of natural processes and people on the distribution, diversity, and abundance of habitats. This provides a starting point for developing conceptual models and restoration scenarios for projected future conditions.

The process by which historical sources are synthesized into a map and comprehensive understanding of landscape process and function relies upon extensive collection, organization, and inter-calibration of various datasets. This involves collection, assembly, compilation, and interpretation of a broad range of historical maps, texts, photographs, and art. Through iterative synthesis and interpretation, we integrate the many disparate sources and datasets to develop an understanding of prevailing conditions in the early 1800s. This process is discussed in detail in this chapter.

Figure 2.1. "W. R. McKean in tules '05". (chapter title page) This photograph was taken during the USGS survey that produced the first series of topographic quadrangle maps for the Delta. (USGS 1905, courtesy of the Center for Sacramento History, Hubert F. Rogers Collection, 2006/028/115)

DATA COLLECTION AND COMPILATION

Reconstructing historical landscapes often requires a broad range of historical data, as a single dataset rarely provides sufficient information for accurate interpretation of complex landscapes (Grossinger and Askevold 2005). Data collection constituted a significant component of this research. We acquired archival data including: 1) maps (e.g., Mexican land grant maps, regional maps, swamp and overflow and reclamation district maps, county surveys, soil surveys, and USGS topographic maps), 2) texts (e.g., Spanish explorer accounts, travelogues, court case testimony, diaries and letters home from early settlers, General Land Office surveys, county histories, and engineering reports), 3) photographs (plan view and oblique aerials and landscape photography) and 4) art (landscape sketches and paintings). The majority of data spanned the period of early Spanish explorers in the late 1700s to the time of the first aerial photography in the late 1930s.

We also drew from contemporary resources, including geology maps, soil surveys, vegetation maps, elevation datasets, and modern aerial photography. Such datasets, while depicting a changed landscape, can often reveal patterns that aid interpretation of the historical landscape when used in conjunction with historical data. For example, modern LiDAR can support the mapping of historical habitat features through interpretation of relict topography. Furthermore, contemporary data allow for comparison to historical conditions and identification of remnant pieces of the historical landscape.

The data collection process began with online searches to obtain publicly available digital material, as well as the development of a database of items to read or acquire at local and regional archives. These data were identified based on ecological and historical keywords. We conducted searches of over twenty websites and electronic databases. We also visited over thirty institutions (Table 2.1), acquiring digital photographs, scans, and photocopies of relevant historical documents.

Through the online and physical data collection process, we reviewed thousands of sources and acquired a subset of those reviewed. Undoubtedly, stones have been left unturned and relevant sources will continue to surface and add to the interpretation of the historical Delta landscape. The local historical record is extensive and additional ecologically meaningful pieces of information are often discovered. Additionally, certain sources only become relevant after substantial synthesis and analysis of other sources have been conducted. In particular, additional details too in-depth for this study, but relevant to specific topics or locations, can be found in sources such as William Hammond Hall's engineering surveys, in newspapers, court cases, and the California Debris Commission mapping.

Once collected, data were processed into readily available formats for mapping and interpretation at the local and landscape scale. We georeferenced selected maps (totaling over 450 individual maps) using

Table 2.1. Institutions from which data were collected for the historical ecology study of the Sacramento-San Joaquin Delta.

Source Institution	Archive Type	Location
Bureau of Land Management	Agency	Sacramento
Bank of Stockton Archives	Local	Stockton
California Department of Parks and Recreation	Agency	Sacramento
California Historical Society	Regional	San Francisco
California State Archives	Regional	Sacramento
California State Library	Regional	Sacramento
California State Railroad Museum Library	Regional	Sacramento
Center for Sacramento History	Local	Sacramento
Clarksburg Library	Local	Clarksburg
Contra Costa Historical Society	Local	Martinez
Dutra Museum of Dredging	Local	Rio Vista
East Contra Costa Historical Society & Museum	Local	Brentwood
Haggin Museum	Regional	Stockton
Isleton Brannan-Andrus Historical Society	Local	Isleton
Isleton Public Library	Local	Isleton
Reclamation District 999	Local	Clarksburg
Rio Vista Museum	Local	Rio Vista
Sacramento County Municipal Services Agency, Survey Section	Agency	Sacramento
Sacramento Public Library	Local	Sacramento
Sacramento Recorder	Agency	Sacramento
Sacramento River Delta Historical Society	Local	Walnut Grove
Sacramento State University – Special Collections and Maps	Local	Sacramento
Sacramento Surveyor	Agency	Sacramento
San Joaquin Historical Society and Museum	Local	Lodi
San Joaquin Surveyor	Agency	Stockton
Solano County Archives	Local	Fairfield
Solano County Public Works	Agency	Fairfield
State Lands Commission	Agency	Sacramento
Stockton Public Library	Local	Stockton
The Bancroft Library, UC Berkeley	Regional	Berkeley
UC Berkeley Earth Sciences Library	Regional	Berkeley
UC Davis Shields Library	Regional	Davis
University of the Pacific	Regional	Stockton
United States Geological Survey	Agency	Menlo Park
Water Resources Center Archives	Regional	Berkeley, now Riverside
West Sacramento Historical Society	Local	West Sacramento
Woodland Public Library	Local	Woodland
Yolo County Archives	Local	Woodland
Yolo County Recorder	Agency	Woodland
Yolo County Surveyor	Agency	Woodland

ESRI's ArcGIS 9.3.1 and 10 software. The earliest aerial imagery available (approximately 1,000 images, taken in 1937) was acquired from the UC Davis Shields Library and the UC Berkeley Earth Sciences Library and orthorectified in collaboration with the California Department of Fish and Game using the Leica Photogrammetry Suite module of ERDAS Imagine 9.2. These were mosaicked into nine datasets that together cover the entire study area.

We read historical texts and selected relevant sections that we transcribed and tagged by geographic area of concern (e.g., Yolo Basin, Mokelumne River) and subject (e.g., riparian forests, hydroperiod, channel geometry). Textual data were compiled from approximately 600 documents into over 400 pages of transcribed quotes. Where possible, text was also linked to specific locations within the GIS (over 800 entered data points). Another valuable dataset is the General Land Office (GLO) Public Land Survey field notes, which provides early spatially explicit ecological information (Box 2.1; Buordo 1956, Radeloff et al. 1999, Collins and Montgomery 2001). We entered and digitized over 3,500 GLO survey points. The database and data entry form were adapted from the Forest Landscape Ecology Lab at the University of Wisconsin-Madison (Manies 1997, Radeloff et al. 1998, Sickley et al. 2000). Non-georeferenceable maps and other images (e.g., landscape photography) were tagged by topic and area of concern. This process prepared data for mapping and allowed for swifter data queries.

CLASSIFICATION

Our mapping utilizes seventeen habitat types, which were based on historical evidence and modern classification systems (Table 2.2). These classes balance the need to provide the detail available in historical sources against the importance of communicating a consistent level of detail across the study area with classes that are relevant to contemporary classification. We drew primarily upon the natural communities identified by the Department of Fish and Game (CALFED 2000c). Although these classes are more oriented toward the aquatic environment, they provide a broad-level basis that translated well to available historical information. Additional classifications were added for historically significant habitat types in the Delta such as willow thickets.

Short explanations of these classes are included in the following sections. Many descriptions draw from contemporary definitions and classification systems (e.g., CALFED 2000b, Holstein 2000, Ornduff et al. 2003, Barbour et al. 2007, Grossinger et al. 2007, Sawyer et al. 2009).

Hydrography

Estuaries are, by definition, transitions between rivers or streams and ocean environments. Estuarine habitat types exist along the gradient between dominant fluvial and tidal processes. Just as the relative tidal influence varies along this spatial gradient, it is also not fixed in time. In the long term, the gradient shifts upstream with sea level rise, and might elongate or shorten depending on the slope of the upstream topography. The upstream position of different salinity regimes and of tidal influence also varies seasonally and from year-to-year for any given season. Because of this complexity, we kept the classification used fairly simple. We classified channels as either "mainstem" or "low order" and either "tidal" or "fluvial." Given that most channels within the study area (except the upland distributaries) were at least somewhat influenced by both tidal and fluvial processes, we identified channels by their probable hydrology, instead of dominating physical process. We classified tidal channels as those that likely experienced bidirectional (tidal) flow during spring tides in times of low river stages. This is different from identifying dominant physical processes in that we may classify a channel connected to a tidal mainstem channel as tidal because we believe it to have experienced the ebb and flow of tides, even though the primary processes forming and maintaining that channel were fluvial.

The "mainstem" channel class includes the rivers (including major streams such as Putah and Cache creeks) as well as the major tidal channels of the Delta. These channels are of high order with large contributing watersheds or are subtidal (i.e., beds below mean lower low water, MLLW) sloughs that delineate the Delta islands.

Fluvial low order channels include distributaries, side channels, swales and other minor channels within the upland Delta edge, channels associated with crevasse splays or other overflow channels dissecting natural levees, and channels within floodplain non-tidal marshes. Channels of this classification are commonly intermittent and dominated by fluvial processes. Such channels often dissipate across alluvial fans or natural levees toward lower-lying wetlands. Fluvial channels dissecting the natural levees of the tidally-influenced Sacramento River generally only flow at high river stages, meaning that their channel beds are likely above the elevations of high tides during low river stages. Channels dissecting natural levees that appear to have carried tidal flow at low river stages are classified as tidal. The numerous side channels and former channel meanders lacing the wetlands of the upper reaches of the San Joaquin distributaries, many of which carry water only during flood season, are also classified as fluvial low order channels. These channels do not carry tidal flows, despite in many cases being surrounded by tidal rivers (e.g., present day Stewart Tract).

Tidal low order channels comprise the sinuous channel networks of the Delta's tidal marshes that usually taper and branch toward the upland edge or drainage divide within a Delta island. The largest of these channels were probably high order, but most are first or second order. Tidal low order channels include both subtidal (beds below MLLW) and intertidal (beds exposed at low tide or beds only wetted at spring tides). Most tidal low order channels are limited to tidal wetlands. Exceptions include the headward reaches of tidal channels that intersect non-tidal uplands.

BOX 2.1. PRIMARY HISTORICAL DATA SOURCES FOR THE DELTA

Successful mapping and interpretation relies upon a diverse set of historical sources. We drew upon a variety of cartographic, textual, and pictorial sources spanning many decades. The summary below provides brief explanations for some sources.

Eliason 1854, courtesy of The Bancroft Library, UC Berkeley

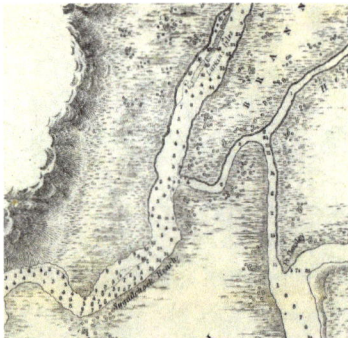

Ringgold 1850b, courtesy of the David Rumsey Map Collection, Cartography Associates

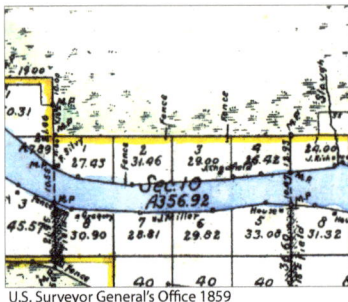

U.S. Surveyor General's Office 1859

Browning 1851, courtesy of The Bancroft Library, UC Berkeley

Jackson ca. 1870, courtesy of the California State Lands Commission

Mexican land grant sketches and court testimony (1840s-1860s). As the Mission system disintegrated, influential Mexican citizens submitted claims to the government for land grants. A *diseño,* or rough sketch of the solicited property, was included with each claim. *Diseños* often show notable physical landmarks which would have served as boundaries or natural resources, such as creeks, wetlands, springs, and forests. Upon California's admittance to the U.S., these claims were often granted through court proceedings, the testimony for which often includes some of the earliest descriptions available of the native landscape.

Expedition reports and mapping (1840s-1850s). Between 1838 and 1842, General Wilkes led the United States Exploring Expedition (referred to as the US Ex. Ex.) which eventually made its way to the San Francisco Estuary. Along with scientific reports concerning the geology and native flora and fauna, the US Ex. Ex. also produced a map of the Sacramento River. One of the members of the party, surveyor and cartographer Lieutenant Commander Ringgold, was commissioned in 1849 by "the enterprising citizens of San Francisco" to create the first navigational charts of the Sacramento and San Joaquin rivers (Ringgold 1852). For the mapping of the San Joaquin, he relied upon the recently published map of C.D. Gibbes (Gibbes 1850b, *Sacramento Transcript* 1850a, Ringgold 1852). The resulting maps from these efforts are particularly valuable because they were made before major modifications and include general depictions of vegetation patterns and river soundings. Few cartographic sources of this caliber exist from this time period.

General Land Office (GLO) Public Land surveys (1850s-1870s). Established in 1812, the GLO conducted the Public Land Survey in the Delta from 1853 through the 1870s, imposing a grid at the resolution of square mile sections on the landscape. Surveyors established survey lines by noting trees and other natural and cultural features. These spatially accurate and detailed field notes taken on the cusp of rapid settlement have been used extensively in historical landscape reconstruction and land cover change research (Buordo 1956, Radeloff et al. 1999, Collins and Montgomery 2001, Brown 2005, Whipple et al. 2011). Unfortunately, many townships within the Delta wetlands were left unsurveyed. Nevertheless, the dataset reveals substantial valuable information for many areas.

Textual accounts (1790s-present). Written accounts can provide a wealth of detailed information, with nuance about landscape dynamics not available on maps. We learn of floods and droughts, seasonal dynamics, depths and widths, water temperature and quality, relative size, species composition, notable features, changes due to land uses, and general character of a landscape that is difficult to visualize in three dimensions. Spanish expeditions provide the earliest accounts; later sources such as diaries of fur trappers, land grant case testimony, newspaper articles, ornithological records, county histories, and travelogues give rich perspectives from early visitors and residents.

Swampland and Reclamation District surveys (1860s-1880s). In the initial granting of swampland to states by the Arkansas Act of 1850, the surveys and subsequent sales of these lands were left to the State. As part of a convoluted history, the State created the Board of Swamp Land Commissioners in 1861 and surveys increased in number. In 1866, counties became responsible for the surveys and reclamation. These decades produced survey maps that can be invaluable in their depiction of conditions prior to significant reclamation efforts, but can be widely variable from one to another in coverage, quality, accuracy, and details depicted.

Paintings, sketches, photographs (1850s-present). Historical paintings and sketches, as well as the earliest photographs, offer a unique perspective of the landscape. Like narratives describing the landscape, this diverse dataset provides details that greatly improve the ability to visualize localized conditions on the ground. These sources can capture conditions in a specific place and time and are often remarkably accurate. Depictions of the Delta prior to substantial modification are extremely rare; those available provide important glimpses of the remarkable complexity and multiple scales of variability in the historical landscape.

Engineering reports (1878-present). The California State Engineer's office, established in 1878, was headed by William H. Hall until 1889. Much of the focus in the early years was directed toward drainage, irrigation, and flood control in the Central Valley. Numerous highly detailed and comprehensive surveys, reports, maps, and sketches were produced under Hall's direction. This body of work provides much of what we know about early conditions (during the hydraulic mining era) of California's streams and waterways.

U.S. Geological Survey 7.5-min topographic quadrangles (1909-1918). Shortly after 1900, the USGS (established in 1879) began surveys in the Delta region that resulted in unique 7.5 minute topographic quadrangles at a scale of 1:31,680 (use of this scale was discontinued in the 1950s). With contours of 10 feet, this scale provides great detail of the early topography of the Delta. Though dramatic changes (including subsidence and the dramatic loss of the wetlands and smaller tidal sloughs) had already occurred primarily from reclamation and hydraulic mining, these provide the earliest detailed, consistent, and comprehensive coverage for the entire region, offering invaluable information on topography and hydrography.

U.S. Department of Agriculture (USDA) soil surveys (1904-1940). Early soil surveys described variability in agricultural viability of regional soils. These maps, and their accompanying reports, are a key source used to infer historical land cover type and extent. Soil types were often mapped based on native vegetation, and the accompanying descriptions of soil properties, native vegetation, and agricultural uses are valuable. Twelve such early 1900s maps and associated reports exist for the area of study. Five were made between 1904 and 1909 and another four between 1930 and 1933, all covering different areas. Two of these were regional compilations done between 1910 and 1920. The final soil survey (1941) was conducted explicitly for the Delta.

California Debris Commission mapping (1893-1913). The Debris Commission was created in 1893 by the Caminetti Act primarily to investigate and address the effects of hydraulic mining debris on navigation and agriculture. Official surveying began in 1905, which led to the creation of highly detailed mapping (scales of 1:9,600 and 1:4,900) of channel bathymetry of the primary channels of the Delta. These maps contain channel cross-sections, profiles, and notes about dredge cuts and other features. While most mapping focuses solely on the channel, some maps include cartographic symbols for marsh and woody vegetation. The Commission developed many of the first flood control plans and projects of the Central Valley.

Historical aerial photography (1937). A Depression-era program to ensure crop stability and soil conservation practices resulted in extensive aerial photographic coverage for much of the country. While the photographs were taken after substantial modification, the photos nevertheless reveal relict ecological features, traces of which are often still present in the landscape. This dataset was particularly useful in detecting signatures of small blind tidal channels still evident in the landscape after many decades of farming. In many places, remnant features such as ponds, riparian forest and vernal pools are revealed. With confirmation from pre-reclamation sources, we were able to map features much more accurately from the signatures in the aerial photography.

McCurry ca. 1910, courtesy of the California History Room, California State Library, Sacramento

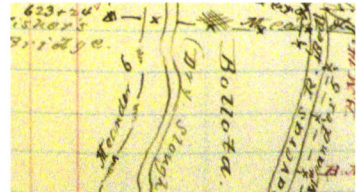
SED 1878, courtesy of the California State Archives

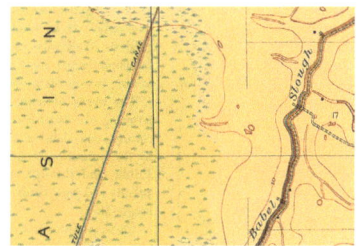
USGS 1909-1918

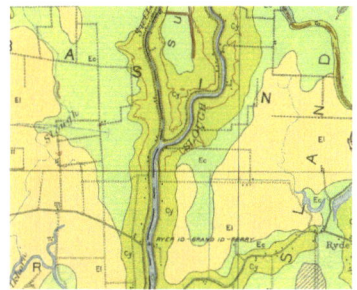
Carpenter and Cosby 1930

Wadsworth 1908b, courtesy of the California State Lands Commission

USDA 1937-1939, courtesy of the Map Collection of the Library of UC Davis and Earth Sciences & Map Library, UC Berkeley Library

Photo by Daniel Burmester

Photo by Daniel Burmester

Photo by Bill Miller

Photo by Bill Miller

Photo by Daniel Burmester

Photo by Bill Miller

Photo by Daniel Burmester

Photo by Bill Miller

Table 2.2. Habitat classification used to map the historical habitats of the Sacramento-San Joaquin Delta.

Landcover grouping	Habitat type	Description	MSCS NCCP Habitat Types (CALFED 2000c)
Water	Tidal mainstem channel	Rivers, major creeks, or major sloughs forming Delta islands where water is understood to have ebbed and flowed in the channel at times of low river flow. These delineated the islands of the Delta.	Tidal Perennial Aquatic
	Fluvial mainstem channel	Rives or major creeks with no influence of tides.	Valley Riverine Aquatic
	Tidal low order channel	Blind tidal channels (i.e., dead-end channels terminating within wetlands) where tides ebbed and flowed within the channel at times of low river flow.	Tidal Perennial Aquatic
	Fluvial low order channel	Distributaries, overflow channels, side channels, swales. No influence of tides. These occupied non-tidal floodplain environments or upland alluvial fans.	Valley Riverine Aquatic
	Freshwater pond or lake	Permanently flooded depressions, largely devoid of emergent Palustrine vegetation. These occupied the lowest-elevation positions within wetlands.	Tidal Perennial Aquatic, Lacustrine
	Freshwater intermittent pond or lake	Seasonally or temporarily flooded depressions, largely devoid of emergent Palustrine vegetation. These were most frequently found in vernal pool complexes at the Delta margins and also in the non-tidal floodplain environments.	N/A
Freshwater emergent wetland	Tidal freshwater emergent wetland	Perennially wet, high water table, dominated by emergent vegetation. Woody vegetation (e.g., willows) may be a significant component for some areas, particularly the western-central Delta. Wetted or inundated by spring tides at low river stages (approximating high tide levels).	Tidal Freshwater Emergent
	Non-tidal freshwater emergent wetland	Temporarily to permanently flooded, permanently saturated, freshwater non-tidal wetlands dominated by emergent vegetation. In the Delta, occupying upstream floodplain positions above tidal influence.	Non-tidal Freshwater Permanent Emergent
Willow thicket and riparian forest	Willow thicket	Perennially wet, dominated by woody vegetation (e.g., willows), emergent vegetation may be a significant component, generally located at the "sinks" of major creeks or rivers as they exit alluvial fans into the valley floor.	Valley/Foothill Riparian

Wildlife Habitat Relationship (WHR)	Representative types from California Terrestrial Natural Communities (CNDDB 2010)	Cowardin et al. (1979)/ USFWS Riparian Mapping System (USFWS 2009)	Hydrogeomorphic classification (HGM) (Brinson 1993)
Estuarine, Riverine	*Azolla (filiculoides, mexicana)* (Mosquito fern mats) Provisional Alliance (52.106.00), *Stuckenia (pectinata) - Potamogeton* spp. (Pondweed mats) Alliance (52.107.00)	Estuarine subtidal, Estuarine intertidal, Riverine	Riverine wetland, surface flow, unidirectional flow and bidirectional flow
Estuarine, Riverine	*Azolla (filiculoides, mexicana)* (Mosquito fern mats) Provisional Alliance (52.106.00), *Stuckenia (pectinata) - Potamogeton* spp. (Pondweed mats) Alliance (52.107.00)	Estuarine subtidal, Estuarine intertidal, Riverine	Riverine wetland, surface flow, unidirectional flow and bidirectional flow
Estuarine, Riverine	*Azolla (filiculoides, mexicana)* (Mosquito fern mats) Provisional Alliance (52.106.00), *Stuckenia (pectinata) - Potamogeton* spp. (Pondweed mats) Alliance (52.107.00)	Estuarine subtidal, Estuarine intertidal, Riverine	Riverine wetland, surface flow, unidirectional flow and bidirectional flow
Estuarine, Riverine	*Azolla (filiculoides, mexicana)* (Mosquito fern mats) Provisional Alliance (52.106.00), *Stuckenia (pectinata) - Potamogeton* spp. (Pondweed mats) Alliance (52.107.00)	Estuarine subtidal, Estuarine intertidal, Riverine	Riverine wetland, surface flow, unidirectional flow and bidirectional flow
Estuarine, Lacustrine	*Azolla (filiculoides, mexicana)* (Mosquito fern mats) Provisional Alliance (52.106.00), *Stuckenia (pectinata) - Potamogeton* spp. (Pondweed mats) Alliance (52.107.00), *Nuphar polysepala* (Yellow pond-lily mats) Provisional Alliance (52.110.00)	Lacustrine	Depressional wetland, surface flow and groundwater, vertical fluctuations
N/A	N/A	N/A	Depressional wetland, surface flow and groundwater, vertical fluctuations
Fresh Emergent Wetland	*Schoenoplectus acutus* (Hardstem bulrush marsh) Alliance (52.122.00), *Schoenoplectus californicus* (California bulrush marsh) Alliance (52.114.00), *Typha (domingensis, latifolia)* (Cattail marshes) Alliance (52.050.00), American bulrush marsh (52.111.00), California bulrush marsh (52.114.00), *Juncus effusus* (Soft rush marshes) Alliance (45.561.00), *Juncus articus* (Baltic and Mexican rush marshes) Alliance (45.562.00), *Salix lucida* (Shining willow groves) Alliance (61.204.00), *Eleocharis macrostachya* (Pale spike rush marshes) Alliance (45.230.00)	Estuarine intertidal persistent emergent wetland. Temporarily to seasonally flooded, permanently saturated.	Fringe wetland, surface flow including tidal, bidirectional flow
Fresh Emergent Wetland	*Schoenoplectus acutus* (Hardstem bulrush marsh) Alliance (52.122.00), *Schoenoplectus californicus* (California bulrush marsh) Alliance (52.114.00), *Typha (domingensis, latifolia)* (Cattail marshes) Alliance (52.050.00), *Juncus effusus* (Soft rush marshes) Alliance (45.561.00), *Juncus articus* (Baltic and Mexican rush marshes) Alliance (45.562.00), *Eleocharis macrostachya* (Pale spike rush marshes) Alliance (45.230.00)	Palustrine persistent emergent freshwater wetland. Temporarily to permanently flooded, permanently saturated.	Riverine wetland, surface flow, unidirectional flow
Valley foothill riparian	*Salix gooddingii* Alliance (61.211.00), *Salix laevigata* Alliance (61.205.00), *Salix lasiolepis* Alliance (61.201.00), *Salix lucida* Alliance (61.204.00), *Salix exigua* Alliance (61.209.00), *Cornus sericea* (Red osier thickets) Alliance (80.100.00), *Rosa californica* Alliance (63.907.00), *Acer negundo* (Box-elder forest) Alliance (61.440.00), *Sambucus nigra* (Blue elderberry stands) Alliance	Palustrine forested wetland. Temporarily flooded, permanently saturated. / Riparian scrub/shrub deciduous.	Riverine wetland, surface flow, vertical fluctuations

Photo by Bill Miller

Photo by Daniel Burmester

Photo by Daniel Burmester

Photo by Marc Hoshovsky

Photo by Jean Pawek

Photo by Christopher Thayer

Photo by Ingrid Taylar

Photo by Jean Pawek

Table 2.2. Habitat classification, continued.

Landcover grouping	Habitat type	Description	MSCS NCCP Habitat Types (CALFED 2000c)
Willow thicket and riparian forest (continued)	Willow riparian scrub or shrub	Riparian vegetation dominated by woody scrub or shrubs with few to no tall trees. This habitat type generally occupies long, relatively narrow corridors of lower natural levees along rivers and streams.	Valley/Foothill Riparian
	Valley foot-hill riparian	Mature riparian forest usually associated with a dense understory and mixed canopy, including sycamore, oaks, willows, and other trees. Occupied the supratidal natural levees of larger rivers that were occasionally flooded.	Valley/Foothill Riparian
Seasonal wetland	Wet meadow or seasonal wetland	Temporarily or seasonally flooded, herbaceous communities characterized by poorly-drained, clay-rich soils. These often comprised the upland edge of perennial wetlands.	Natural Seasonal Wetland
	Vernal pool complex	Area of seasonally flooded depressions, characterized by a relatively impermeable subsurface soil layer and distinctive vernal pool flora. These often comprised the upland edge of perennial wetlands.	Natural Seasonal Wetland
	Alkali seasonal wetland complex	Temporarily or seasonally flooded, herbaceous or scrub communities characterized by poorly-drained, clay-rich soils with a high residual salt content. These often comprised the upland edge of perennial wetlands.	Natural Seasonal Wetland
Other upland	Stabilized interior dune vegetation	Vegetation dominated by shrub species with some locations also supporting live oaks on the more stabilized dunes with more well-developed soil profiles.	Inland Dune Scrub
	Grassland	Low herbaceous communities occupying well-drained soils and composed of native forbs and annual and perennial grasses and usually devoid of trees. Few to no vernal pools present.	Grassland
	Oak wood-land or savanna	Oak dominated communities with sparse to dense cover (10-65% cover) and an herbaceous understory.	Valley/Foothill Woodland and Forest

Wildlife Habitat Relationship (WHR)	Representative types from California Terrestrial Natural Communities (CNDDB 2010)	Cowardin et al. (1979)/ USFWS Riparian Mapping System (USFWS 2009)	Hydrogeomorphic classification (HGM) (Brinson 1993)
Valley foothill riparian	*Salix gooddingii* Alliance (61.211.00), *Salix laevigata* Alliance (61.205.00), *Salix lasiolepis* Alliance (61.201.00), *Salix lucida* Alliance (61.204.00), *Salix exigua* Alliance (61.209.00), *Cornus sericea* (Red osier thickets) Alliance (80.100.00), *Rosa californica* Alliance (63.907.00), *Acer negundo* (Box-elder forest) Alliance (61.440.00), *Cephalanthus occidentalis* (Button willow thickets) Alliance (63.300.00)	Palustrine forested wetland. Intermittently flooded, seasonally saturated. / Riparian scrub/ shrub deciduous.	Riverine wetland, surface flow, vertical fluctuations
Valley foothill riparian	*Quercus agrifolia* Alliance (71.060.00), *Quercus lobata* Alliance (71.040.00), *Quercus (agrifolia, douglasii, garryana, kelloggii, lobata, wislizeni)* Alliance (71.100.00), *Quercus wislizeni* Alliance (71.080.00), *Juglans hindsii and Hybrids* Special stands (61.810.00), *Salix gooddingii* Alliance (61.211.00), *Salix laevigata* Alliance (61.205.00), *Salix lasiolepis* Alliance (61.201.00), *Salix lucida* Alliance (61.204.00), *Salix exigua* Alliance (61.209.00), *Acer negundo* (Box-elder forest) Alliance (61.440.00), *Cornus sericea* (Red osier thickets) Alliance (80.100.00), *Rosa californica* Alliance (63.907.00), *Platanus racemosa* Alliance (61.310.00), *Populus fremontii* Alliance (61.130.00), *Cephalanthus occidentalis* (Button willow thickets) Alliance (63.300.00)	Palustrine forested wetland. Intermittently flooded, seasonally saturated. / Riparian forested deciduous	Riverine wetland, surface flow, vertical fluctuations
Wet meadow	*Lasthenia californica - Plantago erecta - Vulpia microstachys* (California goldfields-dwarf plantain-six-weeks fescue flower fields) Alliance (44.108.00), *Elymus triticoides* (Creeping rye grass turfs) Alliance (41.080.00), *Ambrosia psilostachya* (Western ragweed meadows) Alliance (33.065.00), *Lotus purshianus* (Spanish clover fields) Provisional Herbaceous Alliance (52.230.00), *Juncus effusus* (Soft rush marshes) Alliance (45.561.00), *Juncus articus* (Baltic and Mexican rush marshes) Alliance (45.562.00)	Palustrine emergent wetland. Temporarily to seasonally flooded, seasonally saturated.	Depressional wetland, surface flow and groundwater, vertical fluctuations
Annual grassland	*Lasthenia fremontii - Downingia (bicornuta)* (Fremont's goldfields - Downingia vernal pools) Alliance (42.007.00), *Eryngium aristulatum* Alliance (42.004.00)	Palustrine nonpersistent emergent wetland.	Depressional wetland, surface flow and precipitation, vertical fluctuations
Alkali desert scrub	*Cressa truxillensis - Distichlis spicata* (Alkali weed - Salt grass playas and sinks) Alliance (46.100.00), *Lasthenia fremontii - Distichlis spicata* (Fremont's goldfields - Saltgrass alkaline vernal pools) Alliance (44.119.00), *Allenrolfea occidentalis* (Iodine bush scrub) Alliance (36.120.00), *Sporobolus airoides* (Alkali sacaton grassland) Alliance (41.010.00), *Elymus triticoides* (Creeping rye grass turfs) Alliance (41.080.00), *Frankenia salina* (Alkali heath marsh) Alliance (52.500.00)	Palustrine emergent saline wetland. Temporarily to seasonally flooded, seasonally to permanently saturated.	Depressional wetland, surface flow and precipitation, vertical fluctuations
Coastal scrub	*Lupinus albifrons* (Silver bush lupine scrub) Alliance (32.081.00), *Baccharis pilularis* (Coyote brush scrub) Alliance (32.060.00), *Lotus scoparius* (Deer weed scrub) Alliance (52.240.00)	N/A	N/A
Annual grassland, Perennial grassland	*Lasthenia californica - Plantago erecta - Vulpia microstachys* (California goldfields - Dwarf plantain - Six-weeks fescue flower fields) Alliance (44.108.00), *Elymus triticoides* (Creeping rye grass turfs) Alliance (41.080.00), *Nassella pulchra* Alliance (41.150.00), *Eschscholzia (californica)* (California poppy fields) Alliance (43.200.00), *Amsinckia* (Fiddleneck fields) Alliance (42.110.00), *Plagiobothrys nothofulvus* (Popcorn flower fields) Alliance (43.300.00)	N/A	N/A
Valley oak woodland, Blue oak woodland, Coastal oak woodland	*Quercus agrifolia* Alliance (71.060.00), *Quercus lobata* Alliance (71.040.00), *Quercus (agrifolia, douglasii, garryana, kelloggii, lobata, wislizeni)* Alliance (71.100.00), *Quercus wislizeni* Alliance (71.080.00), *Quercus douglasii* Alliance (71.020.00)	N/A	N/A

Freshwater pond or lake

These occupy topographic depressions that are either perennially or intermittently inundated and that lack abundant emergent marsh vascular plants. Perennial ponds and lakes of the historical Delta generally occupy backwater areas (against natural levees or the upland edge) within the wetlands of low-elevation lands lying parallel to the rivers, or flood basins. These areas probably received very little inorganic sediment. In some locations, large woody debris that caused waters to be impounded may be important in the formation and maintenance of these features. Those within the wetland complex are generally fed by surface water, with groundwater a component particularly in the summer months. Those within the upland ecotone, or zone of transition between perennial wetland and marginal habitats, may be fed by a combination of surface water, groundwater, and direct precipitation. Intermittent ponds and lakes are flooded only seasonally and are usually found at the upland edges of perennial wetlands.

Freshwater emergent wetland

Wetlands that support abundant freshwater rooted vegetation are classified as freshwater emergent wetlands. Salinities lower than 0.5 ppt generally characterize these wetlands (Cowardin et al. 1979). These marshes and swamps are associated with riverine floodplains (lands adjoining a channel that are subject to flooding every one to three years) and flood basins (extensive low-lying regions on the backside of natural levees) as well the upper regions of estuaries. Small freshwater emergent wetlands are associated with low-lying depressions and ponds, small channels, and localized areas of high groundwater. Freshwater wetlands are dominated by plant species such as bulrush or tule (*Schoenoplectus acutus, S. californicus, S. americanus*), cattails (*Typha* spp.), sedges (*Carex* spp.), spikerushes (*Eleocharis* spp.), rushes (*Juncus* spp.), smartweed (*Polygonum* spp.), and the common reed (*Phragmites australis;* Brandegee 1893-4, Jepson 1913, Atwater 1980, Barbour et al. 2007, Hickson and Keeler-Wolf 2007). Vegetation assemblages vary depending on physical drivers. For instance, *S. californicus* was likely more dominimant in the western Delta and along channels given its wind and wave reisistent structure, while the taller *S. acutus* grows in more protected areas like those in the north Delta flood basins (Keeler-Wolf pers. comm.). Particularly in the western-central Delta, this habitat type includes woody shrubs such as willow (*Salix* spp., primarily *S. lucida lasiandra*) and ferns (*Athyrium felix-femina*) to make up a unique plant community, perhaps related to maritime influences (Atwater 1980, Mason n.d., Keeler-Wolf pers. comm.). The wetland species are not precluded by seasonally dry conditions.

Freshwater emergent wetlands can be either tidal or non-tidal. Tidal freshwater emergent wetlands include those areas wetted at mean higher high water during low river stage and comprise what historical records often refer to as tidelands. Non-tidal freshwater emergent wetlands are not directly and predominantly affected by tidal action. However, tides may indirectly affect water table levels in freshwater emergent wetland and hydrological connectivity across landscapes during floods.

Willow thicket

This category includes broad stands of willow (*Salix* spp.), and occasional larger trees (e.g., cottonwood, *Populus fremontii*) that are usually associated with distributary channel networks at the base of alluvial fans and the margins of freshwater emergent wetlands (see discussion of "willow grove" in Goals Project 1999). Often, willow thickets (historically referred to as "sinks," "sausal," or "swamps") grade into freshwater emergent wetland such that the boundary between the two is indistinct. These areas are differentiated from the willow riparian scrub or shrub class because they share hydroperiod characteristics akin to freshwater emergent wetland, withstanding frequent flooding, prolonged periods of inundation, and saturation at or near the surface. They are also not generally linearly oriented along channels, but are larger and more rounded or ovate in plan form and are associated with distributary systems. They therefore occupy lower-elevation floodplain positions relative to riparian forest habitat types. As mapped, this type does not include the willow-fern communities within the freshwater emergent wetland in the western-central Delta.

Riparian forest

Riparian forest, mapped here as either willow riparian scrub or shrub or valley foothill riparian, is distinguished by the predominance of medium to tall woody vegetation adjacent to waterways. Riparian vegetation is usually distinctive due to its lushness as well as its species composition and landscape position. This category includes broad relatively open forests, forests with a dense understory and tall canopy, and riparian scrub or shrub thickets (defined as woody plants generally <10 m in height, usually with two or more stems at the base). In the Delta, these forests are usually associated with the natural levees of the larger rivers. Common species include sycamore (*Platanus racemosa*), Fremont cottonwood (*Populus fremontii*), valley oak (*Quercus lobata*), live oak (*Quercus wislizenii*), Oregon ash (*Fraxinus latifolia*), black walnut (*Juglans californica*), box elder (*Acer negundo*), white alder (*Alnus rhombifolia*), willows (*Salix* spp.), buttonwillow (*Cephalanthus occidentalis*), river dogwood (*Cornus pubescens*), and other bushes (*Rubus ursinus, Rosa californica*) and vines (*Vitis californicus*; Belcher 1843, Day 1869, Jepson 1893, Jepson 1913, Sullivan 1934, Hickson and Keeler-Wolf 2007, Vaghti and Greco 2007). Oaks are found along the higher elevations of natural levees where inundation frequency is low, while willows and other shrub species predominate where flood frequencies are higher and the water table is closer to the surface.

Wet meadow or seasonal wetland complex

Wet meadows and seasonal wetlands are temporarily or seasonally flooded herbaceous communities characterized by poorly drained, clay-rich soils (Goals Project 1999, Cowardin et al. 1979). They are distinguished from freshwater emergent wetlands in part by a lack of dominance of tall emergent monocots such as tules and cattails. These mosaics of moist grasslands generally lie adjacent to freshwater emergent wetlands within the upland ecotone (Goals Project 1999). Characteristic plant species include those found near pools as sparse cover (e.g., goldfields, *Lasthenia* spp.) as well as those found in grasslands (e.g., creeping rye grass, *Elymus triticoides*).This category includes areas that may not satisfy the contemporary state or federal definition of jurisdictional wetland.

Vernal pool complex

Vernal pools are seasonally flooded depressional wetlands underlain by a drainage-limiting subsurface layer and dominated by vernal pool endemic plant species (Holland 1978, Keeler-Wolf et al. 1998, SFEI 2011). These ephemeral shallow ponds are fed primarily by precipitation, though overland and shallow groundwater flow may contribute significantly as well. They are characterized by distinctive and uniquely adapted annual and perennial forbs, such as *Plagiobothrys* spp., *Downingia* spp., *Eryngium, Navarretia, Psilocarphus,* and goldfields (*Lasthenia* spp.; Holstein 2000). Although vernal pools may occur individually, they generally occur together and can be described as a vernal pool complex that includes the surrounding matrix of grassland (Keeler-Wolf et al. 1998). Some vernal pools are characterized by alkali, though the seasonally flooded depressions remain distinguishing features. In the Delta, vernal pool complexes intermix with wet meadow and seasonal wetlands, and alkali seasonal wetland complexes. Consequently, there is some inevitable overlap in these categories.

Alkali seasonal wetland complex

Alkali-associated habitat types typically occur in mosaics of salt-influenced seasonal and perennial wetland types (Holland 1978). They tend to be positioned in the hotter and drier regions of the Delta. They are characterized by fine-grained soils with high residual salt content (0.1% and higher) supporting distinctive, salt-tolerant plant species (Baye et al. 2000, Holstein 2000). Dominant vegetation may include salt grass (*Distichlis spicata*), *Crypsis schoenoides, Eryngium aristulatum, Plagiobothrys leptocladus Pleuropogon californicus,* alkali weed (*Cressa truxillensis*), saltbush (*Atriplex* spp.), alkali heath (*Frankenia salina*), and iodine bush (*Allenrolfea occidentalis*; Ornduff et al. 2003, Sawyer et al. 2009). Distribution and character of alkali habitat types within the complexes vary with salt concentrations, duration of soil saturation, topography, and groundwater depth (Holland 1986, Elmore et al. 2006).

Although well established in the historical record, it was not possible to explicitly map different types of alkali wetlands due to mapping scale and spatial complexity (e.g., perennial alkali wetlands and flats surrounded by seasonal alkali meadow and intermixed with grassland). As mapped, this class represents a mix of alkali habitat types, with varying salt concentrations and inundation frequencies, including alkali meadow (seasonally wet, alkali-affected herbaceous grasslands and forblands), alkali sink scrub (shrub cover of iodine bush, seep weed (*Suaeda* spp.), and Parish's glasswort (*Arthrocnemum subterminale*)), alkali playas (highest alkali intensity of over 1%), and alkali marsh. There was likely great local-scale complexity due to topography, soil, and drainage patterns that is not represented in the habitat mapping. These types also intermix with vernal pool complex, wet meadow and seasonal wetland complex, and freshwater emergent wetland.

Stabilized interior dune vegetation

This type is associated with the relict glacial-age sand dune deposits, which appeared as mound rising above the tidal wetlands like islands in the western Delta. These Pleistocene sand fields were established by winds that blew glacial sands from the Sierra Nevada into dunes (Atwater 1982). They subsequently underwent stabilization and soil profile development, which allowed for the growth of live oaks (*Quercus agrifolia*), forbs, and grasses (Carpenter and Cosby 1939, USDA 1977). More exposed areas likely supported vegetation similar to that of the interior dune scrub associated with these sands today (sometimes referred to as inland dune scrub; Holland 1986, CALFED 2000a, CALFED 2000c, Bettelheim and Thayer 2006, CDFG 2010, Thayer 2010). Such a dune scrub community is found at the Antioch Dunes National Wildlife Refuge, which supports silver bush lupine (*Lupinus albifrons*), the rare Contra Costa wallflower (*Erysium capitatum* ssp. *Angustatum*), and the Antioch primrose (*Oenothera deltoids* var. *howellii*; Howard and Arnold 1980, CALFED 2000a). Some of the larger dunes along the San Joaquin River, including Antioch Dunes, may have been over one hundred feet in height when Spanish explorers encountered them in the later 1700s. The mounds within the wetlands to the east were known to be over 15 feet high (Davidson 1887, USGS 1909-1918, Howard and Arnold 1980). Height was significantly reduced, soils disturbed, and vegetation removed due to mining for the production of bricks and asphalt in the 1880s (Stanford et al. 2011).

Grassland

Grassland is characterized by low herbaceous plant communities, where the soils are rarely saturated and generally have high water-holding capacity and abundant exposure to solar radiation (Holstein 2000). This type encompasses various mixes of annual and perennial grasses and annual forbs (wildflowers), where species composition depends on factors such as climate, topography, and soils (Holstein 2001, Minnich 2008, Sawyer et al. 2009). A variety of different communities were found around in the Delta, transitioning in composition depending on gradients in climate, soils, and topography. The grasslands expressed everything from cold north coastal prairie in the western Delta to valley needlegrass grassland (*Nassella pulchra*) communities in the vicinity of Jepson Prairie. Annual grasslands likely abutted on hills and very well drained sandy soils (Keeler-Wolf et al. 2007). Characteristic species include three-awn (*Aristida* spp.), bunch grass (*Poa*), needle grass (*Stipa*), blue wildrye (*Elymus glaucus*), goldfields (*Lasthenia* spp.), fescue (*Vulpia microstachys*), California poppy (*Eschscholzia californica*), and fiddleneck (*Amsinckia* spp.; Ornduff et al. 2003, Keeler-Wolf et al. 2007, Minnich 2008, Sawyer et al. 2009). Jepson (1893) describes colorful annual herbaceous plants including "Lupines, Clovers, Calandrinias, Platystemons, Baerias [*Lasthenia*], Gilias, Nemophilas and Allocaryas [*Plagiobothrys*]." He also notes that "the shallow streams and pools are edged with handsome Eunani [*Mimulus*] and curious Bolelias [*Downingia*]" (Jepson 1893).

Oak woodland or savanna

The oak woodland or savanna type is characterized by sparse to moderate tree cover (10-25%) with an open understory of herbaceous vegetation (Sawyer and Keeler-Wolf 1995, Allen-Diaz et al. 1999, Davis et al. 2000, Barbour et al. 2007). It is dominated by oak species, including blue oak (*Quercus douglasii*), valley oak (*Quercus lobata*), and coast live oak (*Quercus agrifolia*). The distribution of trees is quite variable, depending on local variation in soil properties, water availability, and topography. Density ranges from thick groves of trees to scattered trees within an open herbaceous plain.

DATA INTERPRETATION

Integration and interpretation of documents produced during different eras, using different methods or techniques, for differing purposes, and with different authors, surveyors, or artists can be challenging (Askevold 2005). Only when compared against each other can these datasets reveal prevailing landscape patterns and processes (Harley 1989, Swetnam et al. 1999). For example, an early 1900s map may be very detailed and accurate in its depiction of a channel, but it is unclear whether the channel existed in the early 1800s without earlier data. Individual sources provide only a single, limited view through which to understand a complex past. Combining, comparing and integrating a wide range of data leads to an improved understanding of historical spatial and temporal patterns of physical and ecological processes. The interpretation of these patterns is guided by an iterative process of source intercalibration and triangulation, where GIS is a central organizing tool. This approach provided independent verification of the accuracy of original documents and our interpretation of them (Grossinger 2005, Grossinger et al. 2007).

The process of source intercalibration and confirmation yields more accurate mapping, as detailed features visible in later sources (i.e., post-reclamation) can often be confirmed by earlier, less spatially explicit, sources. For example, the general alignment of a slough shown on a land grant map is visible more explicitly as a tonal signature in the 1937 aerials, where that signature illustrates the many meanders of the slough which the less precise earlier map does not represent.

Additionally, since many data were created after significant landscape changes had already occurred, it is necessary to understand the contemporary land use context and earlier historical events for accurate data interpretation. For example, we considered the effects of hydraulic mining debris and early levee building when using sources created during those time periods. We calibrated our interpretation of such sources with more general (less geographically precise) pre-modification observations or localized survey points. Inevitably, however, while the process of triangulating sources increases the accuracy of the reconstruction, uncertainties necessarily arise, particularly when few or no early sources are available for confirmation.

It is also important to consider the extensive changes that affected the Delta (and elsewhere in California) shortly after Spanish contact in 1769. Three of these early changes affected the Delta and Central Valley as seen by settlers during the Gold Rush. One is the decrease in native management as the Indian population dropped precipitously due to missionization and epidemics (Cook 1955b). A second early impact is the influence of cattle grazing and the concurrent rapid invasion of non-native grasses (Minnich 2008). A third is the dramatic decline in populations of some wildlife species such as beaver, antelope, and elk that were trapped and hunted extensively beginning several decades prior to the Gold Rush (McCullough 1969).

Data must also be interpreted within the context of ecosystem dynamics on an annual to decadal scale, though it appears many features were remarkably stable. Knowledge of whether data originated during a wet or dry year or at a particular time of year affects our interpretation of them, and subsequently our interpretation of overall "average" or "prevailing" historical conditions. Evidence of lakes within the Yolo Basin in early spring of a flood year means something quite different than observations of lakes during the late summer months during a drought. In addition, we adjusted our understanding of features based on our knowledge of their relative stability through time. For example, some feature types, such as small oxbow lakes or side channels in the south Delta, were more changeable on a decadal scale than many other landscape features (observed through the comparison of different datasets across decades).

A final consideration particularly significant in the subsided Delta of today is that of former ground surface elevations. We did not map historical elevation, but used topography shown by the early 1900s USGS topographic maps to determine tidal extent and to interpolate other habitat boundaries from known points (USGS 1909-1918). Since subsidence has occurred primarily within the central Delta tidal islands where peats were deep, surface elevations at the tidal margin – where peat was shallow or non-existent – have remained relatively stable and could be used for defining habitat type boundaries. Early USGS topographic mapping in the Delta was done prior to standardized tidal datums. The datum for this series of maps is "sea level," which makes it difficult to be confident of the exact elevations within several feet (Atwater pers. comm.). However, this datum can be roughly equated to National Geodetic Vertical Datum (NGVD) 29, which was originally called the "Sea Level Datum of 1929" and was established by measuring mean sea level at 26 tide gauges in the United States and Canada. This translation is supported by 1900s USGS topographic maps that show the edge of mapped tidal wetlands roughly at the 3.5 foot contour line, or the general extent of tidal range in the Delta (Atwater 1982). Unfortunately, we were unable to locate extensive documentation concerning this series of maps.

MAPPING METHODOLOGY

The primary purpose of the mapping process is to represent the diversity and heterogeneity of habitat types at the landscape scale, leading to a better understanding of regional ecological patterns and processes. We aimed to illustrate features and characteristics that could be mapped consistently across the study area. Representing nested scales of complexity within the limits of a two-dimensional map is challenging, however. Some local details and complexity were necessarily excluded from the map, instead described in this report. Consequently, this report is a useful companion for interpretation of the map of historical habitat type extent and distribution produced for this study.

Instead of mapping conditions at a specific point in time, we endeavored to map representative and relatively stable features in the landscape for the early 1800s. The map shows the prevailing dry-season conditions and components of the landforms and habitat types during the period just before major human-induced changes, including reclamation, water withdrawal, and hydraulic mining. Through the synthesis across datasets spanning many decades, we were able to determine those features remaining relatively stable despite inter- and intra-annual climate variability. Additionally, we reconstructed the general patterns of less stable features (e.g., oxbow lakes and ephemeral channels in the south Delta). Thus, while the map does not represent the landscape at a single point in time, the distributional patterns and relative patch sizes can be used to understand the landscape under natural conditions in the early 1800s.

We used a geographic information system (GIS) for source intercalibration, synthesis, and digitizing data layers that represent historical landscape characteristics of the Delta. As a spatial database, GIS allows for the comparison of input of data from many disparate sources and time periods at a single location in space (Fig. 2.2). The relational database component of GIS provides for storage of many attributes about a single feature, which we used to integrate the datasets and document the provenance of our interpretation of the historical landscape. Using GIS, we were able to integrate complex arrays of data by assembling maps and narrative information from different periods, allowing us to assess each data source, more accurately map each feature, and better understand change over time. We used ArcGIS 9.3 and 10 (ESRI) software.

To document the mapping sources and interpretation in the GIS, we attributed each feature with the sources used to map it. Three types were documented: digitizing sources, primary interpretation sources (if other than the digitizing source), and supporting interpretation sources.

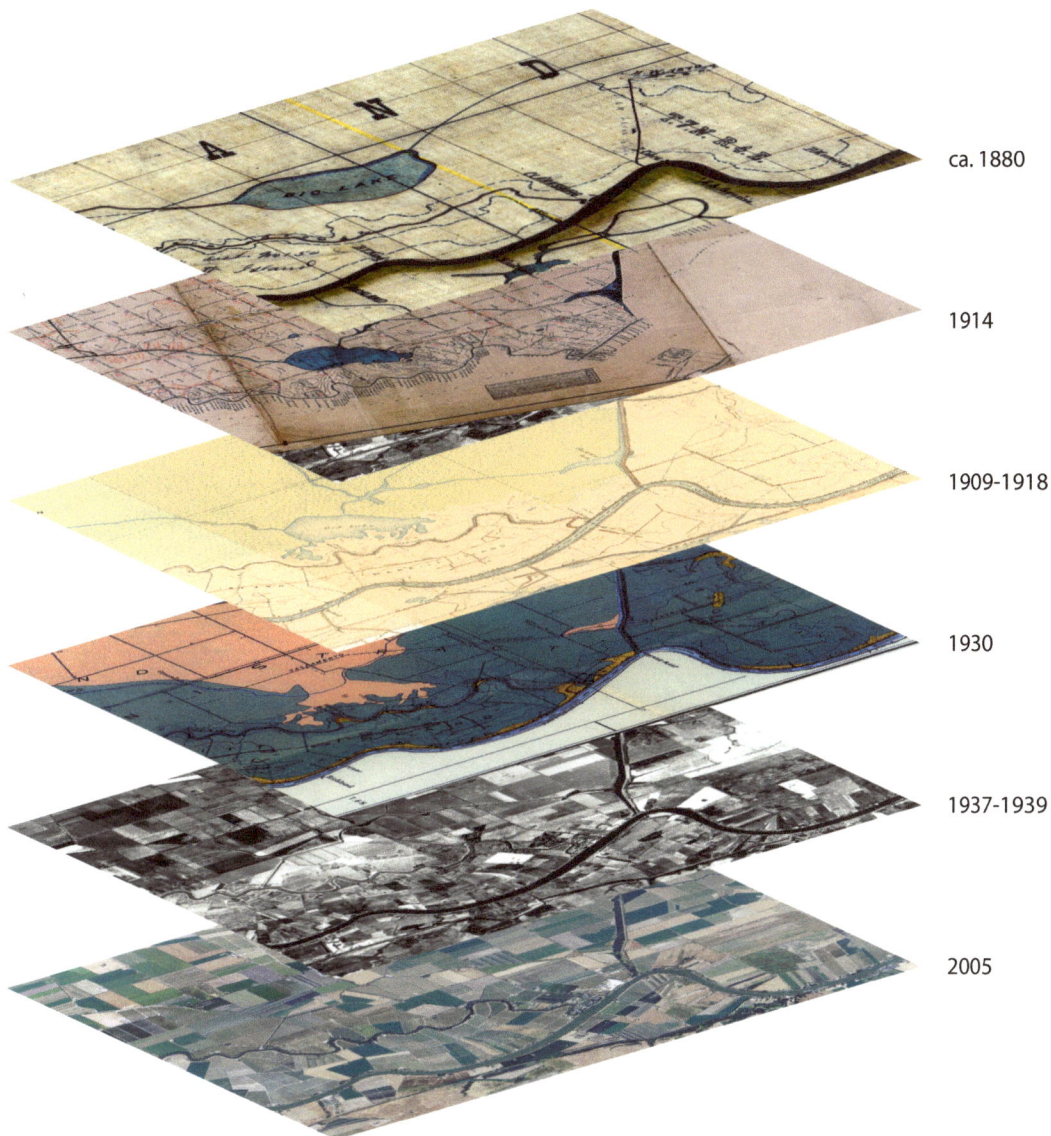

ca. 1880

1914

1909-1918

1930

1937-1939

2005

Where possible, these sources were drawn from varying years and authors. We did not attempt to document every piece of evidence that showed the feature, but those that contributed most to its delineation and interpretation.

We assigned estimated certainty levels to each feature. Our confidence in a feature's habitat type and presence (interpretation), size, and location was assigned based upon the number of kinds and quality of evidence, accuracy of digitizing source, our experience with the particular aspects of each data source, and by factors such as stability of features on a decadal scale (following standards discussed in Grossinger et al. 2007; Table 2.3). Certainty in tidal status was also included for the channel line layer. In cases where features were likely to have shifted positions over relatively short time periods, we assigned lower certainty for location and size. These attributes provide a way to estimate ranges of uncertainty associated with different locations and kinds of feature or habitat type, and allows subsequent users to assess accuracy (Fig. 2.3).

Figure 2.2. Maps assembled from different time periods shown in a geographic information system allows for comparison of features across space and time. (top to bottom: Hall ca. 1880c, courtesy of the California State Archives; Haviland 1914, courtesy of Reclamation District 999; USGS 1909-1918; Carpenter and Cosby 1930; USDA 1937-1939; and USDA 2005)

Table 2.3. Certainty level standards assigned to each mapped feature for the assessment of confidence in interpretation (classification and historical presence), size, location, and tidal status.

Certainty Level	Interpretation	Size	Location	Tidal Status (line features only)
High/ "Definite"	Feature definitely present before Euro-American modification	Mapped feature expected to be 90%-110% of actual feature size	Expected maximum horizontal displacement less than 50 meters (150 ft)	Channel bed definitely within or outside tidal range (<3.5 ft elevation)
Medium/ "Probable"	Feature probably present before Euro-American modification	Mapped feature expected to be 50%-200% of actual feature size	Expected maximum horizontal displacement less than 150 meters (500 ft)	Channel bed probably within or outside tidal range
Low/ "Possible"	Feature possibly present before Euro-American modification	Mapped feature expected to be 25%-400% of actual feature size	Expected maximum horizontal displacement less than 500 meters (1,600 ft)	Channel bed possibly within or outside tidal range (if within, no clear tidal connection)

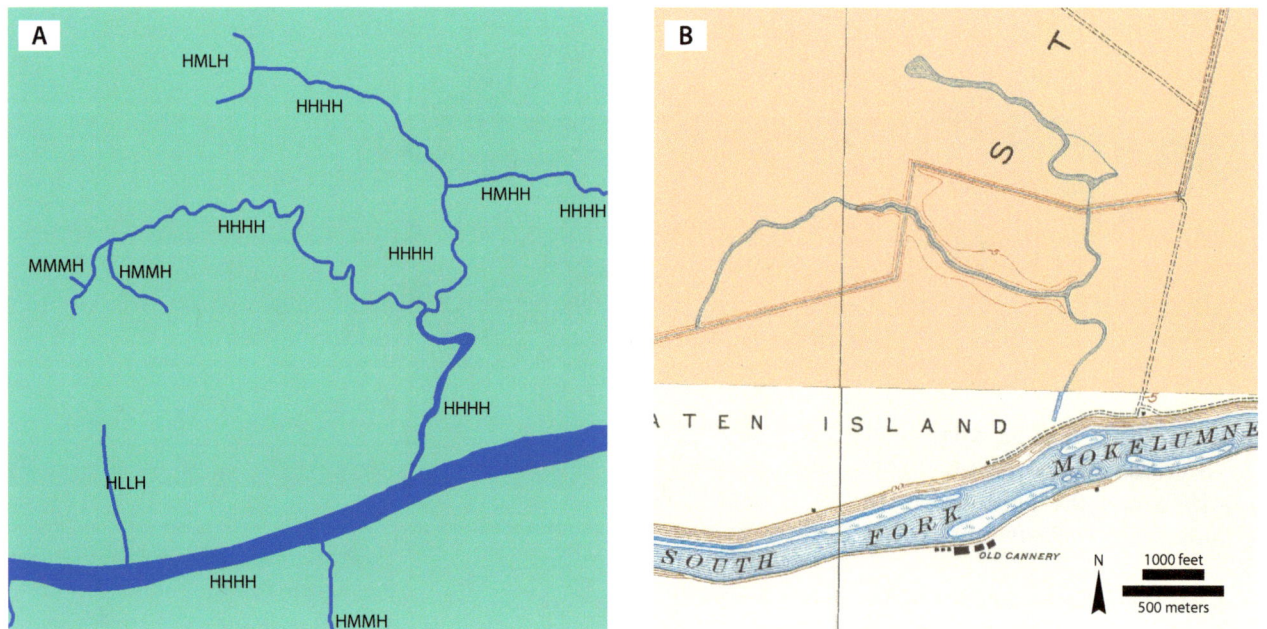

Figure 2.3. Assignment of certainty levels to channels. A channel network (A) off the Mokelumne River within lower Staten Island is shown with certainty levels (in order of interpretation, size, location, and tidal status) assigned to parts depending on sources used to map the features. One of the sources used to map the network, the early 1900s USGS topographic maps, is shown in (B), where not all historical channels are shown, due to reclamation. (B: USGS 1909-1918)

The differences among datasets prevents mapping each feature with the same level of accuracy. While many individually mapped habitat features were assigned high levels of certainty, others were mapped with less confidence. In some cases, a high density of evidence documenting a particular feature or early explicit and accurate detail allowed for high mapping confidence of both presence and extent. However, many individual features are documented by only one piece of evidence and some are not associated with any. In these cases, we inferred conditions based on soil types, topography, hydrology, and general descriptions. Undoubtedly, some features were undocumented in the historical record. It should also be recognized that mapping requires drawing thin boundary lines where the true boundaries are often quite broad ecotonal gradients.

We used a minimum mapping unit of five acres for the primary feature (polygon) habitat layer. This allowed us to capture a large diversity of fairly stable and significant features of the historical landscape. In an effort to accurately portray the heterogeneity within habitat mosaics, but also be true to the accuracy of the historical sources, some habitat types were mapped as complexes that encompassed small features such as small ponds, beaver cuts, and willow patches. Habitat types that were characterized by particularly small features are less likely to be well represented in historical sources and are therefore likely under-represented in the mapping.

Additionally, we captured features less than five acres in size in a GIS layer separate from the primary feature layer. These features tend to be small ponds or oxbow lakes and usually documented by only one source. They are often post-reclamation, which makes interpretation challenging given the degree of change that occurred during the early years of reclamation.

The following sections outline the methods used to integrate and synthesize data in GIS to depict habitat types on the map, both for the purpose of visual representation and for quantitative analysis. We explain the basis for the mapping and describe any important associated uncertainties. Information concerning primary mapping sources and caveats for habitat types is summarized in Table 2.4. For more information on the accuracy of a particular habitat polygon, please refer to the GIS metadata.

Hydrography

Understanding the way water was historically routed through the Delta is critical for determining the relative influence of dominant physical processes (e.g., tides, floods), the nature of flow and hydrologic and ecological connectivity within the system, the character of habitat available to native species, and for selecting various metrics that can describe the landscape, such as channel density or channel edge-to-area ratio. Great effort was placed, therefore, on mapping the channel network of the historical Delta, including the major sloughs that formed the Delta islands as well as the smaller channels – the sinuous dead-end or blind tidal networks of the wetlands and ephemeral distributaries that fed into the Delta (Fig. 2.4). In the GIS, we mapped all channels as line features (features with no width dimension) and also mapped the larger channels as polygon features (features associated with area in GIS). We used a minimum mapping width for polygon channels of 50 feet (15 m). This minimum mapping width was determined based on the infeasibility of accurately mapping polygon features smaller than these due to available data (usually either USGS topographic maps or signatures in historical aerial photography). We found that such widths generally captured the channels that were mapped in regional maps of the Delta in early 1860s and 1870s. In addition, these standards are comparable to the USGS 1:24,000 mapping standards that use a 40 foot (12.2 m) mapping width. We used a minimum mapping length for channels of 165 feet (50 m). We had no minimum mapping width for including channels in the GIS layer of channel lines.

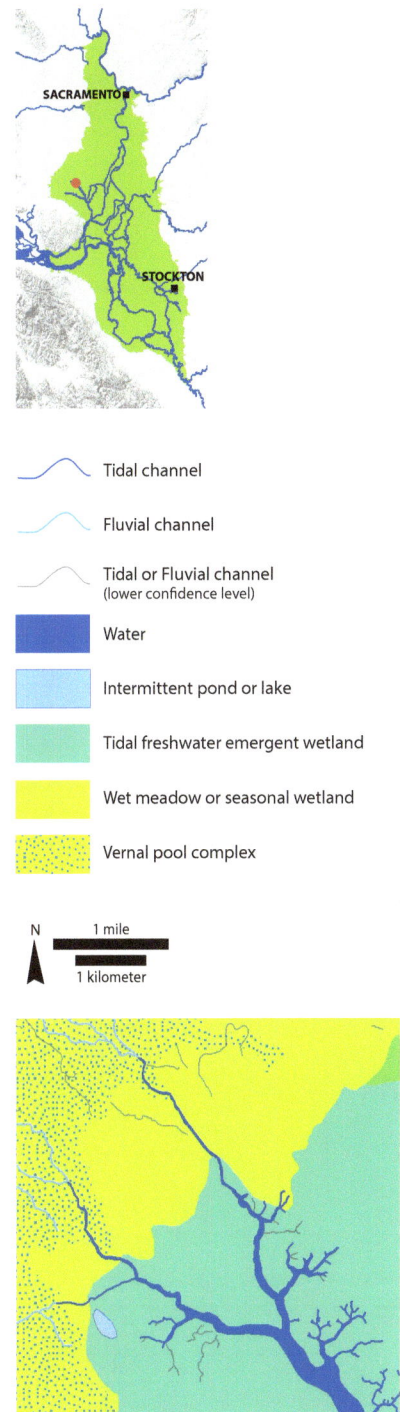

Figure 2.4. Detail of hydrography. In addition to single-line mapping, channels wider than 50 feet (15 m) were mapped as double-line channels (polygons). Channels were attributed as either "mainstem" or "low order" and as "tidal" or "fluvial."

Table 2.4. Primary mapping sources and relevant caveats pertaining to each habitat type.

Data type	Landcover grouping	Habitat type	Primary sources for mapping	Caveats
Line features		Tidal mainstem channel	Early 1900s California Debris Commission mapping and 1937 historical aerial photography. Early (pre-hydraulic mining era) narrative accounts to determine tidal extent.	The maximum extent of tidal influence was not found in historical data in the case of the San Joaquin and several other rivers.
		Fluvial mainstem channel		
		Tidal low order channel	Historical aerial photography and early maps where available. Many features digitized from "ghost" or remnant channel signatures visible within agricultural fields in the historical aerial photography. These signatures are composed of lighter inorganic soils associated with channels, which are detectable against the darker organic peat soils of the historical tidal wetlands.	Given the sources available, mapping most likely does not include the lowest order channels, so these features are likely undermapped. However, particularly in the South Delta, overmapping may have resulted due to the presence of "exhumed" channels in historical aerial photography (ancient channels that are exposed due to loss of peat to oxidation).
		Fluvial low order channel		
Polygon features	Water	Channel	Early 1900s California Debris Commission mapping and 1937 historical aerial photography. Often, boundaries are mapped between the marsh berms (long in-channel islands) created by dredges for levee building. Historical USGS topographic maps used as supporting evidence or as primary mapping source in locations where Debris Commission and aerials were insufficient.	Mapped channels where evidence supported a >15 m width. Blind tidal channel networks (a.k.a. "dead-end") may be undermapped where no sources were available for mapping channel width (i.e., where early sources confirming the presence of the channel are not accurate enough to determine width and where channel is no longer in existence by 1937 historical aerial photography). Such channels, however, are generally captured by the line data layer, which has no minimum mapping width. Due to early levee building, meander cuts, and hydraulic mining, in some locations width may be between 50 and 200% of actual historical width and some meander bends in channels may be missing.
		Pond or lake	Swampland district and reclamation district maps, GLO surveys and plat maps, historical USGS topographic maps.	Minimum mapping unit of 5 acres. Likely undermapped given sparse data in some areas. May include intermittent lakes where sources are lacking.
		Intermittent pond or lake	GLO surveys and plat maps, historical USGS topographic maps.	Minimum mapping unit of 5 acres. Likely undermapped given the particularly sparse data concerning the seasonal characteristics of features.
	Freshwater emergent wetland	Tidal freshwater emergent wetland	GLO survey notes of "tule" boundary and the edges of marsh symbols in early maps. We took the upper limit of tidal freshwater emergent wetland to be roughly equivalent to the 3.5 ft contour in USGS topographic maps, where 0 is "sea level" and assumed to be roughly equivalent to the NGVD 1929 datum (Atwater 1982).	Boundaries extrapolated using topographic contours in places lacking direct evidence. Includes subtypes with much local-scale complexity, including willow-fern swamp complex (Mason n.d.) and a habitat mosaic of tules, grass, and ponds.
		Non-tidal freshwater emergent wetland		

Data type	Landcover grouping	Habitat type	Primary sources for mapping	Caveats
Polygon features	Willow thicket and riparian forest	Willow thicket	To determine the habitat type: textual descriptions of the "sinks," including early settler narratives, GLO surveys, swampland district surveys, Mexican land grant and other regional maps. To determine boundary: historical soil surveys, historical USGS topographic quads.	Considerable error may exist in the size of these features as they were removed early and are rarely explicitly mapped in spatially accurate sources. Small features of this habitat type are likely undermapped.
		Willow riparian scrub or shrub	Early maps and descriptions illustrating transition in vegetation cover relative to natural levee height, historical soil surveys, and GLO survey notes, such as "enter willow."	Given the absence of direct evidence of width in many locations, some mapping relied on a conceptual model of riparian width related to channel reaches built from direct evidence in other locations.
		Valley foothill riparian	GLO survey notes and plats, Debris Commission mapping where available, early maps, historical soil surveys, textual descriptions, and landscape photos and paintings that generally describe the presence and characteristics of valley foothill riparian.	Considerable variability in characteristics exists within this habitat type. For example, evidence suggests that sycamores persisted along the Sacramento, while there is no evidence of sycamores along the San Joaquin. Also, evidence suggests that vegetation varied with the elevation of natural levees, both latitudinally and longitudinally. For instance, the highest parts of natural levees supported large trees, while at the wetland and channel edge, willows and grass were dominant.
	Seasonal wetland	Wet meadow or seasonal wetland	Historical soil surveys, supported by GLO survey notes. Indicative soil descriptions include notes of poor drainage, occasionally overflowed, native cover of annuals. GLO notes mention meadow land or land occasionally overflowed.	Given the sometimes gradual transitions between these habitat types and the reliance on soil mapping, we expect that boundaries are less accurate than habitats with less broad ecotones or habitats that are noted more extensively in historical sources.
		Vernal pool complex	Historical soil surveys describing hog wallow topography and USGS historical topographic maps showing collections of intermittent water bodies.	
		Alkali seasonal wetland complex	Historical soil surveys of alkali concentrations, supported in some places by GLO survey notes mentioning alkali soils or greasewood.	
	Other upland	Grassland	Historical soil surveys describing dry and relatively well-drained soils and GLO survey notes of prairie, good soils, and general absence of trees.	
		Stabilized interior dune vegetation	Distinguished by sandy soils predominantly in the Eastern Contra Costa region where sand mounds rise above tidal wetland. These areas delineated using "sand mounds" shown in early swampland and reclamation district maps as well as elevated land shown in historical USGS topographic maps.	Minimum mapping unit of 5 acres means that many small mounds are not captured in mapping.
		Oak woodland or savanna	GLO survey notes of scattered or heavy timber and associated bearing tree dataset, early narrative accounts of well-timbered land.	

Therefore, this layer includes the paths of the polygon channels as well as an additional level of smaller channels not included in the polygonal habitat features layer.

POLYGON CHANNELS Our goal was to map channels at mean tide in times of low river flow. In developing the polygon network of channels (those channels wider than 50 ft/15 m), we began with the digitized double-line channels from Atwater's (1982) mapping of circa 1850 channels. We then modified the width and orientation and added additional channels based initially on a synthesis of California Debris Commission mapping (see Box 2.1) and 1937-1939 USDA aerial photography.

We determined that the Debris Commission maps and historical aerial photography were superior primary mapping sources for historical channel width as opposed to the majority of earlier cartographic sources. This is because we expected greater channel width error using early sources due to mapping scale differences and lower accuracy. However, using these later sources necessarily meant considering width changes due to reclamation and dredging, as well as the influx of hydraulic mining debris and other channel modifications. We addressed these issues through the process of data interpretation, where we were able to increase our mapping confidence with knowledge of what changes were likely to have occurred where. For example, through calibration with early point observations of channel width, we found that using in-channel island edges in most cases defined the historical edge of the channel (Box 2.2). Despite this, it is likely that some channels may have been wider or narrower in the early 1800s Delta than is depicted in the mapping, depending on a particular channel's history of levee building, cut-offs, and dredging (this is particularly the case for channels where substantial natural levees were absent, which made channel modifications easier to perform). Earlier evidence from GLO surveys and other point data provide some calibration for the channel widths mapped (Fig. 2.6).

We subsequently compared this channel network based largely on post-1900s sources against the larger dataset of available relevant cartographic, survey, narrative, and photographic evidence and made subsequent modifications. Because most early cartographic sources of the Delta mapped only the largest channels, these later data were extremely valuable as illustrations of the primary channel network of the Delta. We used these data to verify that the polygon channel network captured at least the large channels represented in these maps. Most importantly, many early cartographic sources captured the largest of the blind tidal channel networks (i.e., the lower order tidal channels that branch and terminate within the wetland plain). They were the most difficult to map from later sources because many had been dammed, filled in, and farmed. In many cases, these early data supported the use of shifts in tonal signatures in the historical aerial photography to map channel boundaries.

Where the California Debris Commission maps indicated sand or gravel bars (e.g., San Joaquin River south of Middle River), we included these as

BOX 2.2. TRAJECTORIES OF CHANNEL WIDTH CHANGE

Information concerning historical channel width can inform interpretations of Delta hydrodynamics. By comparing historical and contemporary channel widths, scientists can gain a better understanding of how tidal dynamics have changed as a result of Delta reclamation. Whether widths were wider, narrower, or relatively unchanged depends on a number of factors, the most important of which appear to be the presence of natural levees and levee building history and practices.

Because natural levees were the obvious place to build an artificial levee, the widths of waterways with natural levees (e.g., the Sacramento River) have generally not changed dramatically. However, where substantial natural levees were not present – primarily within the more tidally-dominated central Delta – channel widths were more prone to change, and many channels were widened substantially as a result of reclamation. Along these channels, early reclaimers were faced with determining the most stable locations for levees that were the least susceptible to erosion and subsidence, while maximizing farmed area and minimizing levee length. The removal of material to build the levee often resulted in a ditch on the inside or outside of the levee (Tucker 1879a). The early levee building techniques that were employed in any particular location are therefore major determinants of channel change through time. In some cases, the first levees (usually hand-built) were placed on the edge of an island (to maximize farm land area) only to be later moved farther inland onto the slightly more sediment-rich and more stable low natural levees.

There are numerous debates in early newspapers, survey field notes, and published engineering reports about the merits of placing the ditch on the inside or outside of the levee. These ditches served many purposes, including providing building material for the levee and protection against the wash of waves (if the ditch was placed on the outside of the levee). They were often needed simply because the reach of the dredge was not long enough to reach the levee from the main channel. This left narrow strips of marshland along the edge of the ditch and levee. This suggests the origin of the many long in-channel islands present in the Delta today. These distinctive features tend to be most pronounced in early 1900s maps as many have eroded away to tidal flats or bars in the channel today. These features were often used to map width, where the channel-side of the in-channel island was taken to be the historical edge of tidal wetland (Fig. 2.5).

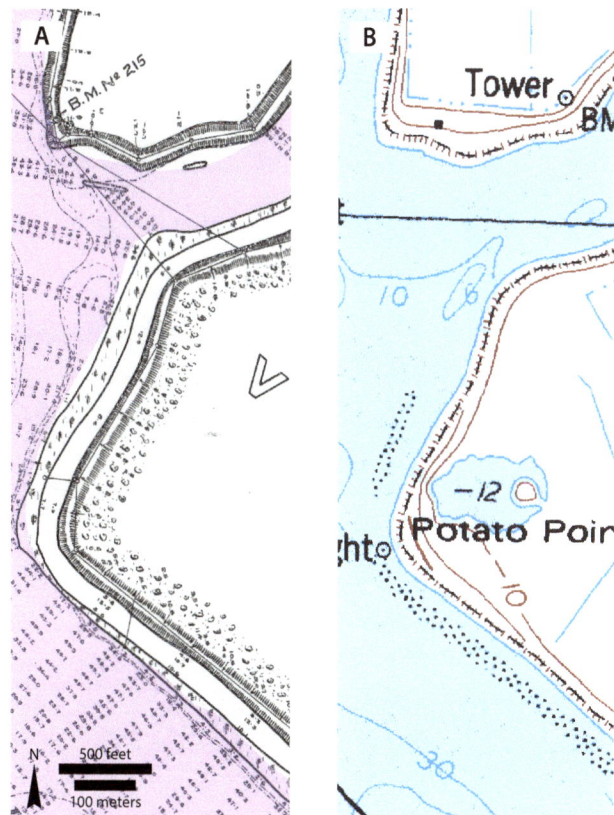

Figure 2.5. Historical channel width mapped based on remnant in-channel islands that outline the pre-reclamation edge of wetlands. The transparent light purple overlay in (A) shows the historical mapping on top of a 1908 Debris Commission map. The artificial levees (shown as parallel sets of hash marks, with a ditch and then a strip of wetland running along the channel edge) are now the edge of islands, as seen in (B). (A: Wadsworth 1908b, courtesy of the California State Lands Commission; B: USGS 1998)

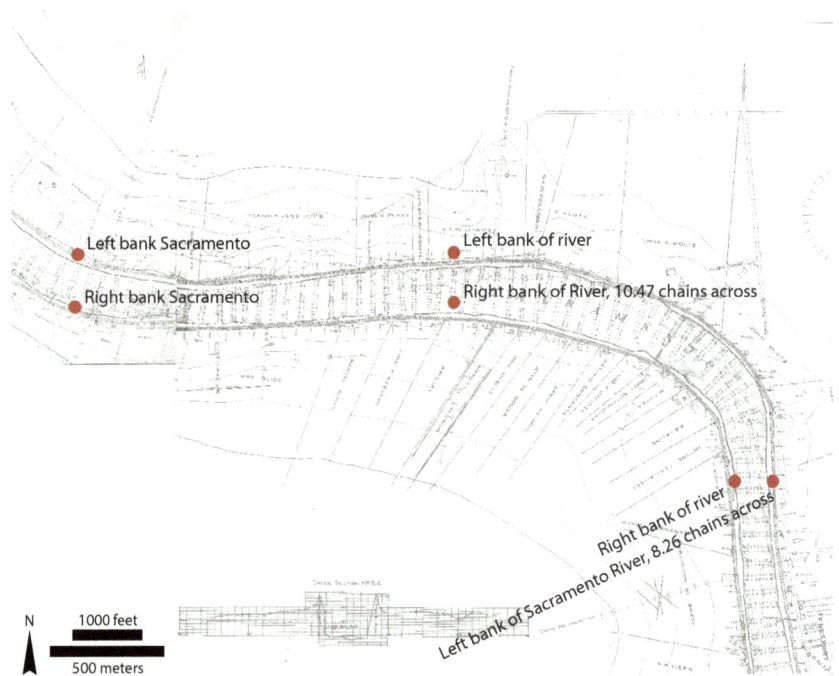

Figure 2.6. Calibration of channel width shown in early 1900s sources using 1858 GLO survey field notes. The locations of the left and right banks of the Sacramento River noted in the GLO survey support the bank locations of the later source. Lewis 1858c; Wadsworth 1908a, courtesy of the California State Lands Commission)

channel as they are part of the active channel bed and we believe that many of these gravel bars could have originated from hydraulic mining debris. We also included tidal flats at the Delta mouth within mapped channels. Many were only exposed at low tides, and such features (particularly at the mouth) were of questionable origin considering hydraulic mining debris. Few high quality data for tidal flats, shoals and bars existed prior to the hydraulic mining era, save for the Delta mouth.

SINGLE-LINE CHANNELS Channels narrower than 50 feet (15 m) wide are represented by single lines (no width dimension). This linework dataset is a complete network in that it includes the "centerline" of mapped polygon channels and connects flowpaths through ponds and lakes where there is an inlet and outlet and also includes channels too small to be mapped as polygons. This network represents our best understanding of Delta hydrography prior to significant modifications beginning in the mid-1800s.

We used the contemporary National Hydrography Dataset (NHD; USGS 1999) vector layer as a basis from which to develop this dataset. We then modified this network such that the planform conformed to the historical sources (e.g., deleting ditches and meander cuts, introducing historical meanders into existing channels, adding historical channels). To account for uncertainty associated with map scale, georeferencing, and accuracy of historical sources, we modified the network only where the historical channel alignment was offset by more than 50 feet (15 m) from the contemporary orientation. We also modified the shape of channels when we estimated the overall historical length of the channel to be greater or less than the contemporary by more than 10%. Using an existing contemporary dataset (NHD) as a starting point avoided the re-digitizing of the main channels of the Delta that have remained unchanged since historical times and provided a comparable dataset from which to analyze the change in the channel network over time. However, using the contemporary dataset also meant deleting many irrigation ditches that mark the Delta today.

The primary digitizing sources for the single-line channel dataset were the 1937 USDA aerial photo mosaics and the historical USGS topographic quadrangle maps. The historical aerial photomosaics were a valuable source despite being flown more than 80 years after reclamation began in the Delta. This is because the lighter colored inorganic sediments associated with the edges of channels often show up quite distinctly against the dark background of the peaty organic soils of the marsh interior (Atwater 1982). These remnant or ghost channel signatures allowed us to include in the dataset channels that were too small to be mapped in the small-scale maps of the late 1800s and were obliterated by the time of the detailed USGS mapping in the early 1900s. Additionally, the aerials allowed us to map more accurate channel shapes than was possible from early maps that confirmed historical presence. These methods were adopted from those used by Atwater (1982) to map historical Delta channels using 1970s-era aerial photography. However, because primary mapping sources used for these smaller channels were post-reclamation, it is likely that smaller sloughs are under-represented. It is also possible an additional class of lowest order channels existed but is not represented by historical sources.

While the majority of single-line channels were mapped based on historical aerial photography, it was not the only source used to map single-line channels. We incorporated additional information, including cartographic sources, GLO survey notes and accompanying plat maps, other textual descriptions, topography, and early soils maps. These sources often highlighted the predominant channel planform of particular locations, which improved our interpretation of signatures in historical aerial photography (or allowed us to complete channels where only parts of them were suggested by aerials). In some cases, these showed channels not depicted in other maps, or illustrated a different channel orientation. Such evidence helped in supporting that the mapping reflected the primary channel patterns of the historical Delta.

As with all other features, this synthesis process was documented by attributing features with data sources and certainty levels. Channels receiving a high interpretation certainty appeared as natural clearly functional channels (physical bed and banks with seasonal or perennial channelized flow) in early reclamation-period sources; or in rare cases had a definite natural form in historical aerial photography that was clearly connected to a channel network established by an earlier source (Fig. 2.7). It can be assumed that channels mapped with high interpretation certainty give the minimum channel present in the early 1800s.

It should be noted that channels mapped within the non-tidal and tidal freshwater emergent wetlands of the upper reaches of the San Joaquin distributaries (i.e., near present-day Stewart Tract) are associated with disproportionate number of channels with lower certainty levels. In

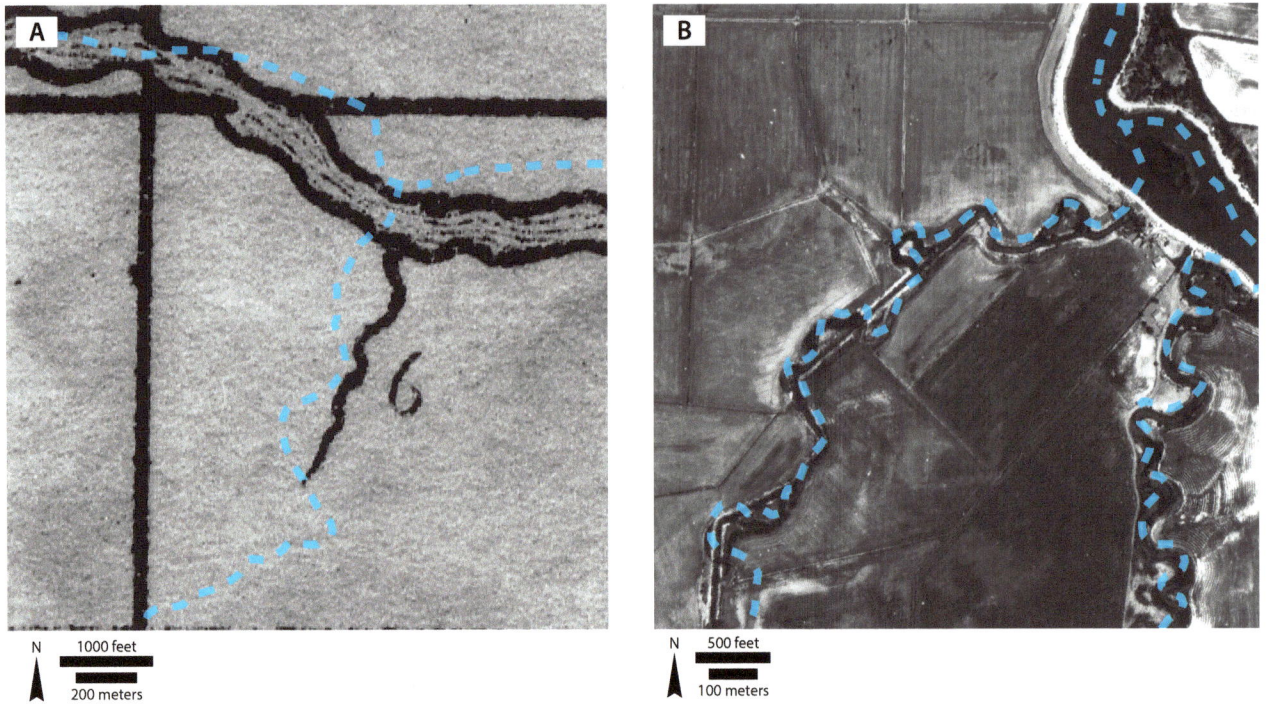

Figure 2.7. Examples of single-line channels assigned with "high" interpretation certainty. In A, a channel digitized off the main river from the 1937 aerial photography is confirmed by a reclamation-era source from 1869. In B, the signatures for two blind tidal channels in the historical aerial photography have a natural form and are connected to a river channel so are assigned a "high" interpretation. (A: Gibbes 1869, courtesy of the Map Collection of the Library of UC Davis; B: USDA 1937-1939)

particular, single-line channel mapping in this area, while reflective of general conditions in the early 1800s, is associated with greater uncertainty with regard to the length and actual physical location of individual features. Sources such as the early USGS and California Debris Commission mapping of the early 1900s show many features that illustrate more frequent changes in channels, such as abandoned channels and oxbow lakes. The mapping effort has captured many of the larger and persistent features, but not all. Accordingly, certainty levels and additional attributes explaining the character of that uncertainty assigned to each feature reflect our confidence in its existence, character, shape, and location in the early 1800s.

The uncertainties associated with mapping the apparently more dynamic system of the south Delta are compounded by complications relating to peat oxidation and subsidence, some of which had already occurred by the time of the historical USGS mapping and 1937 aerial photography. The combination of these factors mean that there are, in fact, several possible eras during which channels visible in the historical USGS maps and aerial photography could have been active (Atwater pers. comm.).

It is possible that many channel signatures in the historical aerial photography are remnants of ancient channels that were actually covered with peat prior to reclamation and were subsequently exhumed as peat oxidized, was burned, removed by wind, plowing, etc. in the process of reclamation (Fig 2.8; Atwater pers. comm.). With the southern Delta's wide zone of shallow peat combined with floodplain dynamics, this is particularly an issue. These channels of questionable origin were labeled as "possibly exhumed." Another possible explanation for some of the signatures is that channel topography in the early USGS topographic maps may be of a channel that was at one time functional (i.e., building natural

Early 1800s channel — Possibly exhumed channel

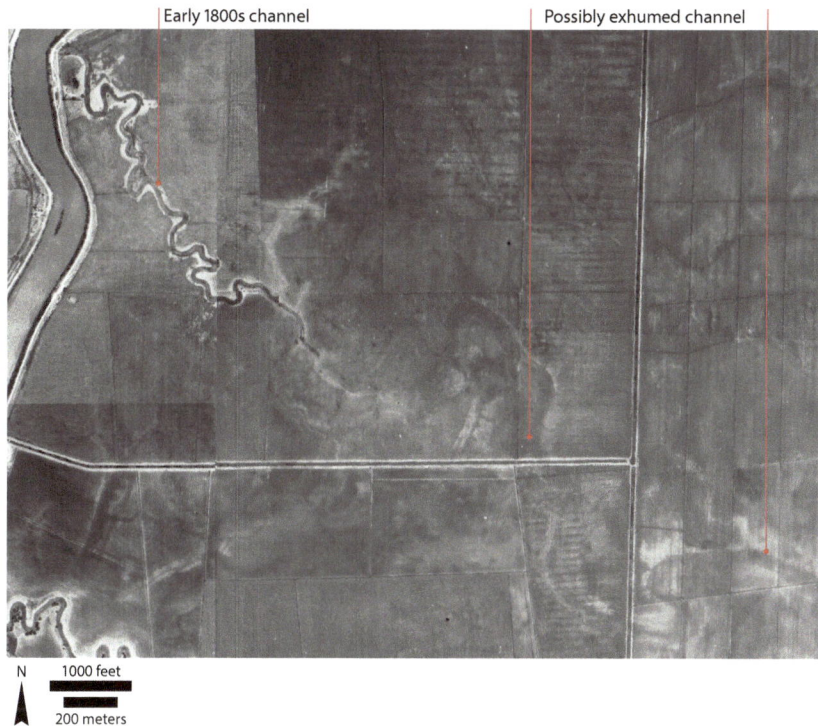

Figure 2.8. A recently active (pre-reclamation) channel that is a high interpretation certainty level channel can be seen (upper left). The fainter, wider, and less sinuous channel signature in the lower right that joins the channel may not have been active in recent history and may be an older channel whose inorganic soils were exposed in the process of reclamation as the overlying peat was oxidized (Atwater pers. comm.). (USDA 1937-1939)

levees), but by the early 1800s was no longer functional. Uncertainty arises with the possibility that the maps show remnant topography of a channel route that was completely abandoned by 1800. A third possibility is that channels were developed post-reclamation either through natural or anthropogenic causes.

Where warranted, we also coded channels as having either intermittent or perennial flow. To assess flow patterns, we modified early USGS flow designations of seasonality with additional information where available from GLO surveys, maps, and textual accounts.

To maintain a consistent depiction of channel density over time, we aimed to match the level of detail captured in contemporary mapping efforts, such as that of the Bay Area Aquatic Resource Inventory (SFEI 2011). For instance, we included those channels visible in the 1937 aerial photography that may have been only small ephemeral channels, swales, or overflow channels, but assigned lower interpretation certainty levels where earlier historical sources were lacking. Many such channels were suggested by topographic patterns in the historical USGS maps.

TIDAL INFLUENCE We based classification of a tidal channel on whether we believed water ebbed and flowed in the channel at least during spring tides during low river flow. We relied primarily on elevations marked in the early USGS topographic maps, where we assumed the approximate extent of tide at low river flow was 3.5 feet (1.1 m) above sea level (see page 66; Atwater 1982). We used this distinction instead of whether the channel was formed and maintained by fluvial or tidal processes because of the challenges associated with determining relative dominance of these processes. The fluvial-tidal interface was, by nature, a zone that moved seasonally along

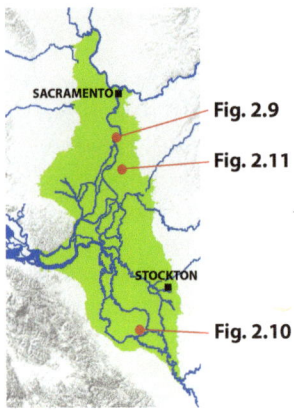

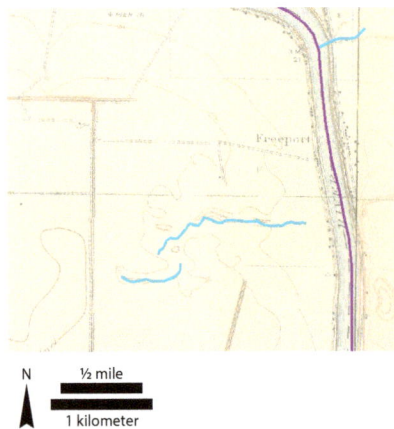

Figure 2.9. A channel shown dissecting the Sacramento's natural levee appears to be within elevations above tidal influence and is therefore classified as a fluvial channel (blue). However, given the possible connection to tidal flow from the Sacramento River, the tidal interpretation certainty was assigned as a "low." (USGS 1909-1918)

the gradient between mostly tidal and mostly fluvial influences. Textual evidence of tidal influence was available for the larger rivers.

All channels within the boundary of tidal freshwater emergent wetlands were classified as tidal, as it was assumed that the bed elevation of these channels would have allowed for at least intertidal flow. Channels associated with bed elevations well above tidal range were classified as fluvial. However, is was very difficult to determine the upstream limit of tidal extent within a channel where tidal influence was very slight or occurred only at the highest tides of the year (e.g., channels that cross the boundary between tidal and non-tidal wetland). Channels that lie at the edge of tidal range make up a relatively small proportion of the total length of channel mapped, however, and therefore do not have a significant impact on overall results.

For the areas where the tidal influence was uncertain, we documented this in the feature attributes. In order to assign channels either "high," "medium," or "low" certainty for tidal status, we followed criteria based on elevation, whether a channel had an established connection with a channel that was definitely tidal, and landscape context (whether it was within a marsh or at the upland edge). For instance, some small channels mapped as connecting to a tidal channel, but traversing a natural levee (above tidal elevation), were classified as fluvial, but assigned a "medium" or "low" certainty level because of the possibility that it intersected the natural levee low enough to receive tidal influence (Fig. 2.9). In addition, some channels connecting to a tidal channel that extended into a non-tidal marsh between 3.5 and 5 feet (1.1-1.5 m) elevation were classified as tidal because we expected their beds were low enough to receive tidal influence and there was an established connection to that tidal influence. However, the lower certainty associated with the non-tidal marsh plain, led to the assigning of a "medium" or "low" certainty level for tidal interpretation (Fig. 2.10).

Freshwater pond or lake

Ponds and lakes, while usually covering only a small fraction of a landscape, significantly affect the process and function of landscapes. Understanding the historical presence and character of ponds and lakes lends insight into dominant physical processes that formed and maintained them and the possible uses of these features by fish and wildlife. We mapped the early 1800s extent and distribution of ponds and lakes that were over five acres in size based on available historical evidence (Fig. 2.11). Given that ponds and lakes are generally quite distinctive and important (recreational use, water supply, etc.), evidence of these features is frequently found in the historical record, both in narrative accounts and early small-scale maps. Mapping allowed for capture virtually all large (on the order of 100 ac/40 ha) lakes. However, ponds (on the order of 5-10 ac/2-4 ha) are likely under-represented, given that detailed historical sources are not available for the entire study area. Ponds smaller than five acres (2 ha) were mapped in the

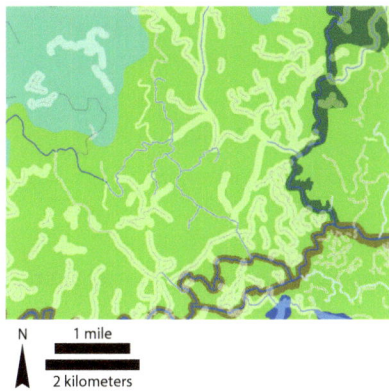

Figure 2.10. Channels within a non-tidal marsh plain (light green) at elevations where channel beds may have been within tidal range are classified as tidal (solid blue). Tidal status certainty levels of "medium" (thin white transparency) or "low" (thick white transparency) are assigned to many of these channels.

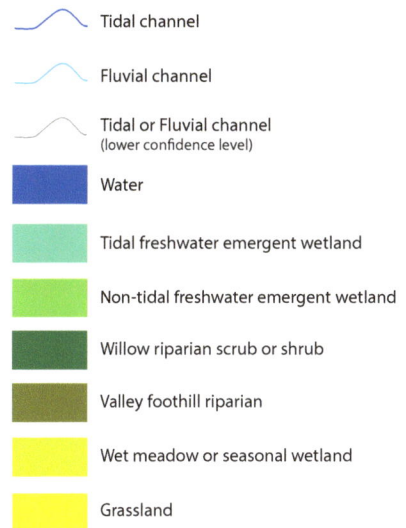

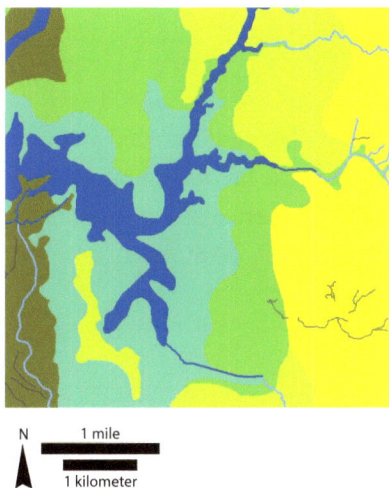

N 1 mile
2 kilometers

Figure 2.11. Detail of mapped wetland features. We mapped ponds and lakes and tidal and non-tidal perennial freshwater emergent wetlands greater than five acres (2 ha).

N 1 mile
1 kilometer

Tidal channel

Fluvial channel

Tidal or Fluvial channel (lower confidence level)

Water

Tidal freshwater emergent wetland

Non-tidal freshwater emergent wetland

Willow riparian scrub or shrub

Valley foothill riparian

Wet meadow or seasonal wetland

Grassland

supplemental small features layer. Such features included the ponds as small as about one acre (0.4 ha) as shown on the early USGS topographic maps. Such small features are rarely confirmed by pre-reclamation sources.

The overall prevalence of ponds and lakes in the historical Delta is well documented, although some ponds and lakes are subject to significant uncertainty concerning size and, in some cases, location. This is largely due to the fact that many early reclamation efforts took place prior to extensive mapping or surveying. Therefore, while interpretation certainty for most features was generally high because the features were confirmed directly by pre-reclamation sources or were large and geomorphically distinctive, other ponds or lakes may have been artifacts of the reclamation process after uneven peat burning and subsequent levee breaks (Thompson pers. comm.).

We did not map ponds that were only generally referred to in the historical record. For instance, some accounts describe numerous small ponds in the dense tule stands of the north Delta, but the descriptions lack enough specificity to map the features. We documented such patterns in the attributes of the larger marsh complex polygons (e.g., the Pearson District freshwater emergent wetland contains many small ponds in winter), as well as in the landscape descriptions within this report.

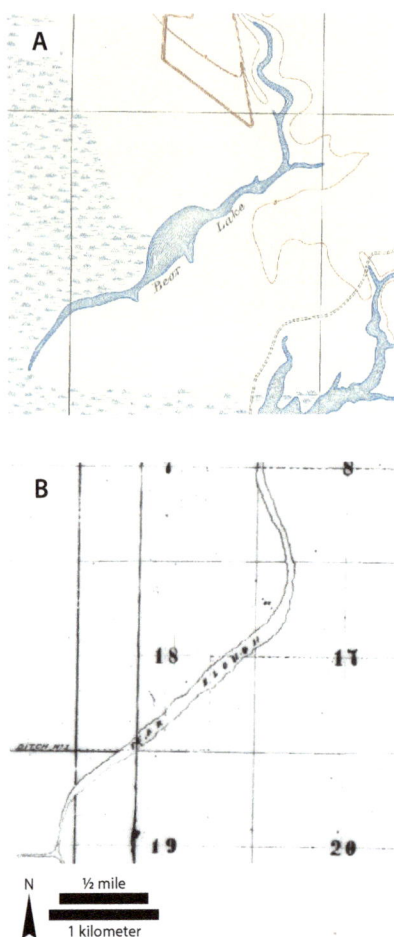

Figure 2.12. Two maps labeling the same feature as a slough and a lake. We mapped this feature as a lake given the narrowing at the downstream end. (A: USGS 1909-1918; B: Reece 1864, courtesy of the California State Lands Commission)

Another complication arose in distinguishing between channels and ponds or lakes. We used the general distinction that ponds and lakes were rounded features or distinct bulges along a channel. In some instances, elongated "sloughs" intersecting the freshwater emergent wetland edge were mapped as lakes, as they were referred to as such in the historical record (Fig. 2.12).

We classified ponds and lakes as tidal or non-tidal and as perennial or intermittent. Tidal or non-tidal status was based on status of the surrounding wetland as well as whether the body of water was connected to a tidal channel. Indication of seasonality came primarily from two sources, the historical USGS topographic maps' symbology for intermittent water bodies and the GLO survey data, where surveyors remarked on whether ponds and lakes they encountered were dry or wet. For instance, in the El Pescadero area (present-day Stewart Tract), surveyor William Norris entered a "dry bed of pond" in October, coinciding with the position of a pond in an early map (Gibbes 1850a, Norris 1851). This pond was therefore classified as intermittent. In cases where no evidence was available for determining the seasonal nature of a body of water, the default was a perennial classification as it was presumed that most accounts and maps would be most likely documenting those features that were persistent year-round.

Freshwater emergent wetland

We mapped the extent of freshwater emergent wetland as characterized by persistent emergent monocot vegetation (dominated by tule) where the land surface was frequently flooded and soils saturated for all or most of the year (see Fig. 2.11). Standard sources used to map freshwater emergent wetland included historical soil surveys, the historical USGS topographic maps (where remnant or re-established freshwater wetlands were mapped), and GLO survey data. Where available, we refined and supported the mapping with additional historical sources that included textual as well as cartographic data. Ancillary data, including LiDAR (CDWR 2008) and Atwater (1982) geologic and 1850 tidal boundary mapping, were used to interpolate boundaries from available historical data.

Selected soil types were used as initial indicators of freshwater emergent wetland. Peat soils clearly indicate freshwater emergent wetland (the buildup of organic material in peat occurs through anoxic conditions created by saturated soils). To identify additional soil types likely to have supported freshwater emergent wetland, historical soil type descriptions were reviewed for descriptions of soil properties, drainage characteristics, native vegetation, and agricultural uses indicative of perennial wetland or former wetlands. For instance, descriptions such as "cocklebur, bur clover, tules, mint, smartweed, and other water-loving plants," "high content of decomposed organic matter," "support a thick cover of tules, sedges, and similar plants" suggest the historical presence of freshwater emergent wetland (Mann et al. 1911, Cosby and Carpenter 1932, Carpenter and Cosby 1934). Using soil types as a first cut at the extent of freshwater emergent wetland was a valuable way to map freshwater wetland using a fairly consistent dataset covering the entire study area.

Uncertainty arises because many of these historical soil surveys were published relatively late (e.g., 1933) and others were mapped at very coarse scales (e.g., 1:250,000). All were performed well after reclamation began (the earliest survey was in 1905). However, although these surveys span almost three decades, general soil characteristics are not likely to change much over time. Also, because of two "reconnaissance" soil surveys from 1915 and 1918 that covered the Sacramento and San Joaquin valleys, we were able to calibrate the more local soil surveys to these general ones where they overlapped. In addition, our interpretations were aided by the numerous Gold Rush-era maps of the Delta region that illustrate the extent of tules, but are too general to be georeferenced and used to map explicit boundaries. These early regional maps give a sense of where the tules were found and then the related soil types could be used to map boundaries, yielding a more accurate approximation of the extent of this freshwater emergent wetland.

We refined the initial soil-based mapping using other earlier historical data. In particular, we used points where vegetation changes (e.g., "entering tule") were noted by GLO surveyors, the mapped freshwater wetlands extant at the time of the historical USGS topographic maps in the early 1900s, and subsequent interpolation of the boundary from known points based on elevation. For instance, because GLO survey data are in the form of points along survey lines, we often used topography to extrapolate habitat boundaries between points, which we obtained from either the historical USGS topographic maps or LiDAR (where topography had not been extensively modified). Given that the location of the tule boundary (freshwater emergent wetland) was an important indicator of "swamp and overflowed land" (Box 2.3), the historical record concerning this boundary is fairly extensive. We also believe, though some confusion arises in the definition of "swamp and overflowed land," that this boundary was rather easily and consistently identified across sources (e.g., where the land became very difficult to traverse or it was difficult to plant crops), as the following quote suggests:

> 1st Question. What is the character of the land covered by Tule on the banks of the San Joaquin adjoining your survey in this case? Answer. It is wet, marshy, overflowed land, generally impassible with here and there patches over which a person can with difficulty wade out…2nd Question. Is the line of Tule as shown on the map distinctly marked by natural features on the ground?… Answer. The line of Tule which I take to correspond with that is quite distinct. (Whiting 1854)

GLO survey data provided some of the earliest, most direct, and spatially accurate evidence of the tule boundary. Field notes such as "to tule" or "leave tulare" were used as direct evidence of the location of the freshwater emergent wetland boundary. "Swamp and overflowed land" is a largely political term used to delineate the boundary between dry land (owned

BOX 2.3. UNDERSTANDING "SWAMP AND OVERFLOWED" LAND

"Swamp and overflowed" (S&O) was the legal term used to identify lands "unfit for cultivation" because they were subject to inundation such that active reclamation (i.e., leveeing and draining) was needed before the land could be farmed. The 1850 Swamp and Overflowed Land Act (the Arkansas Act) transferred these lands from the federal government to the states. The states then sold the land to private landholders at low prices with the stipulation that the land would be reclaimed (Ralston and Broderick 1852). The boundary of S&O land was consequentially significant, as it determined whether the sale of the land would benefit the state or federal government. This was, however, rarely achieved in California because of the timing of surveys and reclamation activities (some surveys were conducted well after initial reclamation) and because the term was subject to interpretation. Additional complications arose from the fact that the State did not resolve how S&O land would be segregated and sold until 1866 (JRP Historical 2008).

It is not surprising that people had very different opinions about what "unfit for cultivation" actually meant, often depending on property interests. To some it meant the natural levees as well as the wetlands because these were "liable to be overflowed at any time during the winter" (CA Swampland Commissioners 1861). Historical accounts reveal much debate over the boundaries settled upon, and in some cases these conflicts resulted in lawsuits. This led to further attempts at defining the meaning of S&O land. In the late 1880s, the courts defined the land as such: "Swamp lands, as distinguished from overflowed lands, may be considered such as require drainage to fit them for cultivation. Overflowed lands are those which are subject to such periodical or frequent overflows as to require levees or embankments to keep out the water, and render them suitable for cultivation" (USDI 1973).

In many cases, there was considerable argument over whether the GLO surveyors were qualified to make these distinctions, whether by surveying in the dry season they overestimated the area of federal land, and whether some land that had already been reclaimed was not being surveyed as S&O land. The question about surveying during the dry season is discussed in the 1854 Report of the Surveyor General, which states that the S&O boundaries were "solely to depend upon the field notes of the U. S. Deputy Surveyors, who, traversing them during the dry season, can scarcely be qualified to judge of their nature" (Marlette 1854). The following testimony appears with many others before the California Swamp Land Committee to affirm that certain land in the vicinity of the Sacramento Basin should actually have been surveyed as S&O land:

> I have seen the whole of said land overflowed… and before any levees had been made on the land reclaimed…At the time said survey was made by the said W. J. Lewis, Deputy United States Surveyor, all of the land returned by him in said survey had been reclaimed and laid dry for a long time by the erection of levees and the closing of inlets from said river. At the time said survey was made by said W. J. Lewis it was impossible for any one to tell what the character of said land was previous to its reclamation. (Denn in CA Swamp Land Committee 1861)

Another testifies that the land once had "a growth of tule upon it; and the timber on the highest part being willow and sycamore, shows that the land in its natural condition and unreclaimed, as swamp and overflowed land" (Greene in CA Swamp Land Committee 1861). Yet another person claimed he had passed through the area in a boat (Hazen CA Swamp Land Committee 1861).

On the other hand, others acknowledged that some legally defined S&O land was not really perennially wet or truly in danger of frequent floods, as this early history of San Joaquin County describes: "There is a large amount of territory classed as swamp and overflowed, that is only occasionally under water, and the most lively imagination could not make of it a swamp" (Gilbert 1879).

Disagreement over the boundaries even occurred between surveyors (some of the earliest surveys in the 1850s were resurveyed a year to ten years later). These re-surveys established boundaries of "swamp and overflowed land" or tule that were different from the original. In some cases these discrepancies could be explained by the expected spatial error in the dataset, while in other instances these were clear differences in interpretation on the part of the surveyor or was an indication of either seasonal fluctuation in the freshwater emergent wetland boundary (as marked by tule) or early reclamation efforts. In one case, a difference between surveys was even noted by the re-surveyor himself:

> At that time of making his [James 1855] survey he was unable to proceed North of the 1/4 section stake on account as he states of deep water and tule swamp. I find that there is a deep slough now dry 8.50 chains North of this point but that sections 13 and 24 are always entirely dry land and subdivision lines should be run. (Lewis 1859b)

As may be imagined, actually defining the boundary on the ground was not easy. Surveyors were required to ask locals about the extent of floods. They also undoubtedly used vegetation changes as an indication of this boundary. However, while some consider "swamp and overflowed" lands to be synonymous with perennial wetland, S&O land was not defined explicitly as such and therefore we considered the possibility that such lands included other ecotonal habitats (Fox 1987b). We compared the "swamp and overflowed" boundary against GLO notes that mentioned entering "tule" as well as against other historical sources and topography before being used as a boundary of freshwater emergent wetland. While we generally found "swamp and overflowed" land boundaries to define the perennial wetlands, there were notable exceptions (Fig. 2.13). The particular biases of surveyors was also explored by examining the relative proportion of the use of the term "tule," "tulare," or "marsh" versus "swamp" or "overflow." Those surveyors who rarely, if ever, used "tule," "tulare," or "marsh" were probably using the "swamp" and "overflow" to mark the boundary of the freshwater emergent wetland (the most distinctive boundary within the landscape). However, those that used all terms were more likely to be making two distinctions, one of the greatest extent of floods ("swamp and overflowed" land that included wet meadows and seasonal wetlands) and another of the "tule" or "marsh" boundary (Table 2.5).

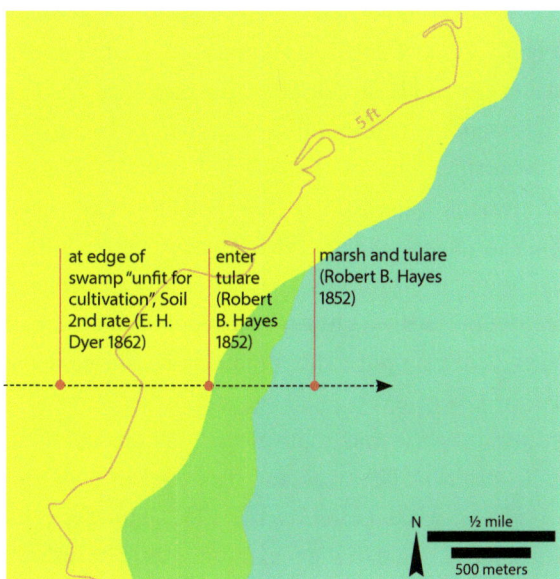

Figure 2.13. Boundary of wetland and GLO field notes. While GLO notes of the margin of swamp land usually coincide with the boundary of freshwater emergent wetland, this map shows that it was not always the case. Here, surveyor Hayes's definitive "enter tulare" is located at a lower elevation within the Yolo Basin than surveyor Dyer's 1862 field note of the edge of "swamp." Dyer here likely includes within his swamp land boundary land that we would classify as wet meadow or seasonal wetland as opposed to perennial freshwater wetland.

BOX 2.3. "SWAMP AND OVERFLOWED" LAND (CONTINUED)

In relation to the legal ownership of swamp
and overflowed land, since the GLO survey was
conducted for the purpose of selling federal
land, it was not necessary to survey the swamp
and overflowed land, except to the extent
needed to establish proper corners on dry land.
Unfortunately, this means that GLO surveys
rarely extended into the wettest portions of
the Delta. In addition, this means that many
lines that were run by the GLO were never fully
surveyed and thus are associated with greater
error in location. Our mapping methods take
these estimated errors into account through
the process of calibrating these data with other
cartographic and topographic evidence and
recording certainty levels.

Table 2.5. Distribution of terms used by GLO surveyors that suggest freshwater emergent wetland.

Surveyor	Tule	Tulare	Swamp	Over-flowed	Marsh
Benson, W.F.	11				
Dyer, E.H.	9	1	41	62	6
Handy, H.P.			15	15	
Hayes, R.B.		35		1	4
Jones, A.H.	14		18	1	2
Lewis, W.J.	126		106	106	14
Loring, F.R.	7			1	5
Norris, R.W.	95		7	1	12
Prentice, J.	12		1		
Ransom, L.	11		10	2	3
Von Schmidt, A.W.	38	2	2	2	
Wallace, J.	2		25	26	

by the federal government) and lands that needed reclaiming (owned by the state). We used such identifications only as supporting evidence for mapping tidal or non-tidal freshwater emergent wetland or wet meadow and seasonal wetland types (see Box 2.3).

TIDAL INFLUENCE Adopting the general approach of Atwater (1982), we assumed that the extent of spring tide at low river stages lay between the zero ("sea level") and five foot contour in historical USGS topographic maps, at approximately 3.5 feet (1.1 m). This elevation is generally understood as the tidal range and is supported by the historical USGS maps showing tidal wetlands extending to about this elevation. In the absence of more information, we assumed the historical USGS datum of "sea level" to be roughly equivalent to the NGVD 1929, which was introduced subsequent to the USGS survey (Shalowitz 1964, Atwater 1982). The delineation of the tidal boundary within the mapped freshwater emergent wetland extent is therefore primarily a physical, as opposed to ecological, definition. Additional evidence supports this approach, including notes about the "edge of tule" by GLO surveyors that generally lie between the zero and five foot contour lines, the maximum elevational limit of tidal marsh mapped in the historical USGS maps, and the upper limit of modern tidal marshes (Fig. 2.14; USGS 1909-1918, Atwater 1980). However, this approach likely results in an estimate of the maximum extent of influence as tidal range varies throughout the Delta and was probably less than 3.5 feet at the wetland edge (see pages 127 and 224).

In addition to the large expanses of non-tidal freshwater emergent wetland in the upper reaches of the north and south Delta that were clearly out of range of tides, we also mapped non-tidal freshwater emergent wetland in other locations. In particular, this occurred at the edge of riparian forest in the northern Delta, where evidence of tule was located above the expected extent of tidal influence. This resulted in a rim of non-tidal freshwater emergent wetland where elevations or distance to tidal source likely prevented the regular influence of tides, but where we had evidence of freshwater emergent wetland.

The boundary between tidal and non-tidal freshwater emergent wetland is not fixed; it fluctuated seasonally and often covered a broad gradient between land that was flooded daily by tides and areas at the margin that were only overflowed at the extreme tides of the year in combination with floods and wind influence. Some areas outside of the tidal boundary were likely indirectly affected by the tidal flow in nearby tidal channels and the tide's general influence on subsurface water elevations (e.g., preventing flow from the upper non-tidal Yolo Basin into the lower tidal portion; Collins and Sheikh 2005, Collins J pers. comm.). Similarly, those areas within the tidal boundary faced different inundation frequencies depending on proximity to the Delta mouth (greater inundation frequencies corresponding with greater tidal range and lower land elevation) and proximity to tidal channels (greater inundation frequencies would be found with areas closer to channels as tidal energy dissipated across the marsh plain). Additionally, some areas mapped as tidal because of their elevations may have been only slightly influenced by tides due to partial or complete isolation by natural levees, or simply because the distance was so great from the tidal source at the nearest tidal channel mouth. Therefore, the boundary should be taken as an approximation of the extent of area wetted by the tides in times of low river flow (Atwater 1982).

In the southern Delta, it was particularly difficult to determine the nature and exact location of the transition between tidal and non-tidal freshwater emergent wetland. The historical USGS topographic maps show a broad zone lying between the zero and five foot contour, making a boundary based on these contours more uncertain than in other Delta locations where contour lines at the tidal margin are closer together. To address this, we used a combination of information about elevation (from both historical USGS topographic maps and patterns in LiDAR, calibrated to account for subsidence), channel planform (whether channels appeared to be tidal or not based on sinuosity and presence of banks, etc.), and soils. We used the soil boundary of peat mapped in the 1905 Stockton soil survey on Union Island to help define where tidal wetland was most likely located (had allowed for peat to accumulate; Lapham and Mackie 1906). Deeper peats were associated with tidal wetlands as these were areas that had experienced slowly rising sea levels that allowed organic material to accumulate over time. At the edge of tidal influence, where soils had only recently been transgressed by tides, these peat accumulations were thin.

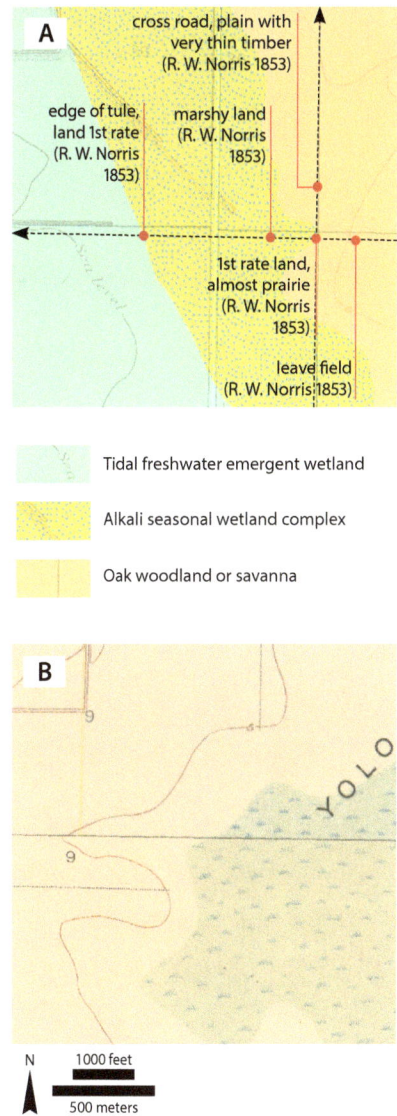

Tidal freshwater emergent wetland

Alkali seasonal wetland complex

Oak woodland or savanna

Figure 2.14. Evidence supporting mapping of the edge of tidal influence. In A, the "edge of tule" falls between the 0 and 5 foot contour line of the historical USGS topographic maps. In B, the mapped tidal tule marsh in the USGS map falls below the 5 foot contour. (USGS 1909-1918)

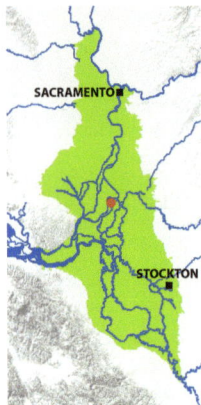

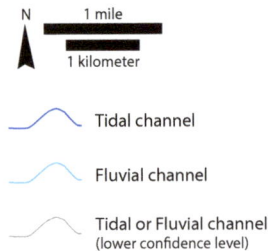

Figure 2.15. Detail of riparian forest mapping. Mapping distinguishes between tall (valley foothill riparian) and medium (willow riparian scrub or shrub) height riparian forest. Where possible, riparian forest width was determined from explicit sources (such as topography along natural levees). Otherwise, a conceptual model was used to assign reaches with a width class.

This boundary generally corresponded with a noticeable drop in channel density (presumably the areas within the tidal ecotone show many channels largely formed and maintained by fluvial processes).

Where the boundary edge adjacent to tidal wetland was upland (usually some form of seasonal wetland), the boundary of tidal extent could be more confidently mapped as the termination of emergent vegetation likely indicating the extent of tidal influence. This is a boundary clearly indicated in a wide range of historical sources. This link between vegetation and tidal extent is expressed by a surveyor describing his efforts to map the regular extent of tides:

> The character of the vegetation growing upon those lands was I considered one of the best tests of its elevation with reference to tides and the location of the drift wood on those lands is another good test of which ordinary tide rises. (Stratton 1865)

Overall, the mapped boundary between tidal and non-tidal habitats should be considered as representing a broad spatial gradient.

Willow thicket

We mapped willow thickets in several locations within the historical Delta. Willow thickets are found primarily within distributary networks at the base of alluvial fans and at the edges of floodplains. These features were mapped in places otherwise occupied by freshwater emergent wetland and were distinguished by data indicating or suggesting willows, thickets, or underbrush. These swamps were not linear features along banks of channels and were thus not mapped as the willow riparian scrub or shrub riparian forest habitat type.

Willow thickets, as mapped, are different from the willow-fern swamp of the central Delta that is discussed in a number of historical sources and described by Mason (n.d.) and Atwater (1980; see page 177). The latter appears to have been part of a matrix of freshwater perennial wetland communities and is not easily mapped as a habitat type separate from freshwater emergent wetland. The willow thickets differ from the willow-fern swamps of the central Delta in terms of landscape position and fluvial influence as well as density and age class distribution of indicative plant species.

Riparian forest

Riparian forests provided a wide array of functions, including shading, sediment entrainment, bank stabilization, allochthonous input, and species support (Collins et al. 2006). Our goal in mapping the historical extent of riparian forests was to capture those areas supporting such functions and estimate overall width (and thus area) and relative tree height across the historical Delta (Fig. 2.15). Understanding the overall landscape-scale pattern of riparian composition and functional width was emphasized over detailed mapping of riparian forest boundary, in part because detailed mapping was difficult given the available historical sources. As with other habitats, our mapping methodology used both direct evidence and extrapolation, which involved data inter-calibration and conceptual models.

Under many scientific and policy-oriented definitions of the term "riparian," most or all of the historical Delta could be considered riparian habitat (Collins et al. 2006, Vaghti and Greco 2007). However, we focused on mapping the likely extent of two major riparian forest habitat types: valley foothill riparian forest and willow riparian scrub or shrub. We focused on these riparian habitat types because of the important functions, such as wildlife support, the produced. Other channel-side areas that some may define as riparian habitat, such as tule mixed with other marsh species or low herbaceous cover, were excluded from the mapping and incorporated into the adjacent habitat type. For instance, although the low natural levees in the central Delta were of higher inorganic content than the surrounding peat, and thus likely supported a wetland species assemblage different from the lower interior island marsh, both areas were mapped as freshwater emergent wetland.

For the purposes of delineating riparian forest in the GIS we developed a conceptual model of potential riparian habitat characteristics in the Delta based on existing scientific literature, calibrated by direct evidence available in the historical record. This was used to map riparian habitat where no direct evidence was available. Our model focuses on the fluvial-tidal transition and accompanying shifts in natural levee size and character as factors controlling riparian width and height (Fig. 2.16). The model captures the transition from fluvial mainstem channels with broad valley foothill riparian forest to low order tidal channels with narrow tule-dominated riparian zones.

Figure 2.16. Graphical representation of riparian mapping conceptual model. By classifying channels by size and by fluvial/tidal influence we could assign likely riparian width and height classes.

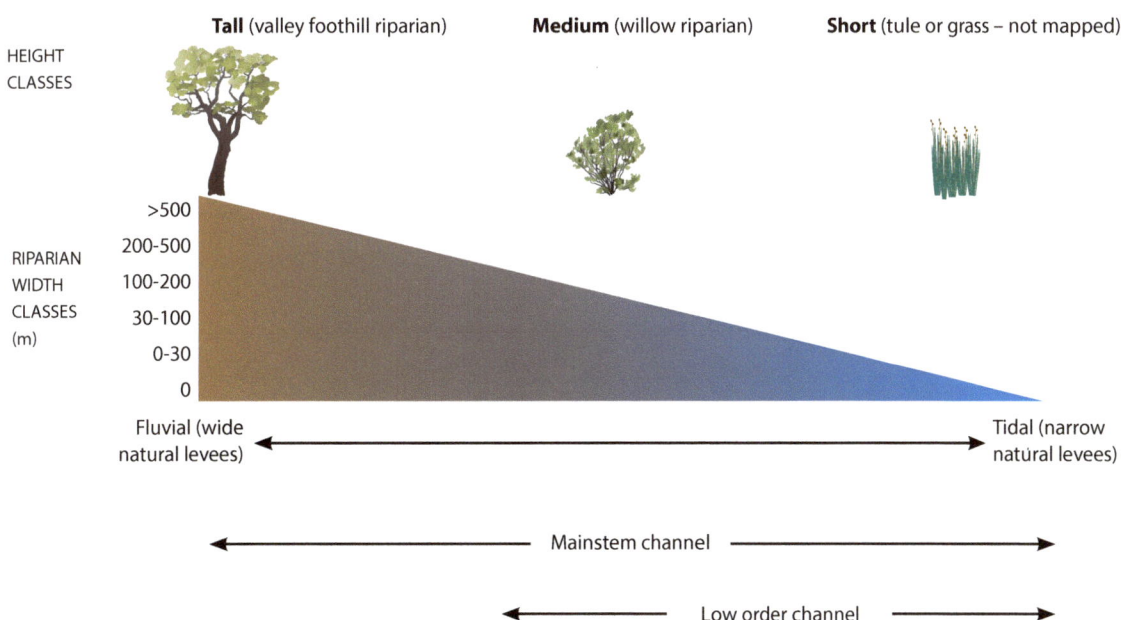

The relationship between riparian vegetation and hydrologic and geomorphic factors is well documented in the literature (e.g., Gregory 1991, Hupp and Osterkamp 1996, Collins et al. 2006, Fremier 2008). Within the Delta, the transition from a fluvial to a tidal landscape and the accompanying shifts in levee height, inundation frequencies, soil type, and soil saturation highlight the connection between vegetation and physical processes. Riparian vegetation structure in the Delta today shifts along elevation and salinity gradients, linked to the fluvial-tidal gradient (Fremier et al. 2008). Historically, large trees such as sycamore and oaks were dominant species on large (high) natural levees along river reaches in the Delta. In contrast, central Delta tidal channels had lower levees and were dominated by tule and other emergent wetland species (Whitlow et al. 1984).

The broad natural levees along the Sacramento River were elevated above tide level and were infrequently overflowed by floods. They contained more coarse inorganic sediment deposited by river floods than the island peat, creating relatively wide zones of oak dominated forest greater than half a mile in some locations and characterized by a dense understory (Thompson 1961). Several early maps document the presence of scrub along reaches where large trees are depicted upstream (more fluvially dominated with larger natural levees) and tule marsh is depicted downstream (more tidally dominated with low natural levees; Gibbes 1850a, Ringgold 1852). Such patterns are still observable today in many locations (Whitlow et al. 1984).

We used our conceptual model for mapping in GIS in combination with available relevant historical data. The conceptual model was especially critical for addressing the necessity of mapping riparian forest width despite there being few historical sources that provided continuous longitudinal information. We developed the following width classes, as defined as a single side of a channel: 0-100 ft, 100-330 ft, 330-660 ft, 660-1,640 ft, and >1,640 ft (0-30 m, 30-100 m, 100-200 m, 200-500 m, and >500 m; Collins et al. 2006, Grossinger 2012). As we considered a reach to map, we also determined the expected average vegetation height using three height classes: short (e.g., herbaceous or emergent vegetation, not mapped), medium (willow riparian scrub or shrub), and tall (valley foothill riparian forest; Hickson and Keeler-Wolf 2007). We also considered the difference between relatively open stand valley foothill riparian forests and those forests with a significant understory component, which was evidenced by descriptions such as "brushwood," "dense barriers of trees and shrubs," and "dense thickets of grapevine and willows" (Belcher 1843, Bidwell and Royce 1907, Belcher et al. 1979). However, separating these forest types was beyond the capacity of the mapping effort and could not be done reliably across the extent of the study area. Forest complexity is described in Chapter Five, pages 285-287.

The conceptual model provided us with a default assignment of width and height of riparian forest depending on location along the fluvial-tidal gradient and channel type. This allowed us to assign a habitat type and define habitat boundaries in the absence of continuous spatially specific information. Most frequently, this involved defining the boundary between woody vegetation of the valley foothill riparian forest habitat type and wetland species of the freshwater emergent wetland habitat type.

Throughout the mapping process we used both specific data and our conceptual model. For example, a reach (perhaps several miles in length) with a single historical cross section showing oak trees extending approximately 330 feet (100 m) from the channel edge would be classed as valley foothill riparian forest and mapped with a 215 foot (65 m; halfway between the width class outer limits of 100-330 ft [30-100 m]) buffer from the channel. In other cases, we were able to map a specific boundary by combining spatially explicit data with extrapolation from topography as well as our understanding of the patterns built into our conceptual model (Fig. 2.17). In cases where only a basic understanding of the width and height of riparian forest existed, we used the conceptual model exclusively to map riparian forest. For example, as the height of the Mokelumne River's narrow natural levees fell to general tide level at the foot of Staten Island, we understood from various textual sources that riparian vegetation transitioned from dense tree cover to scattered willow to predominantly tule (Sherman 1859). We then used the conceptual model to identify this middle reach as willow riparian scrub or shrub habitat type with a width of 100-330 feet (30-100 m), mapped in the habitat layer at a width of 215 feet (65 m).

Reducing the gradient in riparian structure to only two height classes of "tall" (e.g., valley foothill riparian) and "medium" (e.g., willow riparian scrub or shrub) and five width classes necessarily yields a rather unnatural depiction of the gradual transition in riparian structure across the fluvial-tidal gradient. This means, for example, that a mapped continuous zone of 215 foot (65 m) wide willow riparian scrub or shrub may actually represent a situation where the upstream portion of the segment was relatively wide (~330 ft [100 m]) and was occupied by a number of oaks, but at the downstream end the width was narrower (~100 ft [30 m]) and dominated by willow and tule. While it may be shown that trees and brush became less numerous and tule and other marsh vegetation became more common along the banks descending downstream along that reach, the natural, gradual transition in structure and species composition is not conveyed in the mapping. It should be kept in mind that this gradual thinning of structure is not easily captured in a GIS and the abrupt transitions between "valley foothill riparian" and "willow riparian scrub or shrub" should not be interpreted as abrupt vegetation changes or discontinuities in the landscape.

Using a decreasing width class to represent the gradual thinning of large woody vegetation and the concurrent decrease in natural levee height, may in some cases give a false impression of how it looked on the ground, given the challenges of representing changes of 3D structure in a 2D map. Instead of narrower levees as the mapped width implies, the transition was primarily in levee height. So,while the mapping may show a narrow 50 ft

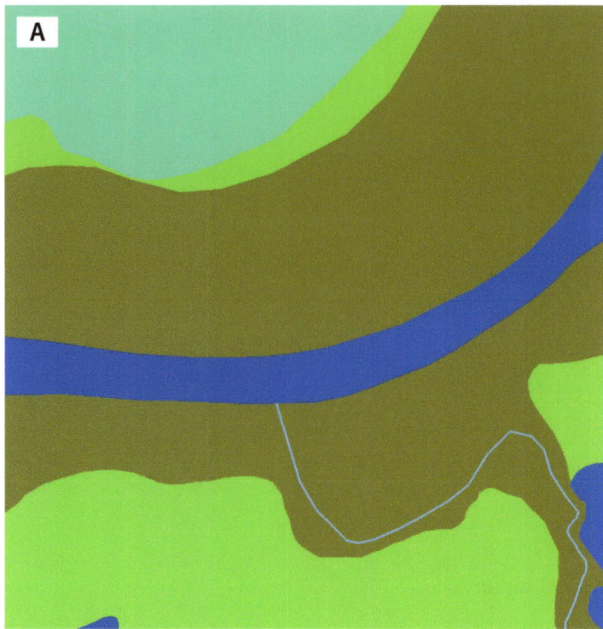

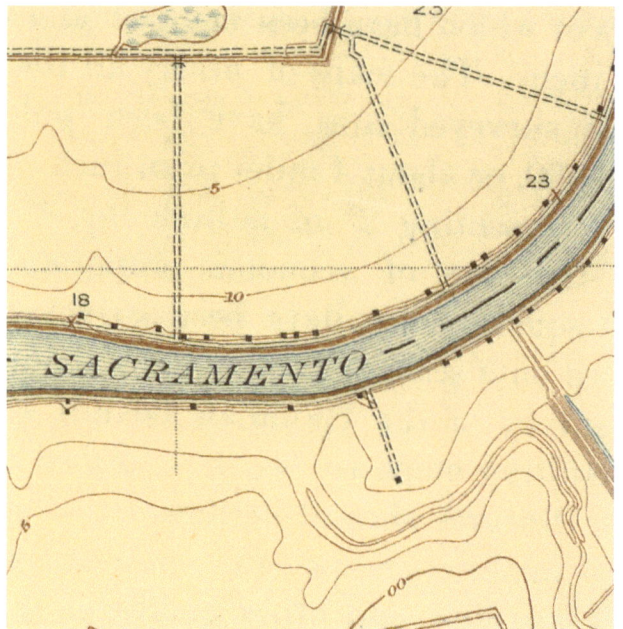

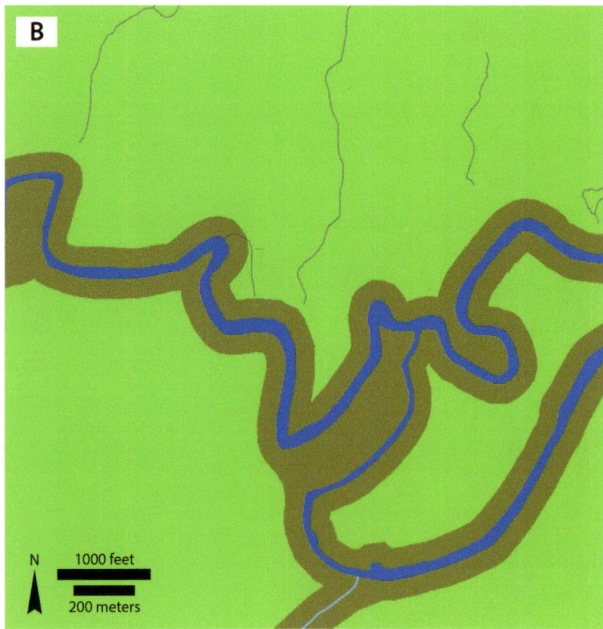

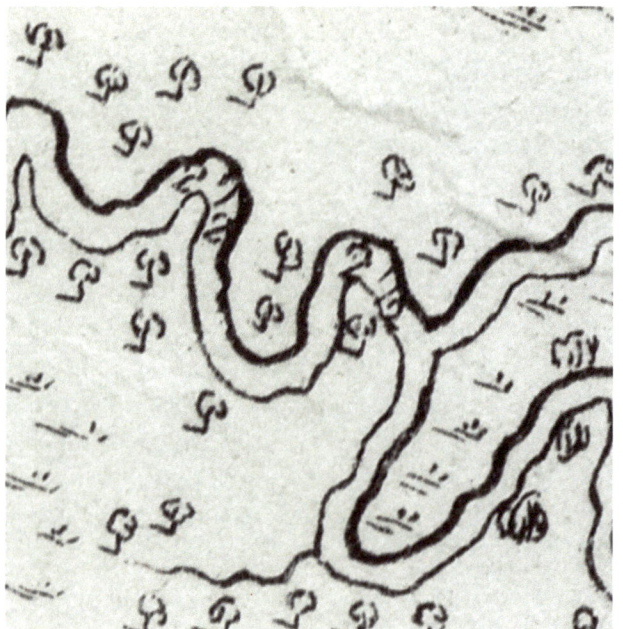

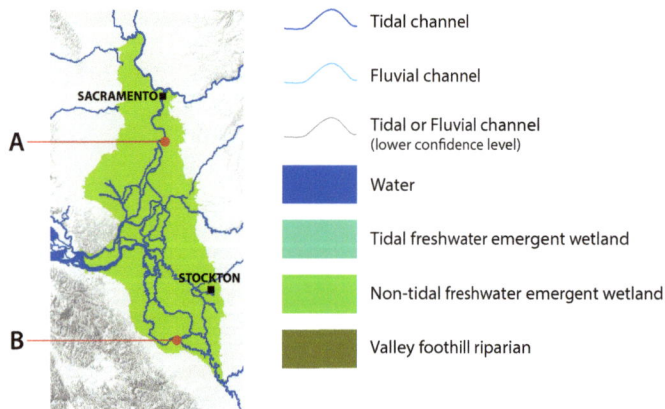

Figure 2.17. Examples of mapped riparian forest based on explicit sources (A) and width classes from a conceptual understanding of riparian width by reach (B). (A: USGS 1909-1918; B: Gibbes 1850a, courtesy of the Map Collection of the Library of UC Davis)

Tidal channel

Fluvial channel

Tidal or Fluvial channel
(lower confidence level)

Water

Tidal freshwater emergent wetland

Non-tidal freshwater emergent wetland

Valley foothill riparian

(15 m) willow strip along a channel close to the tidal Central Delta, it may be the case that sparse clumps of willows and scattered individual trees were found along wider, but low natural levees that were occupied by tules as well as willows.

Wet meadow or seasonal wetland complex

Many perennial wetlands are naturally bordered by seasonal wetlands. As their name implies, seasonal wetlands are characterized by lower inundation frequencies and dry season desiccation (Fig. 2.18). We included areas characterized by herbaceous vegetation cover and a range of inundation frequencies: from those that were seasonally saturated to seasonally flooded (Cowardin et al. 1979, Grossinger et al. 2007, Grossinger 2012). The upland margin of the Delta's freshwater emergent wetlands was frequently characterized as a type of wet meadow or seasonal wetland. We primarily used historical soils maps to determine likely areas of wet meadow or seasonal wetland. For example, one commonly used soil type was adobe clay, described as "sticky when wet" (Nelson et al. 1918). Common descriptions from soil surveys used to identify wet meadow or seasonal wetland are summarized in Table 2.6.

Boundaries provided by soil surveys were adjusted through calibration with other historical sources. In places where freshwater emergent wetland was not already mapped, the GLO notes defining the edge of "swamp and overflowed land" were used as an indication of the edge of wet meadow or seasonal wetland. Often we were able to confirm from the earliest sources the general pattern of areas that tended to be seasonally wet (commonly referred to a "meadow land" or simply defined as being overflowed in the winter) versus those that were drier (with descriptions such as "prairie" or "timbered plain"). Occasionally, an earlier source altered our interpretation of an area (e.g., caused a switch from grassland to wet meadow or seasonal wetland, or vice versa). However, for the most part soil surveys were the most spatially explicit sources available for mapping the boundaries of the wet meadow or seasonal wetland complexes.

Challenges to mapping wet meadows or seasonal wetlands include the often imprecise nature in which they are described in historical sources, their similar hydrology to other seasonal wetland types, the rare depiction of such habitats in maps, and the natural lack of distinct boundaries in the landscape as evidenced by early travelers' rare remarks on changes that would indicate the limits of this habitat type. Finally, wet meadow and seasonal wetlands were complex, intergrading with grassland, ponds, and patches of tule.

Vernal pool complex

We mapped this habitat type (which can be considered a subtype of the wet meadow or seasonal wetland habitat type) where we found evidence of distinctive patterns associated with vernal pools: the presence of seasonally ponded areas associated with clay pans or hardpans. Where the individual

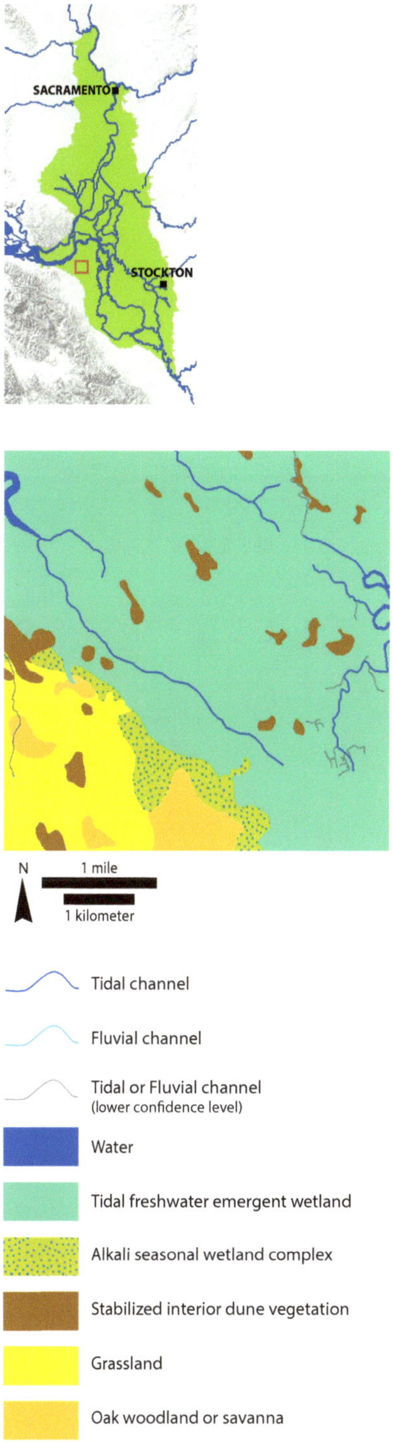
SACRAMENTO
STOCKTON
N
1 mile
1 kilometer
Tidal channel
Fluvial channel
Tidal or Fluvial channel
(lower confidence level)
Water
Tidal freshwater emergent wetland
Alkali seasonal wetland complex
Stabilized interior dune vegetation
Grassland
Oak woodland or savanna

Figure 2.18. Detail of mapped upland ecotone habitat types. These habitat types include seasonal wetlands, inland dune scrub, grasslands, oak savannas, and woodlands.

Table 2.6. Historical soil descriptions indicative of wet meadow or seasonal wetland complex.

Soil Type	Description	Source
Sacramento silty clay loam	"native vegetation consists of weeds and grasses"; "overflow...collect[s] upon the surface for varying periods in the winter months during seasons of excessive floods"	Mann et al. 1911
Sacramento silty clay	"impervious clay subsoil,;"very sticky when wet"; "occasionally overflowed"	Mann et al. 1911
Salinas gray adobe	"extremely sticky when wet"; "readily puddled"	Lapham and Mackie 1906
Stockton clay adobe	"sticky and puddles easily when wet"	Sweet et al. 1908
Hanford sandy loam, poorly drained phase	"saturated with water during the greater part of the rainy season"	Cosby and Carpenter 1932
Stockton clay adobe	"wet during the greater part of the year"; "native cover of annuals -- chiefly wild oats and bur-clover -- with sedges and other water loving plants in the wetter localities"	Cosby and Carpenter 1932
Stockton loam adobe	"heavy, black clay loam adobe"; "natural drainage is frequently deficient"	Lapham and Mackie 1906
Alamo clay adobe	"normally puddled and waxy when wet"; "occupies flat, poorly-drained depressions or low positions in the general region where the upland-plain soils merge with the lower lying basin or lowland types"	Holmes and Nelson 1915
Clear Lake clay adobe	"Occupies depressions or basins"; "developed under poor drainage"; "bur clover, alfilaria, wild oats, foxtail, and other native grasses make a vigorous growth during the rainy season"	Carpenter and Cosby 1934

pools were larger than 10 acres (4 ha) we mapped them separately as intermittent ponds within a larger vernal pool complex. Like most seasonal wetlands, we often relied on soil survey maps and associated descriptions to map their extent. We distinguished vernal pool complex from wet meadow or seasonal wetland using soil types with descriptions such as "hog wallows" (Holmes and Nelson 1915) or "many small bodies occupy local depressions" (Carpenter and Cosby 1934). This type may be overmapped because of challenges interpreting descriptions such as these. Given these issues, vernal pool complex could alternatively be lumped with the wet meadows or seasonal wetland complex class.

Other sources provided additional evidence, including historical USGS topographic maps (mapped intermittent water bodies) and GLO survey field notes (e.g., "enter dry pond"). In addition, the distinctive signatures of seasonal ponds were often quite visible in historical aerial photography. Although vernal pool species such as *Downingia pulchella* are often used to identify vernal pool complexes, historical sources frequently lack such detail, requiring a primary reliance on physical characteristics. The largest and most distinct area of vernal pool complex mapped is the Jepson Prairie in Solano County, much of which persists today. In this case, we were able to use contemporary distribution to calibrate the mapping (Holland 1998).

As with other seasonal wetland types, the edges of vernal pool complexes are broad ecotones. Furthermore, within these complexes, local-scale variability in inundation frequency and vegetation characteristics was high.

Alkali seasonal wetland complex

Like the wet meadow or seasonal wetland complex and the vernal pool complex, this type was a matrix of land cover types with varying degrees of alkali concentrations and inundation frequencies. We identified and mapped alkali seasonal wetland almost exclusively from soil survey maps, using the descriptions in soil surveys that note soils or areas with high residual salt content. Such areas were occasionally confirmed by other historical sources such as GLO notes.

Soil surveys provide two related sources from which we mapped alkali seasonal wetland (Grossinger et al. 2007). The first comes from the soil type descriptions in the survey reports. Soil descriptions commonly used to identify likely alkali seasonal wetland are given in Table 2.7. The second source is direct mapping of alkali performed for several of the surveys. These surveys delineate areas characterized by "growths of alkali weeds and salt grass" (designated with an "A" and red boundary line; Sweet et al. 1908), "alkali present" (designated with "A" and red boundary line; Nelson et al. 1918), or "alkali affected" (designated with "A" and red boundary line; Carpenter and Cosby 1933). Additionally, the 1908 Modesto-Turlock soil survey included a separate alkali map which depicts areas of different concentrations of alkali, with classes of 0-0.1% (background concentration level), 0.1-0.2%, 0.2-0.4%, 0.4-0.6%, 0.6-1%. Any concentration above 0.1% was used to map alkali seasonal wetland complex, which coincides with those areas mapped within the red alkali boundary in the soil survey map. They are characterized: "Except in a few small spots these areas do not at present contain sufficient alkali to be injurious, but with insufficient drainage

Table 2.7. Historical soil descriptions indicative of alkali seasonal wetland complex.

Soil Type	Description	Source
Sacramento clay adobe	"a variable amount of alkali is present"; "a thick growth of alkali weeds and wild grasses"	Mann et al. 1911
Merced sandy loam	"covered by numerous low mounds and hummocks"; "cover of saltgrass and other alkali-tolerant plants"	Cosby and Carpenter 1932
Merced sandy loam , light-colored phase	"a comparatively high concentration of alkali"	Cosby and Carpenter 1932
Fresno fine sandy loam	"alkali salts in small quantities are of general occurrence"	Lapham and Mackie 1906
Fresno loams	"salt grass (Distichlis spicata) the principal native grass, furnishes good pasture," "affected with alkali"	Holmes and Nelson 1915
Capay and Yolo clay loams and clays	"extensive areas of these soils are badly affected by alkali"	Holmes and Nelson 1915
Fresno fine sandy loam	"its puddled and alkali condition"	Nelson et al. 1918
Fresno sandy loam	"generally supports a moderate growth of salt grass"	Nelson et al. 1918
Marcuse clay	"subject to poor drainage and an accumulation of saline salts"; "salt, saltgrass, and pickleweed, greasewood, and other salt-tolerant plants grow"	Carpenter 1939
Solano silty clay	"appreciable salt content"	Carpenter 1939
Alviso clay	"in general it contains large quantities of saline salts"; "it has practically no agricultural value and supports only a growth of saltgrass and other salt-tolerant vegetation"	Carpenter 1939

and other conditions favoring accumulation of alkali, such land may readily become affected to a degree that will interfere with cropping" (Sweet et al. 1908). Unfortunately, this map only covered a small part of the study area on the east side of the San Joaquin in the vicinity of Walthall Slough. The Contra Costa soil survey's mapped alkali areas coincide with concentrations above 0.2%, as all mapped areas falling within concentrations of 0.1% and 0.2% lie within the extent of historical wetland and are thus not mapped as alkali seasonal wetland complex (Carpenter and Cosby 1933). Given the different approaches the soil surveys took mapping alkali areas, inherent inconsistencies are likely present in the mapping as a result.

We mapped according to indications in the soils surveys where earlier evidence did not suggest a different wetland type, such as freshwater emergent wetland. It is possible that some soils may have become more alkaline over time or after freshwater wetlands were reclaimed (reduction in flood frequency – both tidal and fluvial – can allow salts to accumulate in the soils), which may suggest a possible over-representation of alkali seasonal wetland complex. This is an instance where knowledge of land use history bolstered our confidence that some areas became alkaline only after drainage and grazing affected the hydrology and the vegetation cover of an area. In general, however, most regions where alkali seasonal wetland was mapped had at least some localized mid-1800s evidence of alkali, whether from GLO surveyors noting alkali in their description of soils or notes of vegetation such as greasewood, likely iodine bush (*Allenrolfea occidentalis;* Stanford et al. forthcoming).

Stabilized interior dune vegetation
This community occurs exclusively on the eolian (wind-blown) sand deposits and mounds unique to the Contra Costa and western portion of the Delta (see page 186). These areas were mapped based on soil surveys, geology and topography (see Fig. 2.18). The topography was an important consideration because only those sand mounds at elevations above the marsh plain (out of reach of most tides) were expected to have a vegetation community uniquely different from that of the surrounding wetland. Therefore, while there may be large areas of eolian sands, only a portion of those areas are mapped as stabilized interior dune vegetation. Historical sources, primarily reclamation district maps and narrative accounts describing the many small mounds that rose above the marsh surface, offered early confirmation of these areas as unique in the region. We mapped areas larger than five acres, according to our minimum mapping unit. However, many of these features were quite small so the mapping did not capture all of the features which would have been present.

Grassland, savanna, and woodland
Unlike many wetland types, where multiple historical datasets often give detailed information about a feature's location, sources describing the upland habitat conditions of the Central Valley generally contained

less spatially explicit detail. The mapping effort was therefore focused on producing a meaningful representation of general patterns of upland vegetation cover at the landscape scale. The level of spatial resolution and detail in available historical data across the study area meant that we were generally only able to consistently differentiate between areas characterized by few to no trees and those with moderate to relatively dense tree cover. For this reason, areas with few to no trees were mapped as grassland and areas with at least a moderate tree density (approximately 10% tree cover) was mapped as the oak woodland or savanna class.

A primary source of information for these distinctions was travelers' accounts describing where timber (e.g., "scattering timber," "groves of oak") was found in the Central Valley. These accounts are often somewhat general because of the nature of historical narratives (travelers were often not concerned about detailing exactly where they were) and because shifts between grassland, savanna, and woodland are gradual and diffuse, making it difficult to determine where savanna ends and grassland begins. Some of descriptions were location-specific and could be used as direct evidence in the mapping. For example, one traveler leaving Sutter's Fort in 1841 describes first crossing "a vast plain, shaded by enormous oaks" (De Mofras and Wilbur 1937) and another states plainly that "oaks commence" (Lyman and Teggart 1923) at French Camp on the San Joaquin. Overall, we acquired the sense of the landscape to be conveyed through the study of these descriptions. Even broad characterizations were useful: "The east side of the Sacramento and San Joaquin valleys supports the greater part of the groves of scattered trees as contrasted with the west side which is in the main treeless" (Jepson 1910).

Another primary data source used to map vegetation of upland habitats was the GLO surveys, including the bearing tree dataset (trees used to establish the location of a survey corner) and line descriptions (where surveyors describe the general character of the land they have passed over the past mile). Where surveyors were able to find oaks for bearing trees or where they made observations such as "scattering timber" or "timber improving" (Norris 1853), we classified the area as oak woodland or savanna. Although oak removal associated with early settlement had already commenced, it is likely that this cutting was fairly localized and not concentrated in any one location and, therefore, historical patterns of tree density would have still been observed by the surveyors (e.g., Whipple et al. 2011). This early 1860s view of tree cover gives a minimum of the distribution and extent of oak savannas and woodlands in the early 1800s.

As the data discussed above suggest, while we may be confident in the mapping of a particular upland habitat type within a general area (e.g., land surrounding Sutter's Fort), we may be substantially less certain about extent and exact location. In most cases, we used the early narrative accounts and GLO to establish our understanding of an area, but used soil types from historical soil surveys for the actual digitized boundaries. Soil type boundaries often indicate transitions between habitat types as soils were often mapped, in part, from observed changes in vegetation cover. In some cases, soil survey descriptions helped identify whether a particular soil type was grassland or savanna and woodland, but since these twentieth century soil surveys occurred subsequent to extensive tree removal, descriptions of native vegetation cover are often absent. Thus, we generally used these as a confirmation that the description did not contradict with what we had mapped (e.g., we had not mapped oak woodland or savanna on clay soils). Other sources, including maps and landscape and aerial photography, provided additional support.

An important qualification is that considerable variation in characteristics existed within the areas we mapped as either grassland or woodland and savanna. Inevitably, historical documents reveal greater detail than can be represented consistently by habitat mapping. It should be assumed that most grassland supported a few trees (particularly along small watercourses) as well as small mosaics of wetlands and ponds. Similarly, mapped oak woodland or savanna was also locally complex, characterized by patches of dense groves and scattered trees interspersed with open grasslands and small mosaics of wetlands and ponds. For example, within an area near Stockton mapped as oak woodland or savanna, GLO notes of "timber very thin," "good timber" (Norris 1853) and even "no timber" (Wallace 1865) are found. These small-scale patterns depended on local variability in topography, soils, and moisture regime. Though the broad classifications used may obscure some detail, they reveal fundamentally important patterns in the distribution and abundance of major habitat types.

TECHNICAL REVIEW

We sought review from local as well as national experts with backgrounds in ecology, geomorphology, geology, archaeology, estuarine science, geography, and landscape history. Reviewers provided comments on the draft report and many reviewed GIS mapping, aided our interpretation of the data, commented on drafts of graphics, and provided guidance on specific topics over the course of the project.

3. Regional summary

In the early 1800s, a broad expanse of freshwater wetlands met the eye looking east from Suisun Bay. These wetlands comprised the Sacramento-San Joaquin Delta, the upstream portion of the San Francisco Estuary and one of the few inland deltas in the world. The northerly Sacramento River and the southerly San Joaquin River met tidal water and branched into numerous winding and comingling channels within the heart of the Delta. At the Delta mouth, water coalesced into a single broad channel passing into Suisun Bay, San Francisco Bay, and finally the Pacific Ocean (Figs. 3.1 and 3.2). The Delta received an annual supply of water: tides, high groundwater levels, freshwater inflows, and naturally stored water from flood basins all contributed to this highly productive ecosystem (Atwater and Belknap 1980). It remained wet when the rest of California was dry, serving as a refuge particularly during drought.

Before the transformation of wetlands to farms and towns, distinct patterns of native habitats were expressed along the Delta's broad physical gradients. The arrangement of habitats was driven by variations in dominant physical processes. At a fundamental level, the historical Delta habitat patterns and ecological functions reflected the transition between dominant riverine processes upstream and tidal processes downstream. At the Delta mouth, the salinity gradient shifted with inter-annual and seasonal variability. It was also affected by the differences in the hydrologic regimes of the Sacramento and San Joaquin rivers, as well as other systems that fed into it. Landscape patterns were influenced by these and other interacting physical processes and organized within the context of three primary components: the subtidal channels, the intertidal and non-tidal wetlands, and the elevated, infrequently flooded natural levees (Atwater and Belknap 1980, TBI 1998).

Within the Delta, approximately 365,000 acres (147,700 ha) of tidal freshwater emergent wetlands (tule, *Schoenoplectus* spp., dominant) and over 1,000 miles of associated tidal channels occupied the core of the Delta. The wetlands approximated high tide levels as islands ranging in size from a few thousand acres to over 10,000 acres and tracts with an upland edge (Atwater and Belknap 1980, TBI 1998, Thompson 2006). The islands and large tracts slowly accumulated organic matter, which kept marsh elevation rising in pace with gradually rising sea levels. To the north along the Sacramento River, broad zones of tidal wetland graded into non-tidal wetlands occupying flood basins flanking the river behind natural levees. These basins functioned as natural reservoirs for annual overflow from rivers and streams and served to recharge the high groundwater table. They were occupied by unusually dense and tall tule and large lakes. Riparian forests extended far into the tidal wetlands on natural levees along the major rivers and distributaries. Primarily in the western Delta, scattered sand mounds – high points of glacial-age eolian (wind blown) sand dunes – rose above the plain, adding topographic variation and habitat complexity to the flat terrain. In the south Delta, at the margins of tidal influence along

The delta of twenty-five miles in length, divided into islands by deep channels, connects the bay with the valley of San Joaquin and Sacramento, into the mouths of which tide flows, and which enter the bay together as one river.

—FRÉMONT 1845

Figure 3.1. The Delta and Central Valley wetlands as mapped in 1887. The broader wetlands of the Sacramento Valley contrast with the narrower corridor along the San Joaquin River south of the Delta. (Hall 1887, courtesy of the Map Collection of the Library of UC Davis)

N

▆	Water
▆	Intermittent pond or lake
▆	Tidal freshwater emergent wetland
▆	Non-tidal freshwater emergent wetland
▆	Willow thicket
▆	Willow riparian scrub or shrub
▆	Valley foothill riparian
▆	Wet meadow or seasonal wetland
▆	Vernal pool complex
▆	Alkali seasonal wetland complex
▆	Stabilized interior dune vegetation
▆	Grassland
▆	Oak woodland or savanna

Figure 3.2. Oblique view of the historical Delta overlaid on modern aerial imagery in Google Earth. The Delta defines the area where the distributary branches of the Sacramento River from the north and the San Joaquin River from the south meet tide water east of Suisun Bay. The many branches meet at the foot of Sherman Island near Antioch before passing into the Suisun Bay and then San Francisco Bay.

the distributary branches of the San Joaquin River, numerous active and abandoned channels formed by riverine processes laced a floodplain where perennial wetlands were interspersed with intermittent ponds and lakes, willow thickets, seasonal wetlands, and patches of grassland. At the upland Delta margins, the perennial wetlands graded into seasonal wetlands (including vernal pool and alkali seasonal wetland complexes), dry grasslands, and oak woodlands and savannas (Fig. 3.3).

The position of large tidal channels, natural levees, and lakes appears to have remained relatively fixed in place through time. Substantial climatically-driven variability was expressed within these relatively stable patterns, however. As a result, the Delta looked very different depending on the year and season: certain wetlands might be flooded several feet deep by late winter and dry at the surface by the late fall. Species were adapted to the variability, taking advantage of different conditions at different times of the year (Moyle et al. 2010). Such dynamics are important factors in fostering habitat and species diversity (Mason n.d.). Flooding, for instance, provided seasonal connectivity necessary for fish to access the rich food sources of the floodplains and promoted high productivity and nutrient exchange.

While it may be easy to think of the Delta as an unvarying wetland plain – a vast sea of tules, as many put it – it was in fact a place of significant spatial and temporal complexity that provided important ecosystem functions (TBI 1998). This heterogeneity was related to physical gradients that were expressed at different spatial scales, and also related to disturbance regimes and biological interactions. The complex habitat patterns found along the tidal-fluvial continuum led to high levels of habitat connectivity, allowing species to access appropriate environmental conditions at different times in the tidal cycle, season, or year. This more detailed level of complexity was significant in fostering the development and resilience of the diverse historical Delta ecosystem. The historical Delta had the characteristics of a highly productive ecosystem, with temporally and spatially shifting resource availability, physical disturbances, and high degrees of connectivity between different habitats (Moyle et al. 2010).

SUMMARY OF GIS MAPPING

This section summarizes the results of the habitat type mapping of the early 1800s Delta that was conducted over the course of the project (Table 3.1; see Fig. 3.3). We also compare the historical mapping to contemporary conditions in the Delta (Hickson and Keeler-Wolf 2007). This provides an entry into the analysis of landscape patterns and change at the level of major habitat types. The historical spatial datasets produced by this project have been made available for download (www.sfei.org/DeltaHEStudy).

Historical mapping

The historical habitat type map captures the extent and distribution of primary habitat types in a comprehensive fashion across the Delta to determine relative proportions of habitat types and illustrate landscape pattern and process. The mapping does not convey the additional level detail within habitat types, where shifts in factors such as vegetation community or hydroperiod supported significant spatial complexity at the local scale. The three main chapters of this report (chapters 4-6) are intended to provide a deeper understanding of the Delta's historical habitat type characteristics, patterns, functions, and related processes.

RELATIVE WETNESS Within the study area, we mapped 394,400 acres (159,600 ha; 50%) as tidally influenced (open water and wetland). An additional approximately 124,000 acres (50,200 ha; 15.9%) were overflowed every year and kept wet enough year-round through natural surface water storage and high groundwater to support assemblages of ponds and lakes, perennial emergent wetlands, and willow thickets. This is in general agreement with the 380,000 acres (153,800 ha) of intertidal wetlands (not including waterways) and 145,000 acres (58,700 ha) of non-tidal wetlands mapped by The Bay Institute (TBI 1998), the 346,000 acres (140,000 ha) estimated by Atwater et al. (1979), and an earlier estimate of 350,000 to 400,000 acres (140,000 to 160,000 ha) of "fresh water tide lands" in the Delta (Gilbert 1879). The perennially wet features from this study's mapping represent about 40% of the historical wetlands of the Central Valley, as mapped by TBI (1998).

Seasonal wetlands approximating 144,300 acres (58,400 ha; 18.4%) were flooded less frequently and for briefer periods of time. These areas were largely influenced by the smaller upland drainages that spread into distributaries along the alluvial fans before reaching the perennial wetlands along the rivers. However, during extreme floods much of this area was

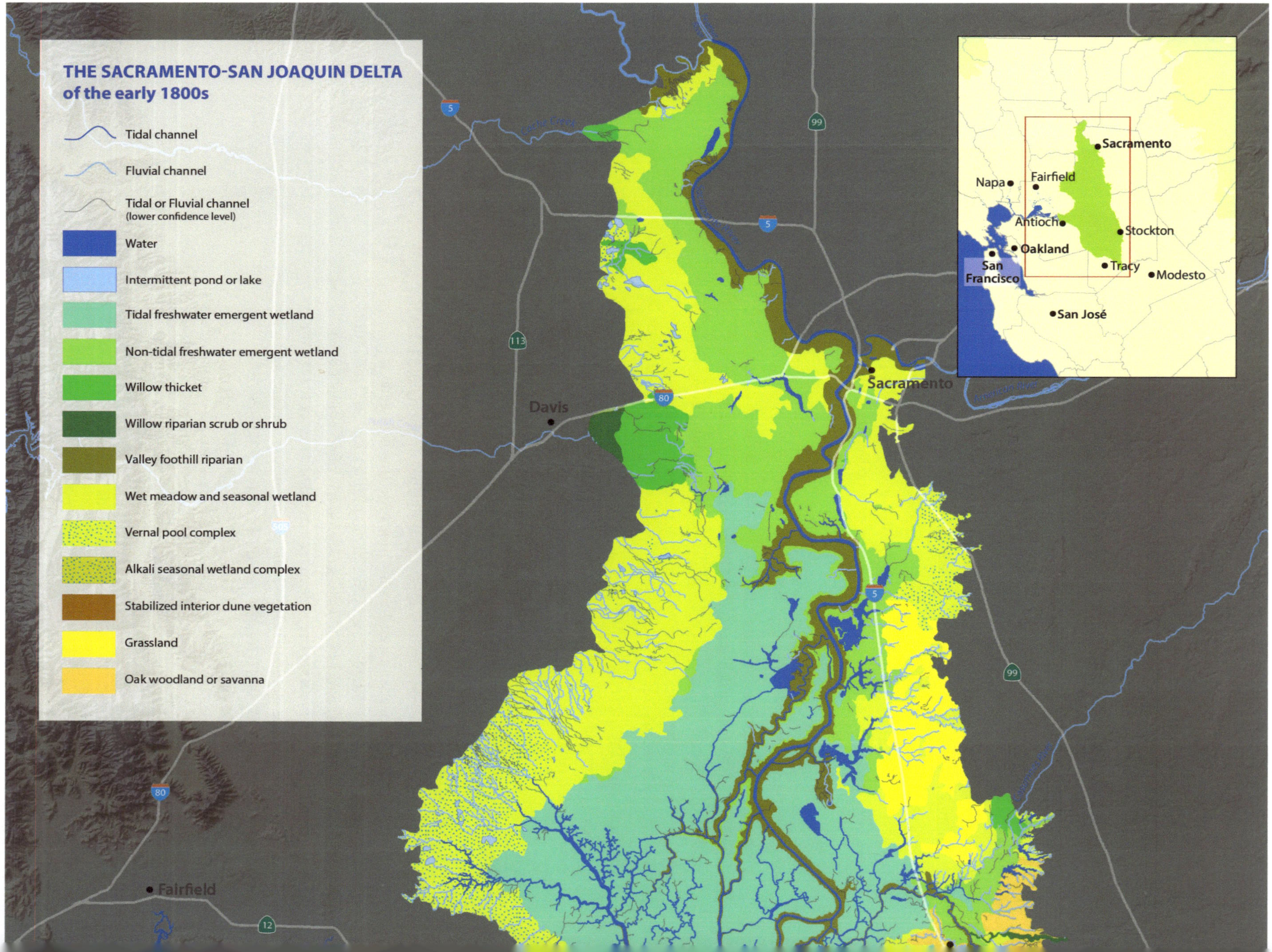
THE SACRAMENTO-SAN JOAQUIN DELTA
of the early 1800s
Tidal channel
Fluvial channel
Tidal or Fluvial channel
(lower confidence level)
Water
Intermittent pond or lake
Tidal freshwater emergent wetland
Non-tidal freshwater emergent wetland
Willow thicket
Willow riparian scrub or shrub
Valley foothill riparian
Wet meadow and seasonal wetland
Vernal pool complex
Alkali seasonal wetland complex
Stabilized interior dune vegetation
Grassland
Oak woodland or savanna
Sacramento
Davis
Fairfield
5
99
113
80
12
Napa
Antioch
Oakland
San Francisco
Stockton
Tracy
Modesto
San José

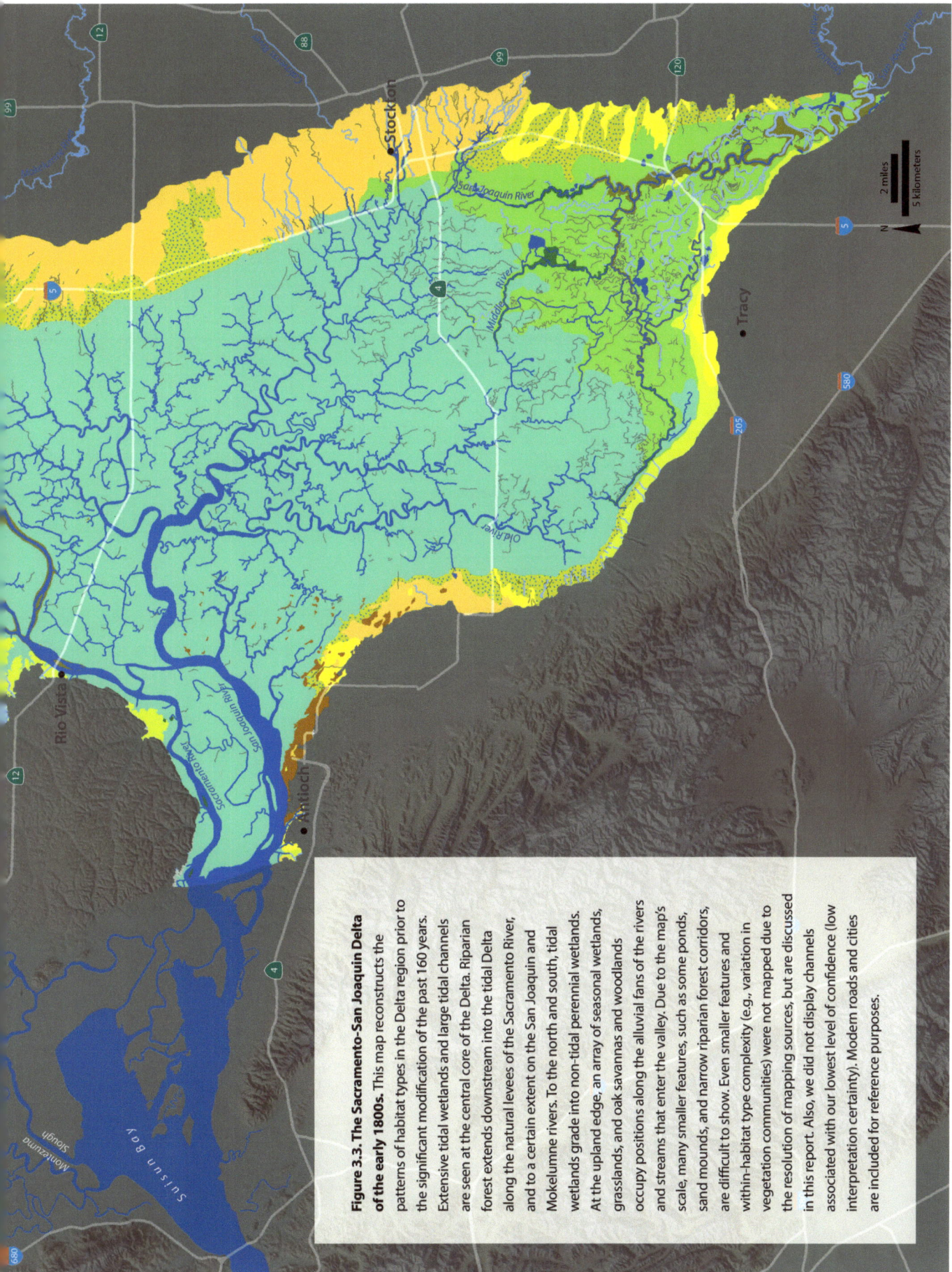

Figure 3.3. The Sacramento-San Joaquin Delta of the early 1800s. This map reconstructs the patterns of habitat types in the Delta region prior to the significant modification of the past 160 years. Extensive tidal wetlands and large tidal channels are seen at the central core of the Delta. Riparian forest extends downstream into the tidal Delta along the natural levees of the Sacramento River, and to a certain extent on the San Joaquin and Mokelumne rivers. To the north and south, tidal wetlands grade into non-tidal perennial wetlands. At the upland edge, an array of seasonal wetlands, grasslands, and oak savannas and woodlands occupy positions along the alluvial fans of the rivers and streams that enter the valley. Due to the map's scale, many smaller features, such as some ponds, sand mounds, and narrow riparian forest corridors, are difficult to show. Even smaller features and within-habitat type complexity (e.g., variation in vegetation communities) were not mapped due to the resolution of mapping sources, but are discussed in this report. Also, we did not display channels associated with our lowest level of confidence (low interpretation certainty). Modern roads and cities are included for reference purposes.

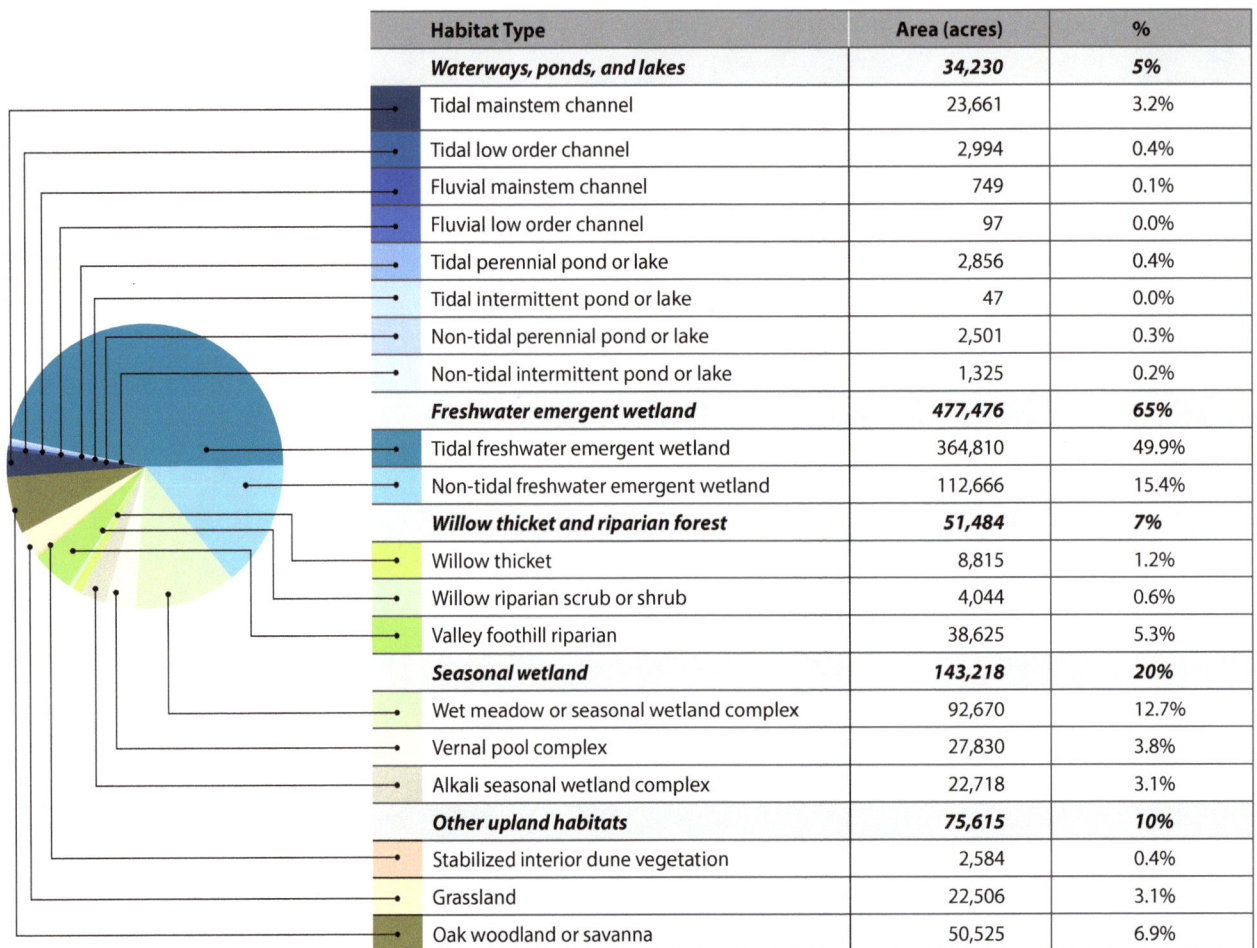

Habitat Type	Area (acres)	%
Waterways, ponds, and lakes	**34,230**	**5%**
Tidal mainstem channel	23,661	3.2%
Tidal low order channel	2,994	0.4%
Fluvial mainstem channel	749	0.1%
Fluvial low order channel	97	0.0%
Tidal perennial pond or lake	2,856	0.4%
Tidal intermittent pond or lake	47	0.0%
Non-tidal perennial pond or lake	2,501	0.3%
Non-tidal intermittent pond or lake	1,325	0.2%
Freshwater emergent wetland	**477,476**	**65%**
Tidal freshwater emergent wetland	364,810	49.9%
Non-tidal freshwater emergent wetland	112,666	15.4%
Willow thicket and riparian forest	**51,484**	**7%**
Willow thicket	8,815	1.2%
Willow riparian scrub or shrub	4,044	0.6%
Valley foothill riparian	38,625	5.3%
Seasonal wetland	**143,218**	**20%**
Wet meadow or seasonal wetland complex	92,670	12.7%
Vernal pool complex	27,830	3.8%
Alkali seasonal wetland complex	22,718	3.1%
Other upland habitats	**75,615**	**10%**
Stabilized interior dune vegetation	2,584	0.4%
Grassland	22,506	3.1%
Oak woodland or savanna	50,525	6.9%

Table 3.1. Acreage summary by historical habitat type within the study area. Habitat types are grouped into classes of open water, perennial wetlands, willow thicket and riparian forest, seasonal wetlands, and upland habitats. Summarized from the mapping performed in this study, these figures represent estimates of the total area of different habitat types in the early 1800s.

overflowed. Several estimates for the larger valleys suggest that such extreme events increased the area usually flooded on a nearly annual basis by a third or more (Fortier 1909, U.S. Congress 1916). For the Sacramento Valley, geomorphologist Kirk Bryan (1923) concluded that 60% was "subject to overflow" prior to reclamation.

WATERWAYS, PONDS, AND LAKES Within the study area of 782,000 acres (316,000 ha), about 34,200 acres (13,800 ha; 4.4%) of waterways and bodies of open water occupied the lowest positions. Most of the water area was found in the form of tidal channels (26,700 acres/10,800 ha; 3.4%). The tidal channel network had a total channel length of about 1,600 miles (2,600 km) (Fig. 3.4). Estimates of possible overmapping due to incorrect assignment of ancient channel signatures in aerial photography suggest this figure could be as low as 1,100 miles (1,800 km), though the strong likelihood of undermapping tidal channel due to lack of detail in mapping sources partly, if not completely counterbalances this issue (i.e., the network likely misses many small, first order, tidal channels).

While the mainstem channels (the primary rivers and sloughs that delineated the tidal islands) were large in size, they only comprised about 27% of the total tidal channel length mapped. This is likely a conservative estimate considering the likelihood of undermapping of small channels.

This proportion is also a proxy for the relative length of channel that was flow-through versus dead-end (mainstem channels connected through to another channel, while low order channels branched and terminated within the wetlands). Also, the mapping illustrates that the tidal channel distribution was not even throughout the tidal wetlands. Most of the tidal channels were concentrated within the central core of the Delta. Moving upstream, sloughs experiencing regular tidal flow became less numerous, though the main river channels continued to be tidally influenced far upstream.

Over 1,000 miles (1,600 km) of fluvial channel were also mapped, with slightly less than half of that length coming from channels found within non-tidal emergent wetlands (e.g., south Delta and Cosumnes Sink). The majority of the remaining fluvial channels consisted of small ephemeral streams located along the upland margin of the Delta and terminating before reaching the tidal wetlands. The rest of these channels were found intersecting the large natural levees of the rivers, which generally only flowed when the rivers were at higher stages.

More than 5,700 acres (2,300 ha; 0.7%) of ponds and lakes were found in backwater locations within perennial wetlands of the Delta. Most of these were found along the margins of tidal wetlands or within non-tidal wetlands. Both the number and the size of ponds and lakes were greatest in the northern part of the Delta, where they occupied the large flood basins. Because of landscape positions away from the tidal core of the Delta, many of those within potential range of tidal influence were likely isolated from substantial tidal action.

Another 1,000 acres (400 ha) of ponds and lakes were mapped within other habitat types, primarily large intermittent features within vernal pool complexes. As we did not capture the smaller features (waterways less than 50 ft/15 m wide and land cover less than 5 acres/2 ha), the area estimates are conservative figures. Aside from these vernal pools, only a few other ponds and lakes were found to be intermittent. It is likely that many such features existed and absence of sources indicating these seasonal features prevented comprehensive mapping (our default classification was perennial).

FRESHWATER EMERGENT WETLAND Perennial freshwater emergent wetlands covered the majority of the study area, consisting of 364,800 acres (147,600 ha; 46.7%) of tidal wetlands within the interior Delta. They graded into an additional 112,670 acres (45,500 ha; 14.4%) of non-tidal perennial wetlands integrally connected to the tidal wetlands. Within these two divisions, inundation frequency varied dramatically. As discussed elsewhere in the report, it appears that only about half of the tidal extent was inundated by twice daily high tides, with the rest ranging from wetted (rather than actually overflowed) by daily tides to only wetted by spring tides (see page 127). Late in the season the surface of the non-tidal wetlands could become dry, although the water table was just below the surface. These and other physical and biological factors supported the diverse vegetation

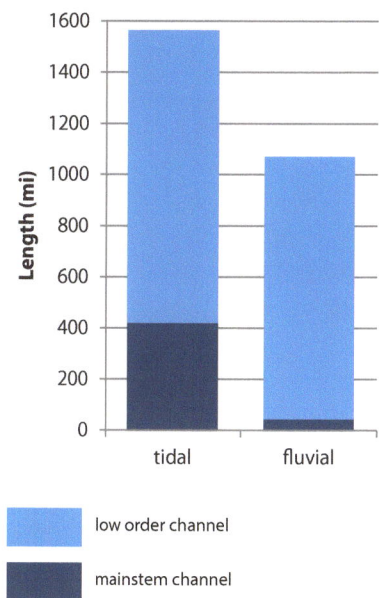

Figure 3.4. Estimates of total length of channel summarized from the mapping of both tidal and fluvial channels within the study area. We were less certain about the tidal status of about 25% of the channel lines, so a portion of the tidal channels might well be more appropriately binned in the fluvial class and vice versa. The tidal and fluvial totals are divided between mainstem (main river channels and connecting sloughs) and low order channels. About half of the fluvial low order channels were found within the south Delta's non-tidal wetlands and are the class of channels most difficult to interpret and thus are associated with the highest uncertainty concerning historical presence. The vast majority of the fluvial low order channels were intermittent streams.

communities found within the wetlands, though tule dominated the vast majority of these habitats. However, particularly within the central Delta in the region of the lower reaches of Old and Middle rivers, willow (*Salix* spp.) occupied a significant portion of this matrix. Its extent is not mapped explicitly. This vegetation community is referred to as willow-fern swamp (Mason n.d., Atwater 1980).

RIPARIAN FOREST AND WILLOW THICKETS Natural levees established by riverine processes extended well into the tidal landscape. These higher lands were occupied by riparian forest, which made up approximately 42,610 acres (17,200 ha; 5.5%) in the Delta, with 36,900 acres (14,900 ha; 4.7%) north of the Delta mouth and 9,400 acres (3,800 ha; 1.2%) to the south. As fluvial influence diminished downstream, natural levee height decreased and riparian forest became narrower and more dominated by water-tolerant species such as willows. To illustrate this shift, we divided the riparian forest into valley foothill riparian along the higher natural levees and willow riparian scrub or shrub at the lower downstream ends, primarily along the lower reaches of the Mokelumne and the San Joaquin rivers. Of the total mapped riparian forest, only 4,000 acres (1,600 ha) was assigned to the willow riparian scrub or shrub class, with the rest classified as valley foothill riparian. This is a conservative estimate of riparian forest as our scale of mapping permits consideration of the larger river channels, but only rarely the many streams, sloughs, lakes, and ponds along the Delta perimeter, many of which were lined with willow.

Once the tidally dominated channels of the central Delta were reached, channel banks were only slightly elevated above the surface of the marsh and so were covered with emergent vegetation. Willows and other wetland associated species persisted as part of the complex of tule and other wetland species along these banks, and thus were included as part of the freshwater emergent wetland class. In the core of the tidal Delta, high salinity did not limit vegetation; rather the high water table, high inundation frequencies, and peaty soils served to exclude tree species along these lower levees.

Willow thickets (i.e., dense woody vegetation occupying floodplains) were mapped associated with the sinks of river and stream distributaries and amounted to 8,820 acres (3,570 ha; 1.1%; see page 294). This type represents large areas of willow-dominated floodplains, where the willows were not exclusively confined to natural levees. This area includes only the large and well defined expanses of willow: patches of willow thickets were found throughout the Delta.

SEASONAL WETLANDS On the order of 143,360 acres (58,000 ha; 18.4%) of various seasonal wetland types bordered the perennial wetlands, including 92,820 acres (37,560 ha; 11.9%) of wet meadow or seasonal wetland, 27,830 acres (11,260 ha; 3.7%) of vernal pool complex, and 22,720 acres (9,190 ha; 2.9%) of alkali wetland complex. These habitat types were associated with less well drained soils, where slopes were gradual, along the upland edge of the emergent wetlands. They were fed by the small intermittent or ephemeral streams emanating from the foothills. The ecotone, or zone of transition, between the tule (freshwater emergent wetland edge) and seasonal wetlands was likely complex depending on local topography and soil moisture regimes. Alkali seasonal wetlands, in particular, were often described as forming a border over a mile wide along the tule (Hilgard 1884).

The majority of the wet meadows and seasonal wetlands (over 94%) – encompassing a diverse range of plant communities, inundation frequencies, and soil types – were found north of the Delta mouth. The gradual slopes west of the Yolo Basin were mapped almost exclusively as wet meadow or seasonal wetland intergrading with vernal pools. Similar patterns were found along the eastern edge of the Sacramento Basin. Identifying this habitat type generally depends on three factors: hydrology, soils, and wetland vegetation. All of the area classified as vernal pool complex (where individual pools may be alkaline, but surrounding soils were not characterized as alkaline) was mapped north of the Delta mouth whereas all of the mapped alkali wetlands were found south of the Delta mouth on the western edge and south of the Mokelumne River on the eastern. This difference reflected the drier conditions in the southern Delta that promoted the accumulation of salts in soils.

OTHER UPLAND HABITATS Sand mound habitats, classified as stabilized interior dune vegetation, occupied 2,550 acres (1,030 ha; 0.3%) within the study area. These habitats occupied the stabilized Antioch Dunes and scattered sand mounds extending into the tidal wetland of eastern Contra Costa County. We mapped 24 sand mounds surrounded by tidal wetland that were over 5 acres (2 ha) in size, amounting to a total of over 250 acres (100 ha). The largest of these features was 27 acres (11 ha). We also identified numerous mounds in aerial photography and topographic maps that were smaller than 5 acres (2 ha) (there were well over 50 individual features) to add over 80 additional acres (32 ha). We mapped only those areas above tidal elevations as this habitat type, but it is likely that the ecotonal boundary surrounding each of these mounds supported a unique assemblage of plant species.

Grasslands and oak woodland or savanna, the upland habitats within the study area, totaled 22,510 (9,110 ha; 2.9%) and 50,560 (20,460 ha; 6.5%) acres, respectively. Grasslands were mapped primarily interspersed with wet meadow or seasonal wetland and vernal pool habitats east of the Sacramento Basin and along parts of the southern Delta edge. The vast majority (92%) of the mapped oak woodland or savanna was found in the vicinity of Stockton and extending north toward the Mokelumne and Cosumnes rivers.

Assessing certainty

Results from the certainty level assignment that was performed for each feature are shown in Figure 3.5. Certainty assessment recorded

Figure 3.5. Certainty levels for habitat mapping. In A, the proportion of high, medium, and low certainty by area is shown for each of the main certainty types (interpretation, shape, and location) for the mapping of polygons. Both the interpretation and location of habitat types was relatively high. Shape was much lower due to the difficulty in assessing the size of large areas of habitat types such as wet meadow or seasonal wetland complex which were characterized by rather indefinite boundaries. The same information is shown for the channel mapping in B. Here, very little of the mapped length is assigned low certainty, under 10% for each type. Interpretation and shape certainty levels are driven, in part, by the challenges associated with interpreting historical aerials and modifications to channel alignment since historical times.

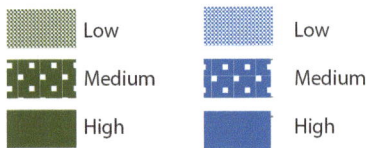

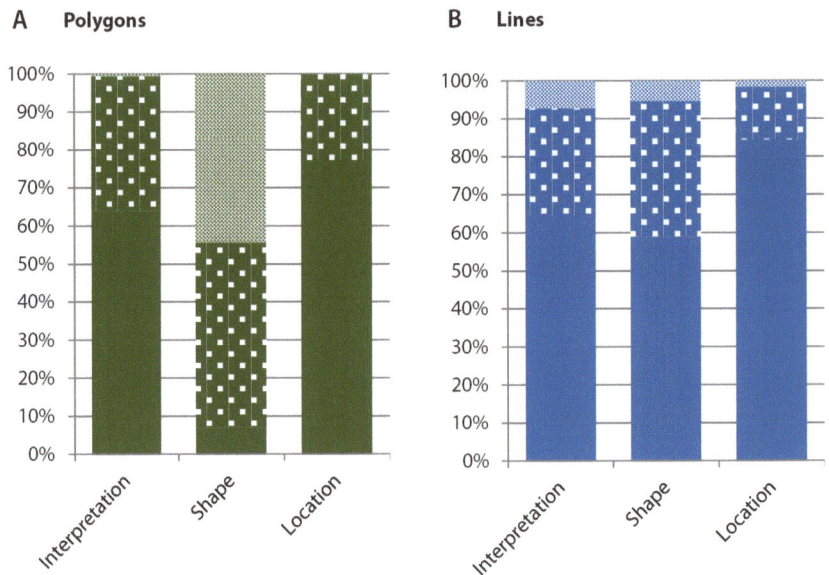

A   Polygons

B   Lines

Low
Medium
High

Low
Medium
High

interpretation (i.e., presence and classification), shape (size), and location (position; see page 49). Overall, confidence in interpretation and location was fairly high, 64% and 77% respectively. The lower certainty in shape (of each mapped feature) reflects the large areas of habitats, primarily around the perimeter of the Delta, where boundaries were challenging to determine. For the channel lines layer (the network along the polygon channels plus the channels narrower than the polygon minimum mapping width), high interpretation certainty accounted for about 64% of the mapped channel length, with high shape certainty at 59% and high location at 85%. Less that 10% of the area was assigned a low interpretation certainty for either mapping layer. The fourth certainty level standard, tidal interpretation, was only included in the lines layer, where 75% of the channel length was assigned a high certainty level for its tidal interpretation.

Figure 3.6 shows the breakdown by habitat type for the interpretation certainty factor for both the polygon (habitat types) and line layers (waterways) of the GIS. This analysis illustrates the variability in confidence of the mapping depending on the habitat type. Generally, lower interpretation levels are associated with those habitat types where few spatially explicit descriptions were available or where interpretation between two types was challenging (e.g., grassland versus wet meadow or seasonal wetland complex). Those habitat types with less than 50% of the area assigned with high certainty includes alkali seasonal wetland complex, grassland, tidal intermittent pond or lake, vernal pool complex, wet meadow or seasonal wetland, willow riparian scrub or shrub, and willow thicket. Habitat types associated with the highest interpretation certainty tended to be the water bodies and freshwater emergent wetland, given the many sources available confirming these habitat types (e.g., descriptions of tule to identify freshwater emergent wetland). Not surprisingly, the similar summary of the channel line layer shows the larger mainstem channels that are well-established in numerous historical sources with nearly 100%

A Polygons

B Lines

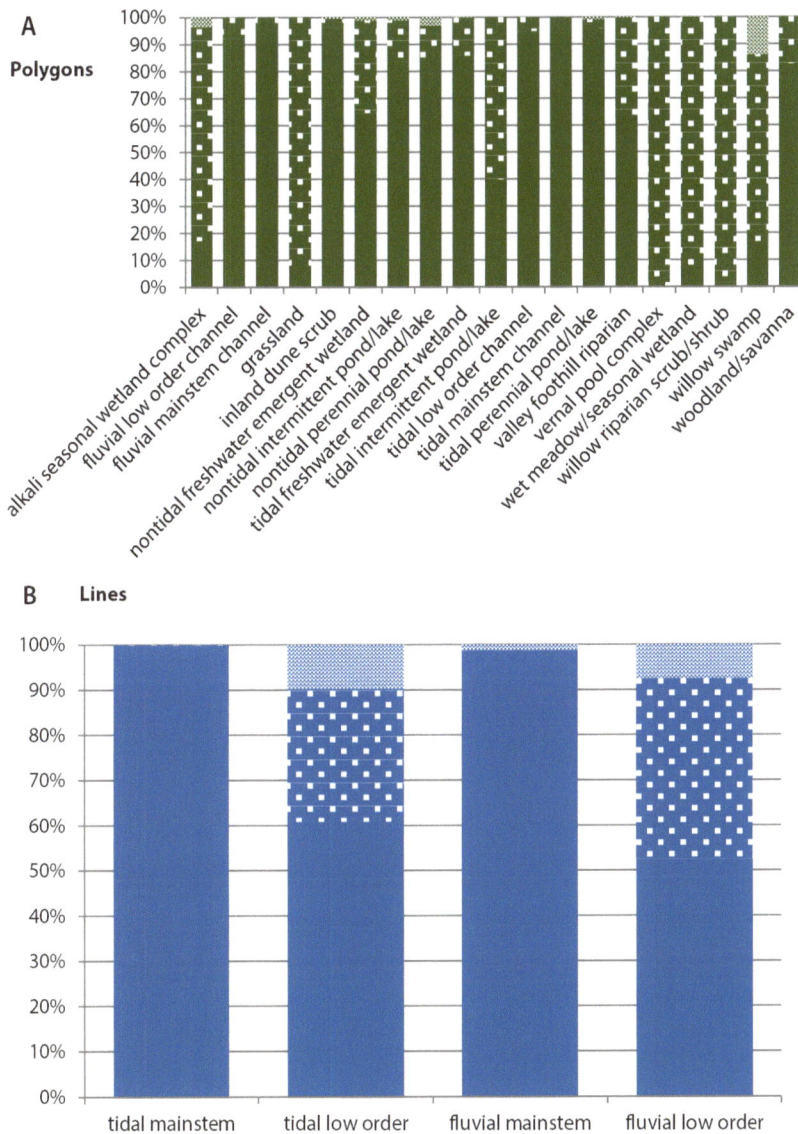

Figure 3.6. Interpretation certainty by habitat type. Certain habitat types were much easier to confirm than others, reflected in the differences in percentages of interpretation certainty (A). Features such as channels and emergent wetlands tended to be easier to determine using historical sources than distinctions between classes such as grassland and wet meadow or seasonal wetland complex. In B, assessment for the channel mapping illustrates the much lower interpretation associated with low order channels, as these were less likely to be shown in early maps of the area and often had to be interpreted using signatures in historical aerial photography.

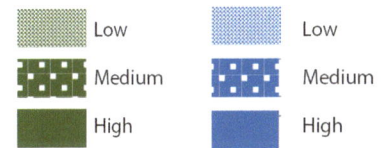

Low   Low
Medium   Medium
High   High

interpretation certainty, while the interpretation of lower order channels was more challenging, mostly due to the difficulties associated with distinguishing the early 1800s channels from the many signatures of ancient channels exposed by exhumed peat in the south Delta (see page 331).

These summaries were performed on the area of features as opposed to the number of features. Larger features were generally associated with lower levels of certainty (e.g., expanse of grassland mapped primarily from soil surveys), and therefore contribute proportionately more to these results than do smaller features.

Comparison to the modern Delta

Today's Sacramento-San Joaquin Delta is one of the most significantly modified deltas in the world. It is consequently challenging to decipher the habitat types and patterns that characterized various locations, imagine the productivity of the Delta's once large complex of wetland and riparian habitats, or analyze the ecological functions that once persisted. The most significant trajectory of change in the Delta region has been the replacement of the historically large expanse of nearly a half a million acres of perennial wetland by an even greater expanse of agriculture and urban development. Another important observation is that much of the existing areas of "natural" habitat types in the Delta – patches of alkali seasonal wetlands, seasonal wetlands, grassland, or willow-lined artificial levees – have been converted from the freshwater emergent wetlands that historically occupied those locations. The remnant natural areas in the Delta today are also often not of the same quality as similar type historically, being significantly compromised in the ecological functions they can provide and often highly disturbed, fragmented, or disconnected from other habitat types.

We compared the historical habitat type mapping to modern extents of vegetation types in recent vegetation mapping of the Delta from the California Department of Fish and Game (Fig. 3.7; Hickson and Keeler-Wolf 2007). Some classes were grouped within both data layers in order to establish a crosswalk between the classification systems of the two mapping efforts (Table 3.2). For the purposes of accurate comparison of acreage, we

Figure 3.7. Land cover change between the early 1800s and early 2000s. The change in land cover is illustrated in bar chart (A) and map (B) form. The dramatic shift from a majority of freshwater emergent wetlands historically to agriculture and urban development today is the most strikingly visible change. The area of open water (including areas of floating aquatic vegetation) has actually increased, in large part due to flooded islands such as Franks Tract and Mildred Island. The early 1800s view is based on the historical habitat type mapping performed in this study. The early 2000s summary is based on mapping performed by the California Department of Fish and Game from field work performed and aerial imagery taken between 2002 and 2005 (early 2000s data: Hickson and Keeler-Wolf 2007).

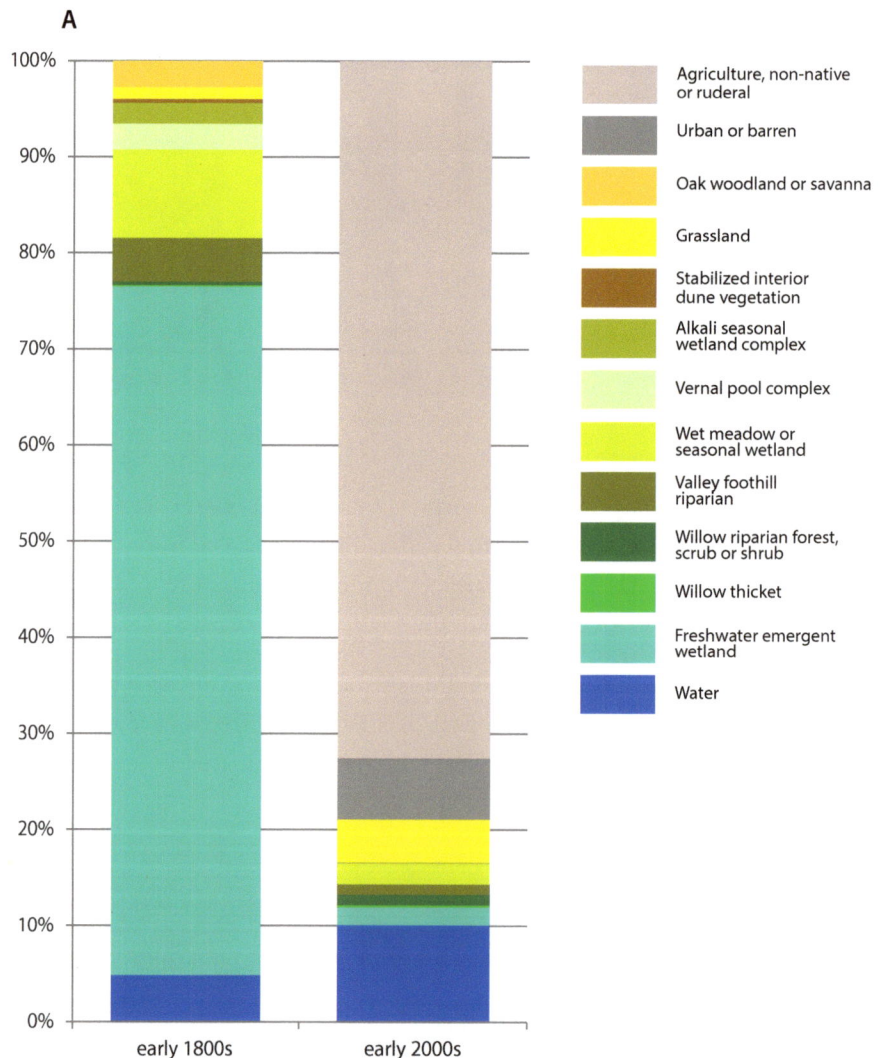

A

only analyzed the mutually mapped area, which totaled 627,300 acres (253,900 ha). As change analysis was not the primary focus of the project, the comparisons discussed here are just some of the many more detailed quantitative analyses that can be performed with the GIS dataset and other information presented in this report. It is expected that further analysis of change, such as evaluating changes in ecological functions, hydrodynamics, and habitat connectivity, will be pursued in subsequent studies.

In comparing historical and modern habitat mapping, we found that only 3% of the former historical freshwater emergent wetland area remains today. While historically 449,420 acres (181,874 ha; 72% of the mutually mapped area) supported perennial wetlands dominated by tule, only 11,590 acres (4,690 ha; 1.9% of the mutually mapped area) is characterized by similar habitats today. Even less of this represents pristine remnant patches of the former Delta: complex tidal wetlands remnant of the early 1800s

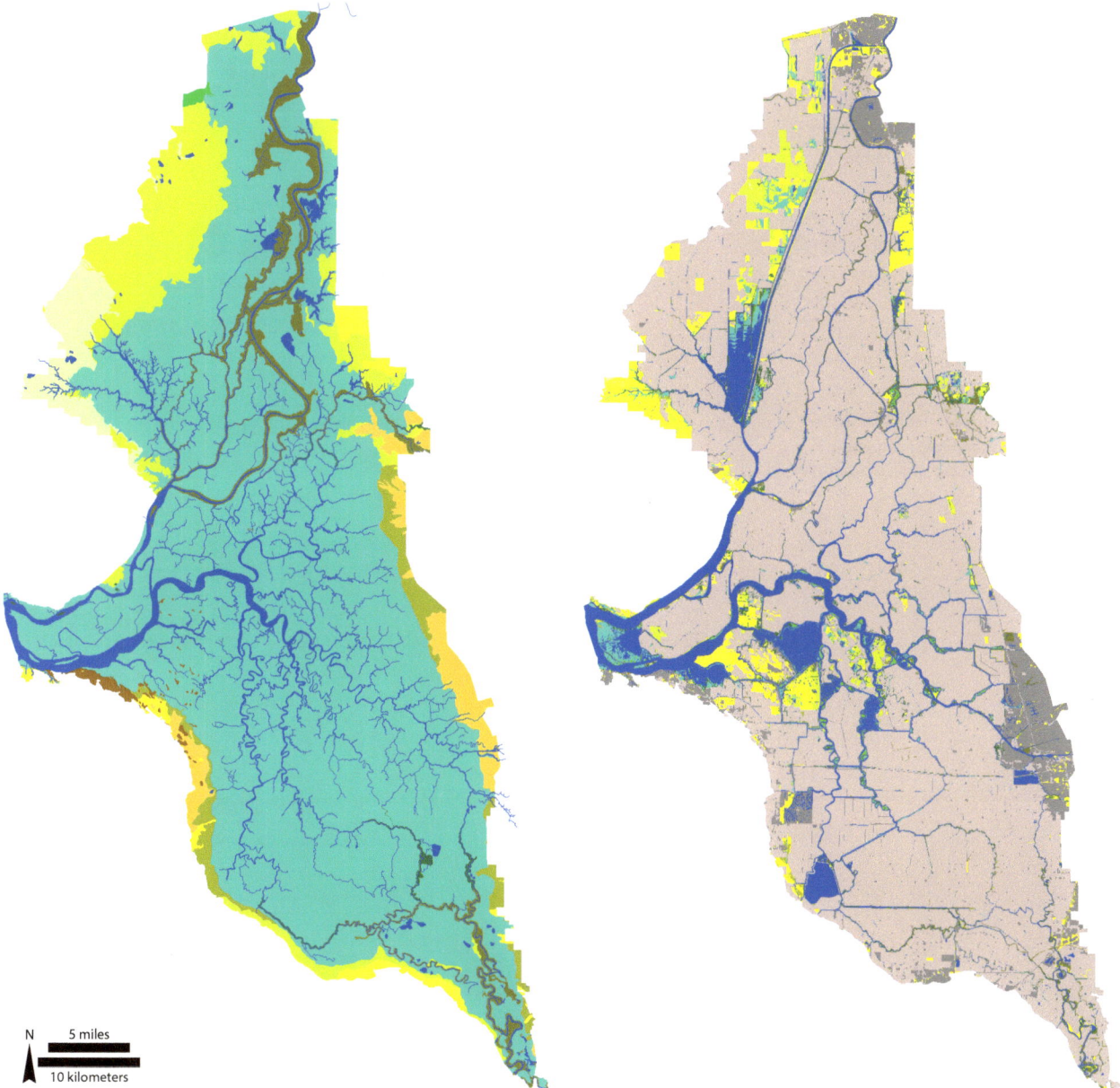

94

Table 3.2. Crosswalk for comparison between early 1800s mapping and recent mapping of Delta vegetation. Classes were grouped for each dataset in an effort to provide the best possible comparison between like land cover classes. Several vegetation types (MAPUNIT) of the 2007 mapping were challenging to associate with a lumped class. For example, "Horsetail *(Equisetum spp.)*" was placed in the "Agriculture, non-native, or ruderal" class given its common ruderal nature on levees. It was also determined that "*Distichlis spicata* - Annual Grasses" should be placed in the "Wet meadow or seasonal wetland" category as opposed to "Alkali seasonal wetland complex," as the area it was extensively mapped in (the Yolo Bypass) holds characteristics more similar to the wet meadow or seasonal wetland type used for mapping the historical Delta (Burmester pers. comm., Keeler-Wolf pers. comm.). Willow dominated communities also posed challenges. We focused on grouping the modern alliances based on the historical habitat classification of whether the willows were part of a backwater swamp community (willow thicket), the dominant species along channel banks (willow riparian forest, scrub, or shrub), or were part of a forest with oaks (valley foothill riparian forest). Finally, the modern types should not be considered to be of the same quality as those they compare to in the historical mapping (today riparian forest is often highly disturbed, wetlands are fragmented and isolated, etc.).

Habitat type for comparison	Historical habitat type
Agriculture, non-native, or ruderal	
Urban or barren	
Water	Fluvial low order channel, Fluvial mainstem channel, Tidal low order channel, Tidal mainstem channel, Nontidal intermittent pond/lake, Nontidal perennial pond/lake, Tidal intermittent pond/lake, Tidal perennial pond/lake
Freshwater emergent wetland	Nontidal freshwater emergent wetland, Tidal freshwater emergent wetland
Willow thicket	Willow thicket
Willow riparian forest, scrub, or shrub	Willow riparian scrub or shrub
Valley foothill riparian	Valley foothill riparian
Wet meadow or seasonal wetland	Wet meadow or seasonal wetland complex
Vernal pool complex	Vernal pool complex
Alkali seasonal wetland complex	Alkali seasonal wetland complex
Inland dune scrub	Stabilized interior dune vegetation
Grassland	Grassland
Oak woodland or savanna	Oak woodland or savanna

MAPUNIT (Hickson and Keeler-Wolf 2007)
Acacia - Robinia, Agriculture, *Eucalyptus*, Exotic Vegetation Stands, Giant Cane (*Arundo donax*), Horsetail (*Equisetum* spp.), Intermittently or Temporarily Flooded Deciduous Shrublands, *Lepidium latifolium - Salicornia virginica - Distichlis spicata*, Microphyllous Shrubland, Pampas Grass (*Cortaderia selloana - C. jubata*), Perennial Pepperweed (*Lepidium latifolium*), Poison Hemlock (*Conium maculatum*), Ruderal Herbaceous Grasses & Forbs, Sparsely or Unvegetated Areas; Abandoned orchards, Tobacco brush (*Nicotiana glauca*) mapping unit
Levee Rock Riprap, Urban Developed - Built Up
Algae, Brazilian Waterweed (*Egeria - Myriophyllum*) Submerged, Floating Primrose (*Ludwigia peploides*), Generic Floating Aquatics, *Hydrocotyle ranunculoides, Ludwigia peploides*, Milfoil - Waterweed (generic submerged aquatics), Pondweed (*Potamogeton* sp.), Shallow flooding with minimal vegetation at time of photography, Tidal mudflats, Water, Water Hyacinth (*Eichhornia crassipes*)
American Bulrush (*Scirpus americanus*), Broad-leaf Cattail (*Typha latifolia*), California Bulrush (*Scirpus californicus*), Common Reed (*Phragmites australis*), *Deschampsia caespitosa - Lilaeopsis masonii*, Hard-stem Bulrush (*Scirpus acutus*), Mixed *Scirpus* / Floating Aquatics (*Hydrocotyle - Eichhornia*) Complex, Mixed *Scirpus* / Submerged Aquatics (*Egeria-Cabomba-Myriophyllum* spp.) complex, Mixed *Scirpus* Mapping Unit, Narrow-leaf Cattail (*Typha angustifolia*), *Polygonum amphibium, Scirpus acutus - (Typha latifolia) - Phragmites australis, Scirpus acutus - Typha angustifolia, Scirpus acutus* Pure, *Scirpus acutus -Typha latifolia, Scirpus californicus - Eichhornia crassipes, Scirpus californicus - Scirpus acutus, Scirpus* spp. in managed wetlands, Smartweed *Polygonum* spp. - Mixed Forbs, *Typha angustifolia - Distichlis spicata*
Buttonbush (*Cephalanthus occidentalis*), California Dogwood (*Cornus sericea*), California Hair-grass (*Deschampsia caespitosa*), *Cornus sericea - Salix exigua, Cornus sericea - Salix lasiolepis / (Phragmites australis), Salix lasiolepis - (Cornus sericea) / Scirpus* spp.- (*Phragmites australis - Typha* spp.) complex unit, Shining Willow (*Salix lucida*)
Acer negundo - Salix gooddingii, Alnus rhombifolia / Cornus sericea, Alnus rhombifolia / Salix exigua (Rosa californica), Arroyo Willow (*Salix lasiolepis*), *Baccharis pilularis* / Annual Grasses & Herbs, Black Willow (*Salix gooddingii*), Blackberry (*Rubus discolor*), Box Elder (*Acer negundo*), California Wild Rose (*Rosa californica*), Coyotebush (*Baccharis pilularis*), Mexican Elderberry (*Sambucus mexicana*), Narrow-leaf Willow (*Salix exigua*), *Salix exigua - (Salix lasiolepis - Rubus discolor - Rosa californica), Salix gooddingii / Rubus discolor, Salix gooddingii* / Wetland Herbs, *Salix lasiolepis* - Mixed brambles (*Rosa californica - Vitis californica - Rubus discolor*), Santa Barbara Sedge (*Carex barbarae*), White Alder (*Alnus rhombifolia*), White Alder (*Alnus rhombifolia*) - Arroyo willow (*Salix lasiolepis*) restoration
Black Willow (*Salix gooddingii*) - Valley Oak (*Quercus lobata*) restoration, Coast Live Oak (*Quercus agrifolia*), Fremont Cottonwood (*Populus fremontii*), Hinds walnut (*Juglans hindsii*), Oregon Ash (*Fraxinus latifolia*), *Quercus lobata - Acer negundo, Quercus lobata - Alnus rhombifolia (Salix lasiolepis - Populus fremontii - Quercus agrifolia), Quercus lobata - Fraxinus latifolia, Quercus lobata / Rosa californica (Rubus discolor - Salix lasiolepis / Carex* spp.), Restoration Sites, *Salix gooddingii - Populus fremontii - (Quercus lobata-Salix exigua-Rubus discolor), Salix gooddingii - Quercus lobata* / Wetland Herbs, Temporarily or Seasonally Flooded - Deciduous Forests, Tree-of-Heaven (*Ailanthus altissima*), Valley Oak (*Quercus lobata*), Valley Oak (*Quercus lobata*) restoration
Distichlis spicata - Annual Grasses, *Distichlis spicata - Juncus balticus*, Intermittently Flooded Perennial Forbs, Intermittently or temporarily flooded undifferentiated annual grasses and forbs, *Juncus balticus* - meadow vegetation, Managed alkali wetland (*Crypsis*), Managed Annual Wetland Vegetation (Non-specific grasses & forbs), Rabbitsfoot grass (*Polypogon maritimus*), Seasonally Flooded Grasslands, Seasonally flooded undifferentiated annual grasses and forbs, Temporarily Flooded Grasslands, Temporarily Flooded Perennial Forbs
Vernal Pools
Alkali Heath (*Frankenia salina*), Alkaline vegetation mapping unit, *Allenrolfea occidentalis* mapping unit, *Distichlis spicata - Salicornia virginica, Frankenia salina - Distichlis spicata, Juncus bufonius* (salt grasses), Pickleweed (*Salicornia virginica*), *Salicornia virginica - Cotula coronopifolia, Salicornia virginica - Distichlis spicata*, Salt scalds and associated sparse vegetation, Saltgrass (*Distichlis spicata*), *Suaeda moquinii - (Lasthenia californica*) mapping unit
Lotus scoparius - Antioch Dunes, *Lupinus albifrons* - Antioch Dunes
Bromus diandrus - Bromus hordeaceus, California Annual Grasslands - Herbaceous, Creeping Wild Rye Grass (*Leymus triticoides*), Italian Rye-grass (*Lolium multiflorum*), *Lolium multiflorum - Convolvulus arvensis*, Tall & Medium Upland Grasses

Delta are limited to relatively small in-channel islands that dot the San Joaquin River (Atwater 1980). There are no remaining large expanses of tidal wetlands in the Delta. A substantial portion of the emergent wetlands today exist as thin strips along the margins of artificial levees. As a result of this fragmentation, habitat connectivity has been significantly reduced.

While wetland has virtually disappeared, the area of open water in the Delta has actually increased: large islands such as Franks Tract and Mildred Island were not flooded historically and some of the San Joaquin River channels have been widened and cutoffs created. Whereas the map shows 30,349 acres (12,282 ha) of channels, ponds, and lakes for the historical Delta, modern mapping includes a total of 63,124 acres (25,545 ha) today (types include areas of floating and submerged aquatic vegetation). The loss of freshwater emergent wetland has greatly impacted the relative proportion of natural habitats in the area, expressed in the ratio between water and wetland. Whereas historically the ratio of waterways, ponds, and lakes to emergent wetland was approximately 7:100, this ratio is now 556:100. Of course, this is a generalization and varies depending on location. This shift has important implications considering that a proportionately much greater area of wetland historically contributed nutrients and organic matter to the Delta waters. One could imagine that water in the Delta once had a much stronger signature of the wetlands than is possible today, impacting food availability for fish and other species as well as overall productivity levels.

Open water habitats are quite different in character today, being generally far deeper (e.g. flooded subsided island) than historical open water features (e.g. shallow lake within tule). They are also often occupied by invasive aquatic vegetation (e.g., *Egeria densa, Eichhornia crassipes*) that disrupt instead of support ecosystem processes like native aquatic plants did in the historical Delta. The shift in characteristics of open water and relationship to marshes means that the majority of the past and present open water features are not equivalent in terms of their ecological function.

The disappearance of tidal wetland has also meant the nearly complete loss of blind tidal channel networks (i.e., the lower order tidal channels that branch and terminate within the wetland plain). These were once the primary method of exchange between the wetland and aquatic environment and the backbone of the complex tidal network, promoting both ecosystem productivity and spatial complexity in habitat conditions. Also, while many of the primary waterways of the historical Delta remain in place today, many have been widened and straightened, and virtually all have been lined by artificial levees. Connectivity between them has increased through connecting canals, meander cutoffs, cross-levees, and dredged channels. This has homogenized conditions (e.g., salinity, temperature, nutrients, flows) and altered tidal and flood routing through the Delta.

The channel network has been altered substantially. Only about 31% of the historical channel is aligned with the modern network and about 27% of the modern network is aligned within 165 feet (50 m) of a historical network

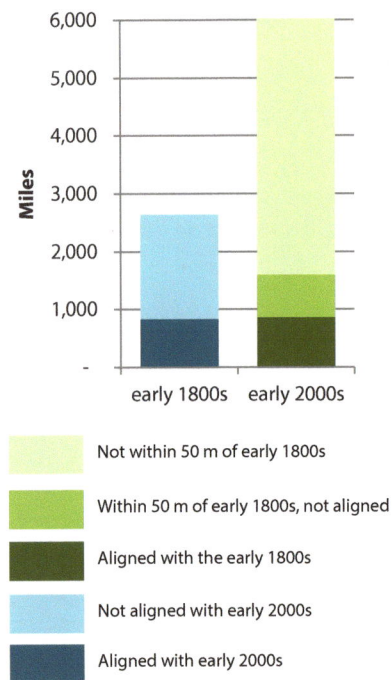

Legend:
- Not within 50 m of early 1800s
- Within 50 m of early 1800s, not aligned
- Aligned with the early 1800s
- Not aligned with early 2000s
- Aligned with early 2000s

Figure 3.8. Alignment of historical and modern networks. The early 1800s channel mapping is summarized based on whether it is aligned with the recent early 2000s mapping by the NHD. About 31% of the total mapped historical length is aligned with the NHD, most of which consists of tidal mainstem channels. The early 2000s mapping is summarized based on whether it is aligned within 50 m of an early 1800s channel, whether they are within 50 m, but not aligned (e.g., cross ditches), and if they are not within 50 m of an early 1800s channel (e.g., new channels). Virtually all of these "new" channels are either canals, ditches or artificial paths found within agricultural areas. (USGS 1999)

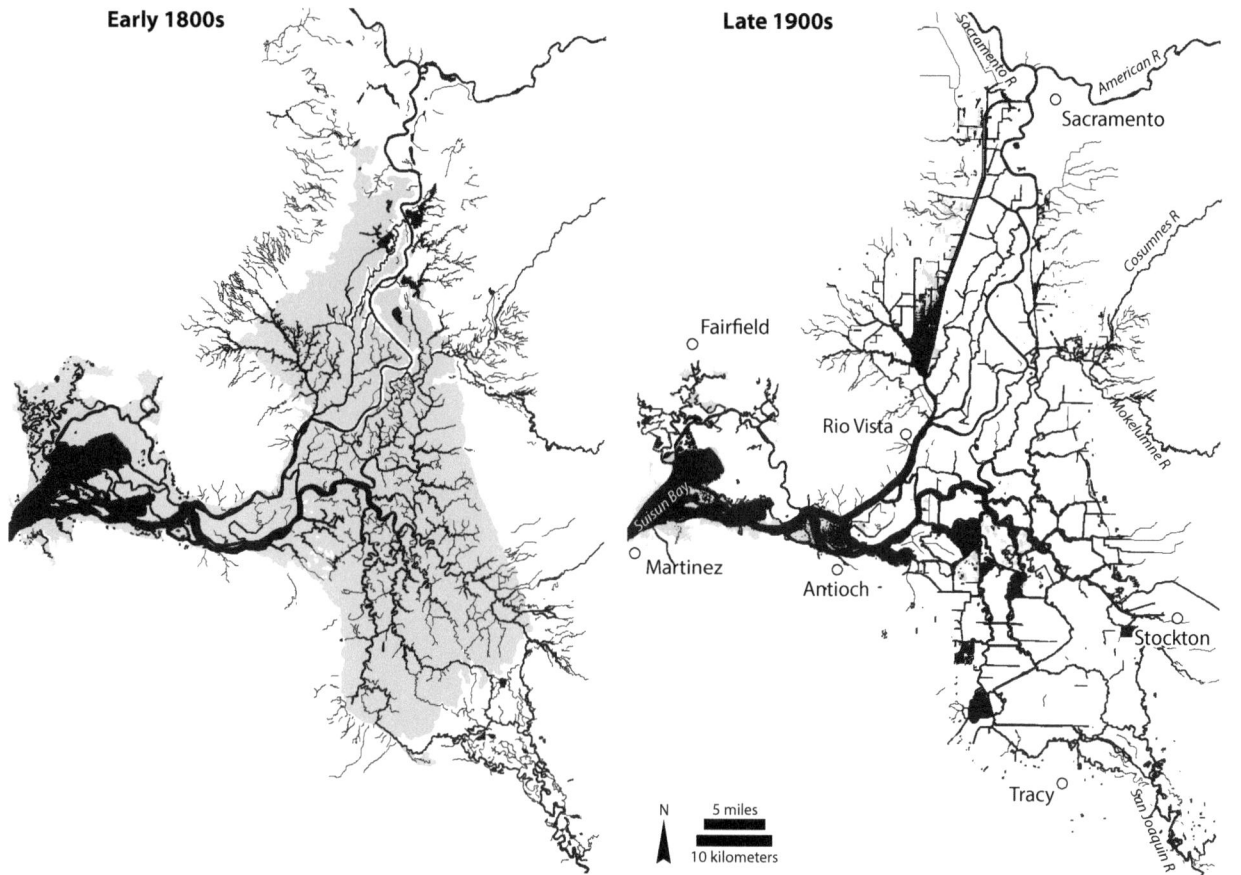

Early 1800s

Late 1900s

N 5 miles / 10 kilometers

(this is because of the thousands of miles of new ditches within agricultural lands; Fig. 3.8). Percent change also varies by channel type: roughly 88% of the early 1800s mainstem channel aligns with the National Hydrography Dataset (NHD; USGS 1999) mapping of the modern Delta, while only about 19% of low order channel aligns with the NHD. This difference relects differing modifications for the historical mainstem channels (which were widened, straightened, and connected) and the low order channels (which were lost to reclamation and to ditching efforts along upland margins). The 81% loss of low order channel includes the loss of about 930 miles of tidal low order channel. The major changes in hydrography are illustrated in Figure 3.9.

Though ponds and lakes are not classified apart from channels and other open water types in the modern mapping used in this comparison and thus cannot be compared quantitatively to historical conditions, it is clear that many of the ponds and lakes that once existed within the wetlands of the Delta have been drained and farmed along with the adjoining wetlands. Many of these former lakes are still present as depressions. Portions of others remain, such as Stone Lake and Lake Washington. Instead of the backwater, more isolated positions of open water historically, today large expanses of open water are found in the heart of the tidal Delta as flooded islands. Most of these water features today are not lined by wetlands extending far beyond the water's edge but are instead bordered by artificial levees rising steeply from the water.

Waterway (includes bays, tidal channels, rivers, lakes, creeks, and intermittent streams)

Tidal wetland (includes brackish and freshwater tidal wetland, late 1900s also includes non-tidal wetlands)

Other (includes natural levees, sand mounds, non-tidal wetland in early 1800s map, and uplands)

Figure 3.9. Comparison of historical (early 1800s) and modern Delta waterways. The map at left shows the complexity of early 1800s Delta hydrography (black) within tidal wetland (gray). The modern hydrography at right shows major differences including channel widening, meander cuts, cross levees, and loss of within-island channel networks and tidal wetland. Wetland (gray) of the late 1900s includes both tidal and non-tidal wetlands. Wetland shown in the early 1800s map are limited to those influenced by tides - non-tidal wetlands continued to the north and south (modern mapping SFEI 2011 (BAARI), Hickson and Keeler-Wolf 2007)

Riparian forest (excluding willow-dominated riparian habitats) has not decreased as dramatically in percentage as the emergent wetland in the Delta. The 6,671 acres (2,700 ha) of valley foothill riparian forest within the Delta today represents 23% of the historical acreage. However, riparian forest extent is far more fragmented, with virtually no wide corridors of riparian forest remaining. Much of the area today was not mapped as valley foothill riparian in the historical mapping. This loss is within the context of an estimated 94-98% loss of riparian forest in the Central Valley since historical times (Vaghti and Greco 2007). The apparent more than two-fold increase in willow-dominated habitat types within the Delta (see Fig. 3.7) is likely reflective of the fact that willows line many miles of artificial levees today where the waterways historically met freshwater emergent wetland. Also, there are some remnant wetland patches on in-channel islands of the central Delta today that are willow-dominated. While the modern mapping captures these and other small patches with its more detailed mapping scale, the historical mapping does not include such patches that would have been present historically within the freshwater emergent wetland type for those areas (see Table 2.2 and pages 42, 68).

Overall, seasonal wetlands have been lost primarily to agriculture and urban expansion. Alkali seasonal wetlands as mapped today are about 4% of the historical extent, a reduction from 13,612 acres (5,509 ha) to 525 acres (212 ha). The loss (99%) of vernal pools is likely overstated given that we mapped the soil types characterized by vernal pools as opposed to individual vernal pools, and recent classification criteria are more strict in definition. For wet meadows and seasonal wetlands, 23% remains (13,490 ac/5,459 ha today compared to 57,696 ac/23,349 ha historically). These seasonal wetlands once greatly expanded the availability of wetland and aquatic habitat for many species at certain time of the year, as well as providing connectivity between wetland and aquatic habitats and the surrounding valley. However, since most perennial wetlands are no longer adjacent to seasonal wetlands, those remaining cannot serve this function. Interestingly, a significant proportion of seasonal wetland types today exist where perennial wetlands were present historically. This relates not only to channelized flow and rivers regulated to limit flooding, but also to lower groundwater tables in many locations. In a few cases, as in the Yolo Bypass north of Liberty Island, a portion of the historical seasonal wetlands are now complexes of perennial and seasonal wetlands.

Stabilized interior dune vegetation, a unique Delta habitat type occupying eolian sands in the region of eastern Contra Costa County, have been virtually lost within the mutually mapped area. As most natural habitats related to this type (e.g., CalVeg types including annual grasses, coast live oak, coastal lupine, coyote brush, soft scrub-chaparral, wet meadow, and willow) occupy highly disturbed land, only 10 acres (4 ha) were left based on the crosswalk established (see Table 3.2).

The decline in grassland, oak woodland or savanna at the Delta perimeter has been significant. Outside the historical boundary of freshwater emergent wetland, the conversion has been nearly complete. The historical extent of the grassland and oak woodland or savanna is now exclusively either agriculture or urban development. There is more grassland in the contemporary mapping (28,077 ac/11,362 ha compared to 7,795 ac/3,154 ha), but much of this is fallow land within the interior Delta that used to be freshwater emergent wetland.

Implications of change

Change in the Delta has not come about through a logical progression of events planned in relation to each other. Rather, layers of only loosely coordinated human modifications and the many unintended and interacting ecological and landscape responses to actions within the estuary and its watershed have led to the Delta as it is today. The Delta is and will continue to respond to the past changes in the future as new land uses and the effects of climate changes emerge (Parker et al. 2011). Although the changes within the last century have been dramatic, with profound implications for ecosystem function, the Delta was substantially altered before the turn of the century. The cumulative effect of modifications has generated a loss of habitat that has facilitated the failing of the once rich Delta ecosystem. It is important to recognize that, without the early loss in habitat, the ecosystem would have been more resistant and resilient to the anthropogenic changes of the twenty-first century.

It is difficult to grasp the magnitude and functional significance of these dramatic historical changes in the absence of cartographic and textual details. This study makes much of this detail available. Understanding how habitats were arranged across the Delta and also what they may have looked like on the local scale in the late summer versus early spring can lend important insight into strategies taken to promote more functional future landscapes.

Though the Delta is irreversibly altered, many fundamental physical processes and landforms are still, to a greater or lesser extent, present. Attaining sustainable ecosystems will require reconnecting pattern and process at a landscape scale, in perhaps different places and scales than what occurred in the historical landscape (Simenstad et al. 2006, Greiner 2010). Restoring aspects of historical landscapes under similar physical processes is a strategy for restoring the habitats that listed species are particularly adapted to, increasing their chances of recovery in the face of stressors (Moyle et al. 2010). The identification of opportunities and viable strategies can be informed by the knowledge of how historical habitat patterns and characteristics reflected their physical context. The historical Delta ecosystem was constantly adjusting to variable conditions, and was therefore more resilient to perturbation. There were few hardened edges, which allowed movement of environmental gradients as well as species in response to physical changes in the system. An understanding of the historical Delta provides valuable information with which to build greater flexibility and adaptive capacity into the Delta. In combination with contemporary environmental research and ecological theory, we can support functional patterns and processes that build an ecosystem more resilient to climate change, land and water use, and related stressors.

Landscape: "heterogeneous land area composed of a cluster of interacting ecosystems that are repeated in similar form throughout"

—FORMAN AND GODRAN 1986

Landscape: "spatially defined mosaic of elements that differ in their quantitative or qualitative properties."

—WIENS AND MOSS 2005

PRIMARY DELTA LANDSCAPES

Much attention in restoration ecology literature is paid to the importance of large-scale process-based restoration in building ecological function over the long term (Hobbs 1996, Bell et al. 1997, Simenstad et al. 2006, Beechie et al. 2010, Greiner 2010). Appropriate strategies are those that address interacting physical processes and how they play out at the landscape scale, such as a tidally influenced floodplain that will, over time, become more tidally influenced with sea level rise (Florsheim et al. 2008). Larger patterns emerge from the research for this project, illustrating how habitats were arranged in response to physical context. The large-scale patterns foster conceptual thinking about landscape characteristics, how they function depending on different suites of physical processes and geomorphic settings, and how they adjust over time in response to physical drivers.

Before significant human modifications over the last century and a half, the landforms of Delta supported complex habitat mosaics arranged in patterns at the landscape scale, where habitats varied predictably in space and time along physical gradients. The Delta was not simply an extensive jumble of patches of different habitat types shifting constantly through time. Landscape patterns were quite stable; the habitats comprising them may have been more dynamic, but the overall patterns reflected physical processes expressed along well defined spatial gradients and primary landforms.

Thinking about the Delta as a series of landscapes is useful for a subregional comparison of pattern, function and process. It is also an exercise fraught with challenges, as there are no clear boundaries between these landscapes and these divisions are different depending upon the characteristic examined. Every landscape of the Delta could be said to have some attributes of other landscapes and it is therefore important to also describe components of the landscape as they fall along the physical gradient of a certain attribute (e.g., tidal influence).

The landscapes of the Delta were an expression of the physical landforms and processes that varied from north to south and east to west. They were governed by many of the same physical processes and shared many of the same habitat types. However, many characteristics differed, including the relative proportion of habitat types, size of features and habitats, vegetation community, hydrologic and habitat connectivity, and landscape position. Although there are multiple levels of complexity, we choose in this report to frame the basic patterns of the Delta as forming three primary landscapes: the tidal islands landscape of the central Delta, the flood basins landscape of the north Delta, and the distributary rivers landscape of the south Delta (Fig. 3.10). The tidal islands landscape characterizes the area between lower Roberts and Union islands to lower Tyler and Staten islands. The flood basins landscape primarily encompasses the Yolo and Sacramento basins and extends into Grand, Tyler and Staten islands. The meandering rivers and floodplains landscape of the southern Delta is representative of areas from upper Roberts and Union Island to the vicinity of the Stanislaus River confluence.

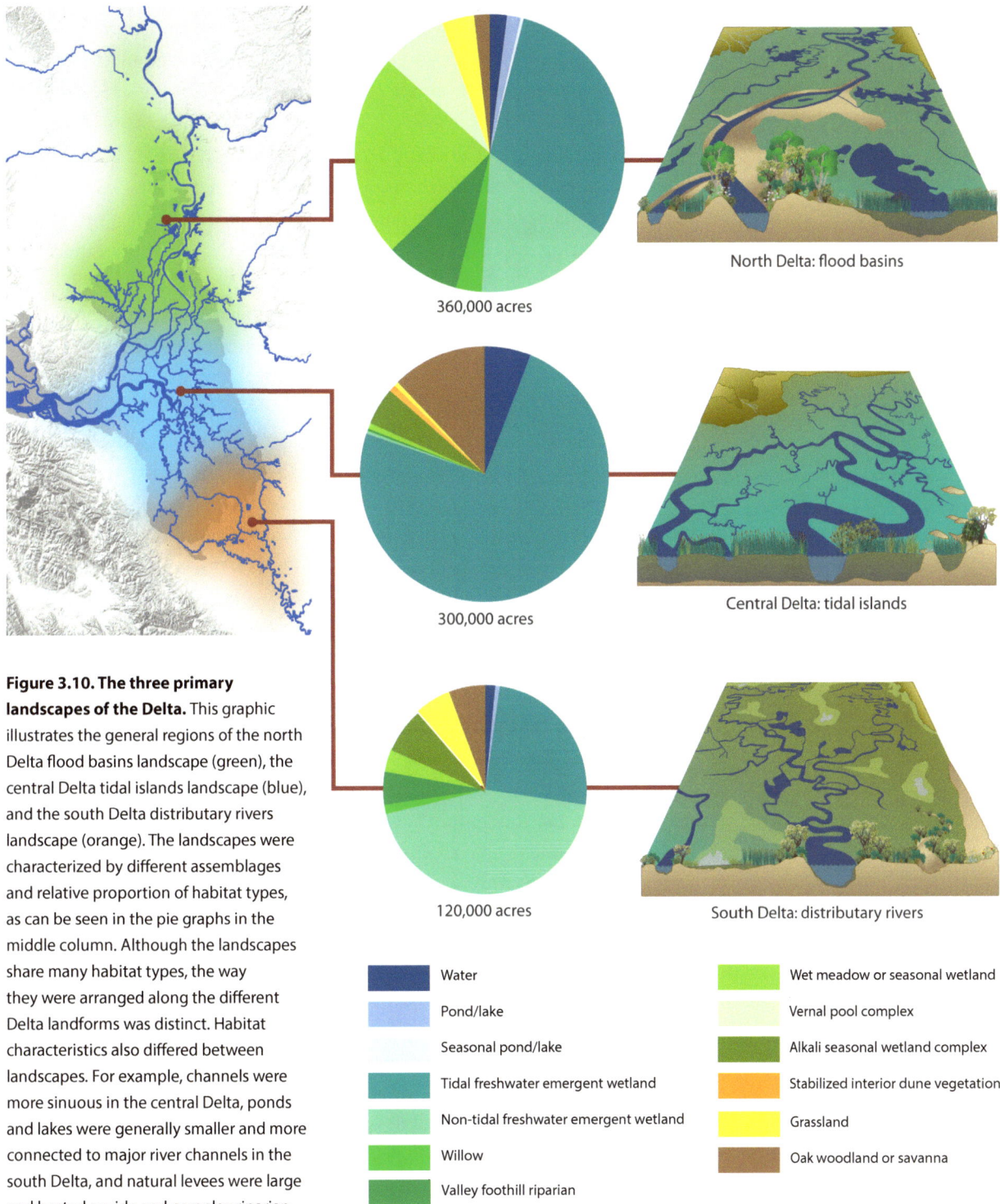

360,000 acres

North Delta: flood basins

300,000 acres

Central Delta: tidal islands

Figure 3.10. The three primary landscapes of the Delta. This graphic illustrates the general regions of the north Delta flood basins landscape (green), the central Delta tidal islands landscape (blue), and the south Delta distributary rivers landscape (orange). The landscapes were characterized by different assemblages and relative proportion of habitat types, as can be seen in the pie graphs in the middle column. Although the landscapes share many habitat types, the way they were arranged along the different Delta landforms was distinct. Habitat characteristics also differed between landscapes. For example, channels were more sinuous in the central Delta, ponds and lakes were generally smaller and more connected to major river channels in the south Delta, and natural levees were large and hosted a wide and complex riparian forest in the north Delta. Conceptual diagrams illustrating these landscapes are shown in the third column.

120,000 acres

South Delta: distributary rivers

Water

Pond/lake

Seasonal pond/lake

Tidal freshwater emergent wetland

Non-tidal freshwater emergent wetland

Willow

Valley foothill riparian

Wet meadow or seasonal wetland

Vernal pool complex

Alkali seasonal wetland complex

Stabilized interior dune vegetation

Grassland

Oak woodland or savanna

About three-quarters of the central Delta tidal islands landscape supported tidal freshwater emergent wetland, composed of a wetland matrix of species, including tule (*Schoenoplectus* spp.), willow (*Salix* spp.), arrowhead (*Sagittaria* spp.), water-plantain (*Alisma* spp.), rushes (*Juncus* spp.), sedges (*Carex* spp.), cattails (*Typha* spp.), reeds (*Phragmites australis*), and lady-fern (*Athyrium felix-femina*). These wetlands were strongly influenced by tidal waters, and were inundated at least by monthly spring tides. Topographic variation was slight and the extensive tidal marsh plain approximated high tide levels (Atwater et al. 1979, Atwater and Belknap 1980). High river stages in the wet season often inundated entire islands several feet deep. The freshwater emergent wetlands were broken into large islands ranging from just a few thousand to tens of thousands of acres. The islands were surrounded by broad subtidal tidal channels that totaled over 6% of the area. Channel banks were low, and numerous small branching tidal channels wove through the wetlands, bringing tide waters to the wetland plain. Channel density and sinuosity in the central Delta was greater than in less tidally dominated northern and southern parts of the Delta. However, related to inundation tolerances of wetland vegetation in fresh water, channel densities appear to have been considerably lower than those observed in brackish and saline marshes of the estuary downstream (Grossinger 1995, Pearce and Collins 2004). Sand mounds rose above the wetland plain in the western portion of the central Delta, where glacial-age eolian sands had not been buried by peaty deposits. Though they amounted to only a small proportion of the overall landscape, these features offered rare relatively dry habitats and topographic complexity within tidal wetland. Alkali seasonal wetlands and oak woodland and savanna habitats were typical upland transitions for the central Delta.

The flood basins of the north Delta lay parallel to the rivers and were influenced by the large-magnitude floods of the Sacramento River that occurred with great frequency, as well as other streams that discharged their annual flows at the basin margins. The floodwaters formed what many referred to as large lakes within the basins; they often extended for many miles and persisted for several months. One defining characteristic was a broad zone of non-tidal freshwater emergent wetland relatively free of channel that graded into tidal freshwater emergent wetland. These wetlands were dominated by dense stands of tules, which numerous accounts state reached heights of 10 to 14 feet (3-4 m). Much of the area of tidal wetlands may have been seasonally isolated from the tides in large part because of the natural levees. Large lakes occupied the lowest and most isolated positions within the expansive wetlands, and few channels penetrated far into the dense emergent vegetation as the wetland transitioned gradually away from tidal influence upstream. Numerous small ponds were found within the tules, many of which may have become partially, if not completely, dry by the end of the season in areas outside the reach of tides. The larger open water features were an important and distinguishing characteristic of the north Delta, although they made up less than 2% of the landscape's area. The basins were bounded by riparian forest along natural levees that extended along the larger rivers. Unlike other Delta landscapes, these forests comprised a significant proportion of the area, close to 10%. In the river's lower reaches this forest became relatively narrow. However, at its widest, the riparian forest of the Sacramento River spread over a mile. For the most part, travelers describe a half-mile wide strip of forest along the river. Seasonal wetlands lined the upland margin. Also at the upland margin of the north Delta – in a few distinct locations – extensive willow thickets occupied the "sinks" of the larger distributary networks of creeks, as well as the Cosumnes River. As can be seen in Figure 3.10, the relative proportions of major habitat types is more even in the north Delta than in the central.

The south Delta (as geographically defined in this report) encompassed an area that was considerably smaller (about 120,000 acres) than either the central (about 300,000) or north (about 360,000) Delta. The three distributary branches of the San Joaquin River were an important influence to the general pattern of the landscape. These distributaries branched into numerous secondary overflow channels within the floodplain, which broadened downstream and merged gradually into tidal wetlands. This complex network of distributary channels with levees of variable height intersected the fluvial-tidal transition zone, likely causing floodwaters to be routed and channelized in ways different from the flood basins landscape of the north Delta. Some of the area between the distributaries was elevated above tidal levels by the sandy deposits left during flood stages. Some parts of the main channels, such as Old River near present-day Fabian Tract, carried large woody debris and were popular salmon fishing grounds for Delta tribes and early explorers. Ponds and lakes were generally smaller and less numerous than in the north Delta, and channels, since they were less tidal, were narrower. Accordingly, less than 2% of the south Delta was open water. Almost three-quarters of this landscape included emergent wetlands, though a much larger proportion of this was non-tidal in comparison to the other Delta landscapes. A broader mix of different habitat types were found within the emergent wetland, including willow thickets, seasonal wetlands, grasslands, and ponds and lakes. In comparison to the flood basin landscape, a greater portion of the natural levee riparian vegetation was composed of willows and other shrubs, and in general the forest was less extensive (only about 5% of the area). Particularly in the most southern extent, the floodplain was occupied by a significant proportion of willows and other trees, reminiscent of the wooded bottomlands of the rivers flowing from the Sierra Nevada and feeding into the San Joaquin. Whereas wetlands and vernal pools made up a significant proportion of the upland edge at the Delta margin in the north Delta, alkali seasonal wetland complex, grassland, and oak woodland and savanna habitat types occupied the south Delta edge.

Local scale

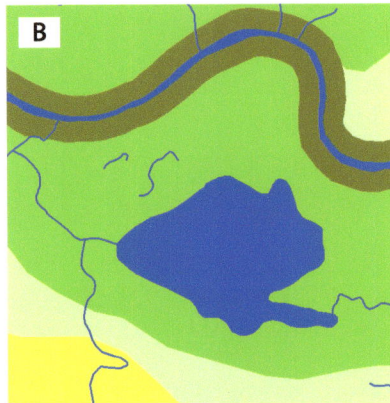

Habitat mosaic scale

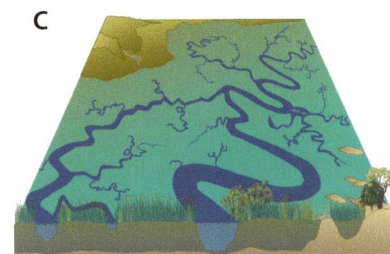

Landscape scale

Figure 3.11. Scales of complexity. The Delta was complex at the local, within-habitat type scale (A). It was also complex in the way multiple habitats fit together to form repeating habitat mosaics (B). At the largest scale, these habitat mosaics formed landscapes that reflected the underlying landforms and driving physical processes (C). We will use these three terms, local-scale, habitat mosaics, and landscape-scale to discuss historical conditions throughout the report. The scales are relative and depend on the location and characteristics being considered. (A: ca. 1910, courtesy of the California History Room, California State Library, Sacramento)

Throughout the report, we will refer to *landscapes* to represent this large-scale arrangement of habitat types upon Delta landforms at different locations along physical gradients. *Habitat mosaics* or complexes comprise groupings of different habitat types that are found within the landscape. Reference to *local-scale* complexity in the report usually refers to the different conditions one might experience on the ground passing through habitat mosaics (Fig. 3.11).

Utility of the landscape perspective

Thinking about these landscapes in a conceptual manner, removed from their exact geographical location, can help support a flexible landscape framework to guide sustainable restoration strategies in the contemporary and future Delta. These larger patterns or landscapes supported historical ecological functions that could not be provided by any single habitat mosaic or type. These patterns are also a reminder that the Delta was not a uniform place, nor was it simply a mixture of habitat types spread throughout the Delta. Rather, the habitat types were arranged in distinct patterns in relation to the Delta's landforms and physical gradients (e.g., climate, elevation, relative fluvial influence). Different landscapes provided different suites of ecological functions, contributing to the Delta's historical diversity and productivity. The landscape perspective facilitates moving beyond restoration efforts focused primarily on reaching overall acreage targets of habitat types or targets based solely upon meeting the needs of single species. It offers insight into how different parts of the Delta can serve different species assemblages and species at different life-history stages. Also, by promoting reconnecting habitat mosaics along physical gradients, this perspective can help the Delta achieve greater fundamental exchange between terrestrial and aquatic systems, nutrient cycling, potential for adaptation through time, and overall ecological function.

Framing the historical Delta in terms of major landscape types provides an avenue for discussing the scales at which functional patterns in habitat types and related characteristics were found in different parts of the Delta. There is no one absolute scale that defines a landscape. One can recognize different scales of complexity, or different landscape units of different sizes, depending on habitat pattern and governing physical process.

Furthermore, the appropriate functional unit may be different for different landscapes. For instance, the central Delta may have a functional unit on the order of a single Delta island. In contrast, the north Delta functional unit might be larger because floodwaters from the upper basin passed across a broad wetland relatively free of channel was integral to the ecological functioning of this part of the Delta. The study of these patterns can help define the meaning of "large" in terms of habitat restoration (e.g., CDFG 2010). As an example, establishing functional processes for north Delta flood basins might involve finding a place where overflows from a river channel could empty into wetlands and be left to drain toward the Delta. On the other hand, supporting south Delta landscape processes would be more reflective of floodplain restoration, where a river should be allowed to move, form side channels, and interact with its riparian forest corridor. Thus, for example, only one side of the river would need to be involved for supporting north Delta basin processes, but corridors along both sides of the river on the San Joaquin might be more appropriate. Such restoration principles are concepts being developed in the ERP-funded *Management Tools for Landscape-Scale Restoration* project, scheduled for completion in 2015.

Describing landscape patterns of the historical Delta is a key step in applying historical ecology to contemporary restoration planning. While we discuss landscape characteristics organized by where they were in the historical Delta, the patterns are inherently driven by their physical and biological processes. The particular historical location of a single channel, lake, or willow grove may, in many cases, be irrelevant as a template for restoration at that location. Instead, knowing how habitat types were arranged in relation to geomorphic position and physical processes allows us to look elsewhere for suitable restoration sites. By comparing characteristics at these broad and conceptual levels, it is easier to translate the thinking into the contemporary and future physical context of the Delta. Using this flexible, historically informed approach, we can better identify appropriate locations to support the functions once supported in the Delta. Certain landscapes or functional components may be possible in different locations from where they were in the nineteenth century Delta; in fact, these landscapes were always adjusting along gradients of controlling factors. One simple application of this idea is moving the tidal islands concept that was historically within the central Delta upslope to locations where tidal elevations are appropriate and floods do not currently provide unacceptable levels of disturbance. The landscape perspective offered by understanding the historical Delta benefits the development and implementation of restoration strategies in the Delta that reestablish functional elements with appropriate scale, location, and connectivity to support native species and increase long-term overall ecosystem health and resilience.

These three primary landscapes are used in this report as a basic framework with which to convey detailed historical information. The intention is to provide an understanding of the landscape patterns and associated processes and functions of the historical Delta. This framework contributes to a basis for defining these landscapes with metrics relevant to restoration, relating them to specific ecological functions, and presenting them in conceptual model form. While outside the scope of this project, these landscape restoration planning tools will be developed in the new CDFG ERP funded project, "Management Tools for Landscape-Scale Restoration of Ecological Functions in the Delta." This project will address questions such as: How can the landscape perspective offered by the study of the historical landscape to inform a strategy going forward? How can the ecosystem's ability to adapt be supported? How can rigidity be removed from the landscape?

Without an understanding of the larger scale, management efforts risk managing the microcosm, instead of addressing structures or processes that exist(ed) or operate(d) at a landscape scale.

—COLLINS 2003

CONSOLIDATED SUMMARY POINTS

This section offers a consolidated briefing of primary findings and implications of the Delta historical ecology investigation. These points can contribute to the discussion of how to support a healthier future Delta ecosystem, one that responds to physical drivers, adapts to change, and provides a suite of functions at the landscape scale.

Overall main points

- **A diverse array of habitat types was found within the historical Delta.** This included deep and broad sloughs (see page 143), small dendritic tidal channels branching into the wetland plain (see page 154), perennial and seasonal ponds and lakes at backwater locations (see pages 253 and 346), extensive freshwater emergent wetlands dominated by tule, willow-fern swamps within the tidal wetland complex (see pages 168 and 351), complex riparian forest with multiple vertical layers (see pages 247 and 357), willow thickets where upland drainages spread at the Delta's edge (see page 294), a range of seasonal wetlands along the perennial wetland perimeter (see pages 193, 301, and 370), stabilized interior dune vegetation occupying the small but pronounced sand mounds of the western Delta (see page 186), and grasslands, oak savannas, and oak woodlands at the Delta margins (see pages 194, 305, and 370). (see Tables 2.2 and 3.1, and Fig. 2.3)

- **The Delta consisted of multiple landscapes.** The central Delta's tidal freshwater wetlands of tule and willow, with its numerous winding channels, looked and functioned differently than the north Delta's broad flood basins, occupied by tule marsh and lakes and bordered by broad riparian forest on the natural levees of the Sacramento River and its distributaries. These landscapes, in turn, were different from the floodplain of the southerly San Joaquin River distributary branches, which was composed of tidal wetlands merging southward into a floodplain wetland interspersed with side channels, lakes and ponds, willows along channels, and patches of seasonal wetland. (see Fig. 3.10 and pages 100, 119, 207, and 309)

- **Landscape-scale habitat patterns were a reflection of the Delta's broad physical gradients and landforms.** Patterns shifted depending on gradients, including tidal to fluvial influence (e.g., flood frequency, duration, magnitude, and extent), brackish to fresh water, low to high elevations, hot to cool temperatures, and peat to clay to loam soils. Landscape-scale patterns reflected the primary landforms of sub-tidal waterways intersecting Holocene peat deposits lying at tide elevation. Supra-tidal natural levees lined the rivers, and small sand mounds rose above the wetland plain. Peat deposits at the wetland edge overlapped the toes of alluvial fans along the Central Valley floor. (see pages 8, 217, 280, 292, 333, 354, and 360)

- **The historical landscapes exhibited gradual transition zones between habitat types that allowed movement and adaption along physical gradients, in contrast to the sharp edges of today.** The river and floodplain, as well as the north-south tidal to fluvial gradient, are largely disconnected today through the leveeing of the main rivers, damming and filling of secondary channels, and reductions in flood flows. The loss of interconnected habitat mosaics, or increase in habitat fragmentation, limits habitat opportunities for species and the ability of the ecosystem to withstand physical and biological stressors. (see page 91)

- **The Delta is unique in its shape.** Characteristics such as the Delta's freshwater character, overall channel planform, and stability of features owe themselves, in part, to the fact that the channels of the Sacramento and San Joaquin rivers meet at the Delta's constricted mouth and flow into the enclosed San Francisco Bay, rather than directly into the Pacific Ocean. (see pages 7 and 124)

- **Temporal variability was overlaid on a less changeable physical template.** Within the context of relatively stable landscape patterns, the Delta experienced droughts and deluge that generated variability in environmental conditions. (see pages 10, 230, and 319)

- **Seasonal variation was expressed differently in different Delta landscapes.** While daily tides and maritime influences muted seasonal differences in flows and water availability within the central Delta, more seasonal variation was evident in the north and, particularly, south Delta. (see pages 8 and 321)

- **A small percentage of the "natural" habitats within the Delta today is remnant of the former landscape.** The majority of the approximately 106,000 acres (42,900 ha) of natural habitat (within the mutual area of the legal Delta and study area) did not exist historically in their present locations. For example, seasonal wetlands are found where perennial wetlands once existed and willow thickets on artificial levees are now present where tidal wetland edges once met water. The Delta has undergone an almost complete transformation, due to land use and water management. (see page 91)

- **Modern anthropogenic modifications occurred early in the Delta.** Changes due to leveeing, agriculture, ditching, clearing of riparian forests, grazing, and other impacts were evident in the 1850s. This affected how floodwaters moved through the Delta and substantially reduced the extent of perennial wetlands. Hydraulic mining debris impacted channel bed levels, among other effects. Most emergent wetlands of the central Delta were leveed and farmed by the 1880s. Habitats of native species were significantly altered or absent over a century ago. (see Boxes 1.2, 5.1, 5.4, 5.7, and 6.3, and page 155)

Habitat characteristics

WATERWAYS

- **Tidal channel planform varied depending on landscape position.** Tidal channels can be binned into three types having different sets of topographic and hydrologic characteristics: they terminated within the tidal marsh plain (blind tidal channels), were met by a fluvial channel from the uplands (e.g., Calaveras River), or transitioned into a non-tidal floodplain occupied by emergent wetland (e.g., upper Union and Roberts islands). Blind tidal channel planform also appears to have differed depending on whether the channels were within an island, connected at one edge with upland habitats, or were influenced by riverine flooding. (see pages 157, 247, and 336)

- **Channel density in the freshwater Delta was apparently lower than that of the brackish and saline wetlands of San Francisco Bay**. These differences likely relate to the presence of freshwater, lower tidal energy, and differences in substrates and vegetation. The trend is evident despite the fact that the historical habitat mapping may not include some of the smallest Delta channels. (see page 161)

- **Most tidal channels appear to have been subtidal**. This follows from the fact that emergent vegetation can colonize below low tide water levels in fresh conditions. (see page 157)

- **Substantial volumes of the riverine flood flows met the tidal Delta from the south through the side channels within floodplains and from the north through flood basins.** While today most water flows through mainstem channels, this was only one of the many ways water historically reached the central Delta. (see pages 212, 230, 240, 319, and 333)

- **Few channels were found in the north Delta.** The broad natural levees of the Sacramento River largely prevented the establishment of extensive secondary or overflow channels extending into the lowlands, as was more common in the floodplain environment of the south Delta. (see Box 5.4 and page 255)

- **Large wood debris jams occupied certain channels.** These features greatly affected habitat conditions and likely promoted the creation of side channel systems and backwater ponds and lakes. Jams appear to have been an ecologically significant structural element of the south Delta in particular. Evidence of woody debris obstructing channels is found on particular reaches of the Old and Middle river branches of the San Joaquin and along the upper tidal reaches of the Mokelumne. (see page 366)

- **The Delta of the early 1800s lacked extensive intertidal mudflats.** This is in contrast to large areas of mudflats found within and adjacent to the more saline tidal marshes of San Francisco Bay. (see page 150)

- **Small tidal channels have disappeared today.** The tidal channels that branched and terminated within wetlands made up over 70% of the historical tidal channel network. While the main rivers and sloughs of the Delta remain today (albeit in modified form), virtually all of the blind tidal channel networks have been dammed and filled in. These features provided connectivity between the marsh and aquatic environment, were characterized by a wide range of environmental gradients (e.g., temperature, tidal range, flow), and provided valuable foraging habitat for aquatic species. (see Figures 3.9 and 4.24, and pages 96 and 154)

- **The dredging and widening of the main channels at the Delta mouth has increased flow capacity.** Historically, features such as channel bars, sinuous channels, and blind tidal channel networks may have had a negative effect on the extent of salinity intrusion. (see pages 136, 141, and 143)

- **Connections between tidal channels caused by cross-levee ditches, meander cuts, and channel widening are likely correlated with the homogenized conditions in the Delta today,** which includes increasing salinity dispersion, temperature, and suspended sediment. Such changes have affected flood and tide routing and reduced the overall diversity of environmental conditions. (see page 146)

PONDS AND LAKES

- **Lakes and ponds were historically largest and most abundant in the north Delta.** Their positions related, in part, to areas most deprived of inorganic sediment supply when floodwaters passed through the basins and to areas isolated due to topography. Many were located in the lower-elevation central core of the flood basin, while others were found along the edge of adjoining riparian forest. Within the wetlands formed by the Cosumnes and Mokelumne rivers, ponds and lakes generally occupied small, short, upland drainages that fed into the floodplain. (see page 255)

- **Numerous shallow small ponds were found throughout the Delta.** Most evidence of such features suggests that they were found generally at tidal margins and within non-tidal emergent wetlands. (see pages 262 and 349)

- **Not all ponds and lakes were connected to the main river in the same way.** Some ponds and lakes were connected via tidal channels (some long, some short), others were connected to overflow channels or intermittent upland streams, and still others appear to have had no substantial connecting channels. (see pages 265 and 348)

- **The shape of ponds and lakes appears to have been quite complex.** Lake edges were marked by inlets and small coves. Some lakes adjoined riparian forest, which influenced the shape of the feature. (see page 262)

- **Ponds and lakes provided slow moving, shallow water habitats.** These would have supported large populations of fish species associated with such conditions, such as Sacramento perch, hitch, Thicktail chub, Sacramento blackfish, and splittail. Such features are uncommon today, and many of these species are now rare or extinct. (see page 268)

- **Biological influences affected vegetation patterns and may have maintained areas of open water.** The use of certain wetland species by indigenous tribes and the consumption of wetland plants by waterfowl, beaver, and other species may have affected vegetation patterns and maintained ponds and lakes. (see Box 4.2 and page 269)

- **Portions of several of the larger lakes persist today.** These include Stone Lake, Beach Lake, Lake Washington, and Beaver Lake. However, they are no longer integrally connected to the Delta through seasonal inundation as they once were. (see Fig. 3.7)

- **Large bodies of water occur in very different landscape positions today, with overall area having increased.** Instead of the backwater, more isolated positions of lakes historically, large expanses of open water are today found in the core of the tidal Delta as flooded islands. These water features are lined by artificial levees rising steeply from the water rather than wetlands extending far beyond the water's edge. The natural Delta today is comprised primarily of water instead of emergent wetlands. (see Fig. 3.10 and page 96)

FRESHWATER EMERGENT WETLAND

- **Freshwater emergent wetland historically dominated the Delta.** About 365,000 acres (147,710 ha) of tidal wetland merged into and integrally connected to a sum of over 100,000 acres (40,470 ha) of non-tidal wetlands at the northern and southern extent. This broad, level expanse was nevertheless diverse, with local-scale vegetation patterns and tidal channel networks. (see page 168)

- **While tule dominated the freshwater perennial wetlands, other species were found within the wetland plain.** Willow-fern swamp was an important vegetation community of the central Delta, particularly associated with islands along the lower reaches of Old and Middle River. Emergent vegetation appears to have been shorter and less dense in the western central portion of the Delta, in comparison to the tall dense tule dominating flood basins to the north. (see pages 176 and 220)

- **Vegetation patterns were not reflective of salinity conditions.** This is unlike the brackish and saline marshes in the rest of the San Francisco Estuary. High salinity levels did not limit vegetation within the Delta; rather the high water table, high inundation frequencies, and peaty soils served to exclude tree species. Although slightly brackish water may have extended into the western Delta at flood tide late in the season and during especially dry years, it does not appear to have affected vegetation patterns far into the interior. (see page 137)

- **Local-scale patterns appear to have been complex in the south Delta floodplain.** In contrast to the extensive tule marshes of the north Delta basins, the wetlands of the south Delta were apparently more broken up by small ponds, patches of tule, willow thickets associated with sloughs, and wet meadows and seasonal wetlands. These patterns related to differences in topographic, edaphic, and climatic variables. (see page 351)

- **Only 3% of the historical freshwater emergent wetland area remains today, which affects the health of the Delta ecosystem at many levels.** For instance, the Delta's waters no longer exchange with surrounding wetlands, impacting nutrient levels, organic matter, and dry-season freshwater input. (see page 93)

- **The Yolo Bypass today occupies the seasonally flooded edge of the historically perennial wetlands of the Yolo Basin.** The wettest core of the historical basin, near Big Lake, lies to the east of the Yolo Bypass. (see Fig. 3.7 and page 233)

RIPARIAN FOREST

- **Riparian forest promoted habitat connectivity.** Tracking the natural levees of major rivers, these forests extended far into the Delta's aquatic and wetland environments. (see pages 282 and 360)

- **A diverse riparian forest was found on the higher natural levees extending into the Delta.** It supported a canopy of valley and live oak, sycamore, cottonwood, Oregon ash, and California walnut. The understory included a number of willow species, alder, buttonbush, dogwood, box elder, buckeye, grape, wild rose, and numerous herbaceous species. These environments contrasted dramatically with the surrounding wetlands, providing dense cover, vertical structure, diverse plant species, food resources, and carbon to the river. The most extensive forests in the Delta were associated with the Sacramento River and its main distributaries. Though narrower and perhaps more dominated by willow, forests along the San Joaquin also extended well into the tidal landscape. Upstream of the head of Old River the forest broadened and extended into the floodplain as well, primarily along secondary channels. (see pages 288 and 364)

- **Riparian forest characteristics reflected natural levee height.** As height decreased, trees such as oaks and sycamores became less numerous, while smaller, more water-tolerant species such as willows became more common. Natural levees decreased in height downstream as tides became a more dominant process. The highest levees were found along the major rivers: width and height generally diminished with decreasing channel size and connection to fluvial processes. (see pages 280 and 360)

- **Riparian forest of the Sacramento River narrowed descending downstream, ranging from widths of approximately 330 feet (100 m) in the lower reaches to about a mile wide upstream.** The San Joaquin and Mokelumne Rivers followed a similar pattern of narrowed width downstream. Compared to the broad expanses of the wetlands adjacent to the forest, the natural levees were comparatively narrow. However, compared to contemporary corridors often only a few trees wide, these forests were of a scale that is difficult to imagine in the modern landscape. (see pages 285 and 362)

- **Natural levees in the Delta, particularly along the Sacramento River, mintained relatively stable positions.** Evidence suggests that, while upstream of the Feather River confluence the main channel of the Sacramento River experienced channel migration, the river channel within the Delta was more fixed in position. The tight river meanders upstream gave way to broad bends downstream with natural levee banks rising relatively steeply on both sides of the river. A pattern of more dynamic river meandering upstream is also reflected in the morphology and vegetation patterns upstream of the head of Old River on the San Joaquin. (see pages 238 and 342)

- **Within-habitat vegetation assemblages shifted depending on the cross-sectional profile of the levee.** Large trees and sometimes groves of oaks with relatively open understory were found along the highest parts of the levee, while at the water's edge more water-tolerant willows, brambles, and vines created a dense border. On the backside of the levee, trees gave way to willows and herbaceous species. (see page 292)

- **The riparian forest edge was complex.** Numerous small overflow channels crossed the levees, flowing only during high river stages. At certain points, forest extended farther into the basins along crevasse splay deposits or other higher elevation landforms. (see pages 223, 251, and 286)

- **Sycamores were plentiful along the Sacramento River, but less noted in the historical record elsewhere.** Sycamores and oaks were the main large trees found along the Sacramento River, while sycamores were rarely mentioned in descriptions of the forest of the San Joaquin and other rivers feeding into the Delta. (see pages 291 and 365)

- **Few large stands of riparian forest remain.** While modern mapping in the Delta contains approximately 6,670 acres (2,700 ha) of riparian forest with large trees, or about 23% of the extent of historical valley foothill riparian forest mapped, there are few places in the area that approximate the complexity and breadth of the mature riparian forest that once lined the rivers. Most mapped valley foothill riparian habitats today along the Sacramento River exist as corridors only a few trees wide along the artificial levees. While these can provide important habitat for species and offer some additional ecosystem functions, the loss of the once broad forest connecting river to wetland basin likely has a significant impact on habitat diversity and ecosystem function. (see pages 98 and 275)

UPLAND MARGIN

- **Gradients in hydrologic, topographic, and soil characteristics produced spatially and temporally variable inundation patterns around the perimeter of the tidal wetlands,** supporting mosaics of seasonal wetland, grassland, oak woodland and savanna, as well as occasional ponds and patches of perennial wetland. (see pages 186, 301, and 370)

- **The upland margin of the Delta's perennial wetlands was typically occupied by seasonal wetlands, aside from willow thickets at distributary sinks.** The treeless seasonal wetlands bordering these lands were often described as a zone less than a mile to over several miles wide along the perimeter. In the southern portion of the Delta, this zone was often affected by residual salt accumulation in the soil (alkali). These lands were occasionally temporarily overflowed by numerous small upland drainages, produced brilliant shows of wildflowers in the spring, and dried out in the summer. (see pages 193, 301, and 370)

- **In some places, the zone of alkali seasonal wetlands at the freshwater wetland edge was an area overlowed during extreme flooding events.** These accumulations of alkali likely related to the irregular flushing resulting from infrequent flood events. These areas were more common in the drier, southern portion of the Delta. (see pages 193 and 370)

- **At the eastern margin, tidal wetlands adjoined oak woodland and savanna habitats on the gently sloping late Pleistocene Calaveras River alluvial fan.** This eastern margin with many trees was a noteworthy contrast to the predominant seasonal wetlands characterizing the ecotone elsewhere. (see page 194)

- **Aside from the larger rivers, most stream systems feeding into the valley were discontinuous; several formed prominent "sinks" at the wetland edge.** The distributaries that spread across their alluvial fans discharged their annual flows across the wetland surfaces, as opposed to directly into the river channels. The wetter systems created "sinks" that were flooded in the winter and were occupied predominantly by extensive willow thickets. Seasonal and perennial ponds and lakes were found within these complexes as well. The disturbances caused by the flooding regime undoubtedly contributed to the complexity of the habitat mosaics within the sinks. Today, these and other discontinuous streams have been ditched, diverted and channelized to prevent overflow. (see page 294)

- **Sand mounds rose above the wetland plain in the western Delta, providing isolated upland habitats.** Some may have reached heights over 15 feet (4.6 m) above sea level; they do not appear to have been larger than about 20 acres. These unique features of stabilized eolian (wind driven) sands likely supported species of plants (e.g. live oaks, silver bush lupine) and animals otherwise not found within the Delta's tidal landscape. More extensive areas were found along the Delta's eastern Contra Costa edge. Substantial areas of eolian sands have been exposed and disturbed in recent times due to peat oxidation and other factors contributing to subsidence in the Delta. (see page 186)

Hydrologic characteristics
- **Floods were integral to landscape form and function.** The north Delta flood basin landforms, including the wetland troughs and the bordering natural levees, reflected the Sacramento River's high magnitude, sediment-laden flood flows. (see pages 212, 230, and 238)

- **Late spring flooding, long duration of the snowmelt hydrograph, and ample tidal and groundwater supply meant that the Delta stayed wet during the dry season.** Wetland and riparian species found ideal growing conditions in the combination of ample water supply, warm summer temperatures, and fertile soils. Many species of the riparian forest grew to be several times larger in size than their counterparts in drier environments. The Delta likely played an important regional and statewide role as refuge and reliable habitat during the dry season and droughts. (see pages 219, 236, and 320)

- **The Delta looked very different depending on the time of the year.** The high flows of the winter and spring would overflow the wetlands annually to a greater or lesser extent. In some years, inundation was several feet deep and extended far into the seasonal wetlands bordering the perennial wetlands. (see pages 232 and 319)

- **Timing of flooding in the Delta differed between the north and south Delta.** The Sacramento River usually flooded during peak rain events in the winter and spring (though snowmelt was also an important contributor to late-season flooding), while the San Joaquin's high stages usually occurred in the late spring and early summer, due to the greater relative influence of snowmelt on its hydrograph. (see pages 10, 233, and 320)

- **The Sacramento River channel was sized to flood into adjoining basins.** Though confined to a single channel for most of its length, the Sacramento River channel did not naturally contain the floods that passed through the Sacramento Valley. Like many low gradient rivers, it overflowed its banks at high stages. (see pages 233 and 238)

- **The flood basins were the historical valley's reservoirs.** The Sacramento River's flood basins and the large lakes within them provided storage of floodwaters in the winter and spring, which reduced peak flows in the channel, recharged groundwater tables, and released water slowly into the central Delta. (see pages 212, 231, 236, and 255)

- **The south Delta marked the terminus of a large riverine system**, the San Joaquin, that regularly overflowed its banks due to spring snowmelt and less frequently as a result of winter storm events. These events filled and connected numerous secondary channels, ponds, and floodplain wetlands. The broad ecotone where the floodplain met the tides permitted floodwater to spread, sometimes inundating the land several feet in depth before it passed to downstream tidal channels. (see pages 313 and 319)

- **A greater percentage of freshwater inflow moved slowly through wetlands historically, suggesting higher water residence times in comparison to today.** Water had more time to interact with the Delta landscape, as broad wetland plains spread and slowed floodwaters, tules retarded flow velocities, large lakes filled, and few channels in the flood basins were present to route flows. Today, a higher proportion of water passes swiftly through leveed river channels and out of the Delta. (see pages 236 and 267)

- **Hydrologic connectivity was high in the historical Delta, particularly during the wet season.** This promoted exchange between the marsh and aquatic environment. Small channels crossing natural levees transported flood flows into the wetlands at the back of the natural levees. At high flow, these waters connected to ponds and lakes and connected through to main river channels or tidal channels in the central Delta. Fish were thus able to access these habitats at certain times of the year. (see pages 135, 265, 271, and 348)

- **Both wetlands and channels were part of the conveyance system that moved and stored water within the Delta region.** Wetlands were connected to channels in a number of ways; these links changed seasonally (e.g., overflow through natural levee low points, return flow through tidal channels, high tide overflow, etc.). (see pages 127, 247, 251, and 330)

- **The pattern of overflow differed in different parts of the Delta.** In the central Delta, daily tides wetted the lands while during floods water spread over much of the area, with the large tidal channels providing flow capacity to carry the waters to San Francisco Bay. As a result, the central Delta was rarely inundated more than several feet deep. In contrast, upstream river stages would rise much higher than in the central Delta, a pattern reflected in the natural levee height. Consequently, these areas, which were less tidally influenced, could be flooded with much deeper water. (see pages 127, 142, 221, 236, 316, and 319)

- **The contribution of discontinuous streams was significant.** While it is important to consider the impact of Sacramento floodwaters on the flood basins, the influence of the annual flows of smaller tributary systems (e.g., Cache Creek) in terms of flood timing, inundation depth and frequency, and groundwater recharge, should not be discounted. (see page 235)

Management implications

- **Consider that native species were adapted to the patterns and processes of the past.** Developing functional landscape units reflective of historical patterns should improve chances of restoration success.

- **Recognize that restored habitats will not necessarily be the same as historical habitats, and will continue to evolve over time.** The many non-native species throughout the Delta, subsidence, climate change, and other large scale changes, will cause future habitats to have many differences from historical habitats, even if they provide function in similar ways.

- **Manage restoration to be reflective of current physical parameters and processes.** Historical habitat reconstruction does not provide a location-specific template for restoration. Instead, by better understanding how habitats reflect physical landforms and processes, more effective restoration can be created that is consistent with the physical gradients within the present-day and possible future. Consider options for managing physical processes to support more functional habitats and leverage restoration efforts by considering physical parameters.

- **Take advantage of physical gradients in the landscape and consider how these may shift in the future.** The Delta is part of the San Francisco Estuary, lying at the upper end of the estuarine continuum. With sea level rise over time, areas at the edge of tidal influence may be intertidal in the future; adequate room for estuarine transgression should be established along these gradients. Tidal wetlands and adjacent natural upland habitats can thus provide a buffer, supporting greater resilience to climate change. By designing landscapes to be reflective of and involving whole physical gradients, there is greater potential to achieve a wider range of habitat characteristics that will provide opportunity for adaptation. This will support continued evolution of plants and animals by maintaining populations at the limits of local habitat conditions.

- **Remove rigidity in the present Delta where possible.** The historical Delta was adapted to shifting conditions along broad gradients. Broad ecotones would better equip the ecosystem to handle the type of future changes expected in the Delta. With the sharp edges and discontinuities in the Delta today, there is little room for the natural adjustments that gave the historical Delta much of the resiliency that is missing in the contemporary system.

- **Recognize what large and interconnected habitats might mean.** The study of landscape patterns can help define these terms more concretely. For instance, supporting basin landscapes may only require one side of the Sacramento River, but requires adequate flood flows. Supporting San Joaquin floodplain processes at the tidal margin may involve allowing the river to meander on both sides of the channel.

- **Employ a landscape perspective and manage toward assemblages of connected habitats,** recognizing that an isolated restoration project will likely provide much less ecosystem benefit than a restoration of the same size and habitat type that is connected to multiple other habitat types. The ecological value of individual habitat types is magnified by their surrounding landscape. Given limited land and financial resources, these considerations are especially important. The landscape perspective helps target broad assemblages of ecological functions, as opposed to specific conditions required for individual species.

- **Promote habitat connection and disconnection in the appropriate places.** The ecological functions of many Delta habitats were provided through the connectivity of features (e.g., side channel habitat connected to riparian forest and backwater ponds and lakes). Improving understanding of historical conditions supports the developing consensus of the importance of floodplain habitat and its connections to riverine processes. At the same time, discontinuities were important (e.g., blind tidal channels, flood basin and river), increasing residence time and heterogeneity. Deciding where to increase and decrease connectivity must be done at a landscape scale and can be informed by conceptual models of the historical landscape.

- **Heterogeneous landscapes are less sensitive to extreme events.** The historical Delta provided a wide array of conditions; places of refuge could be found in times of flood and places with ample water could be found in the dry season.

- **Use Delta freshwater inflows to their greatest potential.** Historically, freshwater inflows encountered and influenced a much broader range of habitats than they do today. Questions about where water should go are valuable in addition to asking how much water is needed. Understanding the role of hydrology becomes more critical when addressing the current and future challenges related to climate change. Such challenges include potentially large floods unknown in recent times related to loss of Sierra Nevada snowpack and large storm events.

- **Different ecological functions can be provided by the same habitat types, depending on the position of those habitats within different landscapes.** In the historical Delta, driving physical processes and habitat connectivity meant that different functions were provided depending on a feature's location. For example, a large lake within a broad wetland flood basin served a different array of functions than a small pond along a side channel system created by woody debris in the river.

- **Recognize that every habitat or function cannot be supported everywhere.** Certain places will provide some functions better than others. Also, certain functions may not be possible, or, may be significantly limited in the contemporary or future Delta. Consider both altered physical conditions (e.g., hydrodynamics) to determine limitations and opportunities identified using the historical perspective. Think in terms of functional landscape units that provide different groups of functions.

- **Match functional targets to the appropriate scale of restoration.** Many desired Delta functions are likely scale dependent, requiring components of certain sizes. Restoration at scales smaller than landscape patterns and processes may not produce the desired characteristics. For example, restoring a functional tidal island may require a restored tidal wetland of sufficient size in order to support a blind tidal channel network. There is a risk that small restoration projects may not achieve desired characteristics. To avoid this pitfall, individual restoration projects should be embedded within a larger vision of a future functional Delta.

- **Think at the large-scale and in the long-term.** Attaining sustainable ecosystems will require reconnecting pattern and process at a landscape scale, in perhaps different places and scales than what occurred in the historical landscape. This should involve re-imagining functional landscapes in new places that leverage existing natural habitats and landforms. Long-range plans should be developed such that individual projects or transformations today can, in the future, become part of an interconnected and diverse complex of both natural and cultural elements that more successfully addresses ecological needs.

SUMMARY

The central Delta encompassed the tidal core of the Delta, where large sloughs with low banks divided the landscape into islands. Numerous channels of varying sizes wove and terminated within the islands, inundating freshwater tidal wetland communities of tule, willows, and other species during spring tides, if not more frequently.

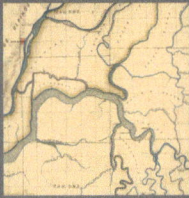

Comparing large-scale patterns (page 124) • Two major river systems, the Sacramento and San Joaquin, meet in the central Delta, and each has its own signature within the large-scale patterns of the region. Hydrologic and geologic differences between the Sacramento and San Joaquin rivers produced assymetry: channels associated with the San Joaquin were generally more sinuous, wider, and had lower banks. Channels associated with the San Joaquin likely contributed more to the Delta's tidal prism (page 125).

Tidal characteristics (page 127) • Approximately 200,000 acres (80,940 ha) of the 365,000 acres (140,710 ha) of tidal wetlands in the Delta were wetted - much of it inundated - by twice daily tides, with inundation depth and frequency greater toward the Delta mouth. The eastern margin was inundated at least by spring tides (page 130). Tidal range in the rivers at Sacramento and Stockton was about two feet (0.6 m) in 1850 (page 129).

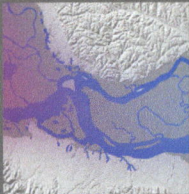

Salinity (page 137) • The gradient between brackish and fresh water fluctuated depending on the year, season, and tide. Freshwater conditions prevailed to the west of the Delta mouth, though high tides late in the season could bring brackish water into the western Delta. Changes in channel geometry and the loss of tidal wetlands may have contributed to the potential for salinity intrusion.

Flood attenuation (page 142) • The wide tidal channels, low banks, and broad wetland plain attenuated the large floods: upstream, near Sacramento, flood heights were over 20 feet (6 m), but only several feet (~1 m) above high tide levels downstream, in the central Delta.

Channels dominated by tides (page 143) • Large winding sloughs branched and rejoined to form the Delta islands (page 143) while networks of smaller tidal channels terminated within the wetland, providing exchange between land and water (page 154). Channel planform varied according to factors such as relative fluvial influence and whether channels were part of an island (page 157). Tidal channel density was lower than that of San Francisco Bay marshes and higher than that upstream in less tidal parts of the Delta (page 161). While most tidal channels ended within the tidal wetland, some connected to non-tidal floodplain channels upstream or to upland drainages (page 164).

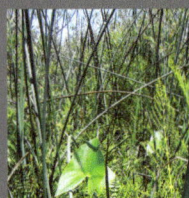

Complexity within the wetland plain (page 168) • The freshwater tidal wetland vegetation communities were unique to the Delta, with different assemblages from the salt-tolerant communities downstream and the riverine communities upstream. Peat soils, reaching depths of 100 feet (30 m) with surface elevations approximating high tide, supported emergent vegetation (primarily tule) and willows. The wetland vegetation in the central Delta was generally shorter and less dense than that upstream in the north Delta (page 176). Willow-fern swamp extended into wetland interiors primarily among the lower reaches of the Old and Middle rivers (page 177).

Upland ecotone (page 186) • Characteristics of the central Delta's upland edge varied in relation to topographic, climatic, geologic, and hydrologic controls. Sand mounds rose like islands above the wetlands in the western Delta, occupied by oaks and grasses as well as dune scrub species on exposed portions (page 188). A zone of alkali seasonal wetlands often lay along the central Delta margin, corresponding with areas inundated by extreme flood events (page 193). An oak studded plain stretched across the alluvial fan of the Calaveras River (page 194).

INTRODUCTION

In the recent past, the central Delta landscape broadened from the westerly Delta mouth to encompass extensive freshwater tidal wetlands interwoven with tidal channel networks (Figures 4.1 and 4.2); Thompson 1957, TBI 1998). This chapter describes this area of roughly 300,000 acres (120,000 ha) historically dominated by tides, with an additional 100,000 acres (40,500 ha) of other wetland and upland habitat types. The more muted tidal landscapes of the north and south Delta, where an additional 65,000 acres (26,300 ha) of tidal wetlands were found, are covered in subsequent chapters. Although exact boundaries are undefinable, we take the general extent of the central Delta to lie roughly from the base of Grand Island and middle of Tyler and Staten islands to the midline of Union and Roberts islands, including both the western and eastern Delta (Fig. 4.3).

Upon close examination, the common conception that the central Delta before reclamation was an unvarying sea of tule resolves into an understanding of a landscape rich in physical and ecological diversity. The landscape pattern was one of multiple tidal islands usually over 5,000 acres (2,000 ha) in area and large tracts of tidal wetland bordered by adjacent upland habitat types (Fig. 4.4). Within that large-scale pattern, factors such as proximity to tidal channels, topographic variability, and biotic interactions produced local patterns and environmental gradients.

Within the central Delta wetlands, sinuous tidal channels of varying sizes branched, rejoined, or terminated to form and dissect the tidal landscape. Topographic relief was slight and the tidal marsh plain elevation approximated high tide levels (Gilbert 1917, Atwater and Belknap 1980, Thompson 2006). The wetlands were composed of a complex mix of freshwater emergent vegetation, willows, and other wetland-associated

Looking toward the northeast we saw an immense plain without any trees, through which the water extends for a long distance, having in it various islands of lowland.

—ANZA AND BOLTON 1930 FROM
APRIL 2, 1776 DIARY ENTRY

Figure 4.1. "A. C. Freese towing the barges, Santa Rita, Ajax and commerce. Schooner is the Rough and Ready." (chapter title page) The flat expanse of the central Delta is captured in this sketch by Ralph Yardley of an 1890 photograph near Stockton. A few willows line the San Joaquin River, likely occupying artificial levees already in place by that time. (see Fig. 1.14; Yardley n.d., courtesy of The Haggin Museum, Stockton)

Figure 4.2. Tidal freshwater wetlands. Tules (*Schoenoplectus* spp.) and other wetland species are seen growing along the edge of Sherman Island in this recent photograph. (photo by Daniel Burmester, September 14, 2005)

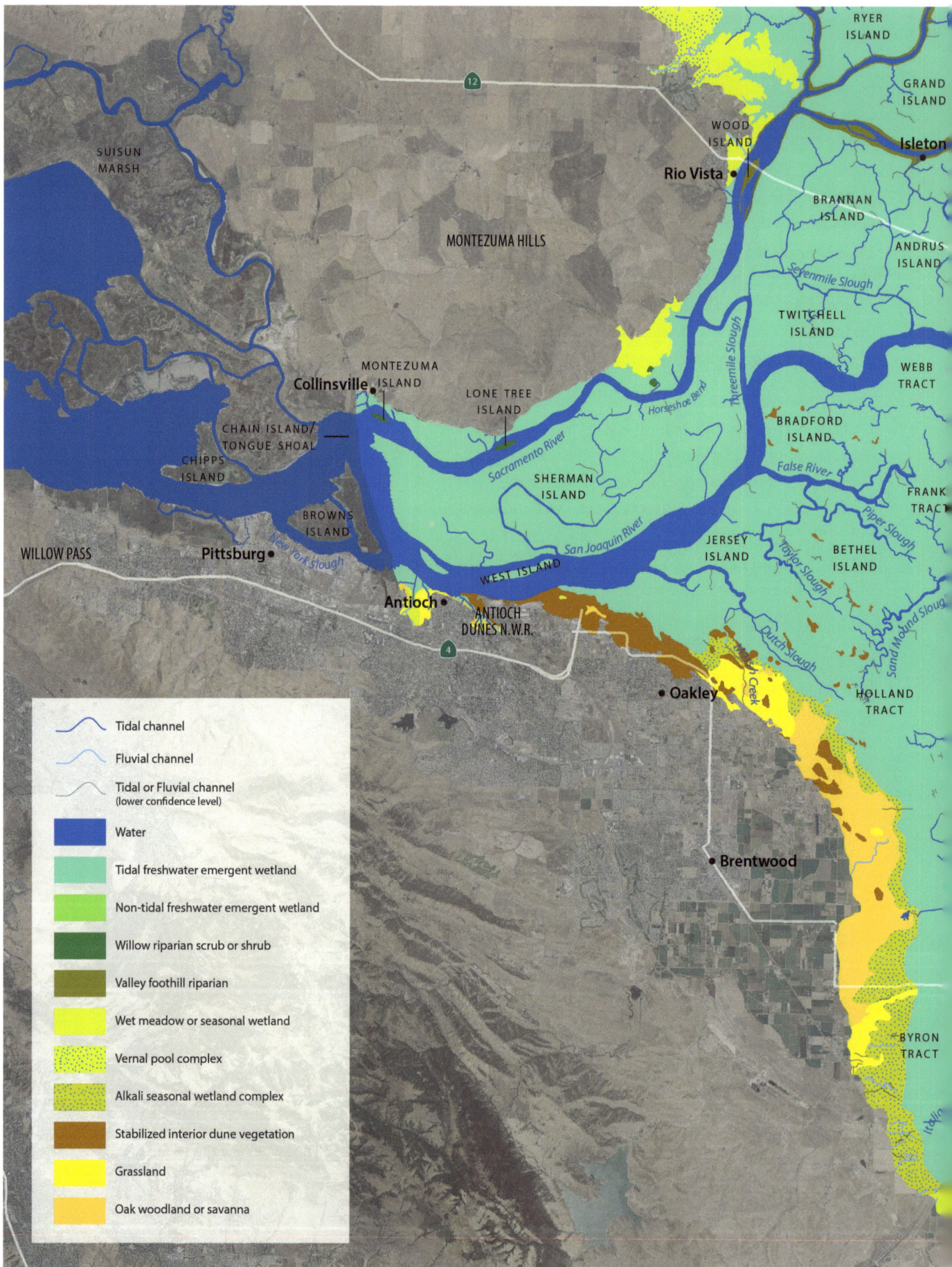

Figure 4.3. Distribution and extent of habitat types within the central Delta tidal islands landscape in the early 1800s. Sinuous tidal channels of varying sizes branched into tidal freshwater wetlands. The larger sloughs rejoined the river channels to form large islands. Tracts of tidal wetlands

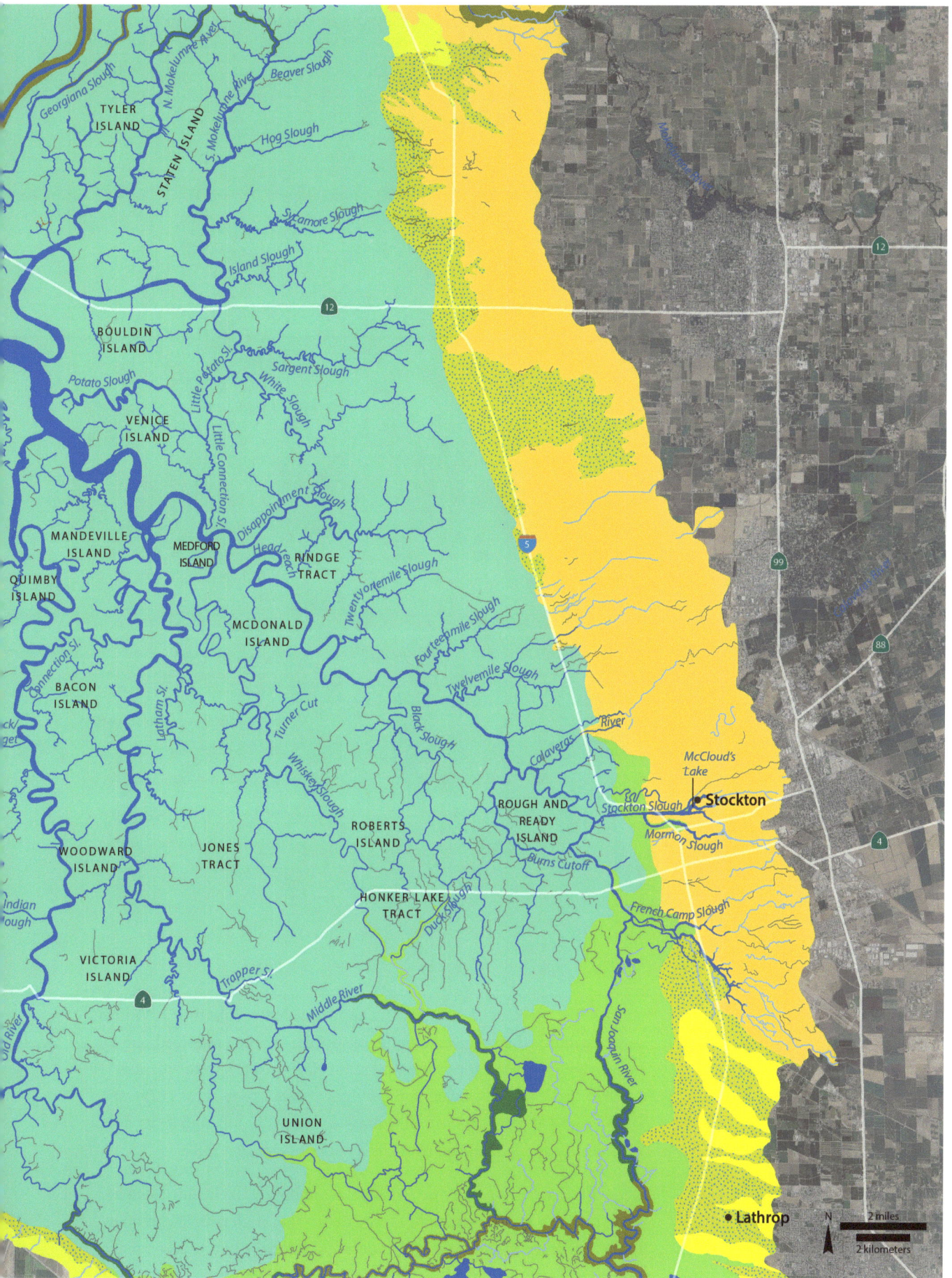

TYLER
ISLAND

Georgiana Slough

N. Mokelumne River

STATEN
ISLAND

S. Mokelumne River

Beaver Slough

Hog Slough

Sycamore Slough

Island Slough

12

BOULDIN
ISLAND

Potato Slough

Little Potato Sl.

Sargent Slough

White Slough

VENICE
ISLAND

Little Connection Sl.

Disappointment Slough

MANDEVILLE
ISLAND

MEDFORD
ISLAND

Head reach

RINDGE
TRACT

QUIMBY
ISLAND

Connection Sl.

MCDONALD
ISLAND

Twentyonemile Slough

Fourteenmile Slough

5

BACON
ISLAND

Latham Sl.

Turner Cut

Twelvemile Slough

Black Slough

Calaveras

River

Whiskey Slough

McCloud's
Lake

WOODWARD
ISLAND

JONES
TRACT

ROBERTS
ISLAND

ROUGH AND
READY
ISLAND

Stockton Slough **Stockton**

Mormon Slough

Burns Cutoff

Indian Slough

HONKER LAKE
TRACT

Duck Slough

French Camp Slough

VICTORIA
ISLAND

Trapper Sl.

4

Old River

Middle River

UNION
ISLAND

San Joaquin River

99

88

4

Lathrop

N

2 miles

2 kilometers

were bordered by an upland ecotone of seasonal wetlands, grasslands, and oak woodlands and savannas.

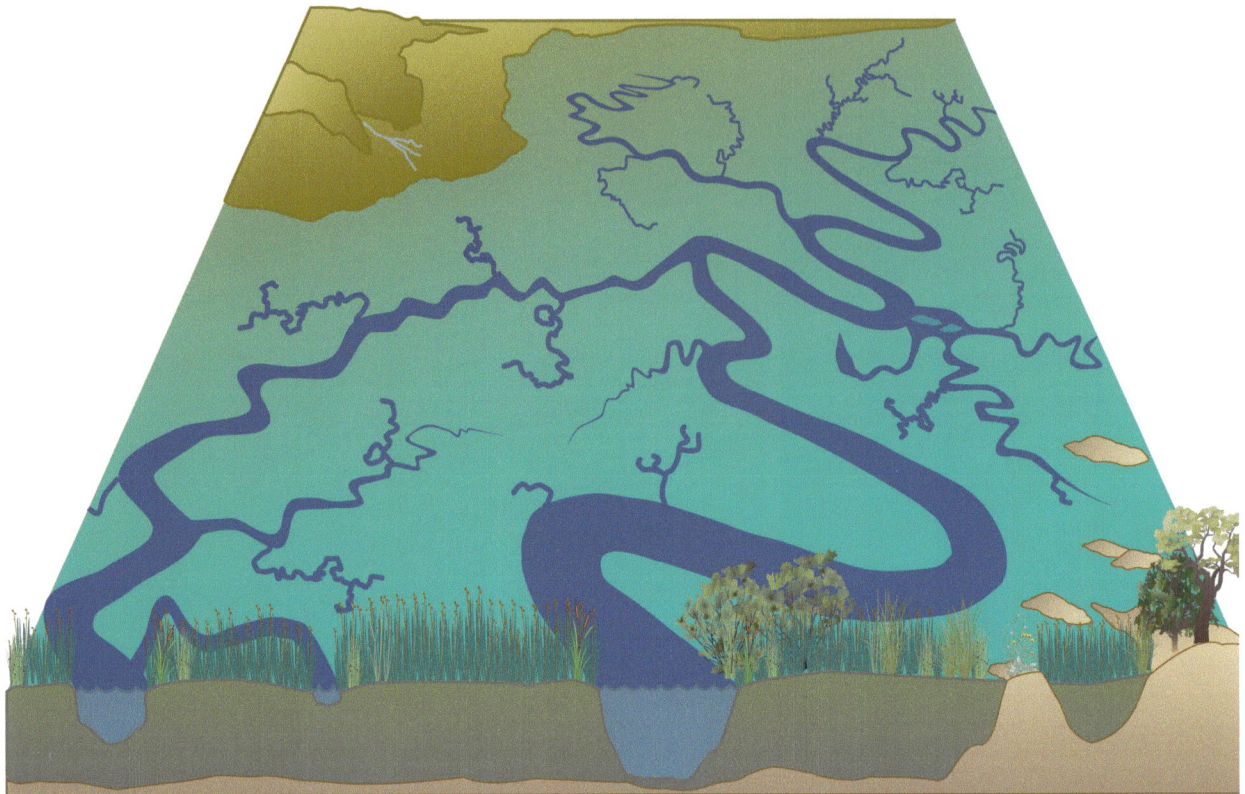

Figure 4.4. Conceptual diagram of the central Delta tidal islands landscape.
Central Delta tidal islands, most well over 5,000 acres (2,000 ha), supported a matrix of emergent vegetation (primarily tule), willows, grasses, sedges, shrubs, and ferns, and were surrounded by broad and deep tidal channels. Channel banks were low and numerous small branching tidal channels wove through the wetland plain, allowing high tides to regularly inundate much of the area. Oaks and herbaceous species with dune scrub in more exposed areas occupied the higher land at the edge and on sand mounds that were interspersed within the wetland plain. Relative proportions of habitat types based on the historical habitat type map produced is illustrated in the pie chart.

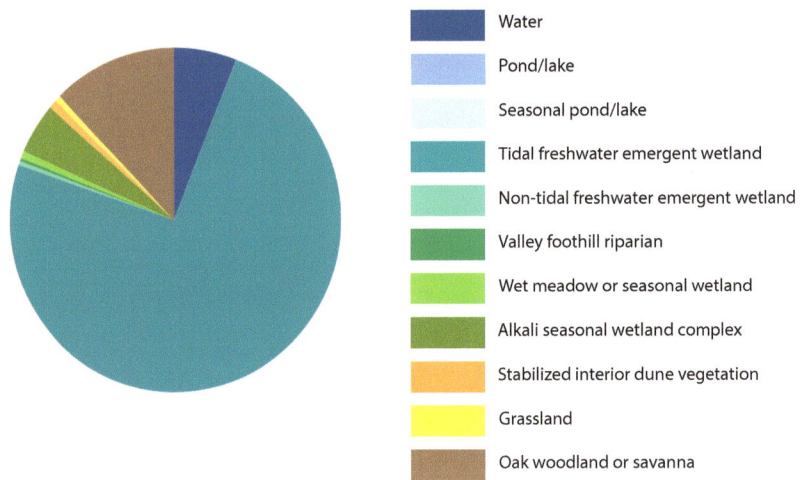

Water
Pond/lake
Seasonal pond/lake
Tidal freshwater emergent wetland
Non-tidal freshwater emergent wetland
Valley foothill riparian
Wet meadow or seasonal wetland
Alkali seasonal wetland complex
Stabilized interior dune vegetation
Grassland
Oak woodland or savanna

species. Tule species (*Schoenoplectus* spp.) dominated, but may have been less ubiquitous than many sweeping historical accounts may lead one to believe: willow (*Salix* spp.), arrowhead (*Sagittaria* spp.), water-plantain (*Alisma* spp.), rushes (*Juncus* spp.), sedges (*Carex* spp.), cattails (*Typha* spp.), reeds (*Phragmites australis*), grasses (*Poaceae* spp.), and ferns (*Anthyrium* spp.), occupied the tidal wetlands (see page 168).

These lands were wetted by twice daily tides and were likely inundated by twice monthly spring tides, if not more frequently. High river stages often inundated entire islands several feet deep, occasionally boosted by a rising tide or persistent strong wind. Flooding associated with precipitation or snowmelt runoff occurred as freshwater inflows backed up through sloughs into island interiors and flowed over the low banks. Upstream of Browns Island at the Delta mouth, freshwater conditions prevailed, as evidenced by early accounts and demonstrated in modern studies of vegetation communities (see page 137; Atwater 1976). The gradient between saline and fresh water moved inland in times of drought and bayward during floods.

Channel density and sinuosity were greater in the central Delta than within the more riverine northern and southern Delta (see page 161). However, channel densities appear to have been considerably lower than those observed in brackish and saline marshes of the estuary downstream, likely related to greater inundation tolerances of wetland vegetation in fresh water (Atwater and Hedel 1976, Grossinger 1995, Pearce and Collins 2004). In contrast to the natural levees found upstream, channel banks were low and regularly overflowed by tides. Following this, the vegetation communities along these banks were occupied by wetland species as opposed to riparian forests.

The upland ecotone was characterized by variable habitat mosaics depending on the hydrologic, topographic, and edaphic characteristics. At the eastern Contra Costa edge, a wide swath of alkali seasonal wetlands, including valley sink scrub, occupied the tidal wetland edge (see page 193; Stanford et al. 2011). Here also, and extending into the wetlands of the western Delta (e.g., Webb Tract), were sand mounds (relict glacial-age dunes) that rose above the land surface (see page 186). Along the eastern margin of the Delta, near Stockton, oak woodlands and savannas came close to the margin of tule, separated by a narrow zone of seasonal wetland. To the north, alkali seasonal wetland complexes formed the transition between tidal freshwater wetland and upland habitat types.

The value of landscape scale restoration to benefit ecological functions can be considered in terms of access of species to conditions along a broad tidal-fluvial gradient, availability of tidal channels at varying sizes with access to food from the adjacent marsh, overall productivity value and high residence times associated with tidal exchange with large areas of marsh, and connection to upland habitats (Simenstad et al. 2000). The historical landscape offered substantial capacity for interaction between aquatic and wetland environments, likely offering feeding and refuge opportunities for native fish as well as reptiles (e.g., giant garter snake) and amphibians. Native fish species would have found the dendritic tidal sloughs that spread through the productive wetlands offering a range of gradients in environmental variables (e.g., temperature, turbidity) at both large and small scales. In addition, the connectivity between wetland and upland habitat types facilitated exchange between these systems.

The following sections provide information about specific topics concerning the historical patterns and characteristics that predominated within the central Delta landscape. First examined are large-scale patterns and then the behavior of tides and floods is discussed. This is followed by sections on the tidal channels, marsh plain, and upland ecotone.

Island Country. At Clarksburg Sacramento River begins to give off a number of minor channels, known as sloughs, which flow independently for a short distance and then unite with other sloughs or with the main river. In this fashion channel after channel leaves the river and returns to it, so that the river may be said to flow into a plexus of channels, each communicating with others and with the main channel and similarly connecting and communicating with the channels of San Joaquin River.

—BRYAN 1923

COMPARING LARGE-SCALE PATTERNS

The Delta is composed of two primary river deltas that meet at a constricted mouth, and it enters San Francisco Bay rather than directly into the ocean. The Delta can therefore be defined as the upstream end of the larger San Francisco Estuary. While tidal processes were primary factors governing how the Delta looked and functioned, other physical factors – particularly riverine inputs – interacted with the tides to shape the Delta landscapes. Hydrologic and geologic differences between the two river systems and their associated major landforms led to asymmetry in the Delta. Interactions with slowly rising tides over a period of thousands of years resulted in different large-scale morphologies between the two dominant river deltas. These are noticeable in a simple visual comparison between the lower reaches of the Sacramento and San Joaquin (Fig. 4.5).

In contrast to the Sacramento River, channels branching from the San Joaquin tend to be more sinuous with less extensive natural levees. This may relate to differences between the natural hydrographs and sediment loads of the two rivers. The Sacramento's annual flows were about three times greater, flood peaks were higher and earlier in the season, and flood water was more laden with sediment (see page 10). Consequently, natural levees built up higher and extended farther into the tidal compartment on the Sacramento, which had the effect of restricting tidal access and slowing flood waters (Atwater et al. 1979). In addition, the soils of the marsh plain varied north to south, grading from inorganic clays in the northern flood basins to organic peat up to 65 feet (20 m) deep in the central Delta (Reed et al. 1890, Atwater et al. 1979). Atwater and Belknap (1980) attribute the accumulation of deeper peat deposits and lack of substantial inorganic sediment accumulation in the central and south Delta (Reed et al. 1890, Cosby 1941) to the absence of large natural levees slowing waters and to sediment more likely falling out downstream where flows meet brackish water. Interestingly, the San Joaquin was also known to transport Sacramento flood flows to the Delta mouth. Partly due to its relatively narrow channel, the Sacramento often delivered floodwaters into the San Joaquin by way of overland flow and connecting channels such as Threemile and Sevenmile sloughs (Bryan 1923, Thompson 2006).

In addition to overall greater sinuosity, the lower reach of the San Joaquin was wider on average (consistently well over 2,000 ft/610 m wide; Fig. 4.6) than the Sacramento, suggesting greater channel capacity. For comparison, before the Dabney Commission of 1908 authorized the straightening and deepening of the Sacramento between its mouth and Rio Vista, the Sacramento was on average around 1,500 feet (457 m) wide and narrowed to around 800 feet (244 m) at Horseshoe Bend (present-day Decker Island). The difference in size was noted by Spanish explorers as early as 1775, when De Cañizares (1909) identified the San Joaquin as the larger of the two upon viewing them at the Delta mouth. In 1817, a member of another early expedition commented on the San Joaquin's greater width in several

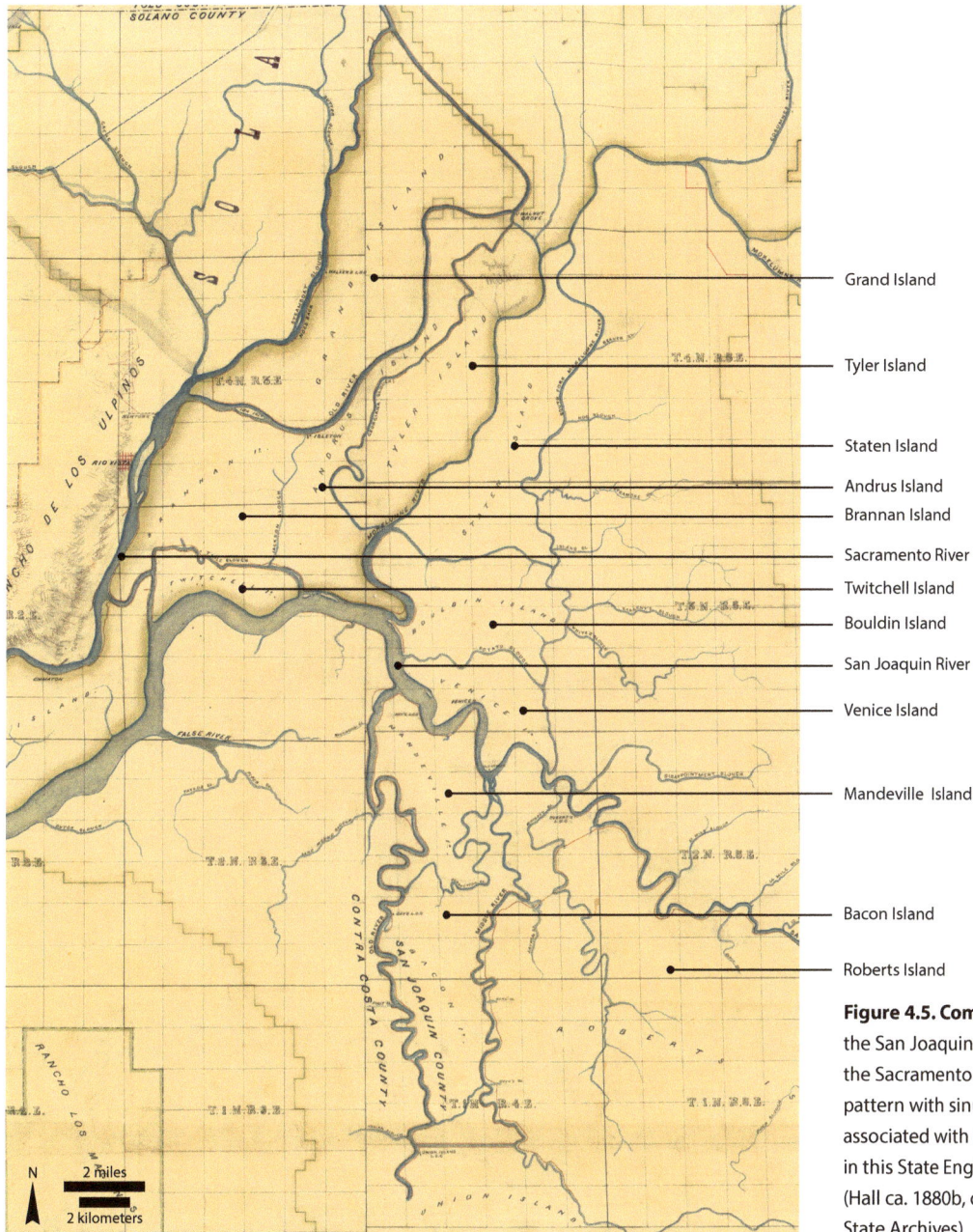

Grand Island

Tyler Island

Staten Island
Andrus Island
Brannan Island
Sacramento River
Twitchell Island
Bouldin Island
San Joaquin River
Venice Island

Mandeville Island

Bacon Island

Roberts Island

Figure 4.5. Comparing the planform of the San Joaquin distributaries to those of the Sacramento reveals the more deltaic pattern with sinuous branching channels associated with the San Joaquin, as shown in this State Engineering Department map. (Hall ca. 1880b, courtesy of the California State Archives)

locations, but comparatively lower flows (likely freshwater flows as opposed to tidal flows; Durán and Cook 1960).

Given channel capacity and related tidal marsh area, the San Joaquin River likely contributed more to the Delta's tidal prism than did the Sacramento (TBI 1998). This conclusion is reached by Gilbert (1917) who determined that "in early years" the San Joaquin River and its wetlands contributed 4.8 billion cubic feet (135 million m^3) to the ebb current at Carquinez Strait while the Sacramento contributed 2.96 BCF (84 million m^3; see page 136). The contribution of the San Joaquin today is still more than that of the Sacramento (about 52%), despite myriad changes (Fleenor pers. comm.). The loss of the large tidal wetlands and damming and filling of numerous channels in the Delta (much of it associated with the San Joaquin) are

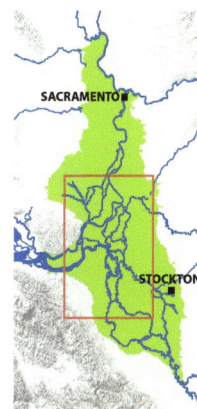

Figure 4.6. The San Joaquin and Sacramento rivers differ in width and sinuosity. Close to the Delta mouth, the San Joaquin is almost double the width of the Sacramento River. Moving upstream along both channels, the San Joaquin becomes a very sinuous with tight bends and much narrower tidal channel, while the Sacramento bends are much wider, reflecting greater riverine influence.

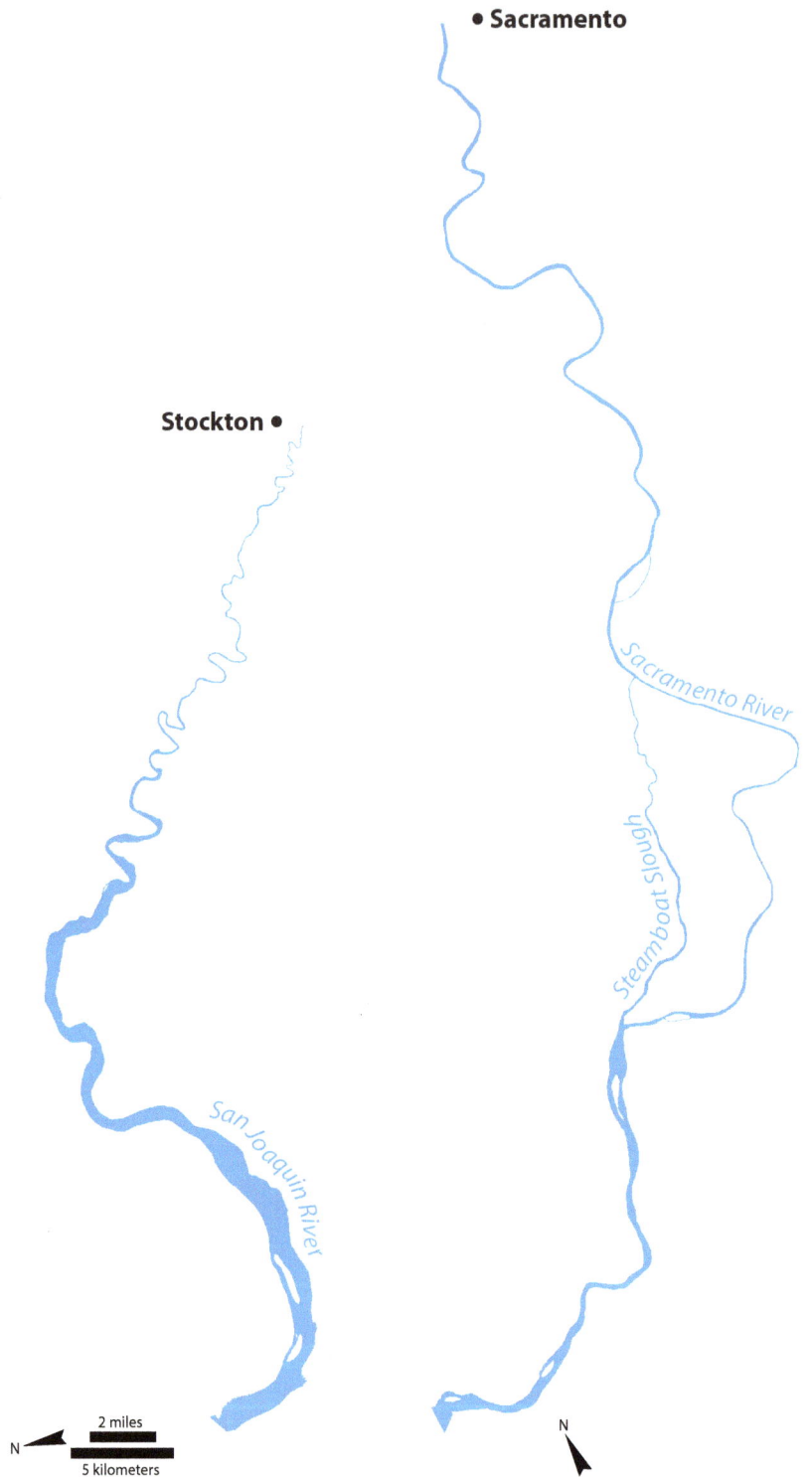

	Channel length (mi)	Direct (mi)	Sinuosity
San Joaquin River	49.0	29.1	1.7
Sacramento River	62.4	40.1	1.6
Sacramento River (via Steamboat Sl.)	55.7	40.1	1.4

likely the primary contributors to this shift (Fleenor pers. comm.). It is challenging, however, to parse out the relative effects of changes (e.g., dredging, reclamation, outflow) to uncover just how important the wetlands and tidal channels of the San Joaquin were to tidal prism (Gilbert 1917, Fox 1987a). Another possible influence on the San Joaquin's planform may be the late spring San Joaquin high flows that arrived to a Delta often already flooded by earlier Sacramento high flows. This may have supported the establishment of a more branching channel network (Fleenor pers. comm.). Overall, these differences suggest two related conclusions: one, that the San Joaquin affected the central Delta landscape pattern more than the Sacramento and two, that tidal processes had relatively greater influence on the formation and maintenance of San Joaquin channels than of Sacramento channels (Atwater et al. 1979).

TIDAL CHARACTERISTICS

Tides have a profound effect on the form and function of wetlands. Tidal action was a dominant physical process in the central Delta: it dictated the frequency with which wetlands were saturated and flooded, kept channels sized appropriately for the tidal prism, influenced tidal channel morphology, reduced riverine flood heights, controlled marsh plain elevations and peat accumulation rates, promoted habitat connectivity, affected species assemblages, and aided the exchange of nutrients and biota through the Delta ecosystem (Atwater et al. 1979). While salt and brackish tidal wetlands are well studied, large freshwater tidal systems are less understood (Odum 1988). Improving knowledge of the historical tidal patterns of the Delta can lend insight into how these freshwater tidal wetlands may have looked and what ecological functions they would have provided.

A large volume of water passed through the Carquinez Strait at each high tide, propagating tides upstream and generating significant currents that traveled through the many winding sloughs. The Delta's tidal lands were inundated by spring tides and some portions were covered twice daily at high tide by several inches of water. During floods, the landscape resembled a large lake several feet deep (Thompson 1957).

Tidal extent and range

Prior to reclamation, approximately 365,000 acres (147,700 ha) of tidally influenced wetland and nearly 30,000 acres (12,140 ha) of water features existed in the Delta. The tidal marsh plain approximated high tide elevations, likely within eight inches (0.2 m) of mean higher high water (MHHW), with estimates based on measurements of remnant marshes and supported by historical records (*Daily Herald* 1869 in Tide Land Reclamation Company 1872, Ferris in Nesbit 1885, Gilbert 1917, Atwater and Belknap 1980, Thompson 2006). At the Delta's core, tidal inundation frequency and depth appears to have been greater than elsewhere in the Delta. Most early accounts state that about 200,000 acres (80,940 ha) or less were regularly overflowed by "ordinary" tides (i.e., daily high tides) or

It is a plain cut by sloughs into a number of islands, whose surface is nearly level, raised but a few inches above high tide and covered with swamp grass and tule.

—SACRAMENTO DAILY UNION 1862

"subject to back water or tidal overflow" (Fassett 1865, Cronise 1868, Whitney 1873). A calculation from an early engineering report states that roughly 160,000 acres (64,750 ha) were "subject to inundation at each high tide, twice in twenty-four hours" (Rose et al. 1895). Some evidence suggests that high tides overflowed the majority of this "regularly overflowed" land by several inches (Higley 1859, Munro-Fraser 1879). However, other accounts reveal that while twice daily high tides wetted the land, extensive inundation only occurred with the frequency of every spring tide, or twice monthly (Belcher 1843, Farnham 1857, Van Scoyk 1869, Day 1869). These more tidally influenced wetlands were found between the latitudes of Clarksburg and Stockton (Day 1869, USDA 1874). The rest of the tidal wetlands were wetted less frequently, likely by spring tides (Fig. 4.7).

Early reports, referring to this core of regularly inundated tidal wetlands as freshwater tidelands, promoted these lands as prime farmland because of the ease with which the land could be irrigated by the tides (*Mining and Scientific Press* 1869, Shinn 1888). Farmers, they claimed, simply had to open the flood gates at high tide: "Fresh water, in any quantity, can thus be brought over these lands, or to within an inch, or a foot or two feet of the surface, as may be wanted, every day of the year, by merely attending to the opening and shutting of the tide-gates" (Alexander 1869, *Sacramento Daily Union* 1873). A similar account gives a sense of the inundation frequency with the statement that "they [the tidelands] can be irrigated, most of them, every day, and all of them once a month, at spring tide" (*Oakland Daily News* 1871 in Tide Land Reclamation Company 1872). However, it should be noted that these boosterish reports may conveniently overlook some of the exceptions to the general conditions they describe.

Tidal and fluvial processes acted upon the Delta's waterways and marshlands to varying degrees, depending on both spatial and temporal factors. Spatially, this physical gradient extended from dominant tidal processes in the western and central Delta to purely fluvial, or riverine, processes at the Delta margins (see Fig. 4.7). This transition included the change from tidal currents ebbing and flowing in the central Delta to the rise and fall of tides caused by water backing up in the river farther upstream. William Hammond Hall's 1880 engineering report describes the manifestation of this gradient on the morphology of the Sacramento River channel. In it, he divides the river into three primary reaches, or "compartments," to designate relative tidal and fluvial influence and illustrate how formation and maintenance of the channel is affected. His "tidal compartment" reach, where "the influence of tidal action predominates over that of the flow from the land drainage in the aggregate… and regulates the dimensions of the channel," extends from the mouth up to the Cache Slough confluence at the foot of Grand Island. The reach that extended from the foot of Grand Island to Sacramento graded from dominant tidal to fluvial processes and "is most variable, and the channel is made and maintained in part by each influence – the one predominating at the time of floods, and the other at times of low water" (Hall 1880). Hall recognized that this compartment would have extended to the Feather River prior to

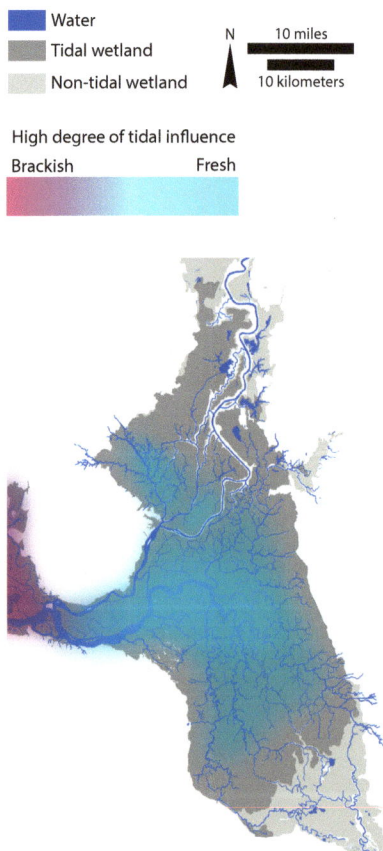

Figure 4.7. Tidal influence, wetlands, and water bodies in the Delta. An overlay gradient illustrates the brackish to fresh gradient that was generally positioned at the Delta mouth in low water. This gradient shifted historically depending on the tides, season, and year. Also illustrated is the decreasing tidal influence eastward and upstream, an expression of the tidal to fluvial gradient. Of the 365,000 acres of tidally influenced wetland mapped (dark gray), only about half of the area in the core central Delta (overlay gradient) is understood to have been overflowed or wetted by tides twice daily. The rest of the area was less frequently inundated by tides.

Water
Tidal wetland
Non-tidal wetland

N 10 miles
10 kilometers

High degree of tidal influence
Brackish   Fresh

hydraulic mining. The confluence of the Feather River is noted by others as well as being a location of distinctive change in the morphology of the river channel, where the quite sinuous and dynamic meanders of the upper Sacramento change into "a series of smooth, large bends in no way suggestive of ordinary meanders" (see page 238; Byran 1923).

Tidal range, and the capacity of tides to influence wetland and channel form and function, diminishes with distance from the Delta mouth. Tidal range between the highest and lowest tides is reported to be on average between 3.0 and 4.6 feet (0.9-1.4 m; Atwater and Belknap 1980). At the Delta mouth, tidal range was between four and six feet (1.2-1.8 m; Abella and Cook 1960, Farnham 1857, Rose et al. 1895). Reports of tidal ranges centered around two feet (0.6 m) at Sacramento as well as at Stockton, 62 miles (100 km) and 50 river miles (80 km) distant from the Delta mouth, respectively (Bryant [1848]1985, McCollum [1850]1960, Wilkes 1845, *Sacramento Daily Union* 1862, Hall 1880, Payson 1885, U.S. Congress 1916, Morgan 1960 in Dawdy 1989, Taylor 1969). Tides were perceptible as far as the mouth of the Feather River on the Sacramento (*Sacramento Daily Union* 1862, Hall 1880, U.S. Congress 1916, CDPW 1931). On the San Joaquin River, tidal range likely became negligible somewhere between Sheppard's Ferry (present-day I-5 overpass on the San Joaquin) and the Stanislaus River confluence (CDPW 1931, Fleenor and Moyle pers. comm.). This was suggested by explorers in October 1811, who reported that tides were slight at the head of Old River (Abella and Cook 1960). Several accounts report regular tide ranges on the Mokelumne at Benson's Ferry (Cosumnes River confluence) of around 3 feet (0.9 m) and spring tides of over 4 feet (1.2 m; Thayer 1859, Thornton 1859, Payson 1885). Tidal influence extended above the confluence with Dry Creek (Gray 1859, Thayer 1859, Van Scoyk 1859, Mendell 1881, Payson 1885). During the dry season on the Cosumnes River, tides evidently reached two miles (3.2 km) upstream from the confluence with the Mokelumne (Gray 1859).

An important point about tidal influence is that the maximum reach of tide on major rivers does not indicate the extent of tidal influence on the marsh plain, given the effect of vegetation, natural levees, and tidal routing through minor tidal sloughs. For instance, knowing high tide elevation in Delta channels does not necessarily mean that water actually rose to that elevation throughout the Delta at each high tide, although this is a fair approximation of the maximum extent of tides in the absence of additional information (Atwater and Hedel 1976, Atwater 1982). This is partly because tidal energy propagates more easily up open channels than across vegetated marsh (due to greater friction). Therefore, particularly where natural levees prevented the direct connection of the tidal river to the marsh plain at most times of the year, tidal influence extended much farther up channels than it did across marsh. This is illustrated by the fact that while water levels rose and fell two feet (0.6 m) with the tides at the City of Sacramento historically, the wetlands within the flood basins on either side of the natural levees were non-tidal. This is not uncommon to tidal systems; such differences

[Sherman Island] The land was then subject to overflow at all high tides, excepting a few spots, where local reclamation had been attempted.

—SAN FRANCISCO TIMES 1869

They are overflowed at spring tides only, before reclamation. The surface of the land, therefore, is a few inches above ordinary high water mark, and 5 feet above ordinary low water. The highest storm tides rise 26 inches above the average level of the land on the creek banks.

—FERRIS IN NESBIT 1885

between tidal limits in channels versus wetlands has been observed in the Pacific Northwest (Collins B. pers. comm.).

Tidal inundation depth and frequency

Diffusion of tidal energy across the large expanse of the Delta's tidal channels and wetlands affected inundation patterns. The frequency, duration, and depth of tidal inundation across the marsh plain was naturally greatest at the western apex of the Delta. Daily high tides appear to have inundated large portions of the western Delta wetlands by several inches, which caught some early explorers off guard. One group camping in the tule in October 1811 found that "the water reached our blankets at the turn of the tide" (Abella and Cook 1960) and another somewhere near the mouth of the Sacramento River in May of 1817 "landed on a small island of tule which at high tide was covered with water, and we had to take refuge upon some places full of brambles to protect ourselves from the water until it receded" (Durán and Chapman 1911). Evidence of daily tidal inundation can also be found in testimony given during the Los Medanos land grant (near present-day Antioch) court proceedings, which offers insight into the pattern of overflow on the marshes at the Delta mouth. When asked to describe how the land along the San Joaquin River was affected by tides, one witness stated that it was "covered daily by water at high tide – nearly every portion of it" (Taylor 1865).

Reported elevations match these descriptions. An 1869 newspaper article reported that the surface of Sherman Island was subject to all high tides (*San Francisco Times* 1869 in Tideland Reclamation Company 1872), being "about six inches below high and from three to six feet above low tide" (*Daily Herald* 1869 in Tideland Reclamation Company 1872). During the spring tides, the depth was far greater: an account from a farmer at Horseshoe Bend on the Sacramento River stated that his two and one half foot (0.76 m) high levee was "about one foot above the spring-tide mark," meaning that the pre-leveed marsh was likely overflowed by a foot and a half (0.46 m) of water at spring tides (Higley 1860).

Extending eastward into the central Delta, sources also support that high tides inundated much of the marsh surface (*Mining and Scientific Press* 1869, Gilbert 1917, Atwater and Belknap 1980, Thompson 2006). Some of the earliest evidence for this comes from Father Ramon Abella's exploration of Old River, where the group decided to sleep in their boats, stating that "there is land but it is flooded" (Abella and Cook 1960). The flooding can be attributed to the tides because this expedition was made late in the dry season, during October 1811. Many decades later, a reclamation document stated that Staten Island, like other central Delta islands, was "swamp over which the higher tides flow" (McAfee 1874). This is affirmed by an 1861 account of the tides on Bouldin Island, which were reported to overtop the low natural levees by six inches (Beaumont 1861b).

That I saw so many islands while Captain Fages and Father Crespí saw only two is no doubt due to the fact that they saw this lake at high tide while I saw it at low tide, which in this Puerto Dulce rises and falls considerably…

—FONT AND BOLTON 1930, REFERRING TO THE REGION AT THE DELTA MOUTH ON APRIL 3, 1776

Even along the eastern Delta margin between the Mokelumne and Calaveras rivers, it is evident that the land was inundated at least by spring tides. Mokelumne land grant court testimony provides several accounts that suggest relatively frequent inundation of the marsh plain, extending virtually to the margin of tule (Fig. 4.8; Van Scoyk 1859). To describe the area generally, one person testified that "at high tides twice a month it is overflowed" (Beaumont 1859a). A more nuanced description is found at the head of Sycamore Slough, where a railroad once extended about three quarters of a mile (1.2 km) into the tules:

> In going in I stepped on the ties of the railroad and when not under water, when under water I stepped on the rails. Holding on to the tules to preserve my balance. This was in the latter part of August 1859. The land on either side of the railroad was inundated, the greater portion of

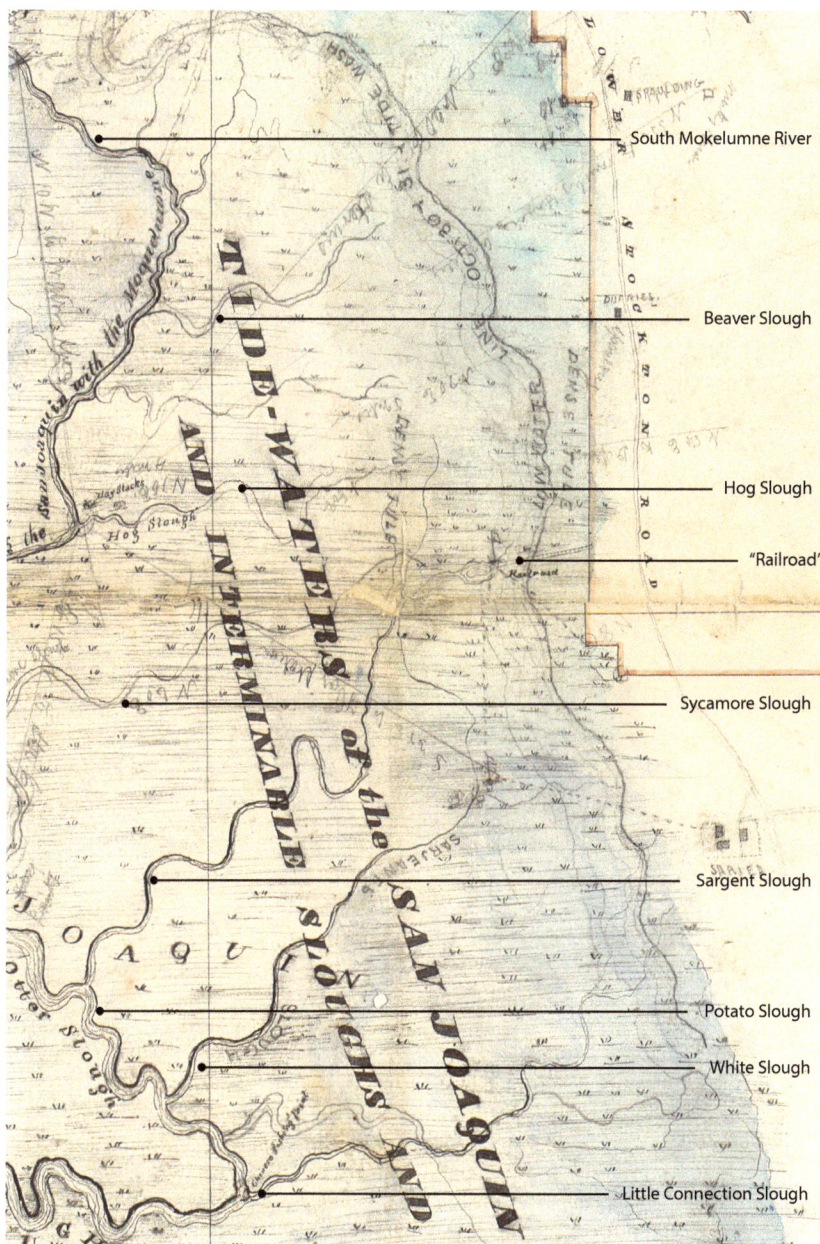

Figure 4.8. Tidal wetlands and sloughs of the eastern Delta. This map, made as an exhibit for the "Sanjon de los Mequelemes" land grant trial, depicts tidal freshwater emergent wetland up to the edge of the land grant boundary (pale orange line). Accompanying testimony clarifies that the land in this region was overflowed by high tides and, more specifically, overflowed at high tide "almost up to the eastern extremity of the railroad" (see label, Sherman 1859). The pencil line on the map of "low water line October 30 & 31 & tide wash" probably illustrates the general extent of overflow at the high tide on those dates to show where the land is reliably wet in the dry season (i.e., low water meaning dry season as opposed to low tide). The line was drawn by surveyor and witness William Watson (1859b) who made it based on four observations taken while walking out through the tules from his boat at the heads of sloughs. Note: some sloughs are mislabed or had different names historically. (Von Schmidt 1859, courtesy of The Bancroft Library, UC Berkeley)

it, almost up to the eastern extremity of the railroad, when the tide was up. At the time the tide was coming in, the land between the western extremity of the railroad and the head of the slough was nearly six inches under water. (Sherman 1859)

When asked about inundation depth at the head of Sargent Slough, another witness reported that upon rowing up the slough seven miles (11.3 km; to a point less than a mile from the edge of mapped marsh), he found "places nearly to the tops of my boots" in November 1859 (Gray 1859). Tides clearly wetted the eastern Delta margin year-round and the edge of tule likely marked the extent of high tides as no major rivers or associated flood basins contributed water from upslope. This boundary lies along the elevation contour close to high tide, or 3.5 feet (1.1 m) above mean sea level, the maximum elevation at which inundation frequency was sufficient to allow accumulation of peat (Atwater et al. 1979, Atwater and Belknap 1980).

Moving away from the core of the central tidal Delta, the combined effect of decreasing tidal energy, increasing land surface elevations, and increasing height of natural levees translated to decreasing inundation depths and frequencies. For instance, one source refers to Bacon Island prior to reclamation as above "ordinary high tides" (History of Bacon Island n.d.). This transition was also apparent within Tyler and Staten islands, where the lower portions had low natural levees and the land was regularly inundated by tides, whereas the upper islands were bounded by natural levees reaching above the extent of tides with interior land wetted only during extreme tides (Thompson 2006).

The upland edge approached close to the river on both the northern Contra Costa and southeastern Solano county shoreline, limiting the extent of tidal marsh at the Delta mouth. Much of the tidal marsh area along the Contra Costa shoreline between Pittsburg and Antioch was found along tidal swales that connected upland drainages. Backwater areas along upland margins, referred to as lagunas or sloughs, were likely only influenced by spring tides (Fig. 4.9; Smith 1866b, Stanford et al. 2011). They were distinguished from high marsh pannes more common to brackish and saline marshes (Leopold et al. 1993, Collins and Grossinger 2004, Grossinger 2012). A number of these features persist in the landscape today, though they are for the most part disconnected from tidal influence.

Tidal inundation complexity at the local scale

Tidal inundation patterns on a marsh plain are affected by proximity to tidal channel and the presence of natural levees (Fig. 4.10). Tidal water primarily accesses a marsh plain through the networks of small tidal channels that extend from the main tidal sources (the major rivers and sloughs that formed the Delta islands) to their termination within the interior marsh. As observed in 1859, "the tide comes up in the sloughs and at their heads flows over into the tule" (Dugin 1859). These smaller channels functioned as the capillaries of the tidal wetland.

Labels on photo:
Sherman Island | trees on stabilized Antioch Dunes | former tidal slough or laguna

Figure 4.9. A former tidal slough or laguna near Antioch is seen in this 1920s-era oblique photograph by George Russell. Early accounts suggest that water flooded these areas during spring tides and was retained in backwater depressions. Residents in the area augmented this characteristic by damming them to retain fresh water through the late summer months (Morse 1888). (Russell ca. 1925, courtesy of the California State Lands Commission)

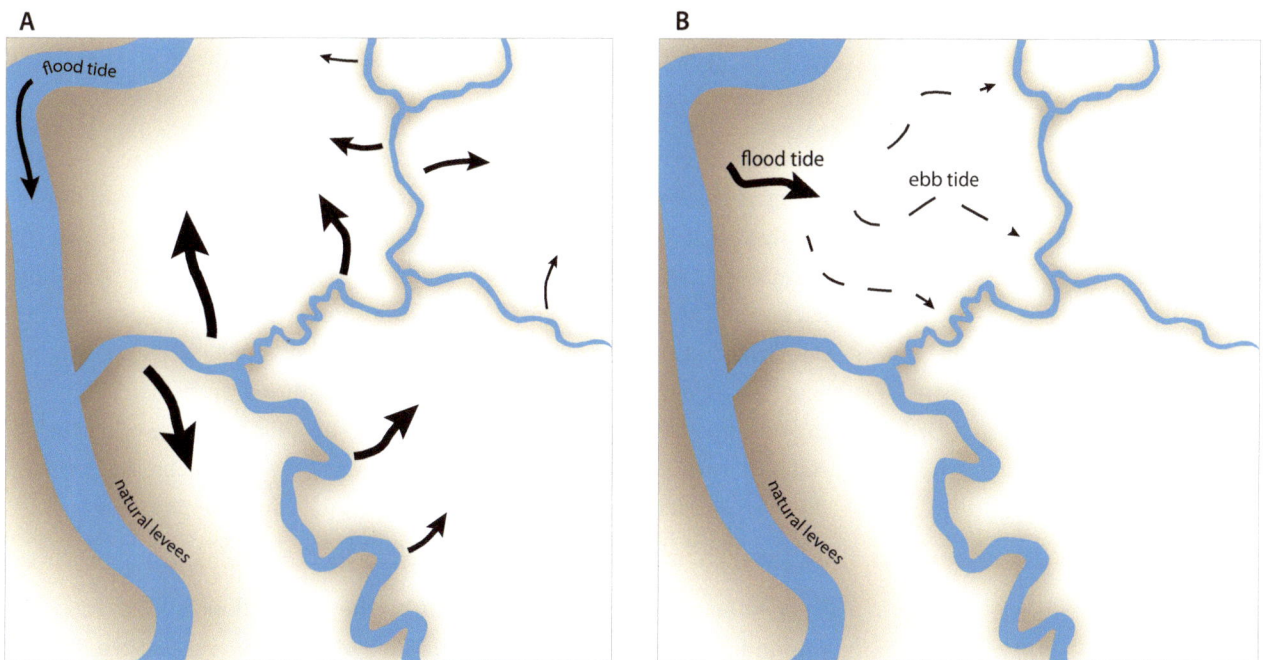

A

flood tide

natural levees

B

flood tide

ebb tide

natural levees

Figure 4.10. Localized tidal inundation patterns related to distance from tidal source (A) and natural levee heights (B). In A, the farther away a point is from tidal source, (e.g., the mainstem channel) the lesser the tidal influence. The size of the black arrows in A indicates the magnitude of overflow. In B, the relative height of low channel banks (natural levees) affects how water enters and exits the marsh plain. Here, at flood tide (solid arrow), water flows over the comparatively high natural levees (brown gradient) of the largest channel (it is also flowing from other channels as well). On the ebb tide, the easiest exodus for that water is by way of the smaller interior marsh channels that have lower natural levees. These topics are discussed in more detail by Leopold et al. (1993) and Collins and Grossinger (2004).

Tidal hydrodynamic complexity is partly driven by the fact that depth and inundation frequencies decrease with distance from a tidal channel due to friction and time of tidal reversal (water may not have the time to spread completely before tide reverses; Gilbert 1917). It also means that a tidal wetland close to the mouth of a tidal slough will be inundated more frequently and with greater depth than one at the same elevation located at the far end of that same tidal slough (Collins J pers. comm.).

Another physical factor affecting tidal hydrodynamics is the presence of natural levees, which tend to decrease in height with decreasing channel order and decreasing sediment input from riverine systems. The patterns of tidal flooding on a marsh plain depend on the relative height of these banks, where even a few inches can have a significant impact. When tides rise high enough, water flows over natural levees into the lower marsh plain, but when tides fall that water must exit at lower points in elevation, found along the smaller tidal channels that have low to non-existent banks.

Within the central Delta, natural levees were generally low and overtopped during high or at least spring tides (Beaumont 1861b, Rose et al. 1895). Where banks were overtopped and where they were not was locally variable, as is suggested by the testimony that banks along sloughs east of the Mokelumne were "overflowed in some places and in some places not" (Van Scoyk 1859). In the absence of fluvial influences, banks of tidal channels generally became lower with distance from tidal source, as is suggested by this reclamation account for the swamp land district (#1) between Taylor and Piper sloughs in eastern Contra Costa County: "as their banks are generally higher than the land on either side, they hold the water until it gets near the head of the sloughs" (Tucker 1879a).

Furthermore, overflow was uneven due to small variations in marsh plain topography. Sherman Day reported that spring tides overflow the "freshwater tidelands," or central Delta islands, "only a foot or so on the lower portions, in hollows, and along the bayous [sloughs]" (Day 1869). The backwater lagunas along the Contra Costa shoreline may have partially owed their existence to this phenomenon, illustrated by two separate accounts: one that explains that "the tides must raise over the banks in order to flood the lands in the rear" (Henderson 1865), and the other that states "at extreme high tide the water gets in and the bank being higher than it is back the water stands and does not run back when the tide goes down…there are low spots in all this ground" (Eddy 1865).

Tidal influence also shifted depending on the season, flood events, and prevailing winds at the Delta mouth. Much of the Delta was flooded for several months of the year, with tides driving the flooding frequency for the remainder of the months (Beaumont 1859a, Mellin 1918). Also, during flood events the high flow of freshwater entering the Delta restricted tidal extent upstream within channels. The disruption of regular tidal patterns by annual floods was described by a witness for the Mokelumne land grant case: "At low water the tide flows out of the Mokelumne River up the sloughs and fills the tules. At high water the water runs over the banks of the river above and flows off into the tules" (Van Scoyk 1859).

After major flood events had passed, but when rivers were at high stages, the greater hydrologic connectivity and higher water surface likely allowed tides to reach greater extents than at the lowest stages in the late summer and early fall. During the periods of low river stages, more of the marsh surface was isolated from the tides. This explains early travelers' accounts of tidal marshes being "dry on the surface" late in the summer and early fall within some parts of the Delta, particularly those at the margins of tidal extent (e.g., south Delta, Yolo Basin; Cronise 1868). Ranchers used these dried out marshlands as pastures for stock, particularly during droughts:

> In very dry seasons, however, the lowness of the river lessens the frequency of the tidal overflow, and this, with the large evaporation, renders the land dry enough for pasturing stock. At such times in the past, large herds of cattle and bands of sheep have been pastured on the tule lands, without any reclamation or leveeing whatever, and considerable amounts of wild grass have been cut and baled under like circumstances. (McAfee 1874)

While it is impossible to know hydroperiod and spatial extent of tidal inundation of the historical Delta precisely, it is clear that it was a heterogeneous landscape at the local scale. Important insights can be gained through improved knowledge of channel planform and relative height of natural levees, calibrated through descriptions of overflow patterns at particular locations.

Tidal current

While there is relatively little information in the historical record concerning tidal currents and flow patterns, several accounts suggest that at low river stages the ebb and flow of tides likely traveled at about three or four miles per hour (*Sacramento Daily Union* 1862, McGowan 1939). Currents varied depending on particular characteristics of sloughs, such as distance to the Delta mouth, blind (dead-end) versus flow-through, sinuosity, width, and depth. For instance, land case testimony describes relatively fast currents in present-day Potato Slough during an August survey: "the action of the tides [was] very swift, running about five miles an hour" (Sherman 1859). Another account, referring to conditions downstream of Stockton on the San Joaquin, referred to the tidal currents as "treacherous" around the larger meander bends (*The Morning Call* 1894).

Additionally, due to the complex planform of the many distributary (or flow-through) channels that delineated the Delta islands, the timing and magnitude of ebb and flow of tide in one channel was quite different from the next. As an example of this, the convergence zone of tides traveling up the two forks of the Mokelumne River was located in the South Fork more than half a mile below the head of the island. According to George Gray, who testified for the Mokelumne land grant case, "the water appeared to stand still; we threw in sticks and pieces to see which way it was flowing – it

Now, at a low stage of the river, the tides ebb and flow through all of the above sloughs and rivers, so that from the head of the Georgiana slough one may float in a skiff, by seizing the tide, north around Grand Island; east, to the Cosumnes; south, to Suisun, or west to the head of Cache slough; and travel in either direction at the rate of three or four miles an hour.

—SACRAMENTO DAILY UNION 1862A

flowed at that time to the north, apparently up stream; we went a little further and we could see it flowing the other way" (Gray 1859). In support of this observation, an exhibit from that trial included a surveyed channel profile showing that the channel depth was shallowest at this location (Fig. 4.11; Watson 1859a).

Tidal prism

Tidal prism – commonly defined as the volume of water between high and low tides – is challenging to estimate for the Delta, given the size and contribution of marsh plain storage. It is even more challenging to compare past and present tidal prisms due to competing historical factors of influence (Gilbert 1917). Tidal prism at Carquinez Strait has likely decreased over the last two hundred years (see page 125; Rose et al. 1895, Gilbert 1917). Gilbert (1917), in his treatise on the effects of hydraulic

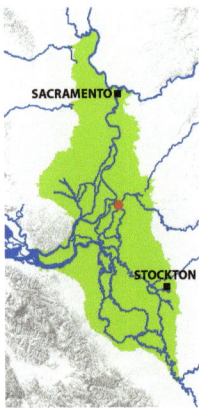

Figure 4.11. Complex tidal flows. This map and profile, Exhibit B of the Mokelumne land grant case, shows the width and depth of the two branches of the Mokelumne River extending 5,000 feet below the head of Staten Island. Soundings were taken at extreme low water in November of 1859. In explanation of his work, Watson described that "commencing at point C on exhibit B on the north slough the tides commence to flow from C to B one hour and ten minutes earlier than from A to B and while it is flow [flood] tide from C to B it was found to be ebb tide from B to A" (Watson 1859b). Shallow waters of a convergence zone (red circle), where tidal flows meet, are seen partway down the South Mokelumne River (B to A). (Watson 1859a, courtesy of The Bancroft Library, UC Berkeley)

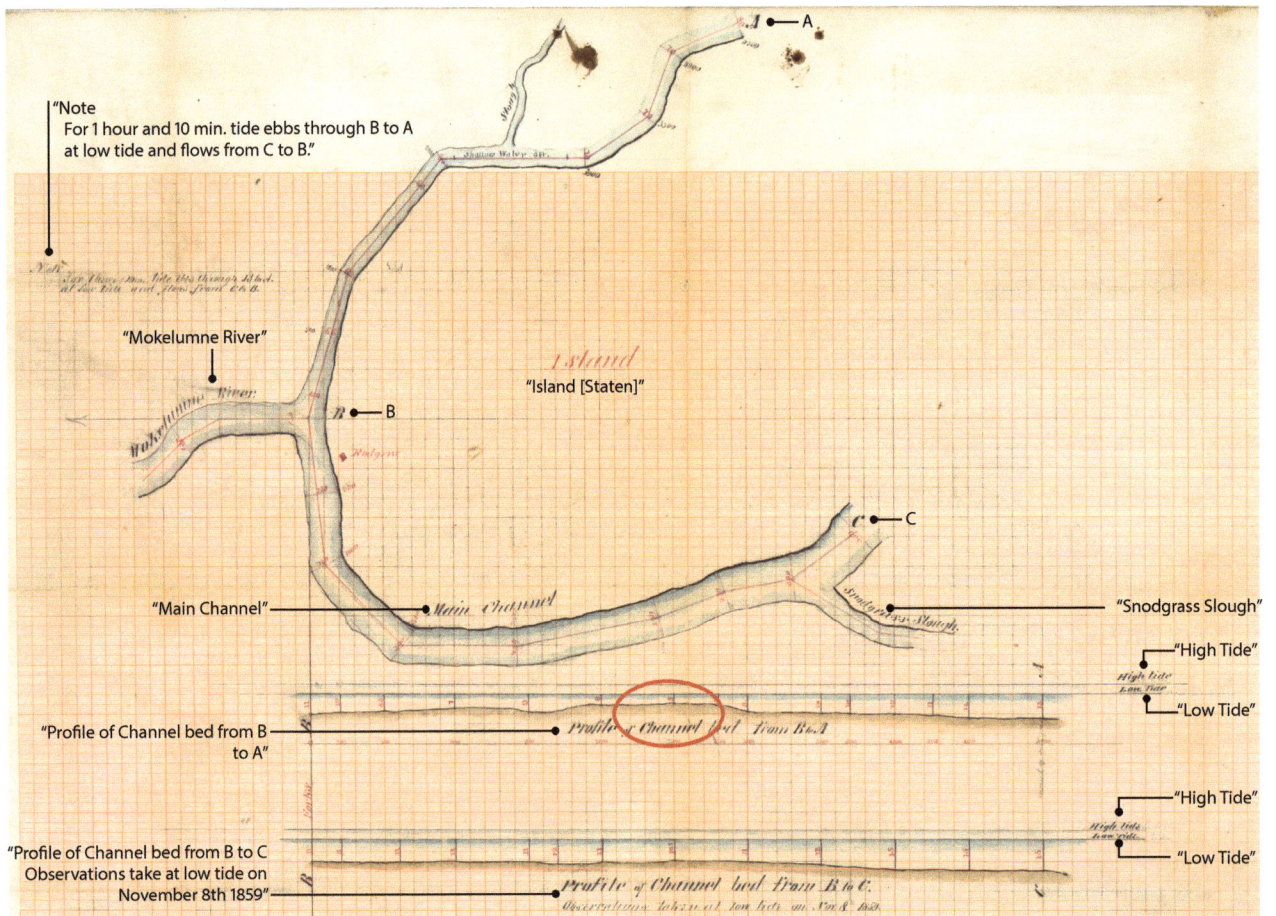

"Note
For 1 hour and 10 min. tide ebbs through B to A at low tide and flows from C to B."

"Mokelumne River"

"Island [Staten]"

"Main Channel"

"Snodgrass Slough"

"High Tide"

"Low Tide"

"Profile of Channel bed from B to A"

"High Tide"

"Low Tide"

"Profile of Channel bed from B to C Observations take at low tide on November 8th 1859"

mining debris, estimated that the historical volume "contributing to the current" at Carquinez Strait by the Sacramento and San Joaquin rivers and marsh totaled 7.76 billion cubic feet (220 million m³), or about 22% of the volume of Suisun Bay. Based on tidal flow estimates, this figure may be closer to 7.39 billion cubic feet (209 million m³) today (CDWR 1993). An important factor negatively affecting tidal prism has been the reduction in the area of marsh plain historically flooded by tides (Thompson 1957, Schoellhamer et al. 2007, Fleenor pers. comm.). Other changes impacting tidal prism include the introduction of deep water ship channels on the Sacramento and San Joaquin, opening of the Delta mouth, flooded islands, channel straightening, and reduction and alteration of outflow (Fleenor pers. comm.). Early engineering reports recognized the impact on tidal volume: "As works of reclamation progress, this flow will continue to decrease, until it is limited to the small tidal flow required to raise the water surface in the waterways alone" (Rose et al. 1895). Objecting to a proposed closure of Cache Slough, an engineer argued that flow into the slough provides "the volume necessary to keep open the channel of the river below, by its passing up and down again" (Young 1880). Channel cuts and channel deepening and widening, however, are counteracting factors that allow for greater volumes of water to pass through a given point (increasing tidal prism). For more detailed discussion of tidal prism in the historical Delta and the rest of the San Francisco Estuary, see Gilbert 1917, pages 71-88.

SALINITY

The San Francisco Estuary has many physical gradients, including the gradient between salt and freshwater at the mouth of the Delta (see Fig. 4.7). Salt content of the water column at any single location will fluctuate according to tidal, seasonal, and inter-annual fluctuations in freshwater inputs. Today, it is also influenced by water diversions, wetland drainage, and channel modifications (CCWD 2010). In the winter when the rivers are at flood stage, large volumes of freshwater push this gradient closer to the Golden Gate, while in the late summer or during droughts lower freshwater inflows allow saline water to extend into the Delta. Also, though flood tides move saline water a certain distance upstream (physical movement of water), tidal influence (energy propagation, expressed as currents or rising and falling water levels) is transmitted much farther up the rivers than the actual extent of saline water upstream.

Historical evidence suggests that freshwater conditions predominated in the early 1800s Delta. There is some evidence of occasional brackish water intrusion at the Delta mouth, though extent is difficult to determine. This was a dynamic gradient that shifted according to a number of interacting temporal factors. Most historical accounts note a transition from saline to freshwater in the upstream portions of Suisun Bay (Farnham 1857, Fox 1987a, CCWD 2010). As an example, a report for agricultural purposes stated that the termination of "salt tide" (presumably salinity levels injurious to crops) was just downstream of Sherman Island (Alexander 1869). In some years, water not suitable for drinking was noted at Antioch during the late summer months, but it is unclear whether brackish conditions extended considerably farther with any regularity. An early history of Contra Costa County described that "the San Joaquin frontage is fresh for ten months out of the twelve, and, in most years, is fresh the entire year; even in very dry seasons it is fresh at low water" (Smith & Elliot [1879]1979). At flood tide and/or dry years, it appears that brackish water could be found at that point. This section (and Table 4.1) summarizes observations made before substantial water withdrawals and other modifications in the Delta (e.g., channel widening and deepening) promoted salinity intrusion during the early decades of the 1900s (CDPW 1931).

While observations clearly suggest that brackish waters did at times extend upriver along Sherman Island under natural conditions, there is little evidence to suggest it regularly penetrated far into the Delta. The first known recorded observations of salinity conditions are those of the Spanish explorers. While it is impossible to determine the exact locations where these observations were made or how the qualitative judgments of fresh or salty might translate into percent salt content, there appears to be general agreement that waters became fresh (to the taste) somewhere along Suisun Bay, usually close to the Sacramento and San Joaquin confluence. The 1772 Fages and Crespí expedition, which marked the first sighting of the Delta by Spanish explorers, found the "water fresh and still" on March 30 after descending from Willow Pass (Crespí and Bolton 1927). An expedition by water was made in August 1775 and, referring to the Delta mouth, pilot Cañizares reported "some rivers empty and take the saltiness of the water which there becomes sweet, the same as in a lake" A map was later produced, the "Plan del Gran Puerto de San Francisco," identifying the islands in Suisun and at the Delta mouth as "Yslas Razas entre agua dulce" or flat or low islands in sweet (fresh) water (De Cañizares 1781, De Cañizares et al. 1909).

Although not from the late summer period when salinities extended farthest upstream, several observations from the Anza and Font expedition in April 1776 are worthy of notice. Suisun Bay is referred to as the Puerto Dulce (sweet port) in this expedition, but Father Pedro Font also reported finding the water "salty, although not so salty as that of the sea outside" (Font and Bolton 1930). Juan Bautista de Anza, likely just east of present-day Antioch, noted that the water of the San Joaquin "was now very fresh, but we noted that it was changeable" (Brown 1998). Although it is unknown whether Anza attributed this changeable nature to the ebb and flow of tide or to seasonal fluctuations, the observation is in general accordance with the idea that the Delta mouth marked a transition. A later account discusses both tidal and seasonal variation with: "water taken from New York Slough on the last of the ebb tide is used by some for domestic purposes all through the year, though it becomes somewhat brackish in the Autumn" (Morse 1888).

Table 4.1. Early textual descriptions of salinity conditions at the mouth of the Delta. Most evidence describes freshwater conditions at the Delta mouth, but there is some evidence for occasional brackish conditions.

Quote	Date	Flow (MAF, Meko et al. 2001)	Location	Reference
"finding the water fresh and still"	1772, March 30	19.5	from Willow pass, "camp this night was probably westward of Antioch" (from footnote)	Crespí and Bolton 1927
"where some rivers empty and take the saltiness of the water which there becomes sweet, the same as in a lake"	1775	18.7	mouth of the Delta	de Cañizares et al. 1909
"Yslas Razas entre agua dulce" [flat or low islands in sweet water]	1775 [1781]	18.7	Islands at the Delta mouth and Suisun Bay	de Cañizares 1781
"the water is unfit for drinking because it is so salty"	1776, April 2	9.1	above Selby, below Carquinez (from footnote)	Anza and Bolton 1927
Puerto Dulce [sweet harbor] "I tasted the water and found it salty, although not so salty as that of the sea outside"	1776, April 2	9.1	Suisun Bay	Anza and Bolton 1927
"it was now very fresh, but we noted that it was changeable"	1776, April 3	9.1	near Antioch	Anza and Brown 1998
"before arriving at the Strait [Carquinez] the water is already salty"	1811, Oct 29	22.8	Crossing Suisun Bay	Abella and Cook 1960
"we found the water perfectly sweet"	1837, Oct 26	14.1	where the Sacramento "becomes a narrow stream" entering its mouth	Belcher et al. 1979
"camped, without water, that of the river being still brackish"	1841, Aug	5.56	likely near Antioch: 11 miles from Suisun Bay, 2 miles north, then 3 miles up the "southeast arm of the Sacramento," which they then find actually leads them to the San Joaquin	Wilkes 1845
"the water being fresh here all the year"	1847	19.8	Rio Vista	Californian 1847
"which if the tides was to wet it the salt would destroy the value of the coal"	1865	18.5	vicinity of New York [Pittsburg] and Antioch	Clayton 1865
"The vegetation is from fresh water"	1865	18.5	vicinity of New York [Pittsburg] and Antioch	Clayton 1865
"Northerly point near the New York where the water is generally so brackish as to be useless for animals"	1865	18.5	New York [Pittsburg]	Stratton 1865
"It is such as is peculiar to both salt and fresh water marshes—Some tule and some salt grass … Sometimes fresh sometimes salt [water]. In summer season high tide would be salt—I have tried the water being in a boat"	1865	18.5	vicinity of New York [Pittsburg] and Antioch	Taylor 1865
"The line of brackish water is at the lower end of Sherman Island…water in the rivers and sloughs above this point rises and falls with the tide and is always fresh"	1869	14.9	foot of Sherman Island	Alexander 1869
"The water along the San Joaquin frontage is fresh for ten months out of the twelve, and, in most years, is fresh the entire year; even in very dry seasons it is fresh at low water"	1879	15.4	vincinity of Antioch	Smith & Elliot [1879]1979
"Natural growth is three cornered tule and sweet grasses. No salt grass or alkli [sic] weed"	1912	11.4	Chipps Island	Unknown 1912

Reports from two other expeditions provide additional evidence of salinity gradients. In 1837, Captain Belcher found "water perfectly sweet" most likely at the entrance to the Sacramento River near the foot of Sherman Island (Belcher et al. 1969). (As an aside, the location of this observation is somewhat debatable: their position was 20 miles (32 km) above the anchorage of their larger craft, the *Starling*, which was 36 miles (58 km) from San Francisco, placing them at about present-day Decker Island on the Sacramento. However, in another part of the text, Belcher states that 20 miles above the *Starling* is where the Sacramento becomes "a narrow stream," which would more likely place them at the foot of Sherman Island (Belcher et al. 1969)). Later, in the particularly dry year of 1841 (estimates show Sacramento River outflows 70% below the 30-year average; Meko et al. 2001), the U.S. Exploring Expedition (referred to as U.S. Ex. Ex.) under Commander Charles Wilkes camped near present-day Antioch where they were "without water, that of the river still brackish" (Wilkes 1845). It is unknown how far these conditions persisted upstream.

Fairly persistent freshwater conditions at the foot of Sherman Island were discussed in the "Fresh water tide land" report of 1869, which enthusiastically assured the Tide Land Reclamation Company that since "the line of brackish water is at the lower end of Sherman Island," then "all their lands are in fresh water" (Alexander 1869). Sherman Day's report to the same company is slightly less conclusive with the statement that the "waters are mainly fresh" at the mouth (Day 1869). While these engineers may have felt encouraged to hail the favorable conditions of the Delta's freshwater tide lands, the comments are in agreement with most early reclamation accounts that describe the ease with which freshwater could be let through ditches at high tides to irrigate crops.

This transition zone at the confluence of the two rivers is also suggested by the salt tolerances of the vegetation present historically (Atwater 1980, Stanford et al. 2011). Evidence of woody vegetation and freshwater emergent species suggests that conditions were fresh on average, as vegetation patterns generally reflect long-term average conditions (U.S. Ex. Ex. 1841, Ringgold 1850a, Durán and Chapman 1911). In addition, fossil records do not indicate that salt tolerant species were prevalent in the Delta within the Holocene period (Atwater and Belknap 1980).

Just downstream of Sherman Island, freshwater vegetation occupied Browns Island (e.g., willow, *Salix lasiolepis*; button bush, *Cephalanthus occidentalis*; alder, *Alnus rhombifolia*; and tule, *Schoenoplectus* spp.), as did salt marsh species, such as salt grass, *Distichlis spicata* (Atwater 1980, Knight 1980). Significant variations in salinity levels have occurred at this site through the millennia. One recent study shows evidence for higher freshwater inflow between 3800 and 2000 calibrated years before present (ca yr BP) and a shift to more saline conditions in the last 2,000 years (Goman and Wells 2000). Another found higher salinity periods between 3000 and 2500 cal yr BP, 1700 and 730 cal yr BP, and 1930 to today (the latter primarily related to freshwater diversions; Bryne et al. 2001). At a finer resolution, periods between 1600 and 1300, 1000 and 800, and 300 and 200 cal yr BP appear to have had relatively high salinity levels (Malamud-Roam and Ingram 2004). The overall trends are supported by core samples taken from Browns Island that reveal a shift away from *Phragmites communis*, a freshwater species, and increased presence of salt grass within the last 1,000 years (Atwater 1980).

The transitional nature of the vegetation is clearly articulated in the Los Medanos land grant testimony. Several accounts mention freshwater vegetation, "wild grass, willow, and tule" as well as more salt tolerant species, "salt grass." (Woodruff 1865). One witness was hard-pressed to state whether the marshes were fresh or not and gave the following statement:

> [Question] Is it such as is peculiar to fresh water or to salt water marshes?
> [Answer] It is such as is peculiar to both salt and fresh water marshes—Some tule and some salt grass…
>
> [Question] How much does the ordinary tide rise and fall at this point?
> [Answer] Perhaps 4 or 5 feet—can't tell exactly never examined
>
> [Question] Is it fresh or salt?
> [Answer] Sometimes fresh sometimes salt. In summer season high tide would be salt—I have tried the water being in a boat. (Taylor 1865)

After about 1920, those who had previously relied on a supply of fresh water through the summer months noted a clear increase in the degree and extent of salt water intrusion (CDPW 1931, CCWD 2010). Much of this change was attributed to greater diversions and impoundments of fresh water upstream. Some of the earliest quantitative information concerning salinity is found in the California and Hawaiian Sugar Company's barge records, which show that boats could find fresh water (<50 ppm) downstream of Jersey Point on the San Joaquin River year round until 1918. In 1919 and 1920, they had to travel much farther upstream late in the summer and fall, after which they ceased to collect water from the river during the dry season (CDPW 1931, CCWD 2010). In the decades that followed, salinity intrusion became more significant, generally extending about 3 to 15 miles (4.8-24.1 km) farther inland over historical conditions, and persisting for longer periods of time (CCWD 2010). A fuller treatment of trends in historical salinity is given in the recent *Historical Freshwater and Salinity Conditions* report by the Contra Costa Water District (2010).

Water withdrawals have not been the sole driver affecting salinity conditions. The effect that channel geometry and tidal wetlands had on tidal excursion and salinity intrusion should not be overlooked (Thompson 1957). All other factors being equal, a relatively narrow and shallow Delta mouth – the historical situation – would have a limiting effect on tidal penetration in comparison to the wider, straighter, and deeper channels in the Delta today (Fleenor pers. comm.). Some have concluded that, depending on timing, greater amounts of freshwater outflows would be needed today to maintain the average historical position of the salinity gradient (Fox 1987a).

Finally, the evidence presented here does not preclude the possibility that brackish (as distinguished by taste) conditions may have extended farther into the central Delta islands historically during extreme droughts. In addition, our focus on early 1800s conditions is a very short time period, geologically speaking. Extended periods of drought and drier conditions over the past 2,000 years (including the cool and dry "Little Ice Age" spanning the 200 years prior to the Gold Rush) likely drove salinity intrusion to extremes unknown in the recent past (Stine 1996, Malamud-Roam et al. 2007). Researchers point to the seasonal and interannual fluctuations in salinity conditions that existed in Suisun Bay and the Delta mouth as an important factor contributing to the heterogeneity of the historical landscape, a landscape where native species are adapted to fluctuating salinity levels (Moyle et al. 2010).

FLOOD ATTENUATION

One ramification of the large tidal channel capacity of the central Delta and its easily accessed expanses of wetland was that floodwaters had a large volume to occupy after passing through the flood basins and narrow upstream riverine channels (State Agricultural Society 1872, USDA 1874). The tule lands of the central Delta were placed in their own class by an agricultural booster because of this, since "the annual floods have no great effect upon them" (Flint 1860). While flood heights at Sacramento reached over 20 feet (6.1 m), flood heights at the Delta mouth attenuated to mere feet above regular high tide levels (Day 1869, Gilbert 1879, Thompson 2006). In reporting flood heights during the infamous 1862 flood, the surveyor general stated:

> I have been told that at the head of Cache slough, at a place called Main Landing [Maine Prairie], the water was ten feet above the ground, which would make it about eighteen feet above low water mark. In the marshes around Suisun City [Rio Vista], the greatest height attained was only about two feet six inches, which would give about nine or ten feet above low water mark. In the islands in Suisun Bay the water did not rise more than six inches above the marsh, and that only at the highest tides. (Peabody in Houghton 1862)

Currents and floods of the freshet season do not have to be leveed against; the waters finding such spreading room in Suisun Bay – which begins at the lower end of Sherman Island – that a rise of fifteen feet at Sacramento is scarcely as much as a foot at Collinsville, Antioch, or Sherman Island.

—*MINING AND SCIENTIFIC PRESS* 1869

Early land reclaimers took note of this fact; the historical record is full of those proclaiming that central Delta islands could be easily reclaimed with levees only three feet high to keep out both high tides and "ordinary" floods (Higley 1859, California Swampland Commissioners 1861, Day 1869). However, they soon began to need even higher levees because of the elimination of land onto which floodwaters could escape, which raised water levels in the channels (Etcheverry 1903-1954, Dillingham 1911). This natural attenuation of floods is likely one of the primary reasons that reclamation occurred early in the central Delta, with Sherman Island being the first officially leveed island in 1869 (USDA 1874, Thompson 1957). This capacity for flood attenuation suggests its significant effect on the way seasonal high flows were transmitted through the San Francisco Estuary, as expressed through increased residence time, increased mixing of tidal and freshwater inputs, and reduced flood peaks.

CHANNELS DOMINATED BY TIDES

A defining characteristic of the central Delta landscape was the fractal-like network of winding channels, sized to accommodate the tidal volume that passed back and forth twice a day and wetted the marsh plain. The channels connected habitats, transporting water, sediment, nutrients, and organisms (Odum et al. 1984). They influenced where, when, and how much water reached the marsh plain (Sanderson et al. 2000). Learning about what these channels were like can generate greater understanding of larger-scale hydrodynamics and the ecological functions of aquatic habitat in the freshwater tidal Delta.

The Delta's tidal channels can be placed into two basic classes: (1) the major subtidal waterways (also referred to here as rivers, distributaries, or mainstem channels) that delineated much of the geography familiar to people today, and (2) the myriad tidal channel networks that branched off into the marsh and terminated in the tules (referred to as blind tidal channels). The historical habitat type mapping suggests that there were about 1,000 miles (1,600 km) of tidal channel within the central Delta, with blind tidal channels making up sixty to seventy percent of that length. Tidal channels wider than 50 feet (15 m) were found within the central Delta wetlands at channel to wetland area ratios of about 1:12. They accounted for roughly 22,500 acres (9,100 ha) out of 28,500 acres (11,500 ha) mapped within the total study area (most of the large tidal channels were in the central Delta).

When rivers meet the influence of tides, they spread into many distributary channels. In the case of the Sacramento-San Joaquin Delta and unlike most deltas, the distributary channels rejoined before passing through the single opening into Suisun Bay, and in so doing, formed the Delta islands (see page 124). It is generally understood that these primary river channels have meandered only short distances during the period of tidal wetland development (Cosby 1941, Atwater and Belknap 1980). This is in contrast to the more dynamic meandering channels of the upper Sacramento and San Joaquin rivers (Bryan 1923). Similar stability in tidal channel planform has been noted elsewhere in San Francisco Bay and other tidal marsh systems (Leopold et al. 1993, Collins and Grossinger 2004).

The following sections address several aspects of central Delta tidal channels, such as channel depth, size, and planform. While not comprehensive, the given sections provides a sense of the defining features of the central Delta landscape.

The rivers and distributaries that formed the islands

The Sacramento and San Joaquin rivers and their main subtidal distributary channels were the primary conduits of flows through the Delta. These channels were generally between 150 feet (50 m) and half a mile (0.8 km) wide. Channels were sized primarily by the volume of tidal flows they carried as opposed to flood flows, and consequently channel widths

It appears highly probable that all the major streams and most of the minor ones have occupied essentially their present positions during the entire period of organic accumulation.

—COSBY 1941

generally decreased with distance from the tidal source. These distributary channels branched off and then met again to form islands that were many thousands of acres in size. The multiple connections and variable channel depths, widths, and sinuosity created hydrodynamically complex flows: the timing of tidal propagation varied, tidal convergence zones were common, and flood flows found multiple paths (see page 127, Bryan 1923).

Overall within the central Delta, the primary subtidal waterways appear to have maintained depths of over 10 feet (3 m) prior to hydraulic mining debris and other modifications (U.S. Ex. Ex. 1841, Gibbes 1850a). Regarding channel depths, only three known navigation charts with soundings were made in the Delta before hydraulic mining debris began filling channels. They include a map produced from the 1841 U.S. Exploring Expedition and Gibbes' 1850 map, which show depths of no less than 15 feet (4.6 m) in the vicinity of Chain Island. These general depths are supported by a third and more detailed 1852 Ringgold chart that includes sounded cross sections. A highly detailed U.S. Coast Survey map (Cordell 1867), produced as the initial waves

Figure 4.12. Maps before and spanning the hydraulic mining era that show soundings at the Delta mouth. (A) An 1841 navigational chart of the Sacramento River shows some of the earliest soundings in fathoms (1 fathom = 6 feet) at the mouth of the Delta. Recorded depths were 15 feet or greater. (B) An 1852 navigation chart shows the locations of shoals as well as soundings (in fathoms). Depths were "reduced to lowest water," which suggests that most of the shoals were covered by several feet of water at low tide (Ringgold 1852). (C) This very detailed 1867 hydrographic map (H-937) of the U.S. Coast Survey shows the depths (in feet up to 18 feet, then in fathoms, at measured mean lower low water) at multiple locations along channel cross sections. (D) Another U.S. Coast Survey map (H-1784) from 1886 shows of the same location (soundings are also in feet), from a period after mining debris had begun entering San Francisco Bay. (E) The 1908 Debris Commission mapping is shown for the same area, where depths are in feet above low tide level. (F) In a recent navigational chart, the primary entrance to the Sacramento River today can be seen lying along the Deep Water Ship Channel between the tip of Sherman Island and Chain Island; navigators were historically advised to travel by way of the other side of Chain Island, next to the right bank of the Sacramento. Maps are shown at different scales. (A: U.S. Ex. Ex. 1841, courtesy of the Earth Sciences & Map Library, UC Berkeley; B: Ringgold 1850b, courtesy of the David Rumsey Map Collection, Cartography Associates; C: Cordell 1867, courtesy of the National Oceanic and Atmospheric Administration; D: Peacock 1886, courtesy of the National Oceanic and Atmospheric Administration; E: Wadsworth 1908a, courtesy of California State Lands Commission; F: U.S. Department of Commerce 1982, courtesy of the National Oceanic and Atmospheric Administration)

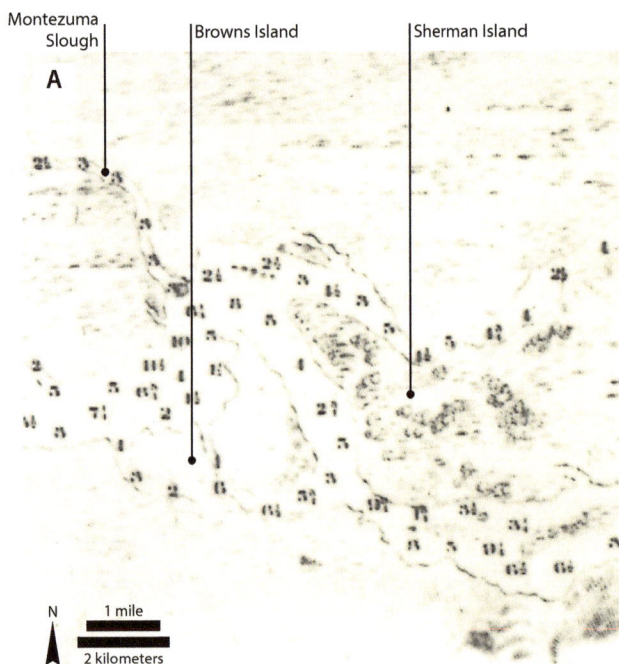
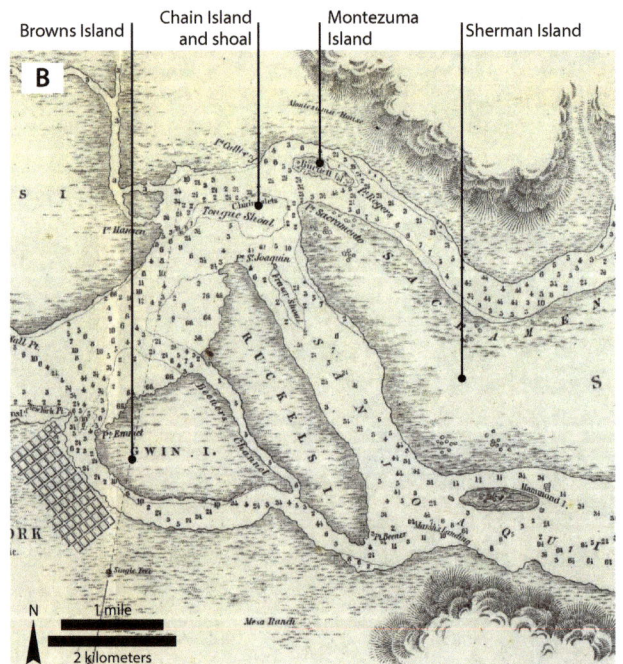

of mining debris were passing into San Francisco Bay, support the depths recorded in earlier sources despite localized changes that are perhaps related to mining debris. Several of these survey maps are shown in Figure 4.12.

Channel width was quite variable longitudinally along a given channel, as well as from channel to channel. The San Joaquin River downstream of the mouth of Old River was the widest channel in the Delta, averaging about a half a mile. Upstream, the San Joaquin branches became much narrower, but were still over 200 feet (65 m) wide. In comparison, Sevenmile Slough, which connects the Sacramento and San Joaquin rivers, was on the order of 100 feet (30 m) wide, while False River, off of the San Joaquin, was well over 500 feet (150 m) wide.

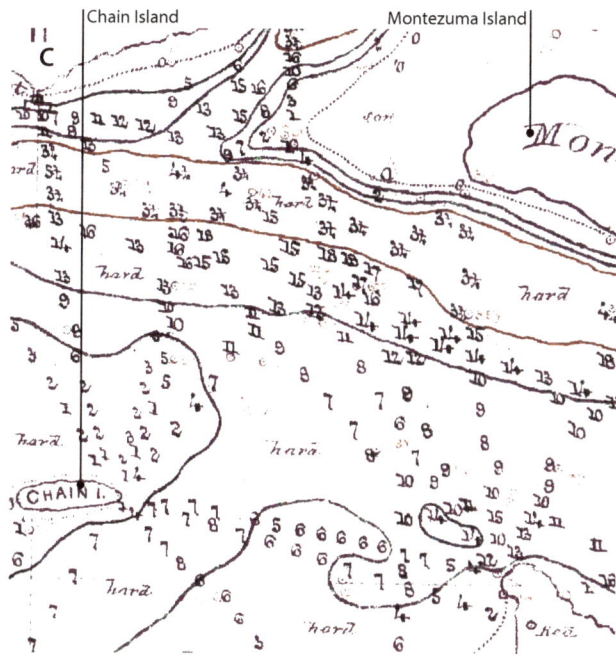

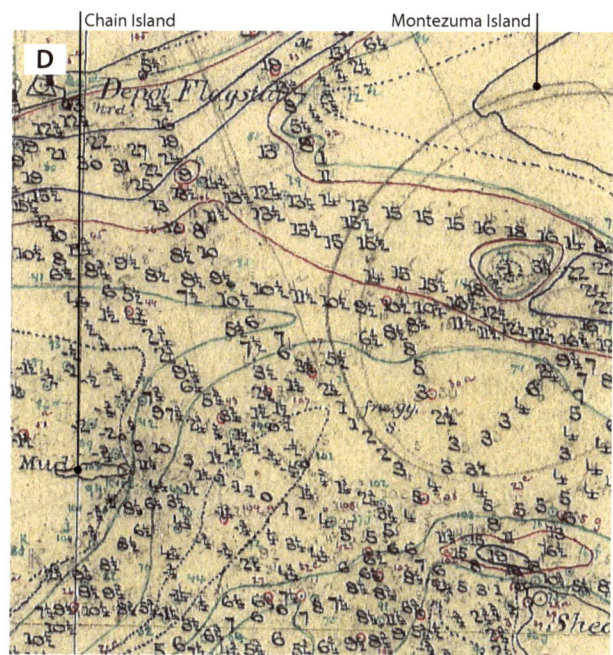

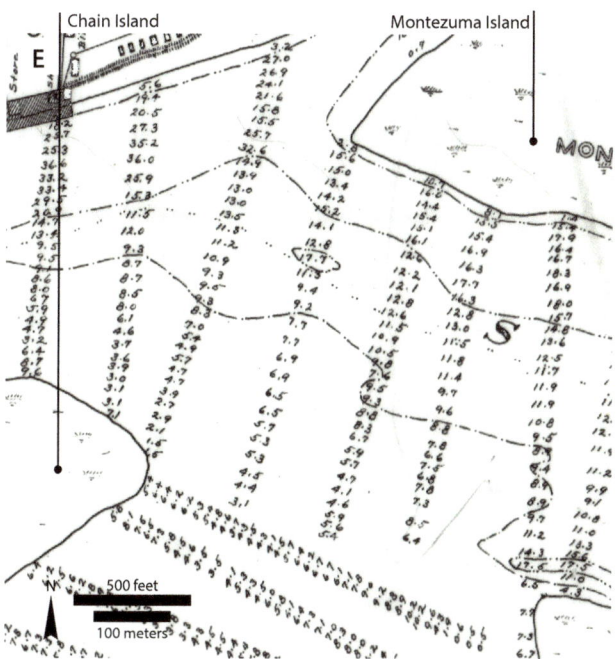

Threemile Slough

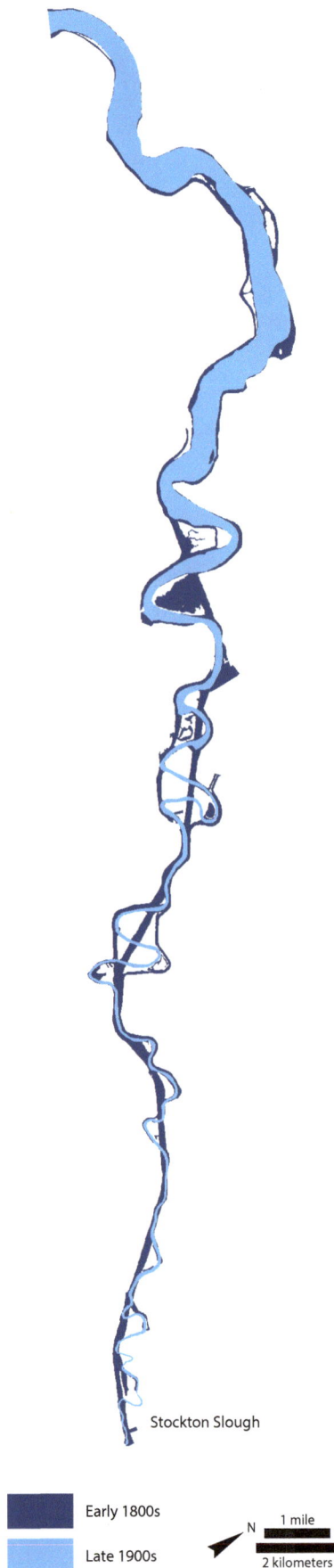

In comparing channel widths to those of today, the mapping conducted shows that many of these large distributary channels of the central Delta (those without natural levees constraining width) are now substantially wider than they were historically. For example, the mainstem San Joaquin River from Threemile Slough to Stockton Slough covered a total of 3,500 acres (1,416 ha), whereas today, the channel encompasses 5,500 acres (2,226 ha), almost a 60% increase (Fig. 4.13). This is partly attributable to increased channel width resulting from levee building practices involving the construction of ditches on the channel-side of levees that later became part of the channel (Fig. 4.14; see Box 2.2). A portion of this increased area is also due to the channel cuts between meander bends.

Most early travelers in the Delta remarked on channel sinuosity, particularly along the San Joaquin (e.g., *The Morning Call* 1894, Kip [1850]1946, Duvall and Rogers 1957; Fig. 4.15). Many diary entries detail the torturous twists and turns, as in Spanish explorer Abella's description: "there are so many twists and windings that at times we circled the compass" (Abella and Cook 1960). An account of a steamboat trip to Stockton explained that the "only method of threading the curves and loops is by running the steamer's nose plump into the tules on this side, which fends her off until she swings around enough to plump her nose into that side" (Smith & Elliot [1879]1979). According to the mapping we conducted, the San Joaquin and the lower reaches of its distributaries were characterized by sinuosity of around 1.6. Sinuosity was lower for the more riverine-dominated channels extending into the central Delta (e.g., Sacramento and Mokelumne rivers). On the other end of the tidal-fluvial and salinity gradient, tidal channels downstream in San Francisco Bay tended to be associated with higher sinuosity, a trend found in other estuaries as well (Garofalo 1980, Grossinger 1995).

However frustrating for those trying to reach a destination, the meander bends served an important role in regulating the ebb and flow of tides. The significance of this was not seriously considered before meander cut after meander cut progressively shortened the distances necessary for boat travel (the distance between Threemile Slough and Stockton was shortened from 35 to 25 mi/56 to 40 km). As a result, tidal flows – and likely salinity – reached farther into the Delta, causing the system to become more homogenous as cross ditches and meander cuts shortened travel distances between points (Fig. 4.16; Enright pers. comm.).

Stockton Slough

Early 1800s

Late 1900s

N

1 mile

2 kilometers

Figure 4.13. Historical and modern channel width comparison. This graphic illustrates the overall wider San Joaquin River channel today between Threemile Slough and Stockton Slough. This increase in width is primarily attributable to ditches made in the process of levee building that have since become a part of the channel. In general, this has added on the order of 350 feet (100 m) to the channel width. Also visible are the many cuts made between meander bends to facilitate travel along the channel. As a result, the channel has become much shorter. The historical channel length was 35 miles (56 km) long in comparison to the contemporary 25 miles (41 km). For most of its length, the river has become wider and straighter.

Figure 4.14. Dredge building levee showing side-ditch. This early 1900s photograph shows a sidedraft clamshell dredge working on a levee, drawing material from a ditch. Over the past century, many of these ditches have become a part of the main channel. (Covello ca. 1900, courtesy of Bank of Stockton Historical Photograph Collection)

In ascending Old River [of San Joaquin] a reasonably straight stretch some 3 miles in length is visible; thereafter the stream is so extremely crooked that straight reaches of even a half mile in length are rare. The steamboat, with her barge, is hardly out of one bend before she is again into another. The river is the most crooked navigable stream that I have ever seen. The river gradually gets narrower, but maintains ample navigable depth.

—US WAR DEPARTMENT 1892

Figure 4.15. Meanders of the San Joaquin are seen in this engraving titled "Night scene on the San Joaquin River – Monte Diablo in the distance" (Hutchings 1862). Rather monotonous vegetation and a lack of large trees along the river is notable, in stark contrast to the tall overhanging boughs that snagged the sails of boats on the Sacramento. Also of interest is a fire in the tules in the distance. (Hutchings 1862)

Figure 4.16. Connecting historically disconnected channels. This map illustrates how the historical channel network (in blue) has been linked up by cross-channels and meander cuts (yellow highlighted areas of the modern channel network, in gray). The cross-channels were created in the process of building cross-levees and established many of the Delta's "tracts," such as Frank's Tract and Drexler Tract. This increased connectivity has altered the routing of tidal water by making travel distances shorter, by and large, between given points.

The localized effects of such alterations are demonstrated in a fascinating 1894 newspaper article discussing tidal dynamics resulting from meander cuts made on the San Joaquin River just below Stockton:

> The San Joaquin River…was simply a long collection of curves, and a steamer had to travel about three miles in a round about manner to make one mile toward its destination. This was, of course, annoying and a great waste of time…
>
> It was not until the river began to fall that it was noticed there was something wrong. It really seemed as if the bottom was coming toward the top… This was puzzling for awhile, and then it was found that in making their calculations for the cuts the engineers had overlooked the effect on the tide.
>
> In the old days, when the river twisted like a snake, the rise and fall of the tide in the bay did not make a difference in the San Joaquin between Stockton and Twenty-one Mile Slough of more than two feet. The reason of this was that the many curves in the stream prevented the water running out as fast as the tide fell. By the time the tide had fallen six feet in the bay the water fell only two feet in the river, and when the tide rose in the bay it caught the flood and the river commenced to rise again. By this natural phenomenon the river was navigable at all hours.
>
> "But now things have changed," said Pilot Arthur Robinson yesterday, "and the water runs through those cuts at low tide as it would out of a tin pan. The tide now falls over three feet at Stockton, and at Twenty-one Mile Slough it falls nearly five feet…

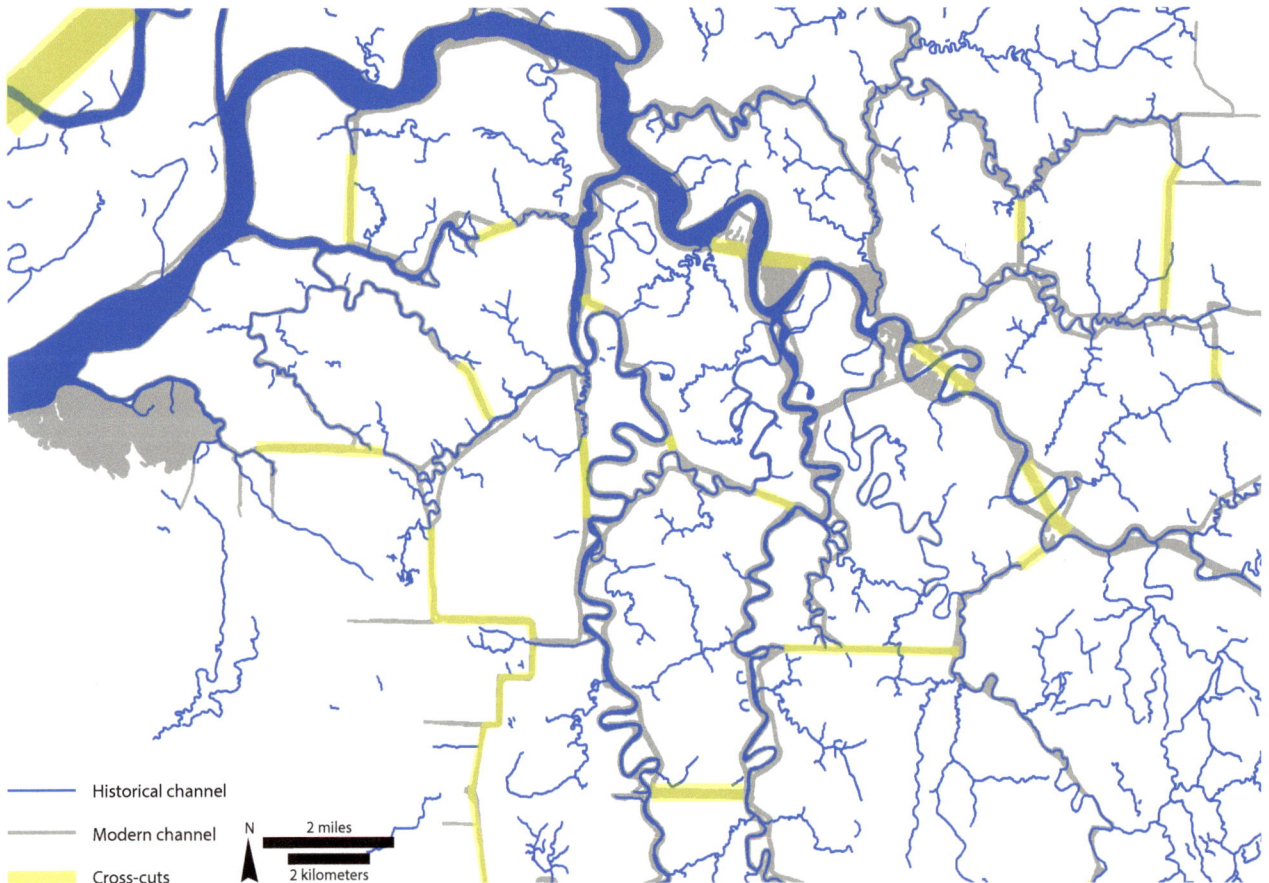

Historical channel
Modern channel
Cross-cuts

"All along the river the effect of the cuts can be seen, as land is uncovered at low tide that has never been before. In some places whole acres are mud flats that used to be covered with water at all times.

"The result of this has caused steamboat pilots trouble all during the summer… In those cuts there is not more than four feet of water at low tide, which is not enough for large steamers. In many spots there is not more than that at high tide." (*The Morning Call* 1894; Fig. 4.17)

LOW-LYING CHANNEL BANKS The height of channel banks followed a gradient from the fluvial to tidal setting: supra-tidal natural levees found upstream transitioned to low banks that lay at general tide levels in the central Delta. These low banks were high in organic content (i.e., composed of peat), owing to the sediment-poor flood waters that reached the central Delta and lack of deposition (Fig. 4.18; see page 134; Thompson 1957, Atwater et al. 1979). According to geographer John Thompson (2006), the central Delta islands' perimeters were "slightly elevated rims." Comparatively, those central Delta channels associated with the Sacramento River had more sediment-rich banks. For instance, low natural levees, or narrow "sediment

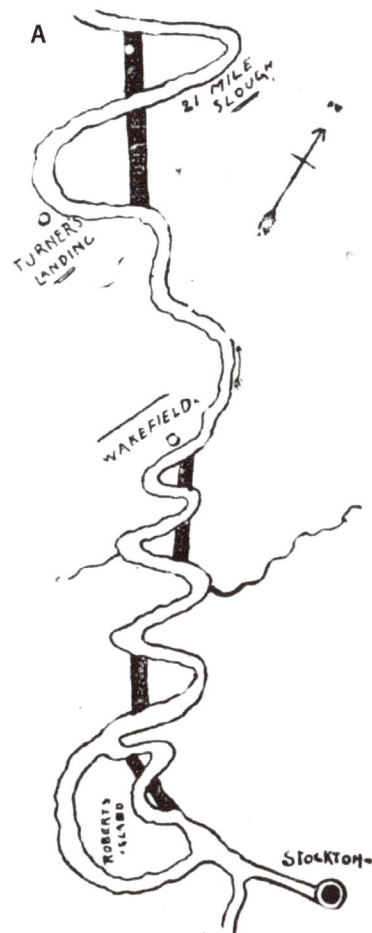

A

Portion of San Joaquin River.
[The heavy lines show how the river has been straightened by "cuts" at the most important points. The two cuts that are shown at the top of the map between Turners Landing and the Twenty-one Mile Slough are the ones that have caused all the trouble, and they are at a point about twelve miles below Stockton in a straight line. The above map is not drawn to scale, but is only intended to give an idea of how the river was straightened.]

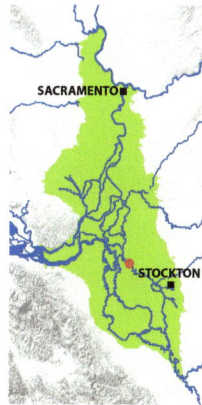

Figure 4.17. Meander cuts on the San Joaquin. In A, an 1894 newspaper graphic accompanying the quoted text illustrates the shortened distance of travel en route to Stockton which inadvertently (according to the accompanying article) caused the tides to flow out more rapidly such that low waters in the channel grounded steamboats. In B, an oblique photograph shows cuts being made by dredges (one visible in the unfinished cut) in the vicinity of Headreach. (A: *The Morning Call* 1894, courtesy of CNDC; B: Covello and Fairchild ca. 1910, courtesy of Bank of Stockton Historical Photograph Collection)

B

Headreach

dredge

land," between 50 and 200 feet (15-61 km) wide were found along the Sacramento side of Sherman Island (Tucker 1879e). The transition to banks characteristic of the central Delta was noted by Tucker (1879e) when he described that the bank height of Middle River "three miles below the cross-levees of Union and Roberts Islands…becomes much less, and the material is partly and generally mostly peat. The banks of the River are but little higher than the adjacent ground farther inland." Field notes describing levee-building along the San Joaquin stated that "there was very little, or no, sediment on that side; it was all soft peat and the hard-pan average 35 feet deep" (Tucker 1879e). On Mandeville Island in the heart of the central Delta, Tucker (1879e) reported that "it is all peat, no sediment." Also, an 1890 profile of a proposed cutoff of the San Joaquin River at Rough and Ready Island illustrates the peaty character of these low channel banks (Fig. 4.19). While reclamation efforts by this time had likely already caused subsidence within the interior marsh plain (particularly noticeable in the section with already high natural levees), the low banks in this illustration are clearly evident. Individuals reclaiming land along the main channels of the central Delta became acutely aware of the peaty banks when the lack of firm footing for artificial levees resulted in their failure.

SHOALS AND FLATS The tidal channel substrate comprised an important element of the Delta ecosystem (Box 4.1). Mud and sand bars, or shoals, were found at the Delta mouth, illustrated in a number of pre-mining-era maps and accounts (TBI 1998). Some have suggested that this historically relatively shallow mouth limited tidal diffusion and salinity intrusion (Fox 1987a). Point bars of mud or sand were also found along the inside of meander bends in some central Delta channels (Atwater and Belknap 1980). Though some of these shallow areas may have been exposed at low tides, most were subtidal. Persistent sand bars exposed at lower water and associated with fluvial processes were found in reaches farther upstream. This picture contrasts with the expansive unvegetated intertidal flats common to San Francisco Bay.

The earliest evidence of shoaling at the Delta mouth comes from explorer de Cañizares' 1775 observations of "sandy bars at their mouths" (De Cañizares et al. 1909). Another Spanish explorer in May 1817 encountered a "shoal" at the mouth of the San Joaquin that could only be crossed at high

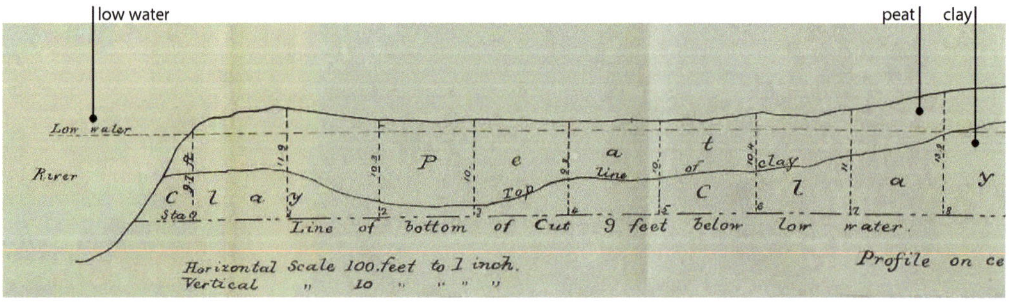

Figure 4.19. An 1890 profile of the San Joaquin River shows the several feet of peat that overlay clay along the slightly elevated channel bank. (Demerill 1890)

A

B

Figure 4.18. Low or absent banks within the central Delta are evident in early photographs. In A, the absence of elevated natural levees along Stockton Slough indicates dominating tidal, as opposed to fluvial, processes. In B, low levees line the San Joaquin River, halfway between its mouth and Stockton (according to the photo's caption). These low levees did not support tall gallery riparian forest like that found on the Sacramento River. The banks do support low scrub, which may be a result of already erected artificial levees. (A: courtesy of The Haggin Museum; B: photo by Gilbert 1905, courtesy of the USGS Photographic Library)

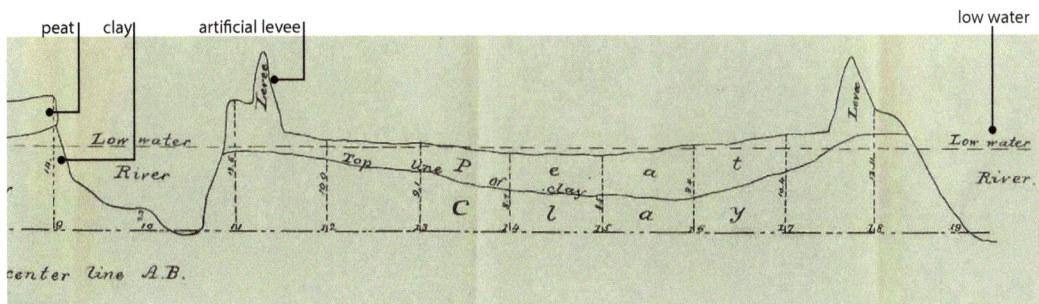

peat   clay   artificial levee   low water

BOX 4.1. EVIDENCE OF MOLLUSKS IN THE DELTA

Mussels and other freshwater bivalves occupy an important position within food webs and riverine function by consuming biomass, producing fine particulate matter, transporting nutrients, affecting substrate composition and stability, and improving water quality (Howard and Cuffey 2006, Howard 2010). Relatively little is known about bivalve species, abundance, and distribution historically in the Delta; in all of California only 400 historical records of freshwater bivalves from 114 collection sites are known, with only a few located within the Delta (Howard 2010). Three genera of freshwater mussels are found in the western US: *Anodanta* spp., *Gonidea angulata*, and *Margaritifera falcata*, with *Anodonta* listed most frequently in historical records (Howard 2010). However, it is likely that mollusks influenced the Delta's tidal channel environments.

Freshwater mollusks were likely abundant in the Delta prior to major modifications of the late 1800s. Archaeological studies of midden sites reveal that these species were a common food source for the numerous tribes of the Delta region (Fagan 2003). An assessment of fish and mollusk remains at several sites in upper and lower river reaches found that mollusks were more abundant in lower reaches (Cook and Heizer 1951). Early anecdotal accounts provide support for the relative abundance of these species. In 1776, freshwater mussel shells delimited the high tide line at the Delta mouth and "shells of snails and turtles" were found east of Byron Hot Springs (Anza and Bolton 1930). Belcher's 1837 expedition up the Sacramento River found "two varieties of mytilus and some univalves" (Belcher et al. 1979), and a report from another expedition four years later stated that "vast quantities of the mussels' shells and acorns" could be seen surrounding the dwellings along the Sacramento River (Wilkes 1845). The report continues, "these Indian had small fishing-nets…they made use of when diving for mussels, and in a short time procured half a bushel of them." This seems to suggest freshwater mussels and clams were bountiful in the Sacramento River (Wilkes 1845). Engineer Grunsky recollected "as many as a dozen clams might be scooped up at one time" from soft mud up to a foot deep on Mormon Slough (Taylor 1969).

tide (Durán and Cook 1960; translated by Chapman 1911 as a "sand bar"). Ringgold (1852) describes "Tongue Shoal" (present day Chain Island) as a "very extensive shoal" at the Delta mouth. In general, these shallow areas apparently were not significant obstacles to navigation, but pilots had to travel with care (Revere 1849, Kip [1850]1946, Ringgold 1852).

While most accounts and surveys indicate that shallow water covered most of these bars at low tide, several records point to the presence of intertidal flats at the Delta mouth. Navy surgeon Duvall noted in June 1846 that the channel was "very much encroached upon by the muddy flats which extend towards it from the dry land for several hundred yards," likely at the base on Sherman Island (Duvall and Rogers 1957). The location and width described here are consistent with the intertidal flats mapped just upstream from Montezuma Island by Ringgold's survey (Fig. 4.20; 1852). These are also shown with greater detail in the 1867 U.S. Coast Survey sheet (Fig. 4.21).

However, there appears to be little evidence of the existence of extensive intertidal flats east of the mouth, like those found in the more saline tidal marshes of San Francisco Bay (TBI 1998). This is likely due to lower

Montezuma
Island | flat

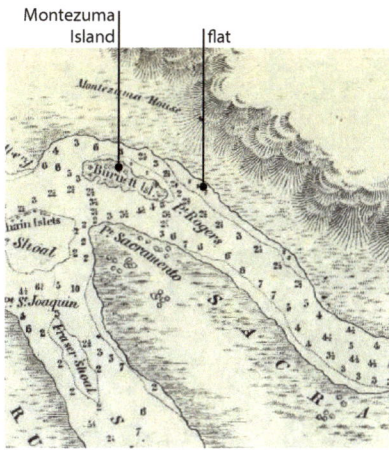

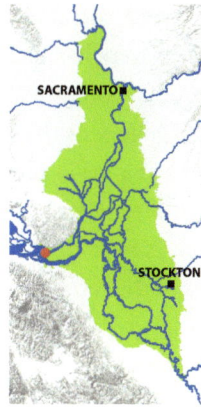

Figure 4.20. Tidal flats at the Delta mouth, 1850. In one of the first hydrographic surveys of the Delta, tidal flats (outlined as stippled lines) are shown bordering the Sacramento River at its mouth and at the tip of Montezuma Island. The absence of soundings in these locations indicate that these areas were bare at low tide. (Ringgold 1850b, courtesy of the David Rumsey Map Collection, Cartography Associates)

Montezuma Island | flat

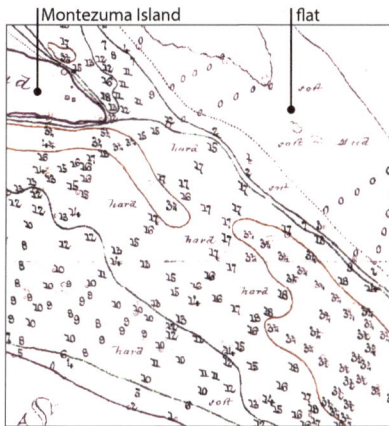

Figure 4.21. Tidal flats at the Delta mouth, 1867. This hydrographic map indicates soft or hard substrate and areas "bare at low tide" (the "0" foot soundings). Although made during the era when hydraulic mining debris was passing through the Delta mouth, the coincidence of flats in this map with those in Ringgold's 1850 survey map suggests that some flats at the Delta mouth were present earlier, though perhaps not as extensively as they were during and after hydraulic mining. Maps are not shown at the same scale. (Cordell 1867, courtesy of the National Oceanic and Atmospheric Administration)

Delta salinity levels (plants could grow lower in the water column) and less erosion from wind and waves (large expanses of open water were uncommon). Small intertidal flats may have been present, though early travelers did not often note them.

One source, Ringgold's 1850 surveys, does however explicitly identify a large tidal flat, or "mud flats" outside of the Delta mouth. These flats are shown in Cache Slough above its mouth and are described in the survey's report: "the waters terminate and waste themselves in swamps and mud flats" (Fig. 4.22; Ringgold 1852). Given the generalized mapping of these channels and accompanying text explaining that the sloughs connecting to Cache Slough were "not navigable except for small boats," it is possible that his expedition did not explore this area extensively and thus made only general observations (Ringgold 1852). For another historical perspective of this area, an account of the first trip of the steamer "New World," describes getting lost and "aground on the shoals of the slough" (Palmer et al. 1881). This later concept of shallow water along channels as opposed to large intertidal flats conveyed by the former Ringgold source may, in fact, be a more accurate description of the Cache Slough vicinity and is consistent with descriptions of bars and shoals elsewhere in the Delta. Layers of mud, generally absent elsewhere in the Delta, were, however, found in cores taken at Lindsay Slough in the 1970s (Atwater pers. comm.), suggesting this area at the base of the Yolo Basin may have been particularly susceptible to the settling of fine sediment.

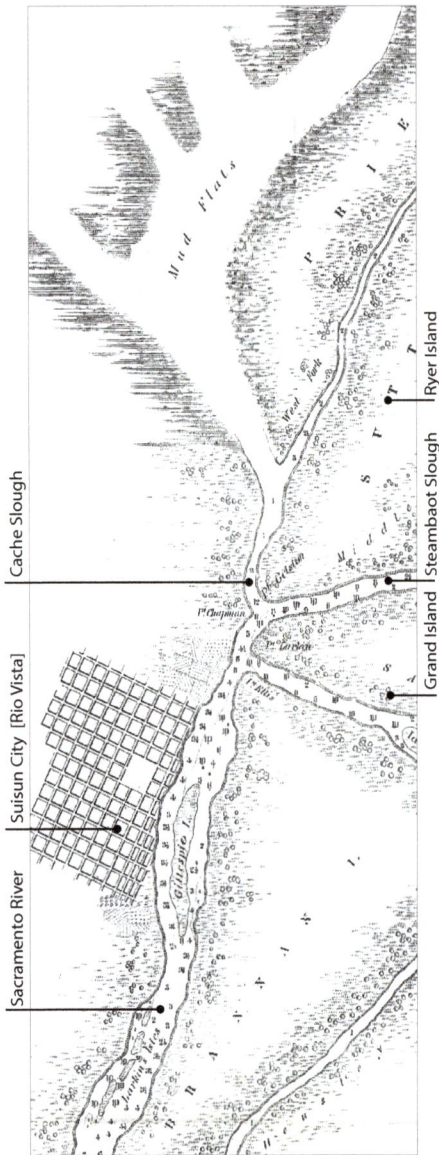

Some of the shoals were associated with small islands within the channel. Several notable islands were located within the Sacramento River channel and occupied by woody vegetation. A pre-hydraulic mining era article in the *Californian* (1847) suggests that Wood Island, opposite Rio Vista, was "densely timbered" (and appropriately named). The dominance of trees on these small in-channel islands contrasted sharply with the more emergent marsh-dominated vegetation of the marsh plain (Fig. 4.23). This difference in vegetation relates to the inorganic sediments deposited on the in-channel islands during floods. Several of the islands, notably Lone Tree and Wood islands, were removed in the early 1900s when the Sacramento was straightened and dredged between the downstream end of Sherman Island and Rio Vista.

Sloughs heading into the tules

Many mid-1800s travelers and residents saw the networks of smaller blind tidal channels that wove intricately across Delta islands in a negative light. Gold miners and others plying the mazelike waterways often would be deceived by invitingly wide channel mouths, traveling for hours before

Figure 4.22. Evidence of mud flats upstream of the Delta mouth. This 1850 chart depicts a large expanse of "mud flats" above the mouth of Cache Slough. Given the focus of this survey on navigation, it is likely that this description overemphasizes the presence of tidal flats and is instead primarily describing an area of shallow channels and small patches of exposed channel margins. (Ringgold 1850a, courtesy of the David Rumsey Map Collection, Cartography Associates)

Figure 4.23. Woody vegetation occupying the islands at the Delta mouth. This 1852 sketch of the entrance to the Sacramento River shows trees occupying the small Montezuma and Chain islands in contrast to the points of marsh near Collinsville and the tip of Sherman Island. This reflects the inorganic sediments accumulated on these small in-channel islands. (Ringgold 1852, courtesy of the David Rumsey Map Collection, Cartography Associates)

discovering that the channels terminated in the tule (Gibbes 1850b). Later, levee builders were annoyed and often humbled by the effort and resources necessary to successfully dam the sloughs (Tucker 1879a).

By the early 1870s, many of these blind tidal channels had been dammed, sluiced, or filled in (Tucker 1879e). Only the largest remained by the early 1900s, and they were usually leveed or had been connected to other sloughs with cross levees and ditches. These cross levees turned land once contiguous to the upland margin into islands. Such areas can today be identified by the term "tract" instead of "island" in their names (e.g., Empire Tract; Thompson 1957). The process of damming sloughs severed the land from tidal flows, although water levels in the islands still responded to the rise and fall of tides. When comparing the historical and modern channel networks, it is evident that although many of the larger distributary channels have persisted (though perhaps wider and straighter), virtually all of the small blind tidal networks are absent (Fig. 4.24). As a result, hydroperiod for most remaining wetlands has been altered, habitat connectivity reduced, and spatial and temporal variability in habitat conditions diminished.

While the primary rivers and the large sloughs that formed the islands of the Delta comprised the network that conveyed water (as well as nutrients and biota) throughout the Delta, the blind tidal channels were central to the exchange between water and wetland. These channels provided significant spatial complexity along the marsh plain, where early observers commented on the remarkable "number and intricacy of the winding sloughs and channels" (U.S. War Department 1856a) and that the tidelands were "cut up by a large number of sloughs" (Beaumont 1859a) that formed a "terraqueous labyrinth" (Bryant [1848]1985). The complexity and variety of channel planform is represented in early maps of the Delta as well as the 1937 aerial photography (Fig. 4.25).

This district is traversed by an interminable net-work of 'slues,' or sheets of shallow water...nearly all of which open broadly and invitingly; but the unwary voyager who trusts to their seeming resemblance to the mouth of either river he wishes to ascend is sure to become involved in labyrinthine mazes, and is not extricated without the exercise of some tact and judgment, the expenditure of a large stock of patience, and peradventure the consumption of all his provisions.

—REVERE 1849

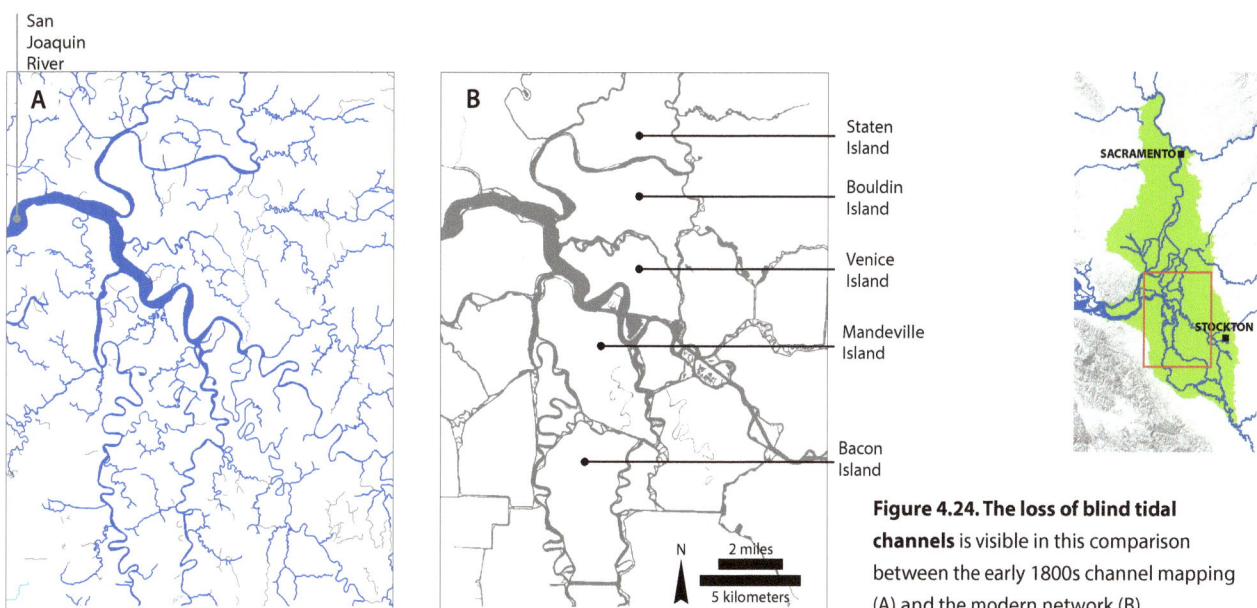

Figure 4.24. The loss of blind tidal channels is visible in this comparison between the early 1800s channel mapping (A) and the modern network (B).

Delta tidal channels were primarily formed and maintained by tidal processes and likely antecedent to the development of dense erosion-resistant emergent vegetation (Crosby 1941, Garofalo 1980, Collins and Grossinger 2004). Though channel positions were generally stable over longer time scales, it does appear that some of the lower order channels may

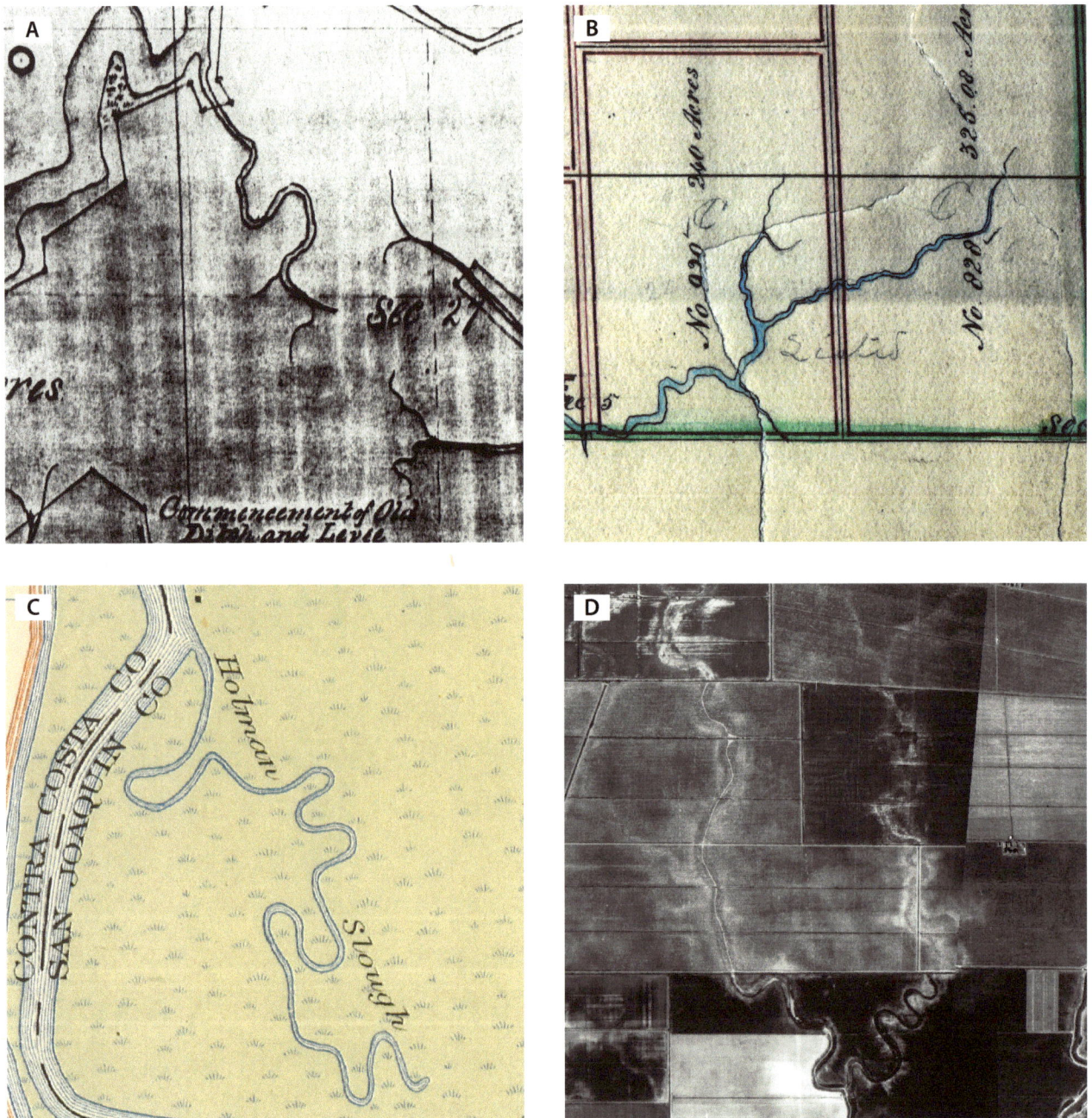

N   1000 feet
     200 meters

Figure 4.25. Low order tidal channels in the central Delta as represented in maps and imagery at the same scale. In A, an 1866 reclamation survey map shows the tidal channels within Venice Island. The end of Sycamore Slough in the eastern Delta is depicted in B. In C, Holman Slough branching from Old River into Bacon Island is shown in the 1913 Woodward Island USGS topographic map. In D, former tidal channels on Ryer Island can be seen in the 1937 aerial photographs; despite several decades of agriculture, the tonal signatures of channel bank mineral deposits and organic peat soils reveal the planform of historical channels. (A: Smith 1866a, courtesy of the California State Lands Commission; B: Unknown ca. 1870, courtesy of the California State Lands Commission; C: USGS 1909-1918; D: USDA 1937-1939)

have occupied former meander bends or routes of main river channels (Fig. 4.26; Atwater 1982).

The tidal channel networks of the central Delta marshlands were akin in form and function to those in the more saline marshes of San Francisco Bay. The branching and quite sinuous channel networks were ubiquitous features of the landscape, delivering water, sediment and nutrients to the marsh plain twice daily. These networks were characterized by decreasing channel width with distance from tidal source and bordered by low to barely perceptible mineral-rich banks, influencing the vegetation patterns of wetland plants, and providing important habitat for aquatic species (Odum et al. 1984, Leopold et al. 1993, Sanderson et al. 2000, Thompson 2006, Hood 2007b). At the same time, there are several marked differences in channel planform between tidal channels in the Delta and San Francisco Bay, including that Delta channels appear to have been wider, less sinuous, and generally associated with lower channel density.

The blind tidal channel networks of central Delta held positions, as described for Petaluma marsh (in San Francisco Bay) by Sanderson et al. (2000), that were either "interior" (branching off a larger blind tidal channel) or "exterior" (branching from a mainstem channel). The main trunks of these networks were quite large; some well over 100 feet (30.5 m) wide at their mouths and navigable for many miles. One of the largest of these, Whiskey Slough of Roberts Island, was described as "navigable for 30 miles above the dam, and it carries a depth of 30 feet of water for ten miles. The dam is at a point where the slough is 42 feet deep and 202 feet wide" (*Pacific Rural Press* 1878). Such channels were often described as maintaining width and depth almost to their heads (Gibbes 1850b, Beaumont 1861b, Tucker 1879e). Smaller channels, less than 30 feet (9.1 m) in width and on the order of 10 feet (3 m) deep, were also numerous. Many appear to have been first order channels. Most tidal channels were apparently subtidal, following from the fact that emergent vegetation can colonize at depths greater than a foot (0.3 m) below MLLW in fresh conditions (Atwater and Hedel 1976). Direct evidence of channel width and depth from surveys and other textual accounts is summarized in Figures 4.27 and 4.28.

Figure 4.26. In the recent past, a blind tidal channel occupied a possible ancient route of the San Joaquin River. The large bends and the wide signature of this channel (illustrated with a dashed blue line in C) suggest this origin, though the actual geomorphic transformation of the channel is unknown. Sediment cores could help address this uncertainty. (A: USGS 1909-1918; B: USDA 1937-1939; C: USDA 2009)

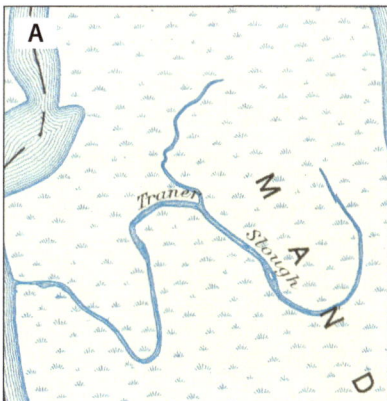

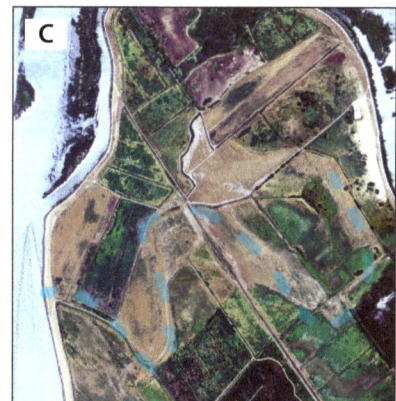

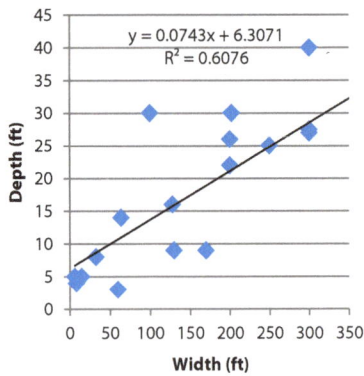

$$y = 0.0743x + 6.3071$$
$$R^2 = 0.6076$$

Figure 4.27. Relationship between channel depth and width drawn primarily from records of sloughs dammed for the purposes of reclamation. Unfortunately, information concerning smaller channels on the order of 20 feet (6 m) wide is scarce since these did not pose significant impediments to reclamation.

The length and sinuosity of channels were also quite variable historically, perhaps related in part to whether channels received significant flood flows from the Sacramento River and whether the surrounding wetland was an island or adjoined the upland margin (Fig. 4.29). Sinuosity was apparently lower overall in comparison to more saline environments of San Francisco Bay (Odum 1988, Pearce and Collins 2004). Most measurements of channel sinuosity fall within the range of 1.5 to 2, but with significant variability in meander belt width and wavelength in relationship to size and location. For example, at the eastern wetland margin (areas not part of islands), the blind tidal channel networks that branched off from the main rivers and sloughs were longer in comparison to channels within island and also appear to have been wider with related large meander belt widths. One of these sloughs, Sycamore Slough, was about 200 feet (60 m) wide at its mouth and seven miles long. Its maximum sinuosity was about 2.4 with an associated wavelength of about 360 feet (110 m) and a meander belt width of about 540 feet (165 m). For comparison, a Bouldin Island channel was found to be approximately 65 feet (20 m) wide at its mouth and 1.8 miles long with a sinuosity of about 1.8. The associated wavelength falls around 170 feet (53 m), with a meander belt width of about 190 feet (59 m).

For many tidal systems, it is useful to think of tidal channel patterns at the larger scale of islands, as it avoids the need to determine contributing marsh area for individual channel networks (Grossinger 1995, Hood 2004, Collins J pers. comm.). This is helpful in the Delta, where dead-end networks on the marsh plain are quite different in character and function from the mainstem channels that delineate the islands (and also the dead end channels at the Delta margins). The following summary is an initial characterization of channel density, a landscape metric challenging to quantify even in contemporary systems. Further research is necessary before explicit use of this information in restoration design. However, even rough estimates of channel density can offer important insights into design considerations, such as the scale of functional landscape units.

Island sizes ranged from around 3,000 acres to over 14,000 (1,210-5,670 ha) for the nine main central Delta islands. The number of mapped primary blind tidal channel networks ranged from four to seven, occurring approximately every three miles (4.8 km) along the main channel (Fig. 4.30). While inconclusive concerning the natural pattern of channel networks in relation to island size, mapping shows that there were few small islands and the islands had relatively few major networks: each of the networks accounted for over 500 acres (200 ha, some over 2,000 ac/810 ha) of its island area. Research has shown non-linear relationships between island area and number of channels, suggesting that greater ecological function may be achieved through the restoration of a single large area over many small areas (Hood 2007b). These observations suggest that, when considering restoration alternatives, the size of the landscape unit should factor into determining expected outcomes.

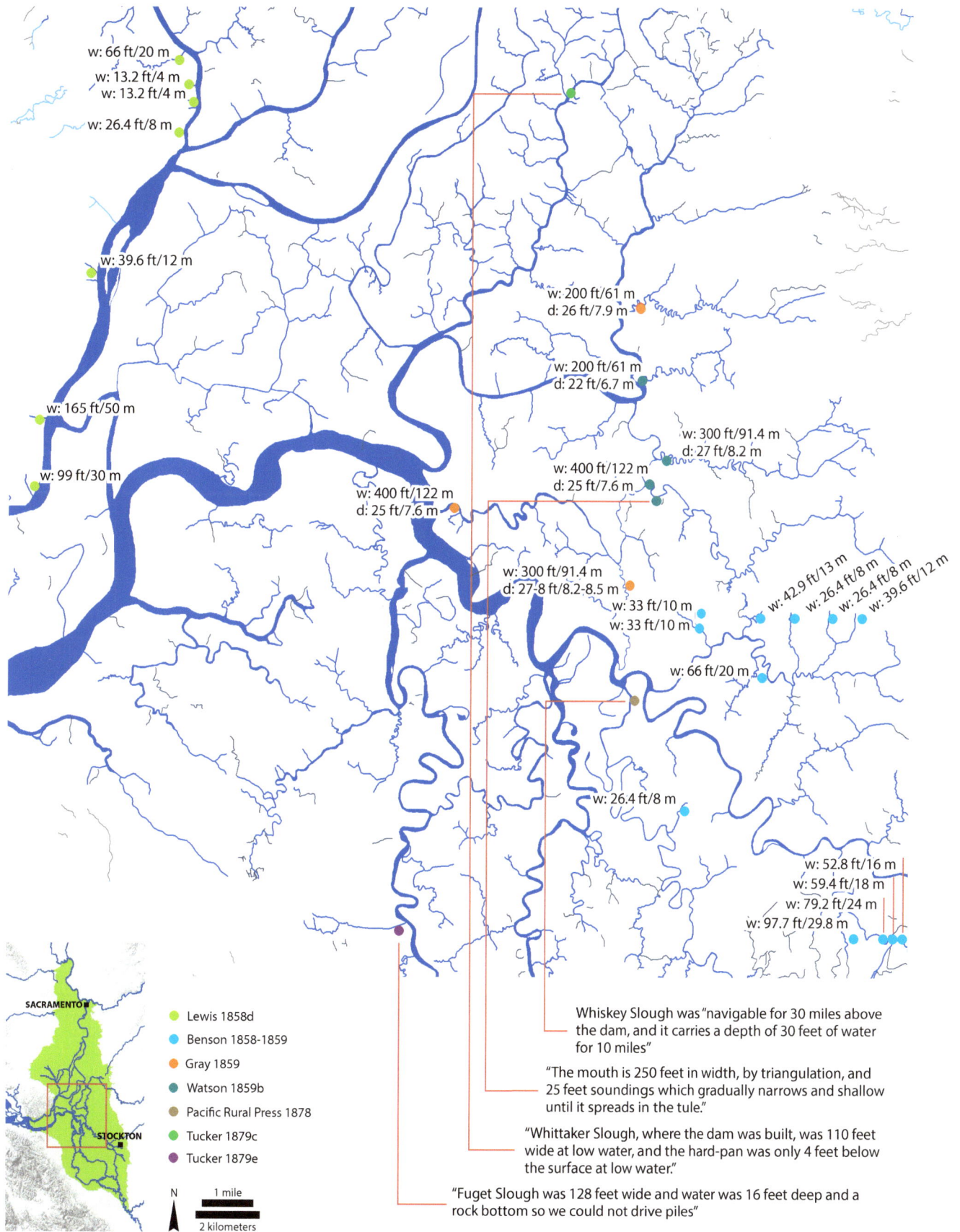

w: 66 ft/20 m
w: 13.2 ft/4 m
w: 13.2 ft/4 m
w: 26.4 ft/8 m

w: 39.6 ft/12 m

w: 200 ft/61 m
d: 26 ft/7.9 m

w: 200 ft/61 m
d: 22 ft/6.7 m

w: 300 ft/91.4 m
d: 27 ft/8.2 m

w: 165 ft/50 m

w: 400 ft/122 m
d: 25 ft/7.6 m

w: 99 ft/30 m

w: 400 ft/122 m
d: 25 ft/7.6 m

w: 300 ft/91.4 m
d: 27-8 ft/8.2-8.5 m

w: 42.9 ft/13 m
w: 26.4 ft/8 m
w: 26.4 ft/8 m
w: 39.6 ft/12 m

w: 33 ft/10 m
w: 33 ft/10 m

w: 66 ft/20 m

w: 26.4 ft/8 m

w: 52.8 ft/16 m
w: 59.4 ft/18 m
w: 79.2 ft/24 m
w: 97.7 ft/29.8 m

SACRAMENTO

Lewis 1858d
Benson 1858-1859
Gray 1859
Watson 1859b
Pacific Rural Press 1878
Tucker 1879c
Tucker 1879e

STOCKTON

N

1 mile
2 kilometers

Whiskey Slough was "navigable for 30 miles above the dam, and it carries a depth of 30 feet of water for 10 miles"

"The mouth is 250 feet in width, by triangulation, and 25 feet soundings which gradually narrows and shallow until it spreads in the tule."

"Whittaker Slough, where the dam was built, was 110 feet wide at low water, and the hard-pan was only 4 feet below the surface at low water."

"Fuget Slough was 128 feet wide and water was 16 feet deep and a rock bottom so we could not drive piles"

Figure 4.28. Spatial distribution of selected observations concerning channel width and depth. This map depicts spatially explicit information concerning channel width derived from various sources. This illustrates the relationship of the size of tidal channel to position.

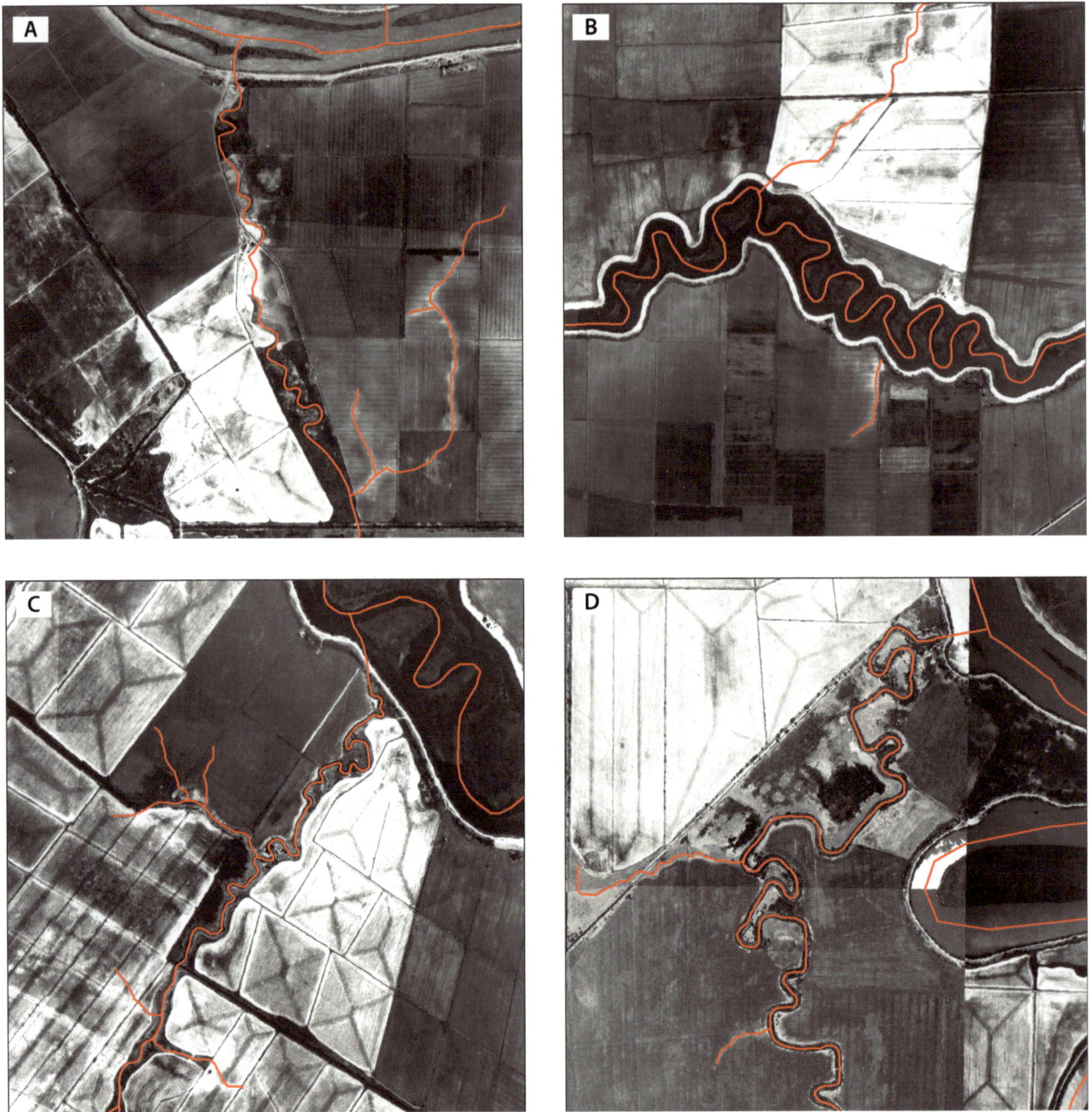

160

Figure 4.29. Examples of blind tidal channel sinuosity. These four networks, all shown at the same scale, represent a range of sizes, planforms, and landscape positions within the central Delta. The aerial photograph in A depicts a Bouldin Island blind tidal channel network. A large and small slough from the eastern Delta are shown in B and C. The channel in D extends westward from Old River on the eastern Contra Costa edge. (USDA 1937-1939)

10

Number of major channel networks

8

6

4

2

0

Venice I

Mandeville I

Bacon I

Webb Tract

Sherman I

Franks Tract

Twitchell I

Bouldin I

Brannan I

-       5,000   10,000   15,000

Island area (acres)

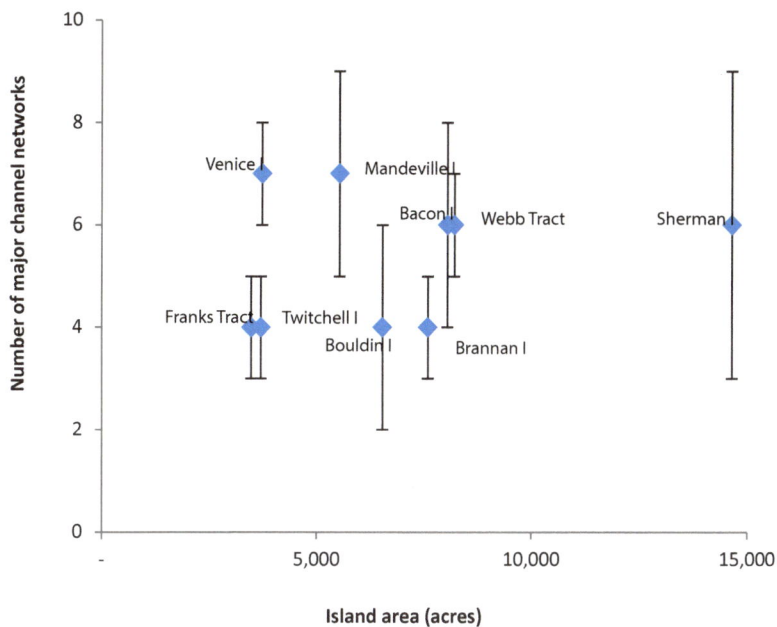

Figure 4.30. The range of channel networks by island area is shown based on mapped historical blind tidal networks on central Delta islands. Uncertainty bars were generated through assessing the number of lower certainty channels that were mapped (e.g., we mapped six high certainty channels on Bacon Island and an additional two low or medium certainty channels, so an error of two was included). Note that for Webb and Franks tracts, the southern boundaries were taken as the connecting line between the two sloughs that have since been connected with a ditch.

Determining a large-scale ratio between island area and number of channel networks is challenging to quantify due to variability in the relative influence of fluvial and tidal drivers. For instance, some islands (e.g., Twitchell Island) appear to have long stretches of over five miles of island edge with no significant intersecting channel networks, while others (e.g., lower Tyler Island along Georgiana Slough) were intersected by relatively large (>50 ft/15 m wide) sloughs occurring less than every mile. A General Land Office (GLO) survey line parallel to and north of Disappointment Slough crossed five sloughs, each between 26 and 42 feet (8-13 m) wide, over a distance of more than three miles (4.8 km; Fig. 4.31).

Although these complex channel networks were ubiquitous features, the mapping produced from the study suggests that tidal channel density was lower than in the brackish and saline marshes of San Francisco Bay. That is, a Delta channel that might have the width and sinuosity of a third or fourth order channel in the Bay might only be a first or second order channel in the Delta. This fits with evidence from studies along salinity gradients in San Francisco Bay and elsewhere that demonstrate decreasing channel sinuosity and density with decreasing salinity (Fig. 4.32; e.g., Garofalo 1980, Odum et al. 1988, Grossinger 1995, Collins and Grossinger 2004, Pearce and Collins 2004). Based on detailed mid-1800s U.S. Coast Survey (USCS) T-sheets, Collins and Grossinger (2004) calculated channel densities around 240 feet per acre (18 km/km^2) in highly saline environments as opposed to around 40 feet per acre (3 km/km^2) in fresher systems. Channel density within freshwater central Delta islands was calculated to be on the order of 12 feet per acre based on the mapping synthesis (0.9 km/km^2; Fig. 4.33). In localized areas, densities were found to be as high as 40 feet per acre (3 km/km^2). These estimates represent a minimum expected tidal channel density for the central Delta. Lower channel densities in the historical Delta are expected given salinity

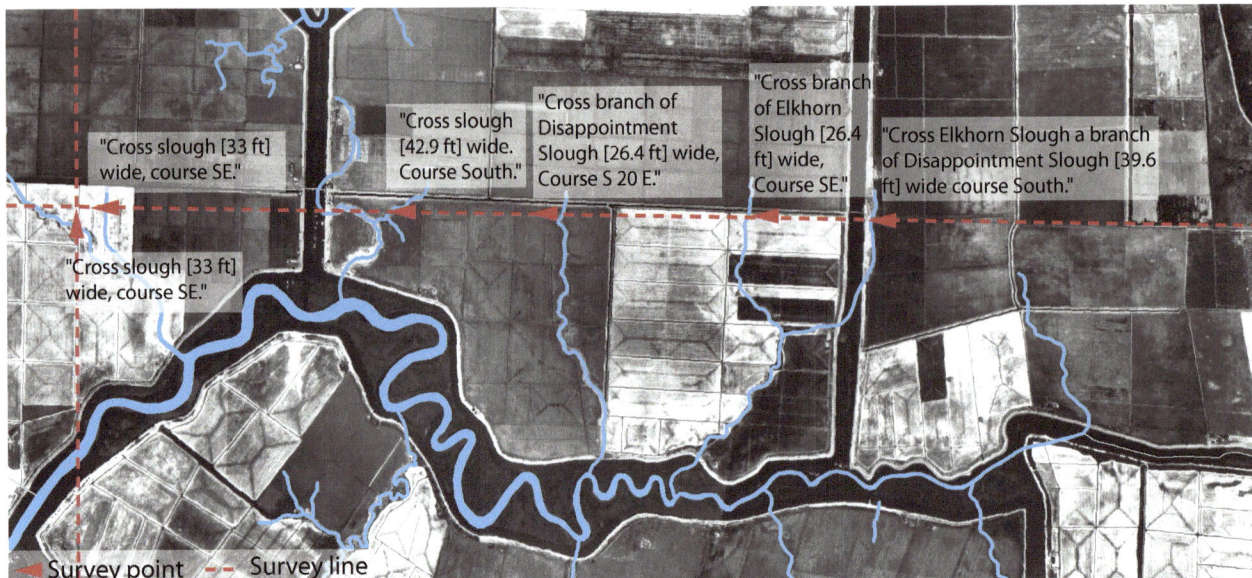

"Cross slough [33 ft] wide, course SE."

"Cross slough [42.9 ft] wide. Course South."

"Cross branch of Disappointment Slough [26.4 ft] wide, Course S 20 E."

"Cross branch of Elkhorn Slough [26.4 ft] wide, Course SE."

"Cross Elkhorn Slough a branch of Disappointment Slough [39.6 ft] wide course South."

"Cross slough [33 ft] wide, course SE."

Survey point — Survey line

N
½ mile
500 meters

SACRAMENTO

STOCKTON

Figure 4.31. The frequency and size of blind tidal channels branching from Disappointment Slough are found in a rare GLO survey that extends within the wetland margin. It should be noted that this survey was conducted in 1878, during a period of extensive levee construction in the Delta. Though records indicate that these tracts of land were not officially reclaimed until the early 1900s, it is likely that some activity was already underway at the time of this survey. (Benson 1878-9; USDA 1937-1939)

Figure 4.32. Channel density by salinity class. The freshest salinity class has significantly lower density than the other classes. There are relatively similar densities across the middle salinity classes. (reprinted with permission from Collins and Grossinger 2004)

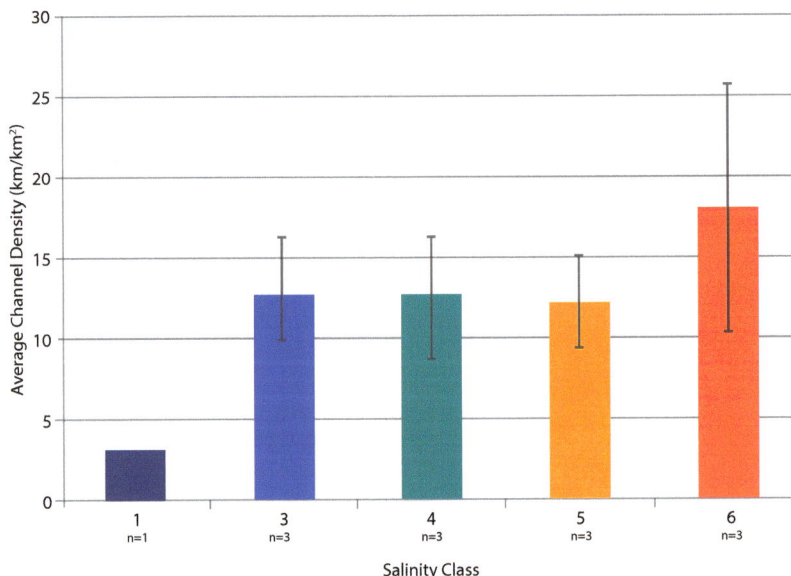

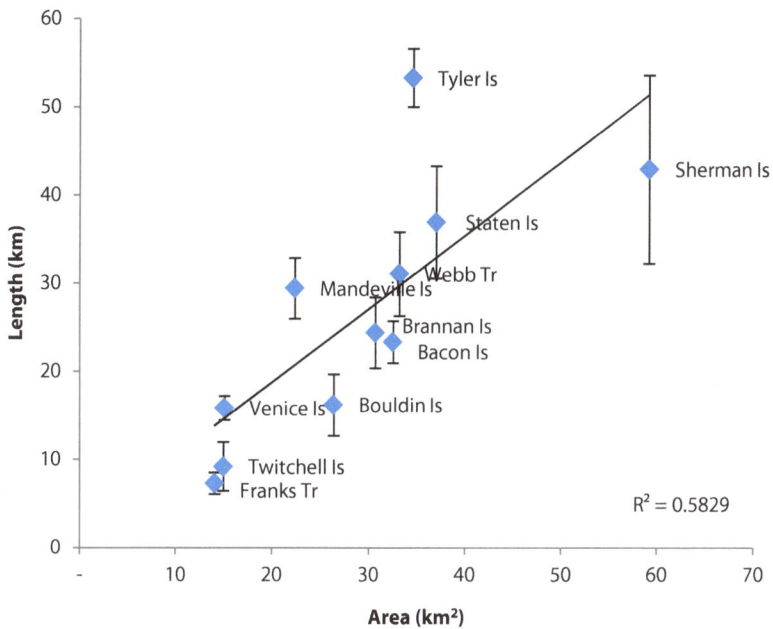

Figure 4.33. Relationship between island area and mapped channel length. Mapped historical blind tidal channels were summed for each of the major central Delta islands. Those that were bounded by substantial natural levees for the majority of their perimeter (such as Roberts, Union, and Grand islands) were excluded. Error bars are derived from the miles of channel mapped that are of "medium" or "low" interpretation certainty, that is, those channels that do not have many lines of evidence supporting them.

tolerances of marsh vegetation (in freshwater conditions, tule is able to grow at elevations below MLLW and would therefore occupy small and shallow channels; Atwater et al. 1979).

A complicating factor for estimating channel density is the uncertainty associated with the level of detail shown in Delta mapping sources. Unlike San Francisco Bay, there is no single comprehensive and detailed data source for historical networks comparable to the U.S. Coast Survey (USCS) T-sheets. To calibrate the level of detail in the mapping from this study with the level of detail found in other sources, we made several comparisons to what similar sources showed in San Francisco Bay. We compared USGS 7.5-minute topographic maps against USCS T-sheets in Suisun Bay and found that the T-sheets usually mapped one channel order more than the topographic maps (i.e., a 3rd order T-sheet channel was shown as a 2nd or 1st order USGS channel). Since most mapping sources we used were closer to USGS scales, one conclusion could be that the mapping does not show the lowest order channels. However, by comparing historical channel density as shown by USGS maps of similar vintages in the brackish Napa River marshlands (Grossinger 2012) and remnant Delta marshes, we found consistently lower densities in the Delta, which supports the conclusion that channel densities were, in fact, lower in the Delta historically than in more brackish and saline marshes downstream.

To further evaluate this relationship using other data, we compared early aerial photography of wetlands in the Delta, Napa, and Alameda at the same scale using imagery from other historical ecology studies (Grossinger 2012, Stanford et al. forthcoming). Signatures of dense networks of narrow sloughs are visible in the reclaimed Napa and Alameda marshes, while fewer comparatively wide and less sinuous channel signatures are seen in the reclaimed Delta (Fig. 4.34). Assuming that this is a comparison of like sources, this would similarly lead to the conclusion that fresher systems (e.g., the Delta) are associated with lower channel density and that channels of the same order are wider. A possible source of complication regarding these observed differences is that the more organic peat soils of the Delta may not show the smaller order channels as clearly as soils in San Francisco Bay. There is also the possibility that different land use histories have caused variable visibility of the historical channel signature. Therefore, it is impossible to rule out the possibility that channels were significantly undermapped.

UPLAND MARGIN The low order central Delta channels could be described as either (1) terminating within the tidal marsh plain (the blind tidal channels discussed in previous sections), (2) connecting to a fluvial channel from the upland, or (3) connecting to a fluvial channel within the non-tidal floodplain wetlands upstream. While most of the blind tidal channels of the central Delta owed their form and function to tidal processes, those that extended into the upland ecotone or non-tidal floodplain were more influenced by fluvial flood flows and depositional processes (Fig. 4.35).

An example of a tidal channel that transitioned to an upland fluvial drainage is Mormon Slough, a main branch of Stockton Slough that intersected the oak woodlands and savannas once common within the vicinity of Stockton (Figures 4.36 and 4.37). The recollections of Carl Grunsky, a prominent engineer in the late 1800s who grew up in Stockton, provide fascinating details. Describing pre-reclamation conditions, Grunsky identified California Street in Stockton as the upper limit of tide on Mormon Slough, marked by a "Rosebush." Just above this point, "a grove of fair-sized oaks grew here within the area that was part of the slough at its high stages" (Taylor 1969). This feature appears to correspond with "Park Is" mapped in the 1850 map of Stockton and shown in Figure 4.38.

In its tidal reach, Mormon Slough was between 100 and 200 feet (30.5-61 m) wide with mud bars exposed at low tide. Grunsky also recalled an interesting tidal phenomenon that can occur when tides meet a channel constriction: "we often watched and even ran from a small-scale tidal bore, perhaps ten to twelve inches high, which would form under certain conditions of wind and rising tide" (Taylor 1969). A pool was apparently used as a fishing spot east of Centre Street, a part of the channel still under the influence of tidal action. The presence of such a feature in this location is indicative of the transitional nature of the channel from fluvial to tidal processes.

The Stockton and Mormon slough complex and French Camp Slough were unique in the extent that their deep tidal channels intersected upland environments, leading to their importance as ports or landings. French Camp was used in the early 1850s as a port along with Stockton before it was apparently blocked by sediment (which may explain differences between French Camp Slough in 1850s-era maps and those of the early 1900s; Tinkham 1880). Other than these networks, most tidal channels

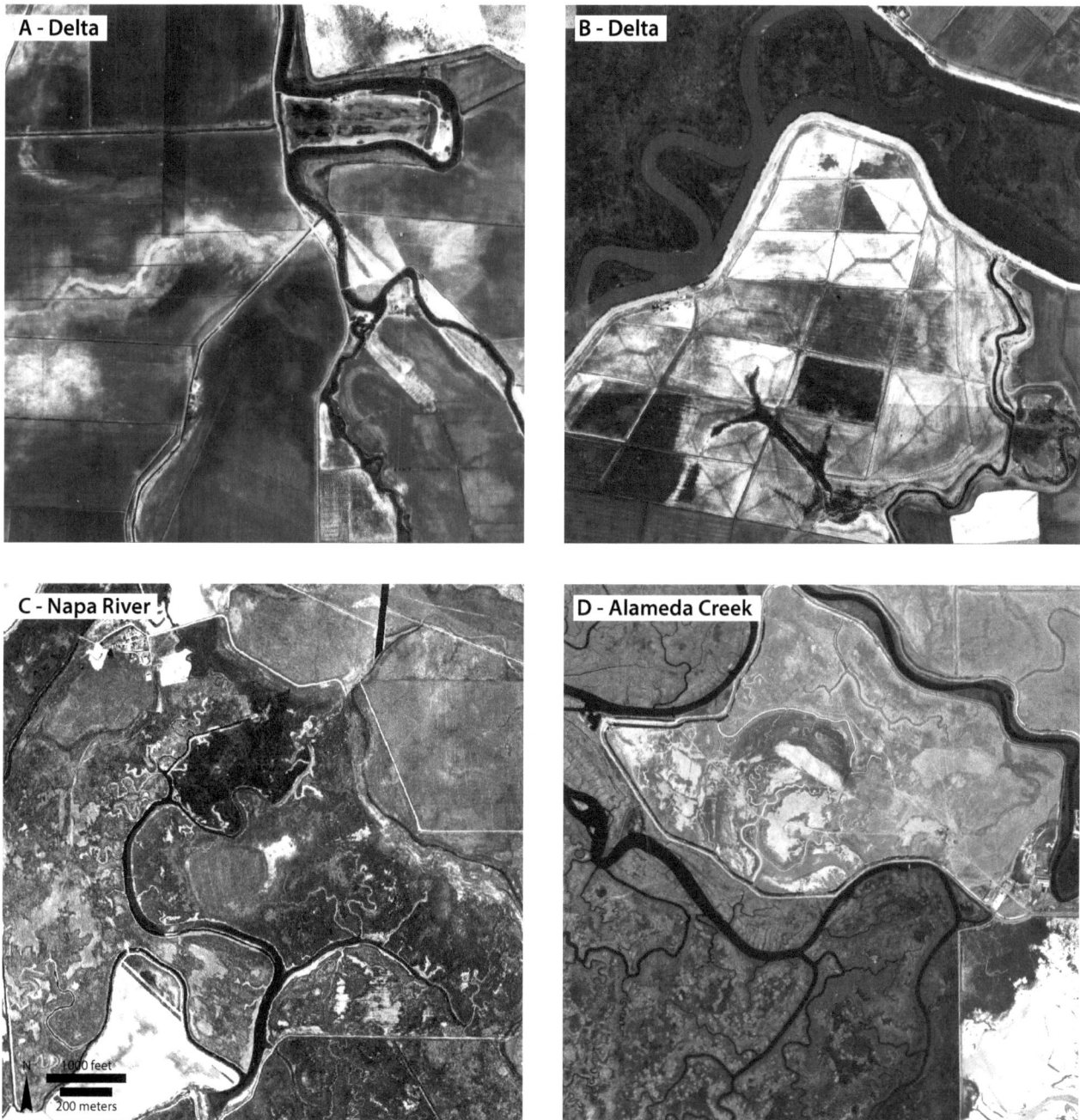

Figure 4.34. Comparing representative channels and remnant signatures of channels in Delta historical aerial photography (A and B) against those in Napa River (C) and Alameda Creek (D) marshes suggests a higher channel density in the more saline systems. However, confounding factors include different land use trajectories and differences in soil types, and it is possible the Delta aerials show less of the historical channel network than do aerials for more saline systems. (A & B: USDA 1937-1939, courtesy of the Map Collection of the Library of UC Davis and the Earth Sciences & Map Library, UC Berkeley; C: USDA 1942, courtesy of Napa County Resource Conservation District and Natural Resources Conservation Service.; D: USDA 1939-1940, courtesy of Earth Sciences & Map Library, UC Berkeley, and the Alameda County Resource Conservation District (ACRCD) and National Resources Conservation Service (NRCS))

Figure 4.35. The Calaveras River at its confluence on the San Joaquin River is shown in this 1905 photograph, likely at low tide. This river had a channel that connected directly to the San Joaquin, despite often becoming dry late in the season above tidal influence. (photo by Gilbert 1905, courtesy of the USGS Photographic Library)

Figure 4.36. Tidal channels of the Stockton harbor with trees shown in the background. The Stockton Channel, Mormon Slough, and McCloud's Lake were the primary subtidal channels of the slough network extending into present-day Stockton. They functioned as tidal sloughs extending into upland habitats that connected to small intermittent streams. (Drawing of Harbor 1852, courtesy of the University of Southern California, on behalf of the USC Special Collection)

Figure 4.37. Mormon Slough is bordered by oaks in this undated ca. 1900 photograph. Unlike most tidal channels, Mormon Slough intersected higher land that supported oak woodlands. (Covello ca. 1900, courtesy of Bank of Stockton Historical Photograph Collection)

Stockton Channel | Mormon Channel | McClouds Lake | California St. | Park Island

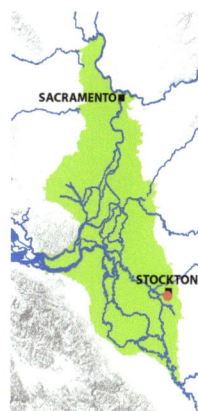

Figure 4.38. Trees within the Mormon Slough channel are symbolized in this 1850 map of Stockton on "Park Island" just upstream of California Street. This street was identified by engineer Carl Grunsky, who grew up in Stockton, as the upper limit of tide. (Brown 1850, courtesy of the Earth Sciences & Map Library, UC Berkeley)

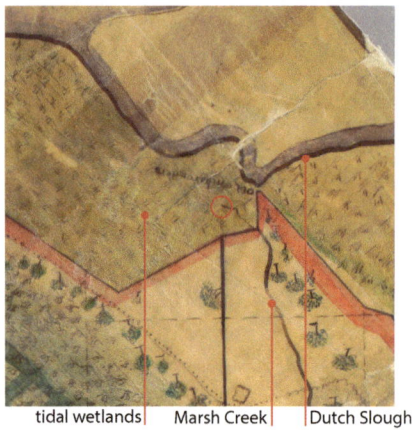

tidal wetlands | Marsh Creek | Dutch Slough

Figure 4.39. Marsh Creek is shown entering the tidal wetlands (in red circle) in this 1853 map, but it does not appear to have connected directly to Dutch Slough. (Whitcher 1853b, courtesy of The Bancroft Library, UC Berkeley)

Marsh – A frequently or continually inundated wetland characterized by emergent herbaceous vegetation adapted to saturated soil conditions.

Swamp – Wetland dominated by trees or shrubs.

—MITSCH AND GOSSELINK 2007

along the eastern wetland boundary south of the Mokelumne River were non-navigable well before the edge of the tule marsh and did not connect with an upland fluvial channel. For instance, testimony from the Mokelumne land case states that such sloughs "headed" in the tule over a mile inside of the wetland edge (Beaumont 1859a). Witness William Watson asserted that he found Sargent Slough's "head in a dense tule with no connection with any landstream" (Watson 1859b).

Most small ephemeral streams draining to the tule lands had insufficient flows to establish direct connections to the tidal wetlands. Instead, they dissipated in the seasonal wetlands that bordered the Delta. This pattern of discontinuous streams spreading across their alluvial fans was common throughout the Bay Area region historically (Grossinger et al. 2008, Grossinger 2012). Marsh Creek may have been an exception, as illustrated in a land grant plat map and accompanying description that the creek "wastes in the Tulare" (Whitcher 1853a, Stanford et al. 2011). However, although the channel seems to have reached the tidal wetland boundary, it does not appear to have connected directly to Dutch Slough via a tidal channel (Fig. 4.39; Whitcher 1853a, Stanford et al. 2011).

COMPLEXITY WITHIN THE WETLAND PLAIN

The Delta's wetlands were among some of the most productive and diverse of the San Francisco Estuary. Studies of this and other estuaries demonstrate that habitat complexity and species diversity is greater at the fresh end of the gradient between saline and freshwater tidal wetlands (Atwater et al. 1979, Odum 1988). Species assemblages within the Delta's wetlands were distinct from the brackish marshes of Suisun as well as from the saline marshes of San Francisco Bay (Atwater and Hedel 1976, Atwater et al. 1979). Positioned at the interface between tidal and riverine systems, Delta freshwater wetland vegetation communities consisted of a combination of species found in the brackish marshes of Suisun and riverine environments of the Sacramento and San Joaquin. Plant communities, landscape position, and patch sizes, rather than any particular endemic species, made the tidal wetlands of the central Delta landscape unique (Atwater 1980). Within the central Delta islands, arrowhead (*Sagittaria* spp.), water-plantain (*Alisma* spp.), rushes (*Juncus* spp.), sedges (*Carex* spp.), cattails (*Typha* spp.), reeds (*Phragmites australis*), as well as low woody plants (predominantly willows, *Salix* spp.), grew alongside the ubiquitous tule (*Schoenoplectus* spp.) and formed freshwater wetlands that included a number of vertical layers (Fig. 4.40; Bryant [1848]1985, *Pacific Rural Press* 1871, West 1977, Atwater 1980, Hickson and Keeler-Wolf 2007, Mason n.d.). They appear to have been more diverse than the wetland interiors of tule-dominated flood basins in the north Delta (*Pacific Rural Press* 1871).

Environmental gradients (e.g., soils, topography, hydroperiod, nutrient availability) affect large-scale species composition differences and vegetation patterns across the marsh plain (TBI 1998, Collins and Grossinger 2004). Biological interactions with species such as beaver also

Figure 4.40. Tule intermixed with willow, lady fern, dogwood, bur reed, and *Sagittaria* in a recent photograph on a non-leveed island north of Franks Tract. (photo by Daniel Burmester, June 20, 2006)

would have affected plant species distribution and abundance (Box 4.2). Because of its freshwater character, the Delta accumulated deep peat soils; soils derived from the productive wetland vegetation (Box 4.3). The freshwater tidal wetlands began accreting organic matter around 6,700 cal year BP when sea levels rose to inundate what is now the central Delta, accumulating at rates between 0.03 and 0.49 cm/yr (0.01-0.19 in/yr; Drexler et al. 2009a). Peat soils, prior to reclamation, were as deep as 65 feet (20 m), in the western Delta (Atwater et al. 1979). Unique to the central Delta, these deep and highly organic soils transitioned to more inorganic clays and loams to the north and south. This transition owes itself to the greater contribution of alluvial inorganic material brought by floods (Fig. 4.42; Reed 1890, Cosby 1941). Though central Delta tidal marsh elevation levels rose primarily as a result of organic matter accumulation, inorganic sediment inputs were important to the marsh development (Drexler et al. 2009a). Today, the peat soils support a highly productive agricultural industry, but the changing land use has meant as much as 26 feet (8 m) of subsidence due to peat oxidation and compaction, the loss in some locations of over 3,000 years worth of accretion (Drexler et al. 2009a,b).

Also related to the relative position of the central Delta at the tidal end of the fluvial-tidal gradient, the island topography was visually quite flat, contrasting with the more riverine-influenced landforms of the northern and southern Delta. Perhaps the most visually striking was the contrast between wetland species lining the low channel banks of the central Delta and the dense riparian forests occupying comparatively broad supra-tidal natural levees just upstream (Thompson 2006). However, slight topographic

Again the soil of the Sacramento islands is to a great extent clay and a late deposit of fine yellow sediment, underlying which is a strata of almost pure decomposed vegetable matter. On the other hand the surface soil of the San Joaquin islands has scarcely any other material in its composition than this decomposed vegetable matter.

—*PACIFIC RURAL PRESS* 1871

BOX 4.2. THE BEAVER FACTOR

As natural ecosystem engineers, beaver can significantly affect hydrology and vegetation patterns (Fig. 4.41). In the early 1800s, the Delta was known for its beaver population. Many people, including men employed by the Hudson's Bay Company and by John Sutter, sought their fortunes by trapping along the Delta waterways (Maloney and Work 1943). The slow-moving water and abundance of tule and willow made the Delta a prime location for the golden beaver (*Castor canadensis*; Skinner 1962). While willows are often thought of as beavers' primary food source, reports suggest that at least in the Delta, tules and other wetland species such as water lilies were their main diet (Grinnell 1937, Tappe 1942).

Since tidal sloughs maintained water levels year-round and did not freeze, Delta beavers may not have built dams as often as beaver in other riverine systems (Grinnell 1937, Bingham 1996). However, some dam-building was likely: a California Department of Fish and

Figure 4.41. Beaver eating cattail tubers along a waterway.
(Grinnell et al. 1937, copyright 1965 by the Regents of the University of California. Reprinted by permission of UC Press)

Game post-reclamation survey of beaver near Prospect Slough found beaver dams (Bryant 1915). Beaver burrowed in banks or built houses. Given the absence of substantial natural levees in the central Delta, beaver likely built houses to escape the tides (Tappe 1942). Construction material included tule, evident in a 1915 description of one hut along Prospect Slough: "One was but 20 yards away from the main slough. This one was well plastered over with mud and tules" (Bryant 1915). An earlier account attests that "the beavers, like true philosophers, have accommodated themselves to circumstances, and build their habitations of rushes, curiously and skillfully interwoven" (Farnham 1857). While numerous in the historical Delta, there is some suggestion that beaver may have become more common post-reclamation (at least in the central Delta region) with the new opportunities for building dry lodges within the recently erected artificial levees (Tappe 1942).

Frequent mentions of "beaver cuts" in the historical record suggest that beavers created their own channels, some of which may have then been captured by tidal processes. If this were the case, tide water traveling up channels encountered more avenues by which to access the marsh than without the beaver modifications, affecting tidal excursion and inundation frequencies and depths. We found accounts of these cuts on Bradford Tract, Jersey Island, Bouldin Island, and Randall Island (Wright ca. 1850a, Tucker 1879a). For instance, reclamation on Bouldin Island required the damming of three beaver cuts, "being from four to seven feet deep" (Beaumont 1861). An early Fish and Game Bulletin described this habit of the beaver: "When a supply of food is situated at a distance from deep water, beavers may dig canals leading to the supply, providing the intervening land is low-lying, level, and easily dug" (Tappe 1942). A long-time resident of Grizzly Island in Suisun agreed that in making the cuts the beaver were "probably going from pond to pond or from a slough to a pond" (Soares pers. comm.). In a more recent study along Sand Mound Slough, Atwater (1980) found narrow channels only about one foot (0.3 m) wide and attributed their presence to beaver. Beaver cuts were apparently quite distinctive from natural sloughs as they were narrower (approximately 18 inches by one account) and straight" (Soares pers. comm.). Though it is clear that beaver inhabited the Delta and impacted its hydrology and habitats, the exact nature and degree remains uncertain.

BOX 4.3. FLOATING ISLANDS

One product of the substantial accumulation of organic material in the Delta was the phenomenon of large areas of vegetable matter breaking loose during large flood events. Some land apparently did not separate entirely, but rather rose and fell with the water. This lighter material floated and was thus referred to as "float land" or "floating islands" (Houghton 1862, Tucker 1879f). They were associated with the region of deeper peat in the central Delta and were sometimes acres in extent and over 10 feet in depth (Hilgard 1884).

Given that they were apparently quite large, they could bear substantial weight. In the historical record, floating islands are discussed with reference to the refuge they provided to livestock during floods (Houghton 1862, Hilgard 1884). For example, records state that floating islands protected all of the livestock during a flood on Venice Island in 1862 (Tucker 1879b).

It is unclear whether the floating islands were a natural phenomenon or occurred because of disturbances such as levee building; no documentation was found prior to reclamation. One 1862 report attributed their presence to vegetation that "had overgrown sloughs and small lakes," which was then separated by rising water from the substrate (Houghton 1862). This was similarly described in a later report: "so rapidly did the rank swamp-land growth add more material to its edges, that not infrequently the peat was formed without contact with the subsoils" (Rose et al. 1895). Documentation from 1879 of "lumps of tule turf" floating at the Delta mouth is made in reference to change in the shoreline at the Delta mouth. This report states that the tule land was "either torn away by the water or cut away by dike-builders" (USCGS 1881). The presence of large floating mats of vegetable matter in the Delta during the early period of reclamation speaks to the Delta's capacity to rapidly accumulate organic material.

differences influenced species assemblages and habitat mosaics. In addition, the absence of natural levees and the multitude of sloughs facilitated the passage of relatively slow-moving floodwaters through the central Delta. Consequently, tidal dynamics primarily controlled hydroperiod, and floods were likely less of a factor than elsewhere in the Delta.

Drawing from the historical record, the following sections convey a sense of the central Delta wetland vegetation communities of the early 1800s. Botanical research performed by Herbert Mason (n.d., 1957) and Brian Atwater (1976, 1980) provides more detailed information concerning native Delta plant communities.

The ever-present tule

Today, it is common to conceptualize the Delta of 1800 as one vast expanse of tule. Certainly, some historical accounts and maps give that impression: the terms tule, tulare, bulrush, and rushes are ubiquitous descriptors in early accounts and in maps. In two of the earliest written accounts of the Delta, Spanish explorer Ramón Abella reported that the banks of what was likely False River were "covered with nothing but tule, and so high that one sees nothing but sky, water, and tule" (Abella and Cook 1960) and an explorer on a different expedition concluded that the branches of the San Joaquin "have no trees" (Sal and Cook 1960). An 1878 sketch by engineer

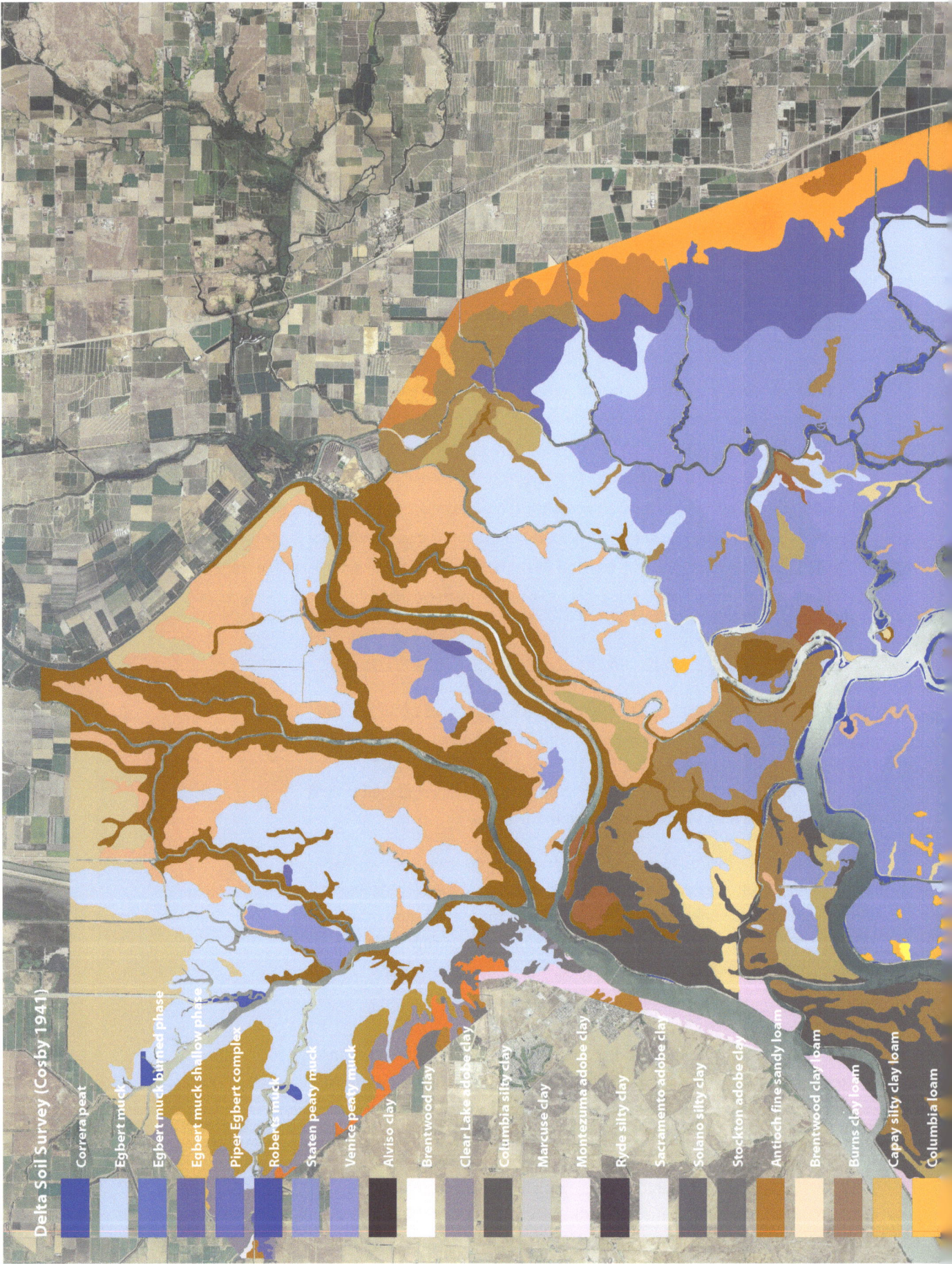

Delta Soil Survey (Cosby 1941)

Correra peat
Egbert muck
Egbert muck burned phase
Egbert muck shallow phase
Piper Egbert complex
Roberts muck
Staten peaty muck
Venice peaty muck
Alviso clay
Brentwood clay
Clear Lake adobe clay
Columbia silty clay
Marcuse clay
Montezuma adobe clay
Ryde silty clay
Sacramento adobe clay
Solano silty clay
Stockton adobe clay
Antioch fine sandy loam
Brentwood clay loam
Burns clay loam
Capay silty clay loam
Columbia loam

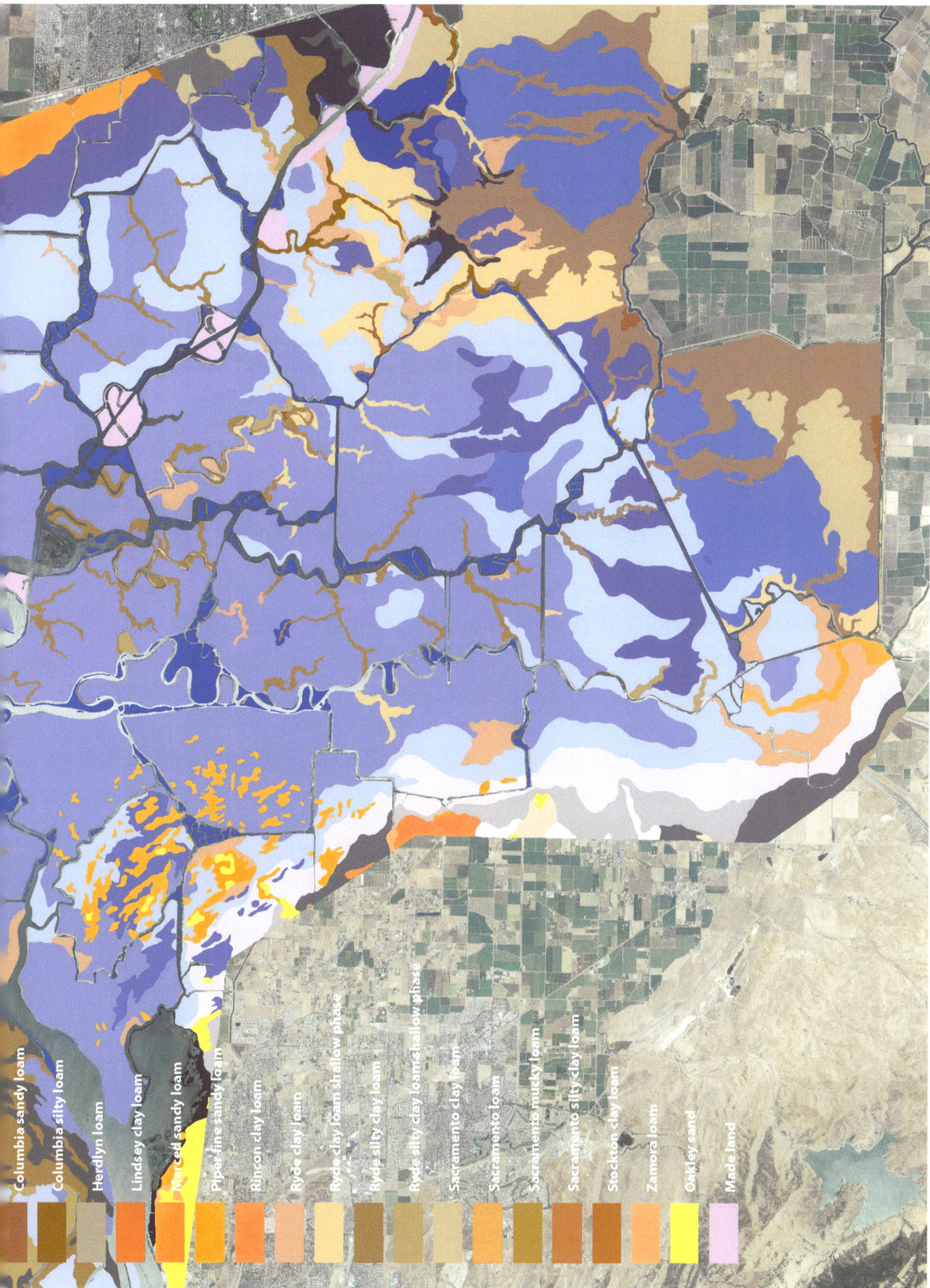

Figure 4.42. The transition from organic peat and muck soils (blues) in the central Delta to clays and clay loams (grays and browns) at the perimeter is visible in the 1941 soil survey of the Delta (Cosby 1941). The map was made based on a digitized version of the 1941 map. (baselayer: USDA 2009)

Columbia sandy loam
Columbia silty loam
Herdlyn loam
Lindsey clay loam
Merced sandy loam
Piper fine sandy loam
Rincon clay loam
Ryde clay loam
Ryde clay loam shallow phase
Ryde silty clay loam
Ryde silty clay loam shallow phase
Sacramento clay loam
Sacramento loam
Sacramento mucky loam
Sacramento silty clay loam
Stockton clay loam
Zamora loam
Oakley sand
Made land

Grunsky, shown in Figure 4.43, offers a similar perspective. The lack of sturdy woody vegetation affected how boats moved up the river to Stockton at ebb tide with no wind.

> In the old days, when it was necessary to make a line fast on shore to heave on, and there were no trees or brush to make fast to but only tule were at hand, we used to take a large armful of tule, and with a long end take a round turn and hitch, then repeat the same with a stake driver to which to make fast the end. (Leale 1939)

While tule clearly made up a substantial portion of the vegetation cover, many other accounts reveal the species complexity.

Exploration of complexity within the Delta's freshwater emergent wetlands begins with the word "tule," which was often used to refer to a range of species. This multipurpose and commonly-used term is often used broadly to encompass the bulrush species found in the Delta (and is used as such in this report), including but not limited to hardstem bulrush (*Schoenoplectus acutus*), California bulrush (*S. californicus*), and, probably less common in the Delta, Olney's bulrush (*S. americanus*). However, the term was also used to refer to all emergent wetland species that inhabited the Delta and was even applied more generally to the lands that were frequently inundated within the Delta (USDA 1874). In one of the first botanical surveys as part of the U.S. Exploring Expedition in 1841, the dominant species at the river mouth is identified as *S. acutus* (U.S. War Department 1856b). Though not in the Delta, botanist Willis Jepson cited O. B. Cromwell as describing three species of tule in Suisun Marsh: the Bull Tule (*B. robustus*) was "tall round tule growing only in soft mud," Common Tule (*S. acutus*), and one referred to as Nut Tule for "the shape of its flower clusters" (*S. californicus*; Jepson 1904).

While myriad historical sources confirm that tule dominated the marsh plain vegetation communities, views of monotonous stands may be somewhat distorted considering the non-specific nature of the term and simple fact that so few individuals actually ventured deep into these wetlands and even fewer wrote about it. Many observations were made from the relatively limited vantage-point of the deck of a sailboat. Consequently, one must look for additional clues in the historical record, along with more recent botanical and paleoecological studies in the Delta, to gain a deeper understanding of the ecological diversity actually present in the historical landscape.

Evidence of "grass"

One aspect of Delta wetland vegetation diversity centers around a potentially taxonomically challenging set of terms used in the historical record: flag grass, flags, coarse grass, long grass, swamp grass, and switch grass. Some historical accounts distinguish tule from grasses or flags, as in this description of the islands being "well stored with long flag grass, and rushes of great size" (Belcher 1843:129), or "covered with swamp grass and tule" (*Sacramento Daily Union* 1862). Along with tule, testimony for the Los Medanos land grant notes that "some grass or tule grass grew on these

The passage up the San Joaquin was a dreary one. The river for the greater portion of the way winds like a tape worm, through low marshy ground, where the tules, (or bull rushes) grow to an enormous height, not allowing us to see out, only by climbing the rigging

—MCCOLLUM [1850]1960

Figure 4.43. Emergent vegetation along the banks of the San Joaquin River. A Grunsky sketch (A) entitled "Delta," shows a line of surveyors among tule and several in boats along the San Joaquin River within the central Delta. A 1905 photograph (B) of USGS surveys in the Delta gives a close-up view of tule dominated vegetation. (A: Grunsky ca. 1878, courtesy of The Bancroft Library, UC Berkeley; B: USGS 1905, courtesy of the Center for Sacramento History, Hubert F. Rogers Collection, 2006/028/115)

lands, with some swamp clover – other kinds of grass, don't know their names" (Brown 1865). Another witness in this same land case calls the grass "dagger grass" and states there are "a good many species of grass" (Clark 1865). In other testimony for the Sutter land case, tule was "mixed with short grass called tule grass" and the vegetation was described as only a foot high (Keseberg 1860). In some cases, the terms are used interchangeably: "to this coarse grass the Indians gave the name of 'tule'" (State Agricultural Society 1872) or "vigorous growth of reeds, of the variety known as 'tules'" (Taylor 1969).

Grass or flags may also refer to the common reed (*Phragmites australis*), cattail (*Typha* spp.), or perhaps sedge species (*Carex* spp.), species which are included in early botanist records and found in remnant wetlands within the Delta today, primarily in the central Delta (Jepson 1901, Atwater 1980). Several studies of peat cores in the central Delta region over the past century have revealed that the common reed was an important component of the central Delta's freshwater wetlands, though relative abundance over time and space is more challenging to determine (Atwater 1980). Soil surveyor Stanley Cosby (1941) reported that the Correra peat of the central Delta region was composed of two layers of vegetable matter – a relatively thin layer composed primarily of tule underlain by a much thicker layer composed mostly of the common reed, though the shift toward tule appears to have occurred prior to anthropogenic changes initiated in the early 1800s. James West (1977), however, found no presence of this lower layer of reed material in his Delta samples. Most recently, attention has been paid to vegetation composition on Browns Island at the Delta mouth, where researchers have correlated a reed-dominated portion of cores with a period 3800-2000 cal yr B.P., where higher flows maintained fresher conditions at the Delta mouth (Atwater 1980, Goman and Wells 2000).

The emergent vegetation of the central Delta, particularly in the western portion, was shorter and less dense than elsewhere. While emergent vegetation in the upper Delta regions (e.g., Yolo Basin) reached heights of well over 10 feet (3 m), it may have been closer to 4 feet (1.2 m) high in the central Delta (Smith & Elliot [1879]1979). It is unclear whether this is a result of species differences or growing conditions (e.g., flood duration and depth of inundation, sediment deposits, and competition from other species). However, vegetation patterns in the Delta today offers insight. The shorter, but more structurally sound *Schoenoplectus californicus* is more common in exposed locations (like the windy and wave-prone western and central Delta) whereas the taller *S. acutus* is found in more protected interior areas (Keeler-Wolf pers. comm).

Historical accounts also suggest that species differences may have been a factor. For example, one newspaper reports that the soils of the Sacramento were composed of tule roots, while the central Delta island soils were "composed in great part of the finer roots of the marsh grasses" (*Sacramento Daily Union* 1873). It is unclear, however, where or whether sufficient soil samples were taken for such a broad statement. Species differences are also suggested in the following account:

> From Collinsville a short distance up the Sacramento and a longer distance up the San Joaquin the land is less solid and formed of peat, most of which will float: upon this soil grows a large amount of grass.

> Above this on the Sacramento and San Joaquin the land is sedimentary upon which the round tule grows rank to the almost entire exclusion of grasses. (Ryer in Tucker 1879c)

Grasses are also discussed in the historical record in the context of cattle grazing, and harvesting the "coarse wiry, heavy swamp grass" sometimes referred to as "tule hay" within the Delta islands (Cronise 1868). A number of early agricultural boosters and county histories highlighted that the "swamps afford good pasturage" (Sprague and Atwell 1870). Livestock were set out particularly in the dry summer months and during drought. During the drought of 1864, "thousands of acres of natural meadows" were harvested "at the mouths of the San Joaquin, Sacramento, and Cosumnes Rivers," with an estimated equal amount left standing (Fig. 4.44; State Agricultural Society 1866). While coarse grass could potentially refer to emergent species such as cattail and the common reed, this connection becomes less clear with the addition of a term like "natural meadows," which implies herbaceous vegetation cover of lower height. Overall, such evidence does suggest that some areas were occupied primarily by species other than tule (Thompson in press).

Explicit accounts of distribution are rare and most discussions of grasses occur alongside those of tule. However, it is possible that large areas of these meadows existed in the central Delta. This is suggested by the hay cutting mentioned above as well as a description in a newspaper article touting the ease of reclamation on Twitchell and Brannan islands because they were "covered with a rich carpet of grass" (*Daily Alta California* 1869). This account is somewhat called into question, however, given other evidence suggesting that "tule, or wet grass land" historically occupied Brannan Island (Shafer 1882). Geographer John Thompson (pers. comm.), an authority on the historical Delta, suggests that the meadows were likely located along more of the elevated sandier and alluvial portions of the Delta, and potentially related to burning by indigenous tribes. Vegetation patterns may also have been affected to some degree by large ungulates, most notably tule elk, which were known to graze at the marsh margin (Burcham 1957, Phelps and Busch 1983). Although we found that many accounts describe grasses within the tules of the central Delta, it is possible that large expanses (e.g., those referred to as meadows) tended to be more located near the wetland margins.

Willows and other associated species

Many less dominant species occupied the wetland complexes alongside the tules. Herbert Mason (n.d.) and Brian Atwater (1980) documented around 40 plant species within native Delta wetland communities. These species, in

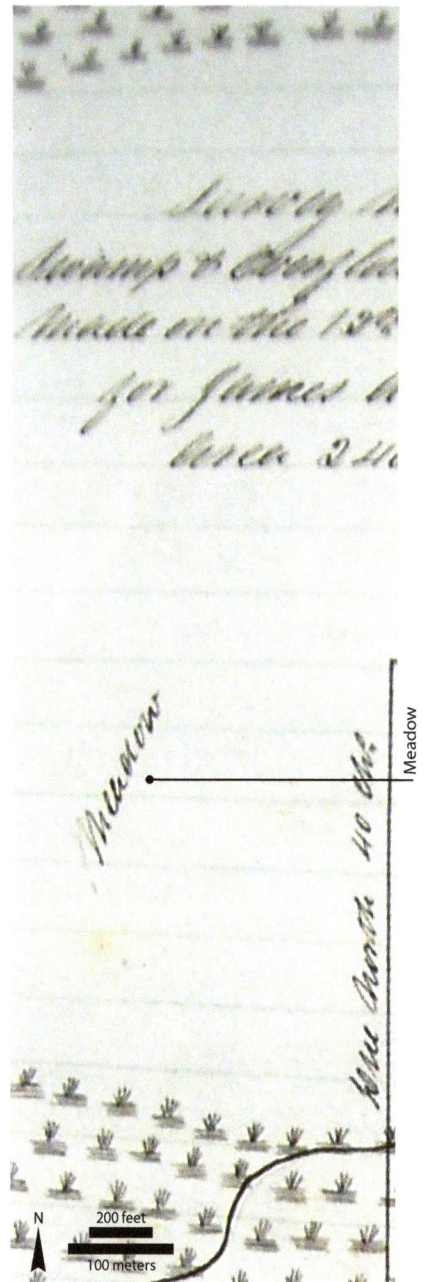

Figure 4.44. An open area within tule is marked as "meadow" inside the eastern edge of the Delta in this 1857 San Joaquin county survey map. (Drew 1856-1857, courtesy of the San Joaquin County Surveyors Office)

addition to those mentioned previously, include Goodding's willow (*Salix gooddingii*), arroyo willow (*S. lasiolepis*), sandbar willow (*S. exigua*), buttonbush (*Cephalanthus occidentalis*), American dogwood (*Cornus sericea*), California hibiscus (*Hibiscus lasiocarpus*), lady-fern (*Athyrium felix-femina*), Mason's lilaeopsis (*Lilaeopsis masonii*), Suisun marsh aster (*Symphyotrichum lentum*), and Delta tule pea (*Lathyrus jepsonii* var. *jepsonii*; Fig. 4.45; Mason n.d., Brandegee 1893-4, Atwater 1980, CNDDB 2010). Some available historical sources hint at this great richness of species. For instance, one observer commented that along the way to Stockton, the "long grass" at the water's edge was "interspersed with some small shrubs, or the flower of the yellow lotus [likely *Nuphar polysepala*]" (Kip [1850]1946). Another, more poetic description of the Delta's freshwater tidelands, includes:

> The islands seem to sink slowly into a wonderful expanse of tule…
> Flowers of royal purple, and deep scarlet, and glorious golden hues
> bloom here in untold profusion; acres of brown-headed "cat-tails" glisten
> in the sun…Sometimes rushes and olive-green canes mingle with the
> tules, and tangle about old barges lying fast in the shallows. (Shinn 1888)

Of particular note, and perhaps unique to the central Delta landscape, was the presence of willows within the matrix of emergent vegetation. Herbert Mason's (n.d.) community profile of "willow-fern swamp," originally described to him by botanist Anson Blake (who grew up near Stockton), likely captures what historical records describe as scattered clumps of willow across many of the central Delta islands at the landscape scale. Mason's quote from Anson Blake and descriptions of the community are as follows:

> "There were extensive willow swamps with a dense understory of Ladyfern."…The vegetation is from 3 to 4 stories. When there are trees they are low. The thickets are from 6 to 10 feet high. Ferns and

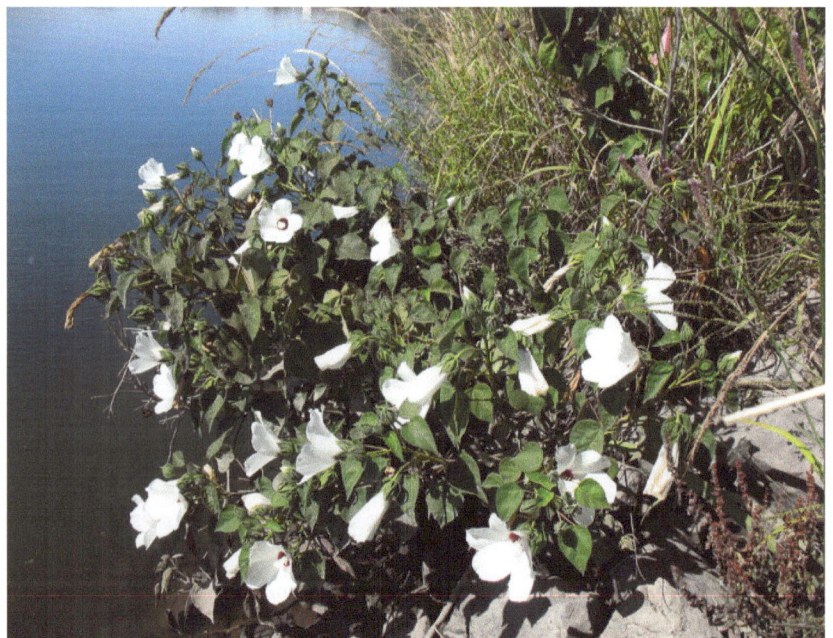

Figure 4.45. The beautiful California hibiscus *(Hibiscus lasiocarpus),* shown here in the vicinity of Latham Slough and Bacon Island, is one of many species that were once part of the rich vegetation community of the Delta's tidal wetlands. (photo by Christopher Bronny, 2007)

undershrubs are below this. Then there is a close bright green ground cover made up of *Tillaea aquatica* [*Crassula aquatica*], *Liliopsis, Hydrocotyle, Utricularia, Eleocharis, Samolus floribundus* [*Samolus parviflorus*] and *Limosella subulata* [*Limosella australis*]. (Mason n.d.)

Evidence of willows mixed with wetland vegetation is found in a number of textual accounts, most of which come from observations made along the San Joaquin River downstream of Stockton (Table 4.2). The willows are frequently identified as bushes or small trees. One photograph of an unreclaimed island gives a sense what these wetland complexes likle looked like from the nearby channel (Fig. 4.46). Though willows are clearly the predominant woody plant, some accounts also mention brambles and alders (*Alnus rhombifolia*).

Several of the earliest maps of the Delta affirm this vegetation pattern at a landscape scale. A circa 1840 map notes "tulares y sauces" (tules and willows) along the San Joaquin in contrast to only "tulares" in the north Delta (Fig. 4.47). In one of the few maps that shows wetland vegetation cover, tree symbols are evenly spaced with marsh symbols in two unreclaimed portions of a Venice Island reclamation map (Fig. 4.48). Another, more spatially accurate map is Gibbes' 1850 survey of the San Joaquin River, where symbols for brushy vegetation are shown in clumps generally within the lower reaches of Old and Middle rivers (Fig. 4.49). Given his focus on mapping hydrography, Gibbes unfortunately did not illustrate vegetation patterns within the central parts of the islands. However, coupling this map with a quote about the lower islands of the San Joaquin helps fill in the picture:

> The banks…appear to be no higher than the centers and are almost uniformly destitute of bushes and have no trees of any size, while the centers of the islands are dotted with bunches of willows, and the tules are thinner and shorter—being mixed with a much greater quantity of coarse grass of different kinds, including now and then patches of California clover. (*Pacific Rural Press* 1871)

Table 4.2. Early descriptions of the central Delta's native vegetation. These accounts highlight the variety of vegetation cover found in the central Delta, of which the "thickets" and "willow bushes" are of particular note.

Quote	Location	Year	Source
"various islands covered with tule rushes and thickets"	from Suisun, along San Joaquin River	1811	Abella and Cook 1960
"the one carrying less water and some small trees"	vicinity of False river	1811	Abella and Cook 1960
"There are a few small trees, like brush, and on the opposite bank also a few other small trees."	mouth of Old River on San Joaquin	1811	Abella and Cook 1960
"island of tule which was flooded when the tide rose and had to take refuge in a bramble patch "	Delta mouth	1817	Durán and Cook 1960
"low marshy ground, covered with rushes and willows"	lower portion of San Joaquin River	1853	U.S. War Department 1856a
"grass and weeds and some small willow bushes"	vicinity of Antioch	1865	Thompson 1865
"tule, alder bushes, few willow some grass or tule grass grew on these lands"	vicinity of Antioch	1865	Brown 1865

Figure 4.46. An oblique view of the willow and tule complex characteristic of central Delta islands is seen in this photograph, captioned, "View of Island Land Before Reclamation." (Yardley Collection, courtesy of The Haggin Museum)

From this, it is reasonable to conclude that the pattern of willows in the map could be extended across the islands.

The willow-fern swamp vegetation community intergraded with other freshwater emergent wetland communities at many scales. The quote above, with its "bunches of willows" and "now and then patches of California clover," conveys the vegetation pattern (*Pacific Rural Press* 1871). Willow-fern swamps added a dimension of woody vertical structure to an ecosystem often considered more narrowly in terms of its emergent wetland species. The size of individual patches likely varied substantially, from only several to potentially several hundred acres in size. Unfortunately, modern analogs within remnant in-channel islands are too small for direct study of such larger landscape-scale patterns.

This vegetation community appears to have been most common within Sherman, Bradford, Webb, Venice, and Mandeville islands. These were areas coincident with areas of cooler temperatures due to the maritime influence and tule fog (see Fig. 1.5). The greater prevalence of these communities in the vicinity of Old River in comparison to the San Joaquin below Stockton is discussed in an 1873 newspaper article:

> A dense growth of tule or flag is the exception rather than the rule. The ground, in its natural condition, is covered with a thick growth of grass and vegetation of less imposing appearance, with here and there an unpretentious patch of tule and an occasional cluster of willows or swamp alder." (*Sacramento Daily Union* 1873).

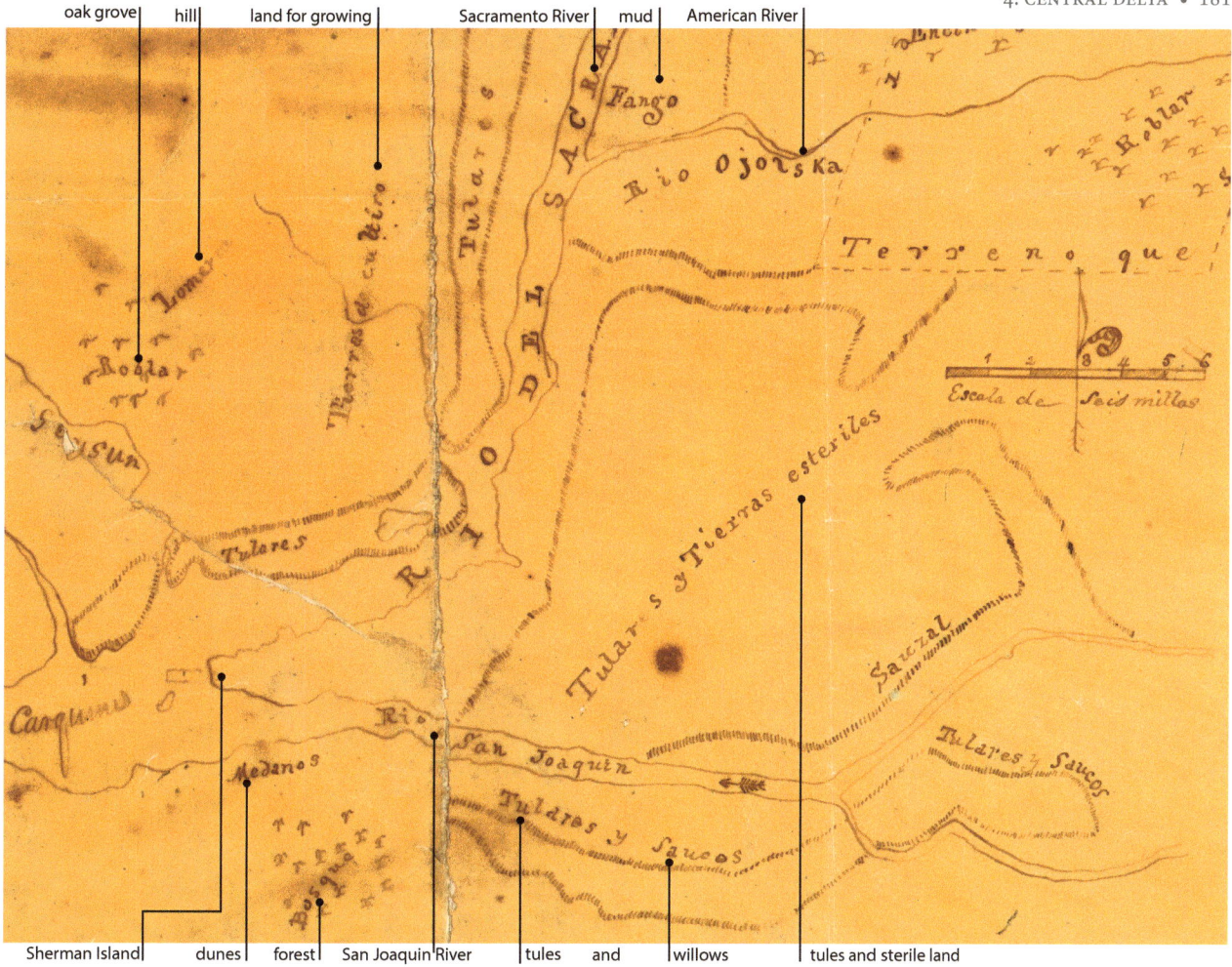

oak grove   hill   land for growing   Sacramento River   mud   American River

Sherman Island   dunes   forest   San Joaquin River   tules   and   willows   tules and sterile land

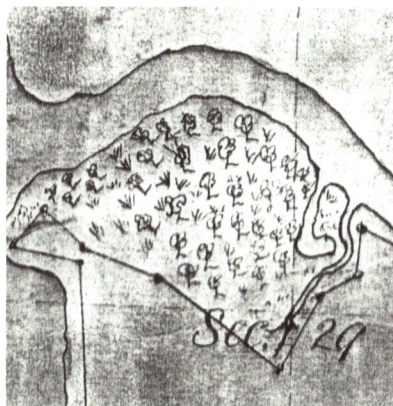

Figure 4.47. In one of the earliest Delta maps, tules and willows ("tulares y sauces") occupy the vicinity of the San Joaquin River in contrast to only tules marked in the Sacramento River and north Delta regions. Though this map covers a large area and is only broadly spatially accurate, this difference between the wetlands of the San Joaquin and Sacramento is clear. (U.S. District Court ca. 1840b, courtesy of The Bancroft Library, UC Berkeley)

Figure 4.48. An even patterning of trees and emergent vegetation symbols is found on this 1866 Swamp Land District map for Venice Island. Though this is of a small area (>50 ac/>20 ha), it suggests that these species were intermixed such that willows were not exclusively associated with channel edges. (Smith 1866a, courtesy of the California State Lands Commission)

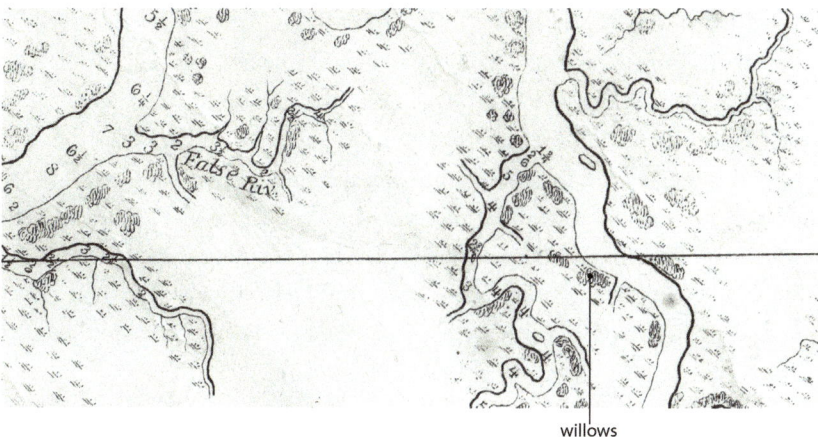

willows

Figure 4.49. Clumps of trees in the vicinity of the Old and Middle river channels suggest the general location of the willow-fern complex in the Delta. This 1850 map was made as a navigational chart and therefore does not include mapping of the interior islands. Based on independent descriptions of these central Delta islands, we hypothesize that the mapped pattern of willows and tule would have continued across much of the islands. (Gibbes 1850a, courtesy of the Map Collection of the Library of UC Davis)

Since no clear defining boundary is evident, however, willow-fern swamp was considered for mapping purposes to be part of the tidal freshwater emergent wetland habitat type. The region in which we understand the willow-fern complex to have been most prevalent is shown in Figure 4.50.

One possible explanation for this vegetation complex's unique presence in the central Delta lies in the different physical dynamics of the Delta landscapes. In his unpublished report on the *Floristics of the Sacramento-San Joaquin Delta*, Herbert Mason discusses a "palustrian continuum" of successional stages of wetland development, where willows become established only during the later stage. The wetlands of the central Delta may have exhibited a later successional stage than those of the northern Delta flood basins. If this were the case, this later stage may have been possible because the central Delta was less prone to disturbance than the more riverine landscapes; floods were less pronounced in flow velocity, depth, and potential duration, and comparatively little sediment found its way onto the marsh plain during these events.

Overall, central Delta vegetation patterns at the local scale were apt to be quite patchy. Patches were apparently not particularly well correlated with small topographic variations within the marsh plain (Atwater 1980), though it is unclear the degree to which patterns were affected by edge-area relationships and other factors relating to channel planform. Smaller patch sizes may relate in part to the absence of strong salinity controls, the relative dominance of tidal processes, the generally level elevation of the marsh plain near high tide levels, and successional processes in the vegetation community (Fig. 4.51). Channel planform may have influenced local species assemblages.: this relationship was demonstrated by Sanderson et al. (2000), who found vegetation patterns in the nearby brackish Petaluma marsh related to channel size, origin, proximity, and location (whether in an "interior" or "exterior" position).

The degree of association of willows with the channel banks of the central Delta islands is somewhat unclear. Some descriptions of Delta vegetation suggest that short woody vegetation preferentially occupied the low banks (Fig. 4.52; e.g., Gilbert 1917, Thompson pers. comm.). Such descriptions capture the observed trend from valley foothill riparian forest to willow riparian forest to scattered clumps of willow shrub (identified as woody plants generally <10 m in height, usually with two or more stems at the base) as natural levee height diminished downstream toward the central Delta. However, once banks decreased to the general elevation of the rest of the marsh plain in the central Delta (where inundation periods, water tables, and soil mineral content was similar to the surrounding wetland), it seems likely that this pattern dissipated. It appears highly unlikely that willows were found exclusively or continuously lining the sloughs in the central Delta, given historical evidence of tules dominating the banks along the river to Stockton and willows standing in clumps within island interiors.

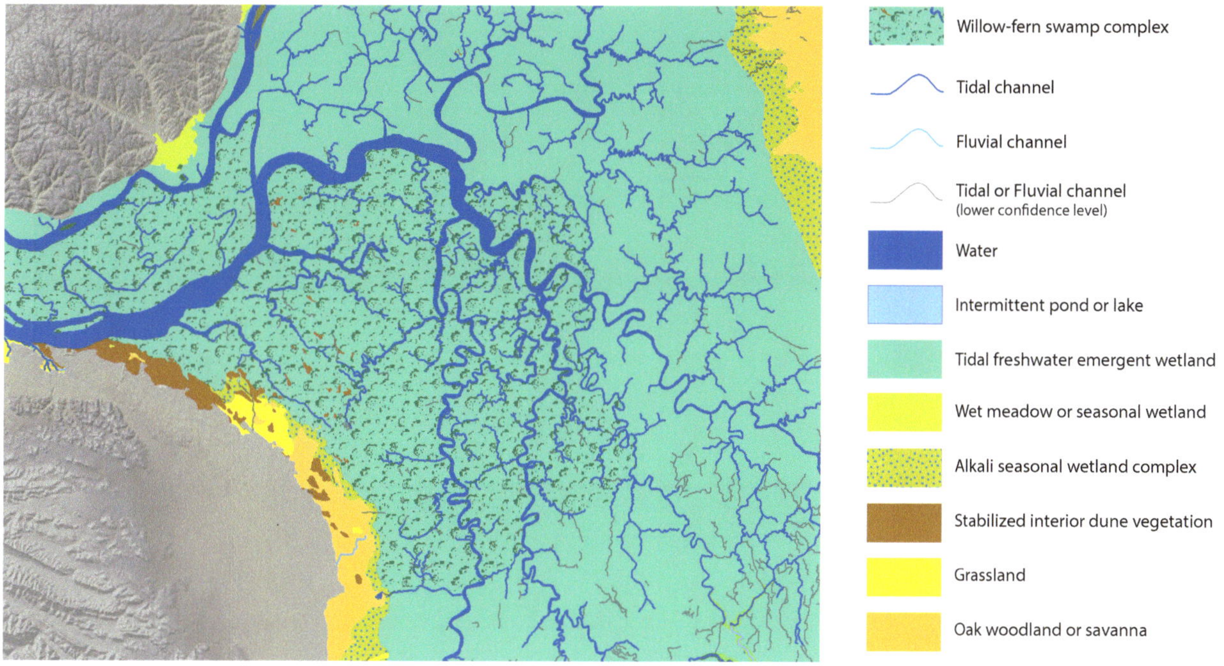

Willow-fern swamp complex

Tidal channel

Fluvial channel

Tidal or Fluvial channel
(lower confidence level)

Water

Intermittent pond or lake

Tidal freshwater emergent wetland

Wet meadow or seasonal wetland

Alkali seasonal wetland complex

Stabilized interior dune vegetation

Grassland

Oak woodland or savanna

It is difficult to determine exactly how far downstream along various channels the relationship between willows and banks persisted. Riparian trees and scrub extended farthest downstream on those channels that received alluvial sediment from the Sacramento or Mokelumne rivers due to higher natural levee elevations. Riparian forest is understood to have largely disappeared around Rio Vista on the Sacramento, upstream of Stockton on the San Joaquin, and within several miles downstream from the head of Staten Island on the Mokelumne, but intermittent sections of willow riparian scrub associated with slightly higher banks continued much further downstream. For instance, two patches of willows were located on the Sacramento at Horseshoe Bend, perhaps a part of that assemblage (Fig. 4.53). At the mouth of Sycamore Slough on the Mokelumne, one man testified that he climbed a willow tree and was able to pick out lines of sloughs to the north (Sherman 1859). The willows along the northernmost sloughs, according to another witness, were apparently so thick that "a boat cannot go through on account of the brush" (Dugin 1859). Channels within the central core of the Delta had hardly any willow: a single willow tree along the San Joaquin en route to Stockton was a landmark used to determine the remaining distance to the city (Knower 1894). One account describes it as the "only tree we saw for an hundred miles" (Kip [1850]1946).

A reminder to use post-1870 vegetation descriptions cautiously is warranted here: the often multiple failed reclamation attempts and associated changes to the elevation, hydrology, and soils of particular areas may have had profound effects on the vegetation patterns within the tidal wetlands that reestablished after each failed attempt. Accounts of willows growing up in places they hadn't been a few years before, or of large areas burned away several feet deep, suggest that some vegetation patterns in the 1870s did not reflect those of the early 1800s. For instance, reclamation

Figure 4.50. The generalized extent of willow-fern swamp complex (shown by the dark, clumped tree symbol) as determined from various sources, none of which explicitly describe the boundaries of this wetland community. Actual boundaries were likely undiscernible, as the presence of willows within the islands gradually became less prevalent moving away from this mapped area.

184

Brazilian Waterweed *(Egeria-Myriophyllum)* Submerged

Broad-leaf Cattail *(Typha latifolia)*

Common Reed

Cornus sericea-Salix lasiolepis / (Phragmites australis)

Generic Floating Aquatics

Mixed *Schoenoplectus* / Floating Aquatics *(Hydrocotyle-Eichhornia)* Complex

Mixed *Schoenoplectus* / Submerged Aquatics *(Egeria-Cabomba-Myriophyllum* spp.) Complex Unit

Salix lasiolepis - (Cornus sericea) / Schoenoplectus spp. *- (Phragmites australis - Typha* spp.) Complex Unit

Schoenoplectus acutus - Typha latifolia

Water Hyacinth *(Eichhornia crassipes)*

500 feet

100 meters

Woody plants, largely *Salix* spp.

Herbacious plants, largely *Schoenoplectus americanus*

Schoenoplectus californicus and robust *S. ocutus*

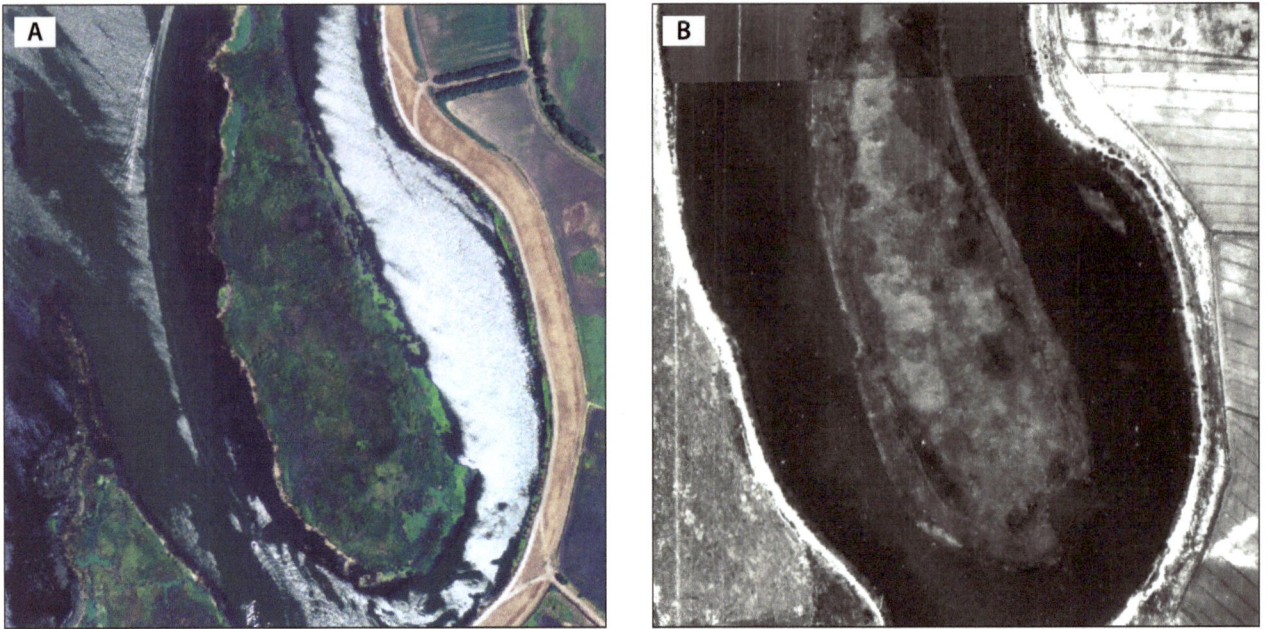

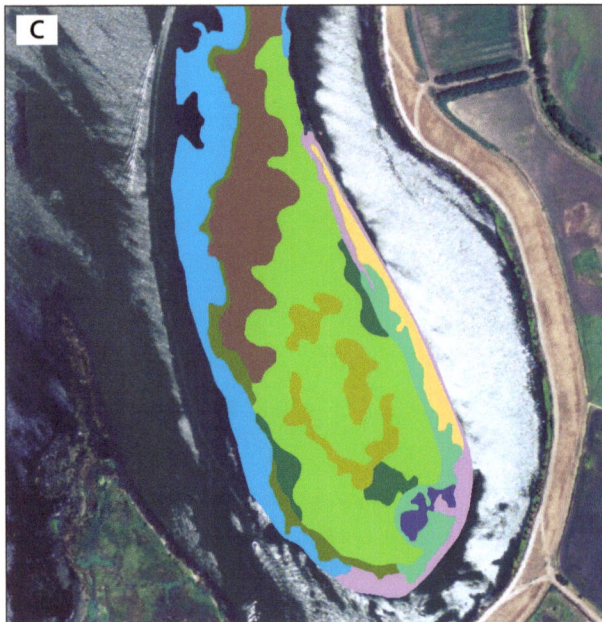

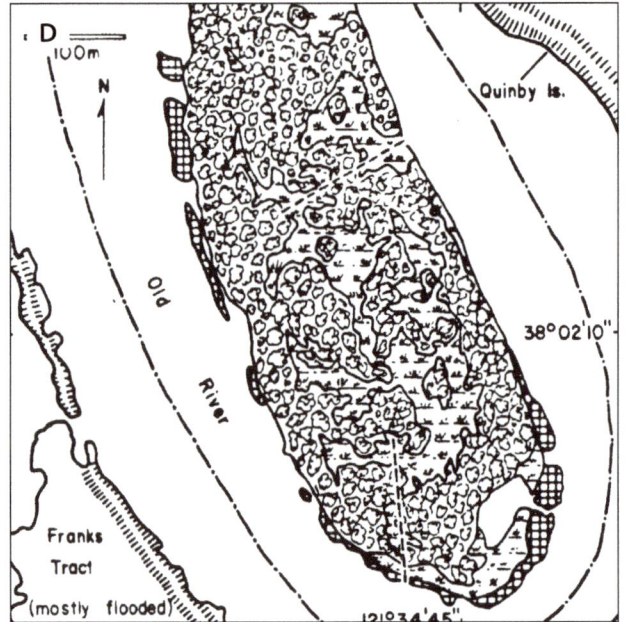

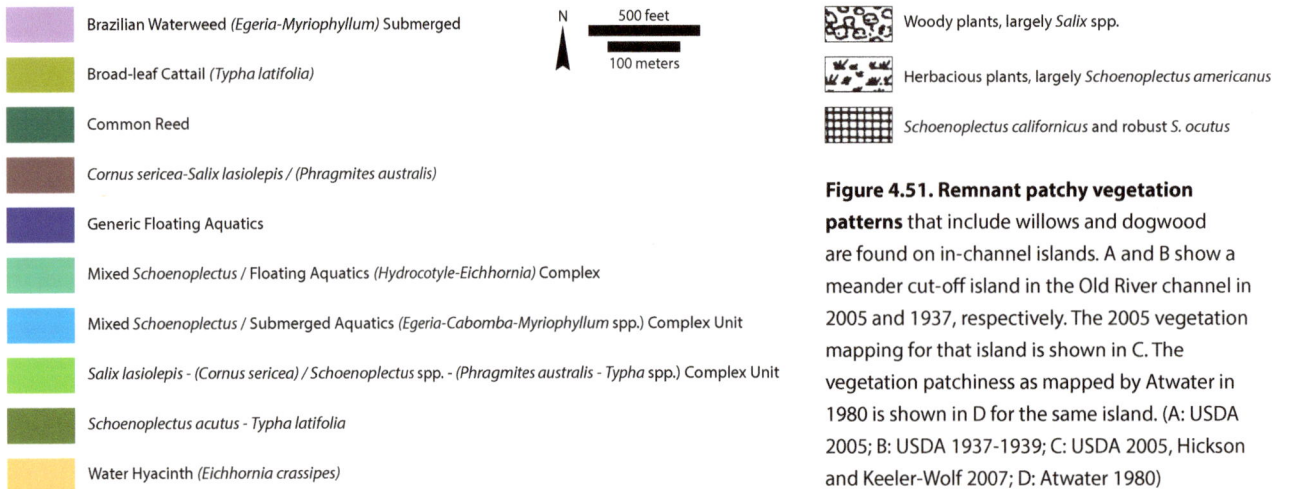

Figure 4.51. Remnant patchy vegetation patterns that include willows and dogwood are found on in-channel islands. A and B show a meander cut-off island in the Old River channel in 2005 and 1937, respectively. The 2005 vegetation mapping for that island is shown in C. The vegetation patchiness as mapped by Atwater in 1980 is shown in D for the same island. (A: USDA 2005; B: USDA 1937-1939; C: USDA 2005, Hickson and Keeler-Wolf 2007; D: Atwater 1980)

Figure 4.52. An early central Delta view with scattered willows along the San Joaquin River. The photograph was taken on the San Joaquin River, downstream of Stockton by the USGS geologist Grove Karl Gilbert. This photograph was later published in his report, *Hydraulic-Mining Debris in the Sierra Nevada* (1917). His caption reads: "A Delta Marsh Bordering San Joaquin River. The foreground shows the dominant vegetation of the tidal marshes where the water is fresh or nearly fresh. The bushes mark the position of the natural levee, here low. An artificial levee may be faintly seen above the rushes. The work of reclamation was in progress at the date of the view, August 31, 1905." The water in the foreground is likely a ditch dug along the channel in the reclamation process. (photo by Gilbert 1905, courtesy of the USGS Photographic Library)

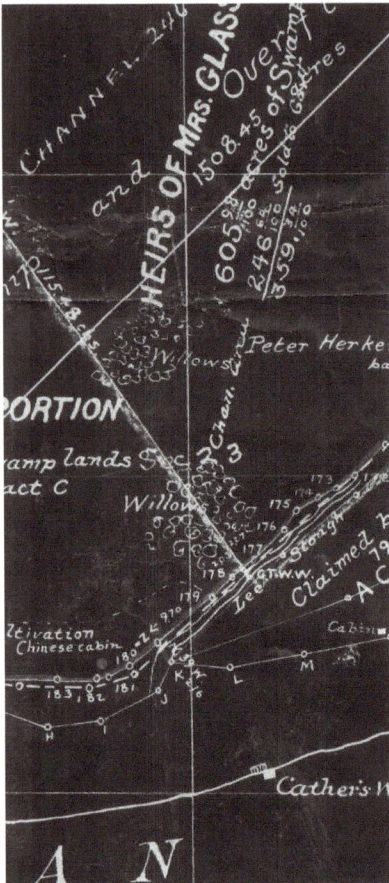

Figure 4.53. Two patches of willows on the order of 10-15 acres (4-6.1 ha) each are shown at Horseshoe Bend on the lower Sacramento River. (Allardt 1880, courtesy of the Solano County Surveyor)

efforts on Bouldin Island were abandoned in 1874 after initial reclamation in 1871. From then until reclamation again began in earnest in 1877, it was reported that "willows and tules sprang up everywhere" (Tucker 1879a). Willows were purposefully planted on early artificial levees as well, and grew quickly (Whitney 1873).

Potential human and biotic-induced modifications

The effects of biotic interactions with the physical environment should not be overlooked as potential drivers of landscape form and function (Box 4.4). While not considered in depth here, native management of the land through burning may have affected vegetation patterns in the Delta (see Boxes 1.1 and 6.1). Some have suggested that the presence of open meadows within the Delta may be a product of this practice. Indigenous peoples also preferentially utilized different wetland species for food, building, and basketry needs (Cromwell in Jepson 1904).

Animals such as beaver and waterfowl likely also changed local vegetation patterns, perhaps to the degree that larger scale patterns were affected. Both beaver and geese enjoyed consuming the tubers of emergent wetland and floating aquatic species, perhaps helping to maintain areas of open water. Geese apparently made fast work of clearing areas, as described by Jepson of geese consuming the pondweed (*Potamogeton pectinatus*) in his trip to Suisun Marsh: "The geese eat the roots and clean out areas of 5, 10 and 20 acres or even more" (Jepson 1904).

UPLAND ECOTONE

The perimeter of the central Delta graded into the extensive riverine wetland landscapes of the Sacramento and San Joaquin rivers to the north and south. Elsewhere, the central Delta landscape graded more sharply into comparatively drier habitats, forming an upland ecotone. The upland ecotone consisted of a variety of habitat types depending on topographic, climatic and geologic controls. At the Delta mouth, wetlands met the steep upland drainages of the Montezuma Hills. Along the more level eastern Contra Costa edge, scattered sand mounds rose above the plain of the tidal wetland and alkali seasonal wetlands bordered the edge. On the eastern boundary, oak woodlands and savannas occupied the alluvial fan of the Calaveras, providing a respite to weary travelers arriving from the treeless plains of the San Joaquin Valley. Moving northward, a matrix of seasonal wetlands formed a broad zone adjacent to woodland, savanna, and grassland.

Sand mounds within tidal wetlands

Along the western Delta margin in eastern Contra Costa County, the flat tidal wetland landscape was broken by numerous sand mounds rising above the plain. These sand mounds were the supra-tidal glacial-age relicts of an underlying sequence of eolian (wind-blown) sand dunes (Fig. 4.54; Atwater 1982). They comprised the distinctive Oakley sand soil unit of the 1939 Contra Costa soil survey (Carpenter and Cosby 1939) and Dehli sand of the more recent soil survey (USDA 1977). This sequence is also associated with

BOX 4.4. RECOLLECTIONS OF MOSQUITOES

One aspect of the pre-reclamation that was Delta particularly annoying to early travelers was the large population of mosquitoes that occupied the freshwater wetlands during the warmer months of the year (Chamberlain 1850, Kerr 1850, Upham 1878, Font and Bolton 1930, Brewer 1974). Diary entries frequently end with notes including "mosquitoes were terribly annoying" in August 1849 (Taylor 1854), "muskitoes [sic] troublesome" in March 1828 (Sullivan 1934), "mosquitoes in swarms" in June 1846 (Duvall and Rogers 1957), and "immense multitude of mosquitoes" in August 1839 (Davis 1889). Many also claimed that they were made miserable by mosquitoes that were the "largest and most voracious" they had seen (Hoag 1882). A particularly colorful account is as follows:

> But your Montezuma mosquitoes should not be named in the same century with those of the San Joaquin. Talk of those dwarfs of Montezuma, carrying brick bats under their wings to whet their bills upon; the mosquitoes of San Joaquin would despise using anything less than an Ohio grind stone! And how they were disciplined! They were well drilled, as we had occasion to know. They would bore through our thick Indian blankets, as if they were as thin as gauze! Swarms of them, as if they were marshaled by a leader, would come out of the tall bull rushes, and attack us sleeping or waking; their warfare was diurnal as well as nocturnal. (McCollum [1850]1960)

This species was ubiquitous and unwelcome within the "swamps and quagmires," as is suggested by a not so glowing diary entry of "continual torture on account of the mosquitoes which attack you in swarms" (Moerenhout [1849]1935). Those attempting to camp close to the river were often driven to higher ground away from the river and tule (Lyman 1848, Clyman and Camp [1845]1928). They may have been less prevalent at the Delta mouth, where "the winds from the bay blew away the mosquitoes" (*Sacramento Daily Union* 1873a). Remarks about mosquitoes were found in diary entries in the spring and summer months. Though reclamation and mosquito abatement efforts have largely removed this characteristic of the Delta found annoying by travelers, it also signifies the loss of rich and voluminous insect life that once was an important food source supporting numerous fish and other species.

the Piper fine sandy loam soil unit that occupies lower elevation positions and was overlain by peat deposits prior to reclamation and subsidence (Carpenter and Cosby 1939, Cosby 1941).

The individual mounds above tidal elevations ranged from less than one to more than 25 acres in size, though the mapping includes only those larger than five acres. Historical USGS topographic maps show that some of these mounds were over 15 feet (4.6 m) above sea level (USGS 1909-1918). Some may have been higher as these maps may not represent early 1800s elevations due to alteration from levee building and reclamation. They were distributed well into the tidal wetlands and were mapped as far northeast as Bradford and Webb tracts and as far south as Rock Slough. One feature was even mapped at the foot of Tyler Island. Because of their size, unique soils, and elevated topography, the sand mounds added local scale complexity to the habitat mosaics at the eastern Contra Costa tidal wetland edge.

A larger region of Oakley sands are found west along the edge of the San Joaquin River. This includes the current Antioch Dunes National Wildlife

Figure 4.54. Sand mounds above the wetland plain. The map (A) shows the plan view of sand mounds larger than five acres in size that were elevated above tidal range. These features, which are relict eolian sand dunes, were historically unique upland features within the context of the surrounding tidal wetland. These features were mapped from various sources, including early 1900s USGS topographic maps, survey maps, 1937 aerial photography, and LiDAR. A conceptual profile is shown in B, adapted from Atwater and Belknap (1980).

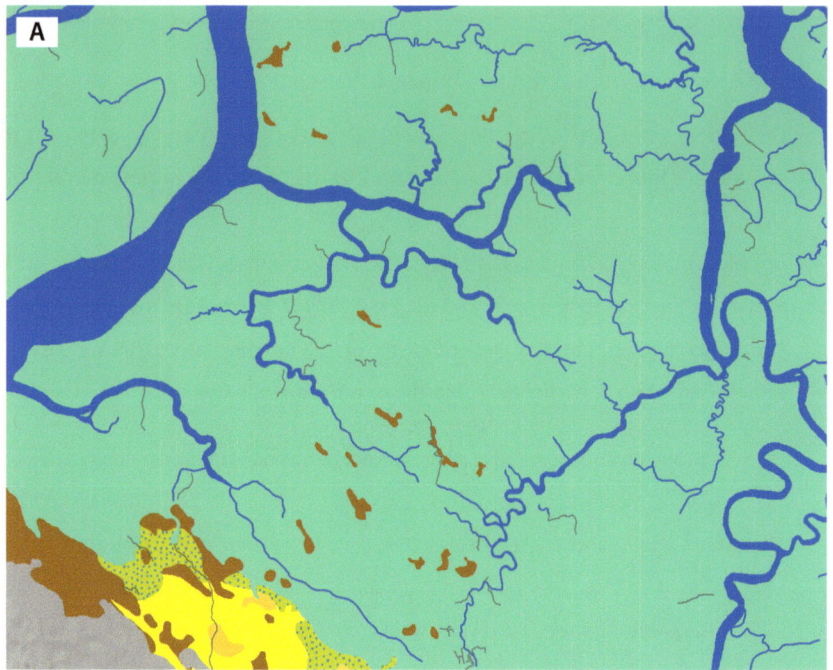

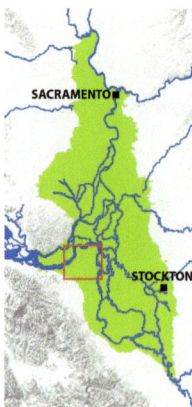

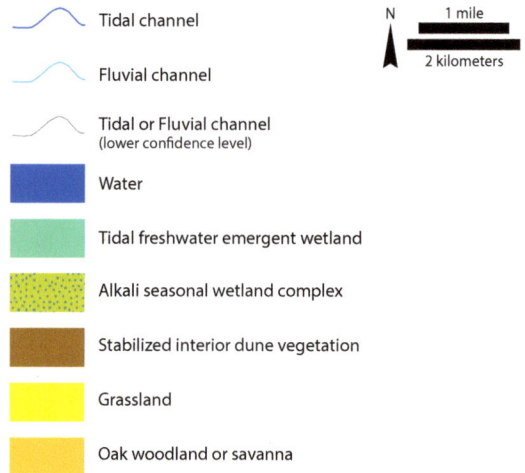

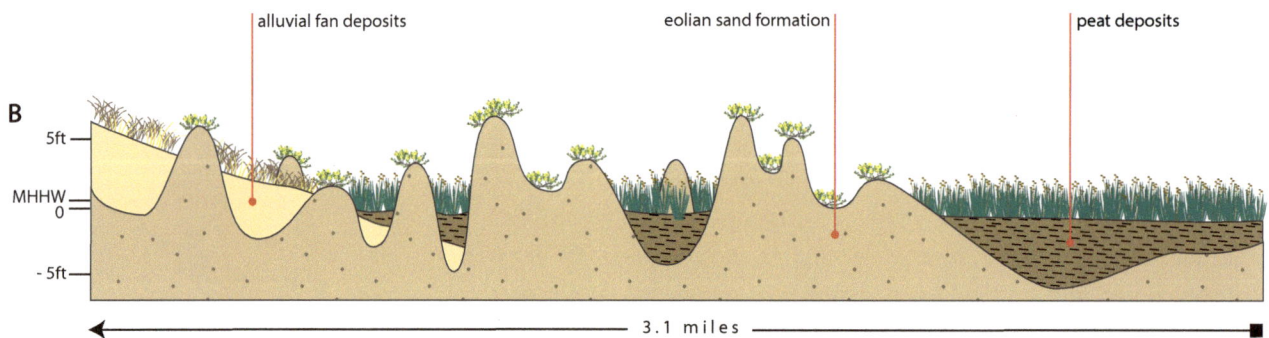

Tidal channel

Fluvial channel

Tidal or Fluvial channel
(lower confidence level)

Water

Tidal freshwater emergent wetland

Alkali seasonal wetland complex

Stabilized interior dune vegetation

Grassland

Oak woodland or savanna

N   1 mile
2 kilometers

SACRAMENTO

STOCKTON

alluvial fan deposits   eolian sand formation   peat deposits

B
5ft
MHHW 0
- 5ft

3.1 miles

Refuge (ADNWR) as well as what was once a prominent 2,800 acre (1,133 ha) expanse of densely vegetated scrub surrounded by more sparsely vegetated areas with oaks in the vicinity of Oakley (most of which is outside the study area; Wackenreuder 1875, Carpenter and Cosby 1939, Stanford et al. 2011). The Antioch Dunes to the west were historically over 100 feet (30.5 m) in height (Davidson 1887).

The relict Pleistocene dune soils of the sand mounds and western edge had stabilized and developed soil profiles that supported live oaks (*Quercus agrifolia*), forbs, and grasses (Wackenreuder 1875, Davidson 1887, Carpenter and Cosby 1939, USDA 1977, Stanford et al. 2011). Historical sources, including early maps, Los Medanos land grant testimony, GLO surveys, and oblique photography, indicate that oaks were associated with these areas (Smith 1866b, Wackenreuder 1875, Russell ca. 1925). An 1887 USCS T-sheet (Davidson) shows tree symbols along the backside of the Antioch Dunes (Fig. 4.55a). Witnesses in the Medanos land case testified that oaks that grew within the tideland boundary and were associated with "a small mound of sand" (Smith 1866b). From the vantage point of Old River, a traveler reported that the "tree-covered mounds look like farm groves of New Jersey" (Smith & Elliot [1879]1979). Oaks are found on some of these features today (Fig. 4.55b and c; Collins J pers. comm.).

More exposed soils were likely occupied by interior dune scrub vegetation (e.g., silver bush lupine, *Lupinus albifrons*), like that found at the ADNWR (Fig. 4.56; Holland 1986, CALFED 2000a, Holstein 2000, Bettelheim and Thayer 2006, Thayer 2010, Stanford et al. 2011). An 1895 botanical account of the Antioch Dunes describes: "sand-hills are brilliant with flowers" and "besides herbaceous plants there are oaks and shrubby Lupines" (Burtt-Davy 1895 in Howard and Arnold 1980). This area is home to the special-status Antioch Dunes evening primrose (*Oenothera deltoides* ssp. *howellii*), the Contra Costa wallflower (*Erysimum capitatum angustatum*), and the endangered Lange's metalmark butterfly (*Apodemia mormo langei*; Fig. 4.57; CALFED 2000a). Common species found today are herbaceous species such as the California croton (*Croton californicus*), slender buckwheat (*Eriogonum gracile*), and valley vinegar weed (*Lessingia gladulifera*; Thayer pers. comm.). Due to uncertainty in spatial extent of vegetation communities, the relict stabilized dune soils above tidal elevations were classified as stabilized interior dune vegetation.

This region was well known for the sand mounds. One of the large sloughs heading west from Old River is Sand Mound Slough, named for the mounds that were found at its head (Fig. 4.58; *Sacramento Daily Union* 1873, Contra Costa Board of Supervisors 1875). Another place named after these features was Sand Mound Ranch (established before 1879) of Bethel Island. The mounds were used by early occupants as homestead sites, as well as for material to shore up sinking peat levees along nearby waterways (Tucker 1879a). The larger dunes were mined for silica. Consequently, the heights of many mounds have been substantially reduced. Some examples of this can be seen comparing the historical USGS topographic maps and

Figure 4.55. Oaks found on sand mounds.
The 1887 T-sheet (A) depicts oaks growing along the Antioch Dunes. Several trees, likely live oaks, along with a residence, persist on a sand mound (outlined in red) on Bethel Island (B). An 1870s-era lithograph (C) illustrates several trees occupying mounds in the vicinity of Old River as well the mounds' topographic distinction. This illustration from about the time of reclamation in the area as well as pre-reclamation textual descriptions suggests that these oaks were present when the surrounding landscape was still tidal wetland (A: Davidson 1887, courtesy of the National Oceanic and Atmospheric Administration; B: USDA 2005; C: Smith & Elliott [1879]1979)

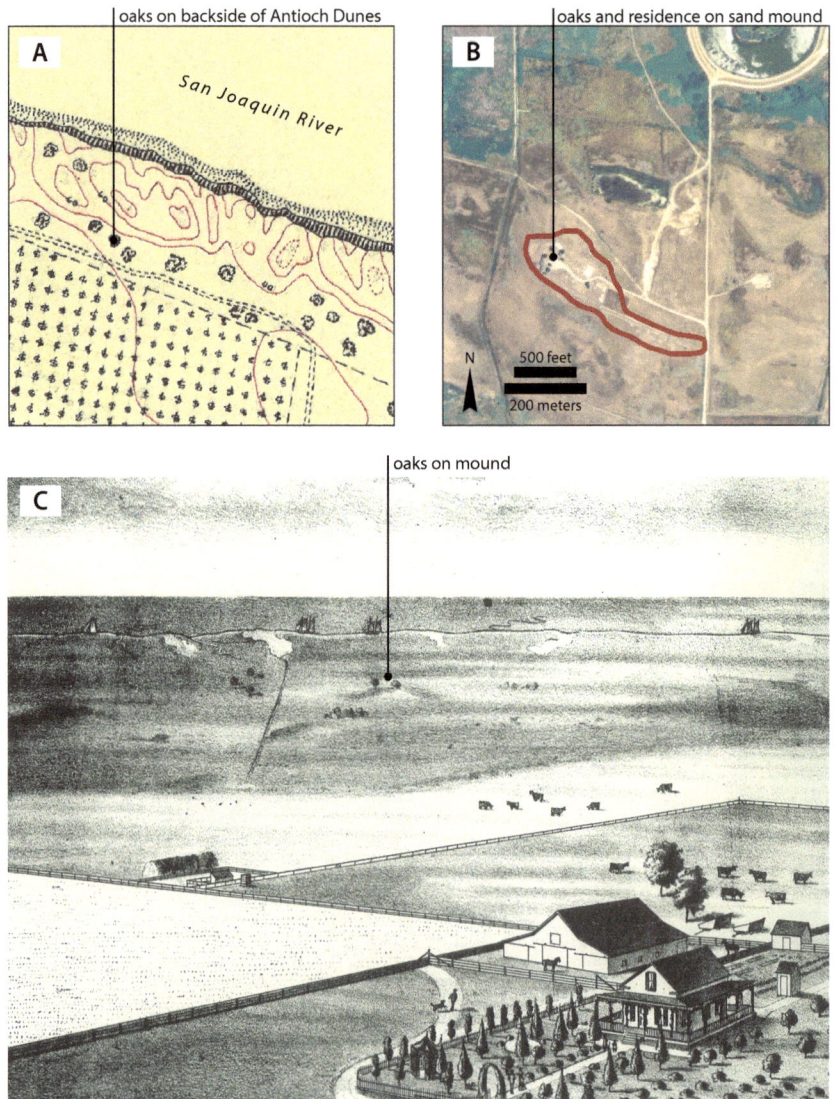

Figure 4.56. **Interior dune vegetation** is shown in a recent photograph. (photo by Christopher Thayer, 2011)

Figure 4.57. The Antioch Dunes evening primrose is found within the Antioch Dunes National Wildlife Refuge, part of a unique vegetation community occupying stabilized eolian dune sands in eastern Contra Costa County. (photo by Ruth Askevold, October 1, 2009)

contemporary LiDAR (USGS 1909-1918, CDWR 2008). It is interesting to consider that where the peat of wetlands overlying the sandy soils has oxidized and been removed in the process of reclamation, the sandy soils have been "exhumed," or exposed at the surface, since reclamation began (Cosby 1941). As a result, areas of exposed sandy soil (mostly Piper fine sandy loam) have increased since historical times, although these newly exposed areas lie at elevations well below sea level.

Sand mounds were not the only mounds in the Delta, however. Sherburne Cook (1960b), who studied the Delta's indigenous tribes, wrote of two types of mounds:

> 1) small, scattered mounds formed of residual calcareous sand (the so-called 'sand mounds') on the summits of which the Indians established their villages; (2) true habitation mounds, perhaps originally situated on a slight elevation, but built up by midden deposits to a height of several feet.

The mounds of anthropogenic origin, or middens, were apparently about 300 feet (90 m) in diameter, about the size of a city block (Belcher et al. 1979). The spatial distribution of historical descriptions suggests that the human-constructed mounds were prevalent in the north Delta along natural levees, but they were also associated with the higher ground of sand mounds (Cook and Elsasser 1956). Both mound types were inhabited by tribes according to anthropologist Nils Nelson (1909), who described sand dunes "rising like islands through the surrounding peat" and noted that they "furnish evidence of having been more or less permanently occupied by the aborigines."

Vegetation cover on true habitation mounds was likely quite different from that of sand mounds. The variety of vegetation cover that is discussed in the historical record suggests that vegetation cover on the Delta's mounds was not consistent across the study area and depended on soils and history of occupancy by indigenous tribes. Given that many village sites were abandoned by the early 1800s (as a result of epidemics, wars, and mission influence), it is difficult to tell to what degree the vegetation communities on individual mounds of the late nineteenth and early twentieth centuries differed from those in the early 1800s. The human origin of some of these mounds, most of which appear to have been located north of the central Delta islands, implies that they would have been highly managed areas. Mason (n.d.) associated the California walnut with indigenous sites. On the single mound Mason bontanized near the town of Locke, he found California walnut (*Juglans californica*), oaks (*Quercus agrifolia* and *Q. lobata*), Oregon ash (*Fraxinus latifolia*), box elder (*Acer negundo california*), and willows (*Salix lasiolepis, S. laevigata,* and *S. exigua*), among other species. A somewhat whimsical but intriguing description by a noted naturalist of "old Indian mounds" states that they are "clothed with short grass…Sometimes, in a place like this, springs of freshwater flow from the grassy knolls, and willows and blackberry vines grow there, with the yellow grindelia and the graceful wild asters" (Shinn 1888). There are also descriptions of large oaks growing on these human-made mounds in the mid-1800s (Houghton 1862). This species assemblage is more similar to the riparian forest of nearby natural levees than the stabilized interior dune vegetation of the eastern Contra Costa sand mounds.

Those looking out across the Delta in its pre-leveed state often remarked that the small mounds dotting the interior Delta (also referred to as knolls or hillocks) provided the only topographic relief to meet the eye. Descriptions like these are found for both sand mounds and the artificially

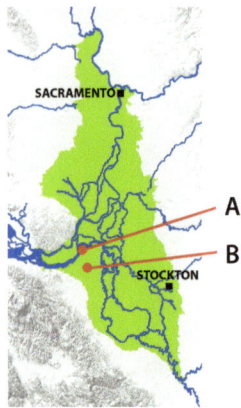

Figure 4.58. Two maps show sand mounds rising above the elevation of the central Delta landscape; several are occupied by homesteads. In (A), mounds of varying size – some larger than ten acres – are shown at the end of Sand Mound Slough. The reclamation map for Bradford Tract (B) depicts similar mounds adjacent to the San Joaquin River. (A: Contra Costa Board of Supervisors 1875, courtesy of the California State Lands Commission; B: Brown 1901, courtesy of the California State Lands Commission)

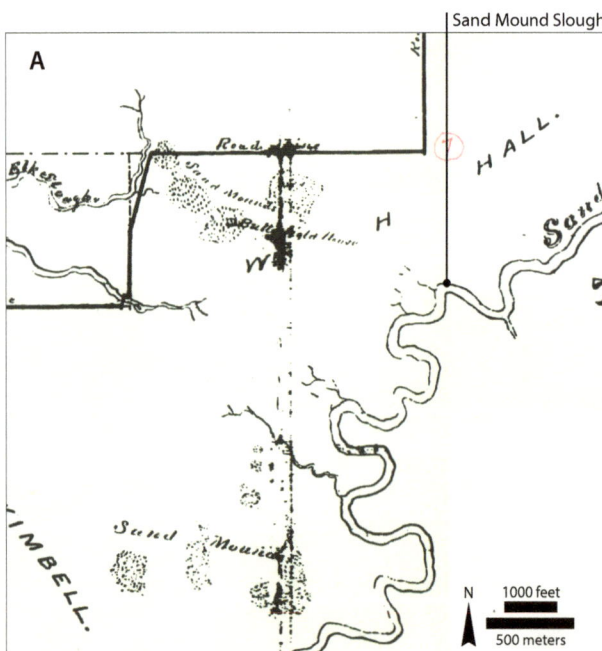

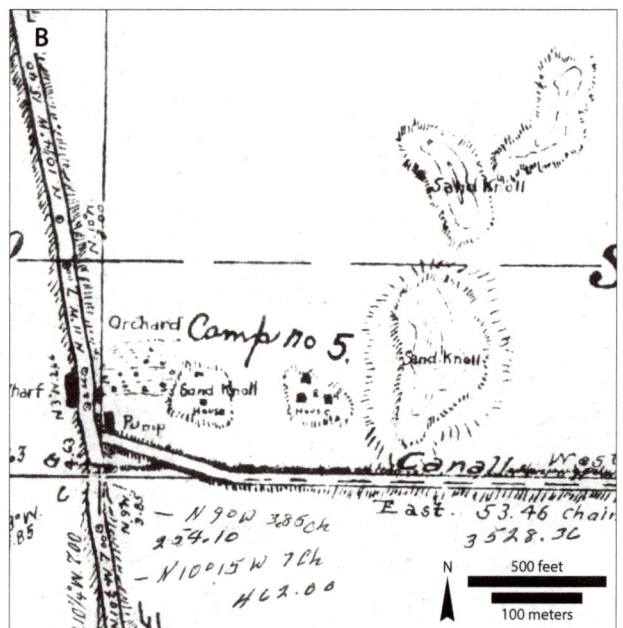

constructed habitation mounds. The landscape could be viewed from atop the mounds when the rest of the land was overflowed: "many mounds of earth on these great savannas built unknown ages ago by the Indians, from which to gaze over these surpassing regions, and to view in safety the rush of the spring-floods covering the country far and near" (Farnham 1857). An early report attested that the only features that were not prone to flooding east of the Sacramento River were the "artificial Indian mounds, which would not amount in the whole to more than ten acres" (California Swamp Land Committee 1861). A Yolo County history recounted that the flood of 1852 inundated the Delta north of the Montezuma Hills and west of the Sacramento except for the mounds (Gregory 1913).

The presence of the sand mounds scattered about the perennial wetlands created isolated upland habitats and supporting local-scale complexity. The mounds supported numerous species of plants and animals that would have otherwise been unable to persist within the Delta's tidal landscape. For instance, tule elk were reported to find valuable protected breeding and foraging habitat within these secluded refuges (Hulaniski 1917). In such close proximity to aquatic and wetland areas, the mounds would have provided excellent habitation sites with a rich abundance of food and other materials available to the indigenous tribes occupying them. Since the mounds were elevated above tide levels and most spring floods, they offered some of the few refuges for terrestrial species in the Delta during high water (Belcher et al. 1979, Swan [1848]1960). This characteristic was also advantageous to settlers and their stock: "these peculiar elevations, lifting from the surrounding plain, were never submerged and were the refuge resorts of stock and frequently people in the vicinity during the floods" (Gregory 1913). Early settlers on Roberts and Union islands were known to take refuge on mounds during floods (Tucker 1879c).

Alkali seasonal wetland complex at the edge

In addition to scattered sand mounds, alkali seasonal wetlands characterized much of the land adjacent to the tidal wetland boundary of eastern Contra Costa. These complexes were comprised of a variety of alkali habitats, including small brackish ponds, perennially wet alkali marsh, alkali flats, alkali sink scrub, and seasonally inundated alkali meadow (Stanford et al. 2011). These were arranged along moisture gradients and characterized by salt concentrations ranging from 0.2 to 1.5% (Carpenter and Cosby 1939). This zone along the edge of the freshwater emergent wetland encompassed over 5,000 acres (2,023 ha; ~20% of the alkali seasonal wetlands mapped within the study area) and varied from less than a quarter mile to about a mile (0.4-1.6 km) in width (Fig. 4.59).

Alkali sink scrub (also known as valley sink scrub) is a subtype of the mapped alkali wetland complexes. It was found along more than seven miles of this western wetland margin and was particularly associated with Marcuse clay soils (Carpenter and Cosby 1939, Stanford et al. 2011). The soil survey describes native vegetation cover of "pickleweed, greasewood,

Left the river in good season and departing gradually from its timber – came into large marshes of Bulrushs…The earth was in many places strongly impregnated with salt – came into hills.

—BIDWELL [1842]1937. TRAVELING WEST FROM THE SAN JOAQUIN TOWARD MARSH'S ADOBE IN EASTERN CONTRA COSTA COUNTY

and saltgrass" (Carpenter and Cosby 1939). GLO survey notes confirm the early presence of scrub in this area, describing "thin scrub or greasewood," which likely refers to iodine bush (*Allenrolfea occidentalis*; Fig. 4.60; CNDDB 2010, Stanford et al. 2011). This area represents perhaps the northernmost range of what was a common habitat type in the upper San Joaquin Valley (Coats et al. 1988).

These alkali wetlands at the Contra Coast edge are representative of a narrow transition zone between the freshwater emergent wetlands and upland habitats found in other localities at the Delta periphery (Sprague and Atwell 1870), primarily in the southern Delta:

> The flood plains or tule lands of the streams are commonly bordered by more or less interrupted belts of land impregnated with an unusual amount of soluble salts or "alkali," which, during the dry season, bloom out on the surface. (Hilgard 1884)

In many locations this zone, characterized by evaporative salt residues and distinct plant communities, demarcated the edge of extreme overflow during flooding events. For instance, a band of alkaline soils along the eastern edge Delta's tidal wetlands between the Calaveras and Mokelumne rivers were mapped in the 1930 Lodi soil survey (Carpenter and Cosby 1932, Cosby and Carpenter 1932). The recent soil survey of the area describe soils that are mildly or moderately alkaline (USDA 1992). This area generally coincides with areas noted as "subject to overflow" by GLO surveys, but is distinguished from areas identified as "swamp" (Wallace 1865a). One GLO field note includes "land sandy alkali" in a description of the land (Handy 1864). Also, testimony from the Mokelumne land case includes multiple witness descriptions of an area extending for about half a mile east of the western line of the grant (which was defined as the edge of tule) that was subject to overflow, but not vegetated by tules (Beaumont 1859a, Gray 1859, Sherman 1859). These zones of alkali lands were not likely characterized by the same vegetation community, however. The alkali scrub in the western edge of the Delta in Contra Costa County appears to have been unique to the Delta edge. Interestingly, GLO surveyors also noted a number of bearing trees within the alkali seasonal wetland complexes we mapped, indicating local complexity that we were unable to include at the resolution of the mapping.

Calaveras River alluvial fan

On its eastern edge, the central Delta tidal island landscape was adjoined by oak woodlands and savannas on the gently sloping late Pleistocene Calaveras River alluvial fan, which extends for over ten miles north from French Camp Slough (Purcell 1940, Atwater 1980, Atwater 1982). An early history of San Joaquin County described a "literal forest of white and live oaks" in the vicinity, where Stockton was home to "thousands of these trees" (Tinkham 1923). Within the area upslope from the tidal wetland edge (extending to the

"No evidence of alkali grass"

"Asparagus. Slight evidence of alkali"

"Saltgrass"

"Greasewood and pickleweed"

"Greasewood, bare spots"

"Greasewood, bare spots"

"Spots of milo and pickleweed"

"Saltgrass, occasional clumps of pickleweed"

"Bare spots, pickleweed, salt sage"

tule

alkali sink scrub

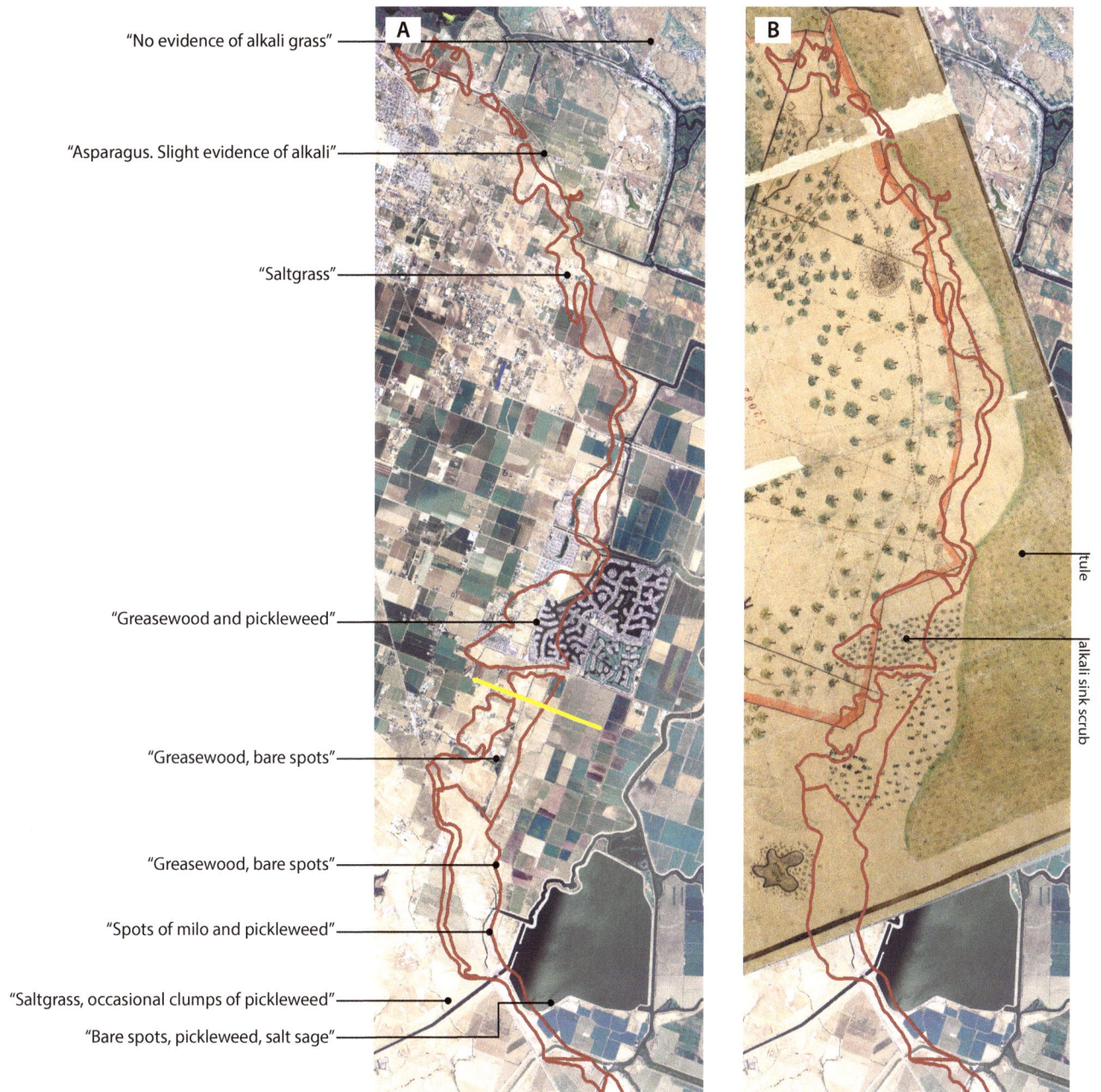

Figure 4.59. Alkali seasonal wetland was found along the western boundary of tidal freshwater wetland in a zone between a quarter to one mile wide. Nine soil survey sample points from the 1939 soil survey are shown against mapped alkali seasonal wetland complex (A) with supporting Los Meganos land grant *diseño* with a vegetation pattern apparently representative of alkali sink scrub (B). The Delta's freshwater wetlands lie on the east side of the red alkaline area, adjacent lands of eastern Contra Costa County to the west. The yellow line in A shows the location of the GLO survey line that is reconstructed in Figure 4.60. (A: USDA 2005 and Carpenter and Cosby 1939, B: Whitcher 1853b, courtesy of The Bancroft Library, UC Berkeley)

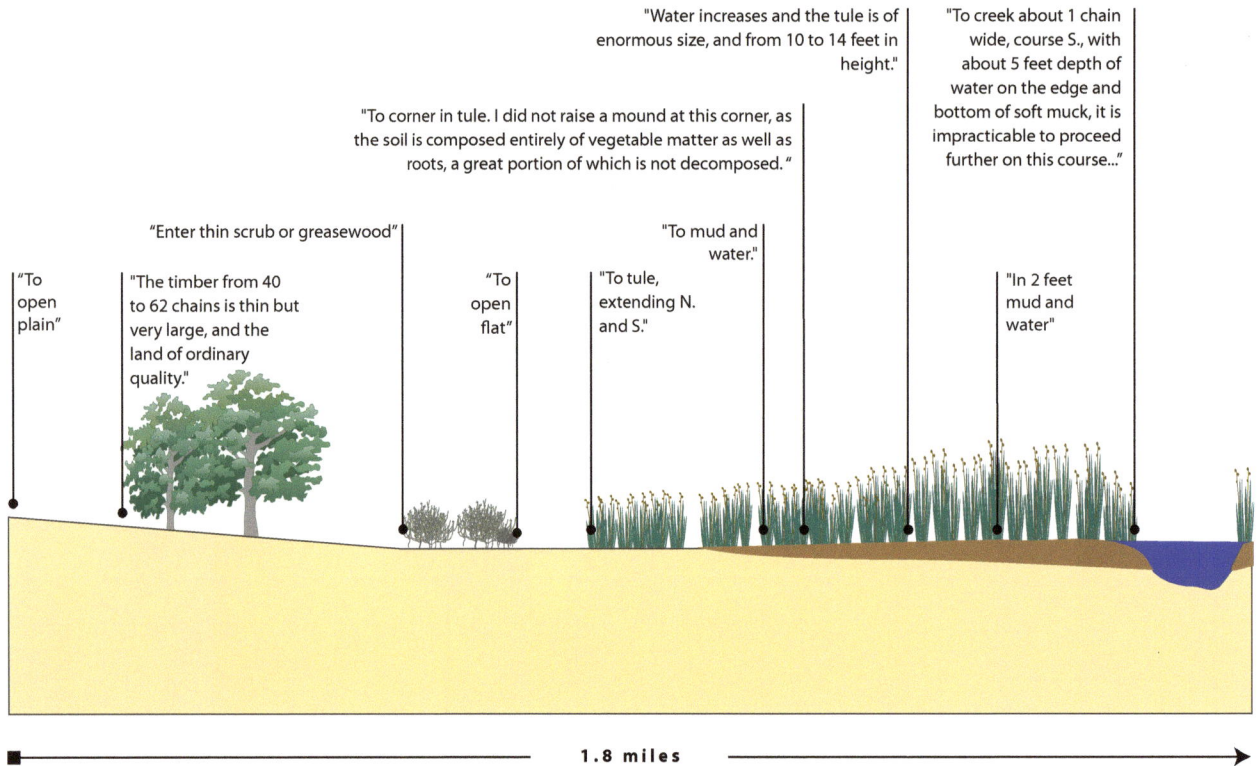

"Water increases and the tule is of enormous size, and from 10 to 14 feet in height."

"To creek about 1 chain wide, course S., with about 5 feet depth of water on the edge and bottom of soft muck, it is impracticable to proceed further on this course..."

"To corner in tule. I did not raise a mound at this corner, as the soil is composed entirely of vegetable matter as well as roots, a great portion of which is not decomposed."

"Enter thin scrub or greasewood"

"To mud and water."

"To open plain"

"The timber from 40 to 62 chains is thin but very large, and the land of ordinary quality."

"To open flat"

"To tule, extending N. and S."

"In 2 feet mud and water"

1.8 miles

Figure 4.60. The transition from oak savanna to alkali seasonal wetland to freshwater tidal wetland found along Ralph Norris' GLO survey line exemplifies the eastern Contra Costa-Delta upland ecotone. Its location is shown as a yellow line in Figure 4.59. (Norris 1851a)

We had made about two leagues in that direction, still over this sandy and ungrateful soil, we finally came into another zone and continued on our way through a well-wooded country, covered with pasturage and as beautiful as one could hope to see.

—MOERENHOUT [1849]1935

25 foot elevation contour that defines our study area boundary), we found that the southern and northern boundaries of this fan closely align with the extent of oaks described by the historical evidence. French Camp Slough, at the southern extent of the alluvial fan, marked the beginning of the oaks, which were a respite for many weary travelers from the San Joaquin Valley (Moerenhout [1849]1935, Lyman and Teggart 1923, Hilgard 1884, Cook 1960b). Some of the earliest cartographic evidence of the prominence of oak woodlands and savannas lying between French Camp and the Calaveras is found in the *diseño* maps of the Los Franceses land grant (Fig. 4.61). At the northern extent of the fan, about five miles north of the Calaveras, the alkali seasonal wetlands discussed in the previous section commenced. However, oak woodlands and savannas did continue farther to the east on the Hanford sandy loams of the 1932 Lodi soil survey (Fig. 4.62; Handy 1864, Wallace 1865a, Cosby and Carpenter 1932).

The spatially explicit GLO survey dataset corroborates the presence of oak woodlands and savannas. It includes multiple bearing trees (used to establish survey corners) at most section and quarter-section points, though the trees were usually quite far away from any given survey point (on average, over 200 ft/61 m away; Fig 4.63). These distances contrast with bearing trees in the southern Santa Clara Valley, which were on average over 100 feet (30.5 m) away. Field note descriptions of "scattered timber" that accompany many of these points also suggest low density oak woodlands and savannas (Grossinger et al. 2008, Whipple et al. 2011). Over 130 trees were recorded by the GLO within the Calaveras alluvial fan in the study area, and about half were less than two feet in diameter (Fig. 4.64).

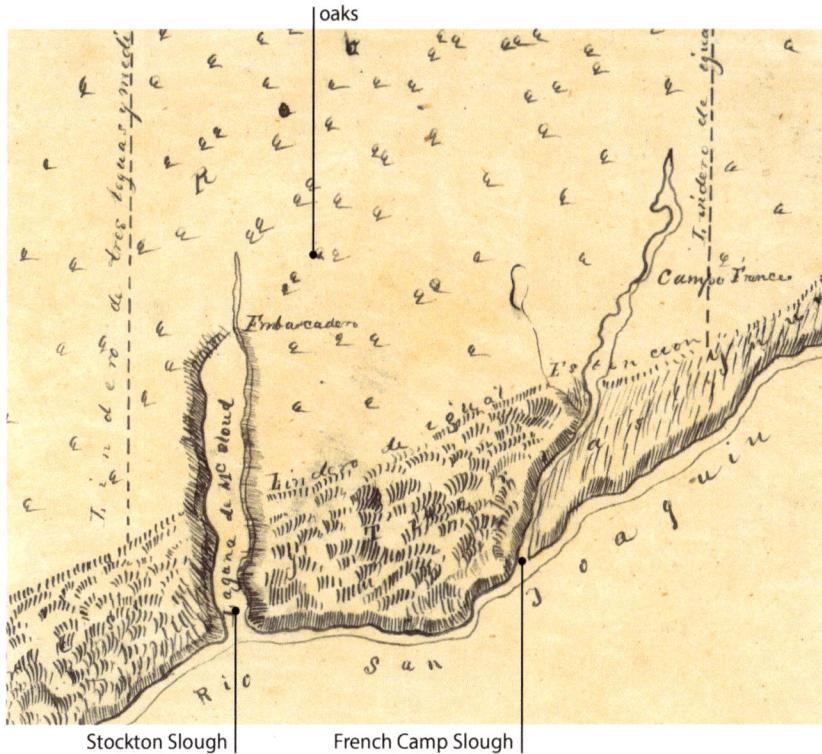

oaks

Stockton Slough | French Camp Slough |

Figure 4.61. Oaks in the vicinity of Stockton. This circa 1840 Los Franceses land grant map covers the land between Stockton Slough ("Laguna de McCloud") and French Camp Slough, and shows oaks extending to the edge of the tidal wetland. Multiple sources support the prevalence of oak woodlands and savannas in this part of the Delta's upland ecotone. (U.S. District Court ca. 1840c, courtesy of The Bancroft Library, UC Berkeley)

Figure 4.62. Oak savanna landscape near Lodi is shown in this 1907 photograph, taken as part of the USGS surveying efforts. (USGS 1907, courtesy of the Center for Sacramento History, Hubert F. Rogers Collection, 2006/028/121)

This may indicate a fairly young age distribution. Other GLO field note descriptions within this area are summarized in Table 4.3. More detailed analysis using bearing tree distances could help approximate historical density (Whipple et al. 2011).

Physical conditions, including groundwater levels and soil characteristics associated with current and former waterways, were appropriate for this notable establishment of oaks. Loams were found closest to the myriad old routes of waterways. Explorer John Frémont, passing through in 1844, noted "lines and groves of oak timber, growing along dry gullies" (Frémont

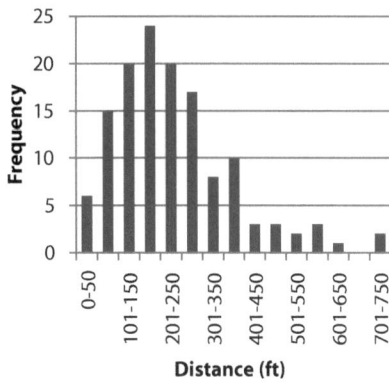

Figure 4.63. Distance class frequency distribution for the 134 oaks recorded in the GLO field notes occupying the alluvial fan of the Calaveras River. These distances are summarized from distances measured to trees nearest survey points. Relatively few trees are found close to the survey points, suggesting that these were not dense woodlands.

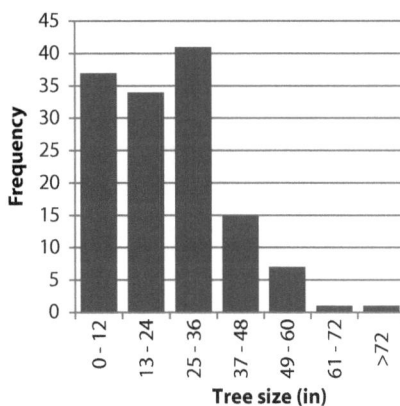

Figure 4.64. Size class frequency distribution for the 134 oaks recorded in the GLO field notes occupying the alluvial fan of the Calaveras River. Over 80% of the trees are no more than three feet in diameter, with a fairly even distribution in the three lowest size classes.

1845). Although historical soil surveys mapped clay adobes for a significant portion of the area (a soil type usually associated more with seasonal wetlands and wet meadows than oaks), they also note native cover that includes "occasional trees or small groves of valley oak" (Nelson et al. 1918). Local-scale differences in soil properties would have influenced the patterns at which these trees occurred.

At the landscape scale, the oaks were found standing alone and in groves, with a low herbaceous understory (Fig. 4.65). The Calaveras alluvial fan is one of few areas on the Delta periphery where historical accounts use words such as "fine oak park," "like a park," "beautiful groves of oak," "open groves of handsome trees," "covered with clumps and groves of oaks," and other terms commonly used to describe California's inland valleys (Moerenhout [1849]1935, Fremont 1849, Taylor 1854, Sal and Cook 1960). A landscape view is conveyed in engineer Grunsky's boyhood recollections:

> Our oaks covered the site of the city and extended far to the north, south and east. They were wonderfully graceful, giving a parklike character to the landscape. Their fascination never lost its charm for me. They did not stand in compact forest, but were isolated or in small picturesque groups and were of goodly size, generally from two to four feet in diameter and from sixty to a hundred feet high. There was no underbrush on the plains – just the iridescent green of grass interspersed with flowers, with here and there a pure golden patch where wild mustard or sunflowers had taken over. (Taylor 1969)

Unlike the treelessness of much of the Delta's ecotone, oaks were found within a mile of the tidal wetland edge. From the San Joaquin River, one traveler described the view: "As you approach Stockton, the uplands, oak-openings and glades of timber, begin to approach the river" (McCollum [1850]1960). This proximity is demonstrated by a GLO line between the Calaveras River and Stockton where two bearing trees are marked within 600 feet of the edge of tule.

Norris also notes that the boundary of tule was bordered by a thin strip of "marsh" almost 650 feet (200 m) wide (Fig. 4.66; Norris 1853b). It is unknown whether this ecotone "marsh" community identified by Norris was unique to this area. This pattern may have been common elsewhere at the tidal edge. For instance, an 1886 USCS descriptive report for Suisun Bay discussed the challenges inherent in determining the high water line based on the edge of tule. For their mapping purposes, they used "the inner edge of the tule in cases where there was a line of demarcation between tule and marsh grass" (Morse 1888). Such ecotones presumably represent lower wetland vegetation found at the edges of remaining tidal freshwater wetlands today. Since the spatial resolution necessary to map such an ecotone was more detailed than the sources available, we did not include it in the mapping.

As Stockton became a boom town during the early years of the Gold Rush, pressure on nearby oak woodlands became severe (Fig. 4.67). Within Charles Weber's Campo de los Franceses land grant, encompassing the area of

Table 4.3. Oaks found at varying levels of density in soils deemed high quality along GLO survey lines in the vicinity of Stockton.

Quote	Citation
"Soil 1st rate. Timber portion, large timber."	Norris 1853b
"Land level and first rate. Scattering oak timber."	Wallace 1865
"1st rate land and timber scarce."	Norris 1853b
"1st rate soil - fair timber."	Norris 1853b
"1st rate soil and timber improving."	Norris 1853b
"1st rate soil. An occasional tree."	Norris 1853b
"Land first rate and good timber."	Norris 1853b
"1st rate soil in ordinary timber."	Norris 1853b
"Soil 1st rate. Timber very thin."	Norris 1853b
"on edge of swamp land…Land level and first rate. Some timber."	Wallace 1865

present-day Stockton, trees were cut for building houses as early as 1844 (Smith 1853). The pressures on the riparian forests of the Sacramento River (see Box 5.7) expanded to the plains within reasonable distances of shipping ports. The loss of trees was noticeable as early as 1859: "the timber for any purpose but fire-wood, has nearly all disappeared" (Higley 1859).

ARROYO DE LAS CALAVERAS Unlike most of the Sierra Nevada rivers, the Calaveras River was an intermittent stream in its lower reaches, a characteristic noted by numerous travelers. The Calaveras was deemed a "non-navigable channel" in contrast to rivers like the Mokelumne and the Stanislaus (Figitt 1859, Stockton Commercial Association 1895). On June 14, 1848, a diarist noted that the Calaveras was "a stream of clear cool water a few yards in breadth" (Lyman 1848). At a location farther downstream where the Calaveras had begun to spread into multiple distributary channels, a GLO surveyor referred to a "dry bed of slough" when crossing the Calaveras in early June (Norris 1853b). By August the river was usually dry, with some water remaining in pools (Carson [1852]1931, Taylor 1854, Fugitt 1859). The seasonal nature of the Calaveras was attributed to its lower-elevation headwaters, with relatively little summer snowmelt water to sustain flow (Carson [1852]1931, Beaumont 1859b). In the winter, the Calaveras became a "deep and rapid river" (Carson [1852]1931). Nearing the vicinity of Stockton, its numerous distributary channels transported much of the winter floodwaters out onto the plain, which was noted as a potential obstacle for growing grain (Carson [1852]1931, Long in Houghton 1862, Hilgard 1884).

The Calaveras likely followed a pattern common to many seasonal streams historically as they entered their alluvial fans at the base of the foothills, that of spreading into numerous distributary channels. Textual descriptions suggest that the Calaveras may not have been well defined in parts of its lower reaches:

Nevertheless, for the first time in several days, we slept in a bed – the bed of Calaveras River, and in the deepest hollow of its gold-besprinkled sands. The stream, which in the spring is thirty feet deep, was perfectly dry, and the timber on its banks made a roof far above, which shut out the wind and sand, but let in the starlight

—TAYLOR 1854. AUGUST 1849, TRAVELING FROM STOCKTON TO THE MOKELUMNE RIVER

Figure 4.65. Views of oaks in the vicinity of Stockton. The map in (A) shows an oblique view of Stockton in 1870 shows scattered oaks beyond the city. Similarly, oaks can be seen in a photo (B) with the caption "Stockton, looking S. from Jas Littlehale's Tower" in the fields beyond what is likely Mormon Slough. (A: Britton & Co. 1870, courtesy of The Bancroft Library, UC Berkeley; B: photo by Batchelder ca. 1876, courtesy of Bank of Stockton Historical Photograph Collection)

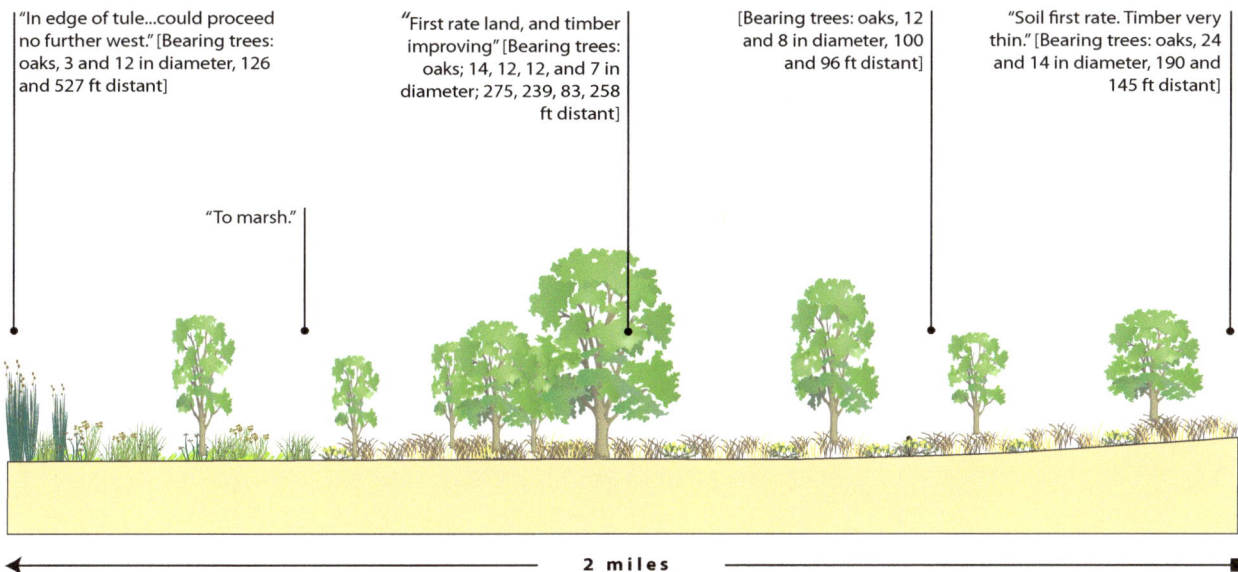

"In edge of tule...could proceed no further west." [Bearing trees: oaks, 3 and 12 in diameter, 126 and 527 ft distant]

"To marsh."

"First rate land, and timber improving" [Bearing trees: oaks; 14, 12, 12, and 7 in diameter; 275, 239, 83, 258 ft distant]

[Bearing trees: oaks, 12 and 8 in diameter, 100 and 96 ft distant]

"Soil first rate. Timber very thin." [Bearing trees: oaks, 24 and 14 in diameter, 190 and 145 ft distant]

2 miles

Figure 4.66. Upland ecotone near Stockton traversed by GLO surveyor Ralph Norris in 1853, above. Oaks were found close to the tule margin, apparently within or just alongside a thin strip of "marsh" at the tule margin. This line is located between Stockton Slough and the Calaveras River, shown in Figure 4.68.

Figure 4.67. Evidence of wood cutting in an undated photograph at the Port of Stockton. The photograph's caption reads: "Channel St. looking in a westerly direction from El Dorado St. New York Hotel right center. Oak wood in foreground is for fuel for steamboats (cordwood fuel for river boats)." (courtesy of The Haggin Museum, Stockton)

At the point where the North boundary strikes the Calaveras, it is about 35 feet wide and 10 feet deep. About 1 mile below this point the Calaveras divides, and loses itself in a Tule swamp, and for about two or three miles no trace of a River can be found. (Beaumont 1859b)

This and other sources indicate that the point at which the Calaveras divided into its distributary channels falls outside of the study area boundary. Within the study area, most of the early land case maps depict a defined Calaveras River channel extending from the edge of the study area down to the tule margin (Fig. 4.68; Unknown 1859, Unknown 1854, Beaumont 1858). One map includes sites of permanent water in the channel up to three miles (4.8 km) from the edge of tule, and patches of tule persisted in "spots and strips a yard or two wide for a mile or so up that [Calaveras] river" from the boundary of the Delta's tidal wetlands (McQueen 1859). Describing the Calaveras close to where it enters the tidal wetlands, a witness for the Los Franceses land grant case stated that "the Calaveras begins to spread out and ceases to be a river below where the Old Sacramento road crosses" (Buzzell 1859). Although the historical habitat type mapping depicts a primary flowpath, it should be recognized that this channel may not have been a true riverine channel and certainly did not possess the straightened channelized form it takes today in its lower reaches.

N
1 mile
1 kilometer

Tidal

Non-tidal

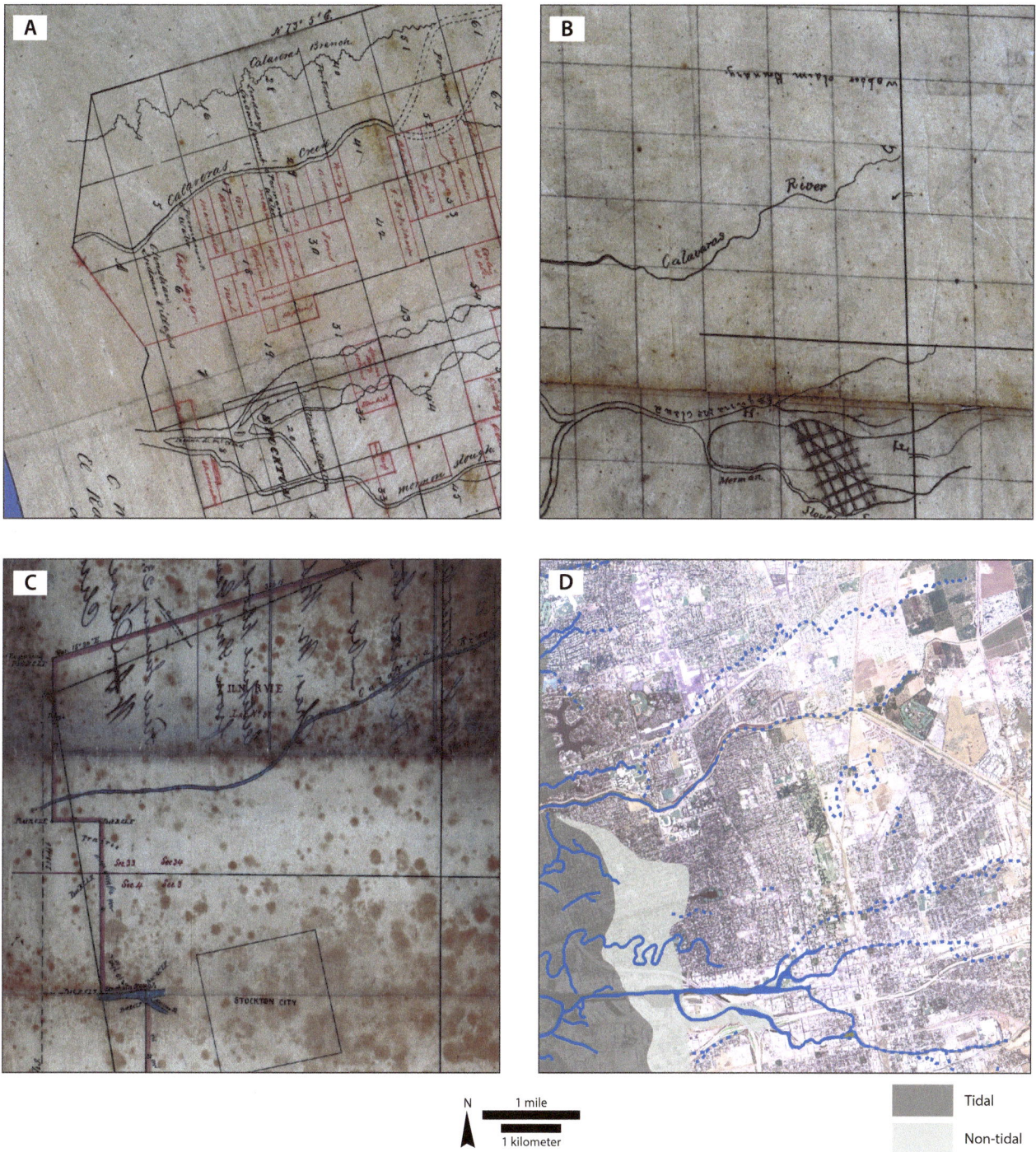

Figure 4.68. Mapping the Calaveras River. A land case map (A) for the Los Franceses grant illustrates the numerous distributary channels along the Calaveras alluvial fan as they near the Delta's perennial wetlands (flowing east to west). The dashed sections of channel, just outside of the study area boundary, likely correspond with surveyor Beaumont's (1859) description of where "the Calaveras divides, and loses itself in a Tule swamp" and "no trace of a River can be found." Two other land case maps, one from Beaumont's 1858 survey (B) and another from 1854 (C) depict defined channels of the Calaveras. The 2005 NAIP imagery is shown in (D) with the location of GLO line (2) shown in Figure 4.66. (A: Unknown 1859, courtesy of The Bancroft Library, UC Berkeley; B: Beaumont 1858, courtesy of The Bancroft Library, UC Berkeley; C: Unknown 1854, courtesy of The Bancroft Library, UC Berkeley; D: USDA 2005)

SUMMARY

Wetlands dominated by tule stretched north along the Sacramento River and its distributaries, gradually transitioning from tidal to non-tidal freshwater wetlands. The expansive wetlands were interspersed with large lakes but few channels, bordered by riparian forest along the rivers, and merged with seasonal wetlands at their upland margins.

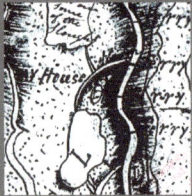

Basin morphology controlling habitat patterns (page 212) • Low-lying basins extended between gradually sloping natural levees and upland alluvial fans. The north Delta's two main flood basins, the Yolo and Sacramento, were over 40 and 20 miles (64 and 32 km) long, respectively, and several miles wide. Soils were less organic (page 217) and the tule grew taller, denser, and in more homogeneous stands than in the central Delta (page 219). Natural levees between river and basin were over 20 feet (12 m) high near the Feather River, gradually decreasing in height downstream (page 221). Tidal access was primarily limited to downstream outlets (page 224).

Seasonality and flow (page 230) • The Sacramento River naturally overflowed during high flow, spreading water into the basins and reducing flood peaks, where water flowed parallel to the river before re-entering through southerly outlets (page 232). Extreme floods could extend more than a mile beyond the edge of tule (page 232), which often occurred in late winter and spring in response to rain events (page 233). More flow could pass through Delta basins than the river channel in large flood events (page 235). Land could be inundated several feet deep (page 236) with large areas sometimes remaining flooded for several months, into summer (page 237).

Channels at the fluvial-tidal interface (page 238) • The Sacramento River flow capacity was a fraction of flood flows (page 240), width was about 600 feet (183 m) near Sacramento, depths were usually over 7 feet (2.1 m), and sand bars maintained distinct positions (page 241). On the Mokelumne, the north fork was wider and deeper than the south fork (page 244). Channels within wetlands were concentrated in the downstream, more tidal portions (page 246). Other channels that bisected natural levees carried flood flows into the wetlands (page 251).

Lakes and ponds of the wetlands (page 255) • The largest lakes of the Delta (page 258) were found in the flood basins outside of or at the upper limits of tidal influence (page 259) and surrounded in part or completely by large expanses of wetlands (page 263). The lakes and ponds were filled by and connected to the river via flood flows, many became isolated in the dry season (page 265), and some lacked direct channel connections (page 267). In addition to pondweed, yellow pond lily grew in the shallower lakes and ponds (page 268).

Riparian forest extent and composition (page 274) • Structurally complex riparian forests grew on natural levees of rivers and distributaries. Riparian forest width and composition tracked the natural levee height and width as it diminished downstream (page 280), becoming minimal at Rio Vista on the Sacramento River and just below the head of Staten Island on the Mokelumne (page 282). Width was generally about half a mile wide (0.8 km; page 285). The forest was composed primarily of oaks and sycamores, with an understory of willow (page 288), often with transitional vegetation at the edge (page 292).

Sinks at distributaries (page 294) • Many smaller rivers and creeks became distributaries and dissipated at the wetland edge. Putah and Cache creeks (page 295) and the Cosumnes River (page 298) had sufficient flow to support large thickets of willow and other riparian species, referred to as "sinks."

Upland ecotone (page 301) • Perennial wetland margins primarily consisted of seasonal wetlands, temporally overflowed by upland distributaries and occasionally by extreme Sacramento River floods. Vegetation assemblages varied considerably with topography at this edge (page 302). Vernal pools were common, including localized alkali areas (page 304). Points of higher land sometimes intruded into the wetland edge (page 305).

INTRODUCTION

As tidal influence became less significant to the north, the tidal islands of the central Delta historically gave way to a landscape where fluvial processes played a significant role in shaping habitat form and function (Figures 5.1 and 5.2). Low basins, or flood basins, – often several miles wide and tens of miles long – lay parallel to the Sacramento River's natural levees. They were inundated by floods in the wet season and supported perennial wetlands. The north Delta, as it is defined in this report, includes the roughly 360,000 acres (145,687 ha) located generally within the 25 foot contour and extending north from Cache Slough, Grand Island, and upper Tyler and Staten islands, to the Feather River confluence on the west side of the Sacramento River and to the American River on the east side (Figures 5.3 and 5.4).

The basin landforms reflected the Sacramento River's high magnitude sediment-laden flood flows. Natural levees outlined the basins along the rivers and distributaries, gradually building up in elevation with sand and silt deposits from periodic overtopping and supporting dense riparian forests. In contrast, the lower-lying basins behind the natural levees accumulated the finer sediments that produced clays and soils with organic content.

In this landscape, the fluvial-tidal interface interacted with distinct topographic and geologic environments to produce habitat mosaics arranged along the broad physical gradients offered by the basin morphology (CDFG and YBF 2008). At the large scale, zones of tidal freshwater emergent wetland in the southern portions of the basins graded northward into non-tidal freshwater emergent wetland. This transition was masked by dense tule (*Schoenoplectus* spp.) growth. Numerous channels lacing the wetland plain in the more tidally-influenced areas, adding landscape complexity. The channel form and natural levee height depended on whether flood flows were received directly from the major sediment bearing rivers (high natural levees) or indirectly from the flood basins where most sediment had already settled out (low natural levees). Where tidal influence was slight or non-existent in the upper basins, few

In the basin formed by the natural levee of the streams and the high grounds to the rear, lie these vast bodies of swamp lands. The original body of water contained in them is constantly fed by the annual overflows, the back water of the sloughs connecting with rivers, the fresh water streams which flow into the tules and the constant absorption through the porous soil.

—BREWSTER 1856

Figure 5.1. Waterfowl in flooded lands of the Sacramento Valley (chapter title page). (photo by William G. Miller, Cole~Miller Photography, December 31, 2011)

Figure 5.2. Riparian forest along the Mokelumne River. Forests described as jungles occupied the natural levees of the major rivers and distributaries. These landforms owed themselves to the riverine floods that passed through the valley. (photo by Daniel Burmester, August 24, 2005)

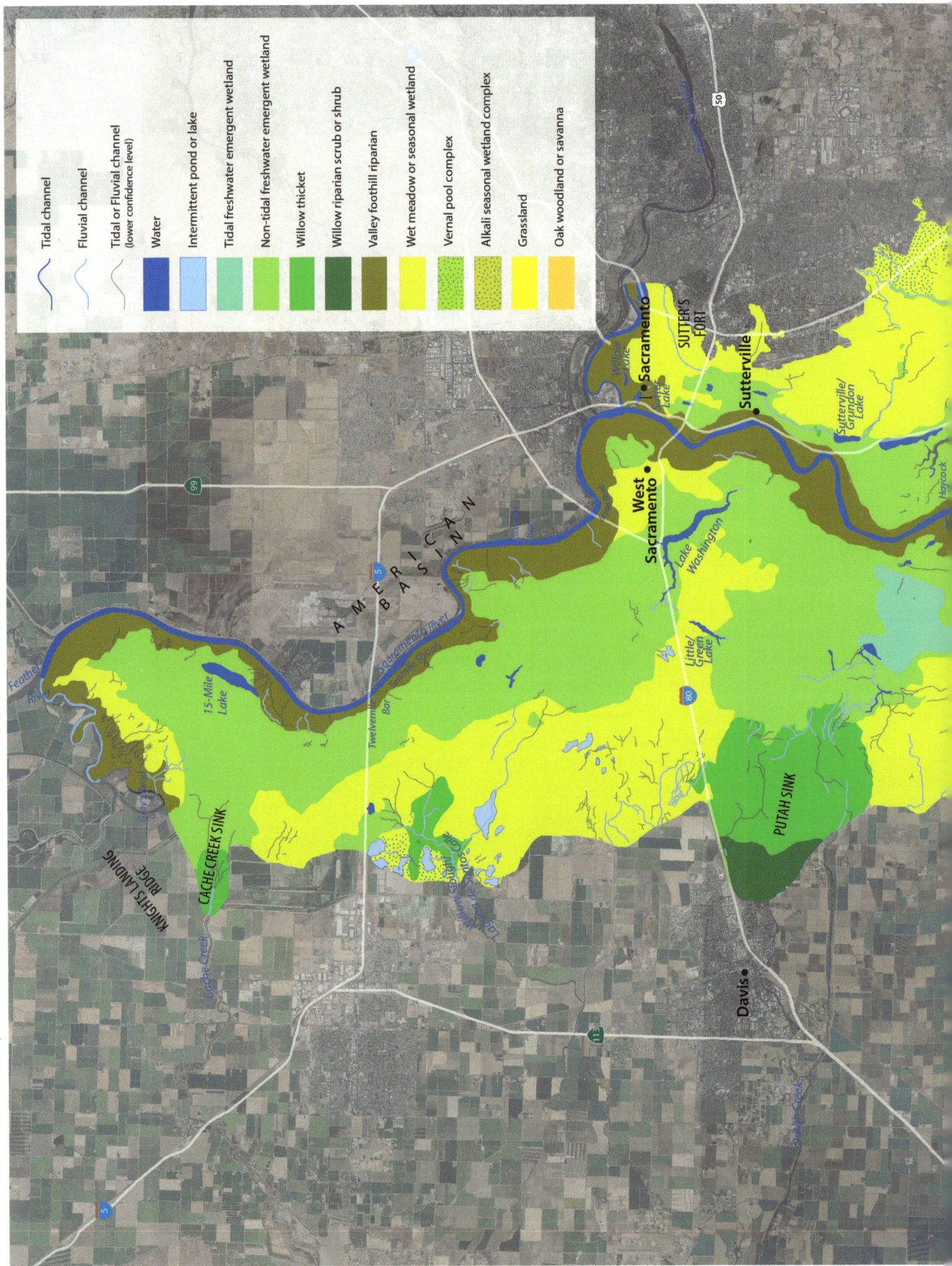

Legend:

- Tidal channel
- Fluvial channel
- Tidal or Fluvial channel (lower confidence level)
- Water
- Intermittent pond or lake
- Tidal freshwater emergent wetland
- Non-tidal freshwater emergent wetland
- Willow thicket
- Willow riparian scrub or shrub
- Valley foothill riparian
- Wet meadow or seasonal wetland
- Vernal pool complex
- Alkali seasonal wetland complex
- Grassland
- Oak woodland or savanna

Figure 5.3. Distribution and extent of habitat types within the north Delta basins landscape in the early 1800s. Large low-lying basins occupied by tule-dominated wetlands lay between the upland plains and the riparian forests of the Sacramento and Mokelumne rivers and their

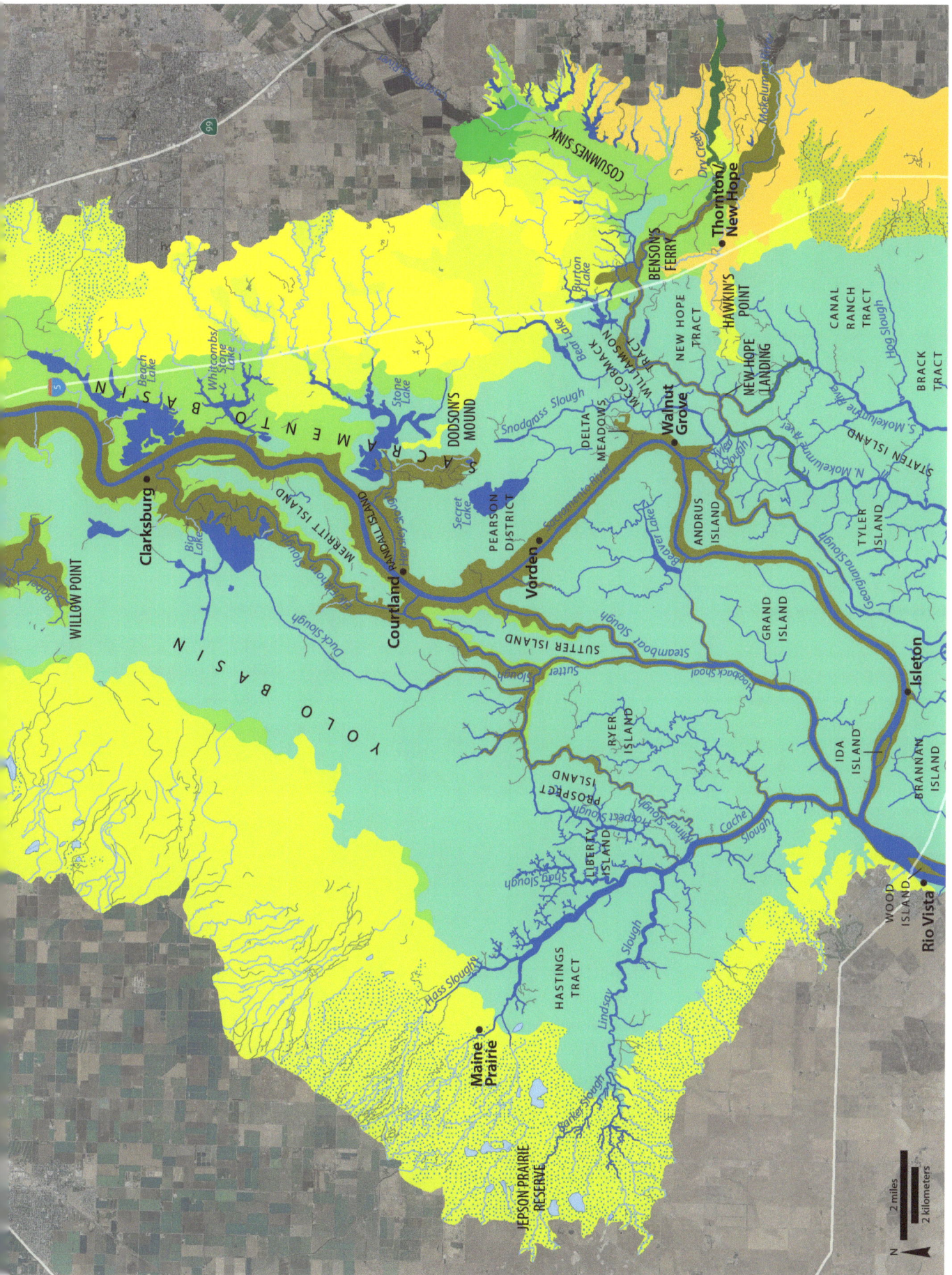

distributary channels. Tidal and intermittent overflow channels coursed through the wetlands and lakes and ponds occupied the lowest and most isolated parts of the basins. Habitat types were arranged in predictable mosaics according to the landforms and physical gradients of the north Delta. (USDA 2009)

Figure 5.4. Conceptual diagram of the north Delta flood basins landscape. Flood basins of the north Delta were greatly influenced by the flooding regime of the Sacramento River as well as other streams that regularly overflowed into the low-lying basins running parallel to the rivers. Large lakes occupied the lowest and most isolated positions, and few channels penetrated far into the dense emergent vegetation wetland plain as it transitioned gradually away from tidal influence upstream. The basins were bounded by riparian forest along natural levees and seasonal wetlands of the upland margin. The relative proportions of habitat types based on the map are illustrated in the pie chart.

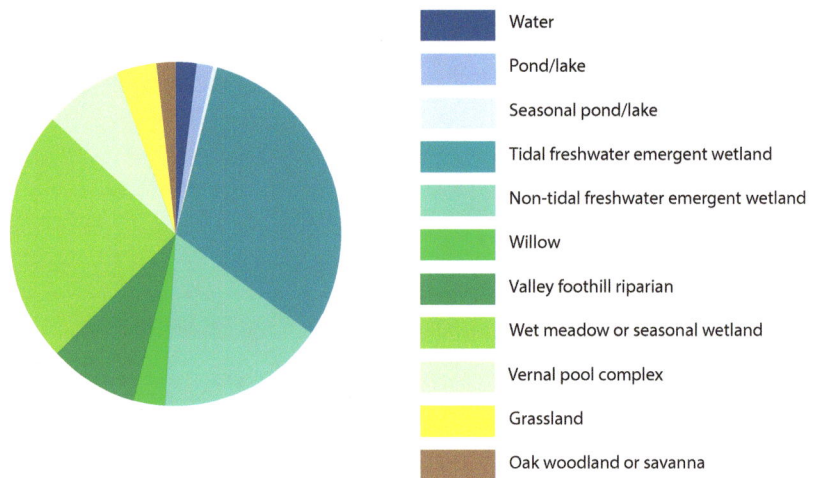

- Water
- Pond/lake
- Seasonal pond/lake
- Tidal freshwater emergent wetland
- Non-tidal freshwater emergent wetland
- Willow
- Valley foothill riparian
- Wet meadow or seasonal wetland
- Vernal pool complex
- Grassland
- Oak woodland or savanna

channels were present to break up the dense tule-dominated vegetation (Board of Swamp Land Commissioners 1864-5, Jepson 1893). Instead relatively shallow perennial lakes and ponds occupied low-elevation, backwater positions (Browning 1851, USGS 1909-1918). Many parts of the flood basins may have been at least seasonally disconnected from tidal influence as a result of the intervening natural levees.

Riparian forest and upland habitats bordered the basins. The riparian forests along the river's natural levees varied in width, ranging from narrow strips at the tidal end of the spectrum and along the smaller channels to well over half a mile in width on the broad natural levees of the Sacramento River. They comprised the downstream extent of the once extensive and often wider diverse gallery riparian forests of the Sacramento Valley. Dominated by oak (*Quercus* spp.) and California sycamore (*Platanus racemosa*) in the canopy and willow (*Salix* spp.) in the understory, these forests offered rich habitat complexity at the tidal wetland edge. Occasional crevasse splays broke through the natural levees and brought deposits of inorganic sediment out into the basins, adding complexity at the edge by permitting the extension of this forest further into the wetlands. At the upland margin, perennial wetlands of the flood basins transitioned to less frequently inundated and less saturated seasonal wetlands (including alkali and vernal pool complexes) and were occasionally intersected by stream distributary "sinks" supporting willow thickets.

The Sacramento River channel alone was insufficient to carry most winter peak flows, being naturally sized to pass much of its flood flows through the basins. In addition, annual flows from the smaller systems, including all of those draining the western Coast Range, passed through flood basins before entering the river proper (Bryan 1923). Only larger rivers such as the American and Mokelumne connected directly to the Sacramento. At times, flows through the basins were greater than those of the Sacramento. Water passed more slowly through these wide floodways than it did through river channels, reducing peak flows entering the Delta (Gilbert 1917, Thompson 1957, TBI 1998). The floodwaters formed what many referred to as immense lakes or inland seas; inundation extended for many square miles during high flows and persisted for weeks, if not months. Nevertheless, parts of the basins did become dry at the surface late in the season, evidenced in part by the numerous tule fires observed during the early settlement period.

The north Delta landscape supported rich assemblages of species within a highly productive ecosystem. Hydrologic connectivity and the depth, duration, and frequency of flooding would have directly affected the ability of aquatic species to utilize the flood basin habitats and to access upper watersheds, such as Putah and Cache creeks. Lakes were likely frequented by migratory fish during high flows, when river and lake were functionally connected. Other fish associated with slower moving waters, such as thicktail chub, lived there year-round. Both resident and migratory waterfowl also used these lakes, grazing on submerged aquatic vegetation and floating-leaf aquatics. The lakes, marshes, and surrounding seasonally inundated lands made the Delta an important stop along the Pacific Flyway (Garone 2011). The high residence time associated with the water retention in these lakes and in the basins as a whole provided substantial capacity for aquatic food web development and nutrient exchange between the marsh and aquatic environment. In the riparian forest, a diverse and abundant array of birds occupied the many available niches. The forest also provided opportunity for terrestrial species such as elk, grizzly bear, and smaller mammals to access the wetland and aquatic environments of the tidal Delta.

The following sections discuss the historical habitat patterns and characteristics of the early 1800s north Delta landscape.

Into these low lands or basins thus formed empty numerous creeks from the foothills of the Sierra Nevadas and coast Range of mountains, and the overflows of the rivers at high stages of water keep them full during the wet seasons, and generally well into the Summers. Hence the soil becomes wet and swampy, and all vegetable growth coarse and rank.

— STATE AGRICULTURAL SOCIETY 1872

BASIN MORPHOLOGY CONTROLS OF HABITAT PATTERNS

The flood basin is the primary geomorphic unit which framed the historical habitat patterns of the north Delta (CDFG and YBF 2008). These basins formed as Pleistocene alluvial fans to the east and Holocene alluvial fans to the west encroached on the valley and as the Sacramento River developed its natural levees since the last glacial period, leaving lower areas in between (Dawson 2009, Gutierrez 2011). The Yolo Basin to the west of the Sacramento River and the Sacramento Basin to the east fall within the study area and are the two southernmost basins of a series that continues north up the Sacramento Valley (including the American, Colusa, Sutter, and Butte basins in addition to the Yolo and Sacramento basins). A geomorphic term in use today, "flood basin" was first applied in the Sacramento Valley in state engineering documents of the late 1800s (Rose et al. 1895). Scientists used the term extensively in the early 1900s, defining these distinctive floodplain environments as a "natural trough or depression" (Mann et al. 1911), a "system of settling basins" (Gilbert 1917) and "broad but shallow troughs between the low plains and river lands" (Bryan 1923). Over several miles, the gradually sloping natural levee deposits of the rivers and encroaching alluvial fans of upland drainages bounded these "troughs" (Fig. 5.5; Brewster 1856, McGowan 1961, CDFG and YBF 2008).

The downstream extents of the basins were tidal, at approximate high tide levels (*Sacramento Daily Union* 1862a). Here, the margin of perennial wetlands marked the tidal limit. As the elevation of the basin plains increased gradually northward, non-tidal perennial emergent wetlands continued up the basin (Fig. 5.6). The elevations increased to over 10 feet (3.0 m) above mean sea level in the Sacramento Basin near the American River (a fall of about 7 in/mi/11.1 cm/km) and to over 20 feet (6.1 m) in the Yolo Basin near the Feather River (a fall of about 5 in/mi/7.9 cm/km; USGS 1909-1918). An early Yolo Basin survey west of Sacramento recorded the lowest point about one and a half miles (2.4 km) from and about 13 feet (4.0 m) below the river bank and noted that the height difference decreased downstream (Mathews in Houghton 1862). On the Sacramento Basin side, the lowest areas were reportedly 15 feet (4.6 m) below the river bank (Sanford 1860).

The basins were much longer than they were wide: the Yolo Basin extended over 40 miles (64.4 km) and the Sacramento Basin over 20 miles (32.2 km), but both were only a few miles in width. The Yolo Basin began at Knights Landing Ridge (a ridge of Holocene alluvium built by an ancestral course of Cache Creek) and ended at the mouth of Cache Slough. The Sacramento Basin began south of the American River and drained into Snodgrass Slough and the Mokelumne River. It was described as "an irregular narrow basin" that had "a very irregular eastern border" (Bryan 1923). As defined by the mapped extent of historical freshwater emergent wetland, we estimate that the Yolo Basin

To the east the surface slope is toward the Sacramento River and on the west side the basins merge into the upland plain without distinct boundaries. The surface of the region is flat. Shallow sinks occur in which surface water remains until dispelled by evaporation.

—MANN ET AL. 1911
DESCRIBING THE YOLO BASIN

(excluding Ryer Island and marshland south of Cache Slough) was about 73,600 acres (29,780 ha; about 52% within tidal range at the downstream end), while the Sacramento Basin (extending down to the mouth of Snodgrass Slough) on the east side of the river was both narrower and shorter and occupied about 32,700 acres (13,230 ha; about 44% within tidal range; Fig. 5.7). The basin boundaries cannot, however, be precisely defined because vegetation and hydrologic characteristics shifted within broad ecotones that looked and functioned differently depending on the time of year.

The basins were isolated from the river by the natural levees along the Sacramento River and its primary distributary channels (e.g., Steamboat Slough, Elk Slough; Atwater 1982, Thompson 2006). Unlike in most floodplain environments, water levels in the basins did not directly respond to those in the adjacent river channel (Sanford 1860, CDFG and YBF 2008). Once water entered the basin at high river stages, it was unable to return to the river the way it came due to the natural levees (Brewster 1856, Control of Floods 1916, Gilbert 1917). Groundwater levels were high, usually at or within several feet of the surface (Holmes and Nelson 1915, Unknown 1919, TBI 1998). The basins were kept wet year-round by these stored floodwaters, as well as by inputs from upland streams that drained into the wetlands, seepage from adjacent river channels, the high groundwater levels, and tidal flows in some places.

Floods were quite important to the form and function of the landscape. Fluvial depositional processes formed and maintained the basins over time (Bryan 1923). Flood waters from the Sacramento River and the smaller drainages spread across the basins as broad, slow-moving sheets of water, which gradually drained and evaporated after flooding (TBI 1998). Geomorphologist Kirk Bryan (1923) described the lacustrine-like basin deposition as "that resulting from standing water rather than running water," creating broad relatively homogeneous flat topography (Fig. 5.8).

The downstream connection to tides was an important feature of the Yolo and Sacramento basins. Because of this interaction, the basins were a part of the Delta's upper deltaic plain, a term used to describe the outer limits of a delta where fluvial processes dominate over tidal influence and where the surface has only relatively recently come under tidal influence (Coleman 1976, Brown and Pasternack 2004). The landforms and associated habitats were affected by the tides accessing the flood basins through tidal channels as well as by the annual floods that filled the depressions, deposited sediment and woody debris, and established hydrologic connections. These bidirectional processes played out at the landscape scale and have been largely disconnected from most parts of the Delta today.

The flood basins and their bordering natural levees comprise a landscape unlike that of the central Delta. Numerous accounts note the stark contrast between the "pure tule swamp" along the central Delta channels and the

The flood basins can be defined either as the tracts actually covered by water during the highest known floods or as the flat areas between the sloping low plains on one side and the river lands on the other, occupied by heavy soils and commonly having either no vegetation or a strictly swamp vegetation. Under either definition the boundaries of the basins are indefinite and usually are transitional in character.

—BRYAN 1923

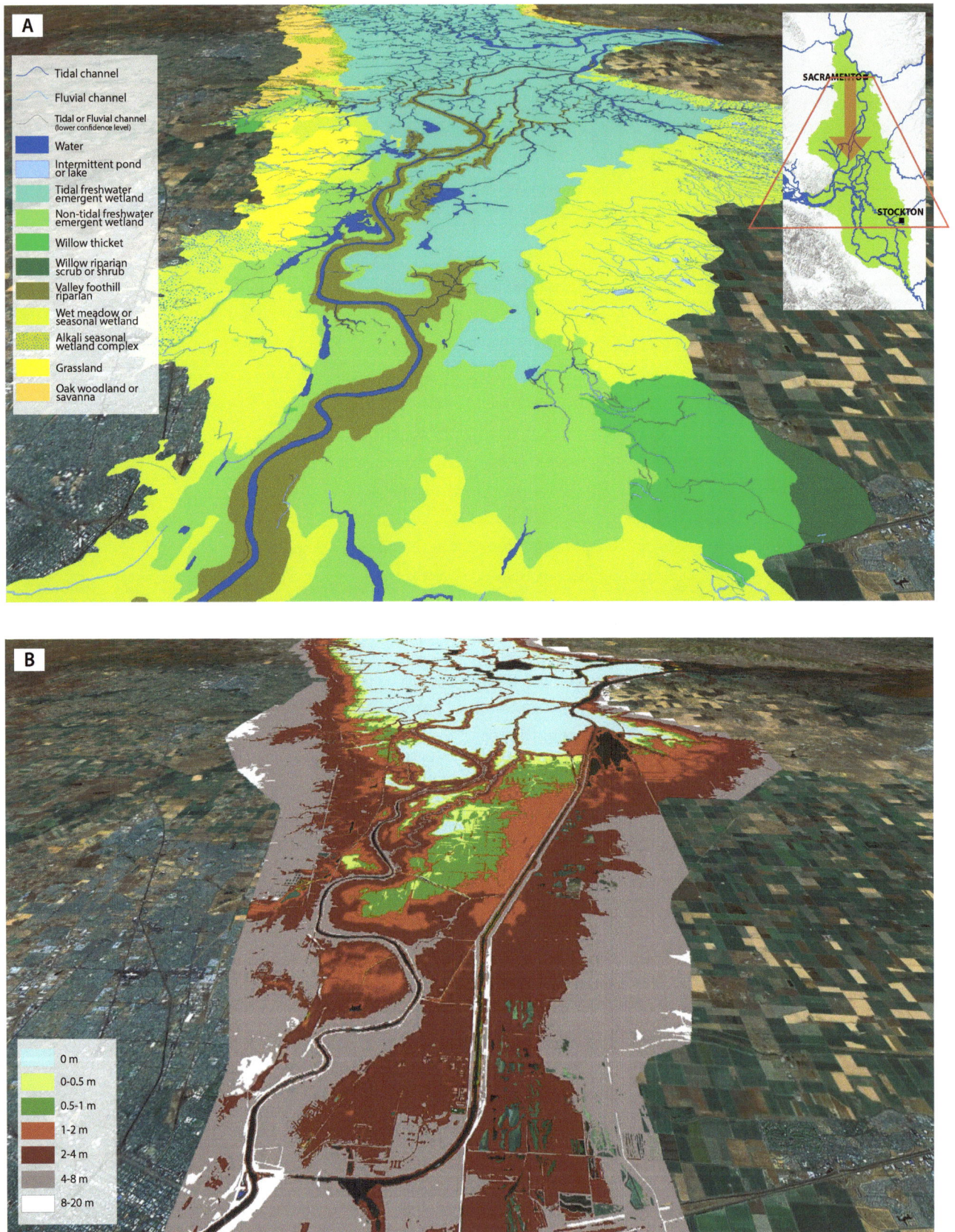

Figure 5.5. Oblique views of the north Delta landscape looking southward show the basins (freshwater emergent wetlands) lying alongside the natural levees of the Sacramento and Mokelumne rivers and their distributaries. The distribution of historical habitat types in (A) illustrates the relationship of landscape patterns to the major Delta landforms and topography. In (B), 2007 LiDAR imagery is displayed with warm colors depicting land above tide level and cool colors below, illustrating the gradual shift toward lower elevations in the central Delta. Also, the natural

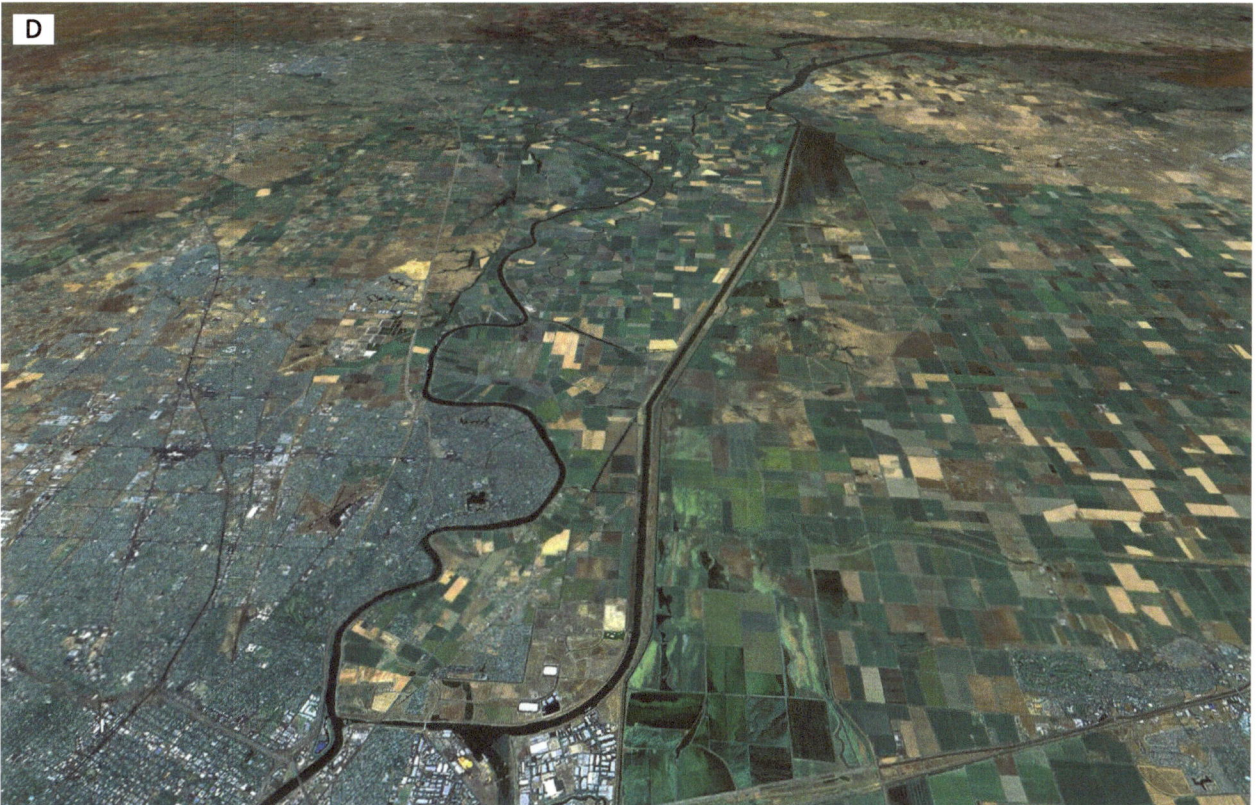

levee topography and adjacent, gradual topographic gradients are clearly visible in the landscape today. These extend as fingers of land along channels and above tide levels well into the historical tidal Delta. Changes, such the construction of the Yolo Bypass, flooding of Liberty Island, and urban expansion, are visible comparing the aerial imagery from 1937 (C) and 2009 (D). (B: DWR 2007, C: USDA 1937-1939, D: USDA 2009)

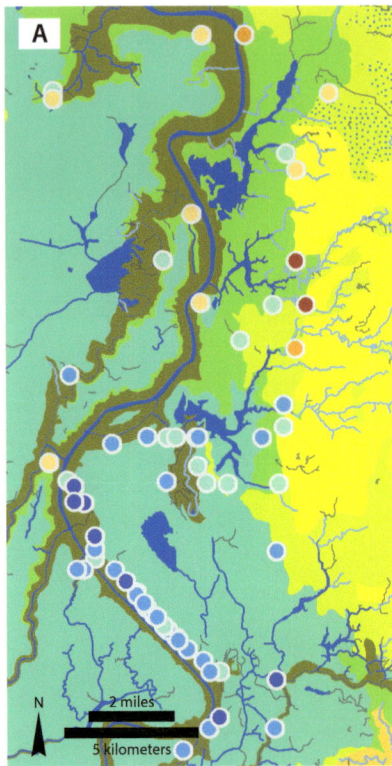

A

Elevation (ft above sea level)

- 0-2 ft
- 2-4 ft
- 4-6 ft
- 6-8 ft
- 8-10 ft
- 10-12 ft

B

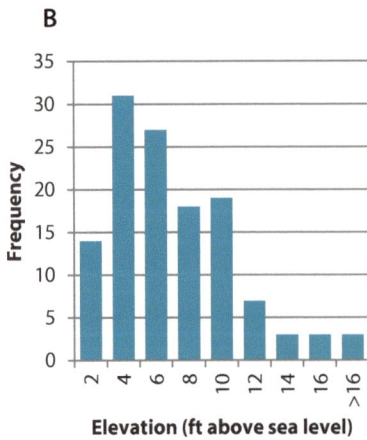

Frequency vs Elevation (ft above sea level): 2, 4, 6, 8, 10, 12, 14, 16, >16

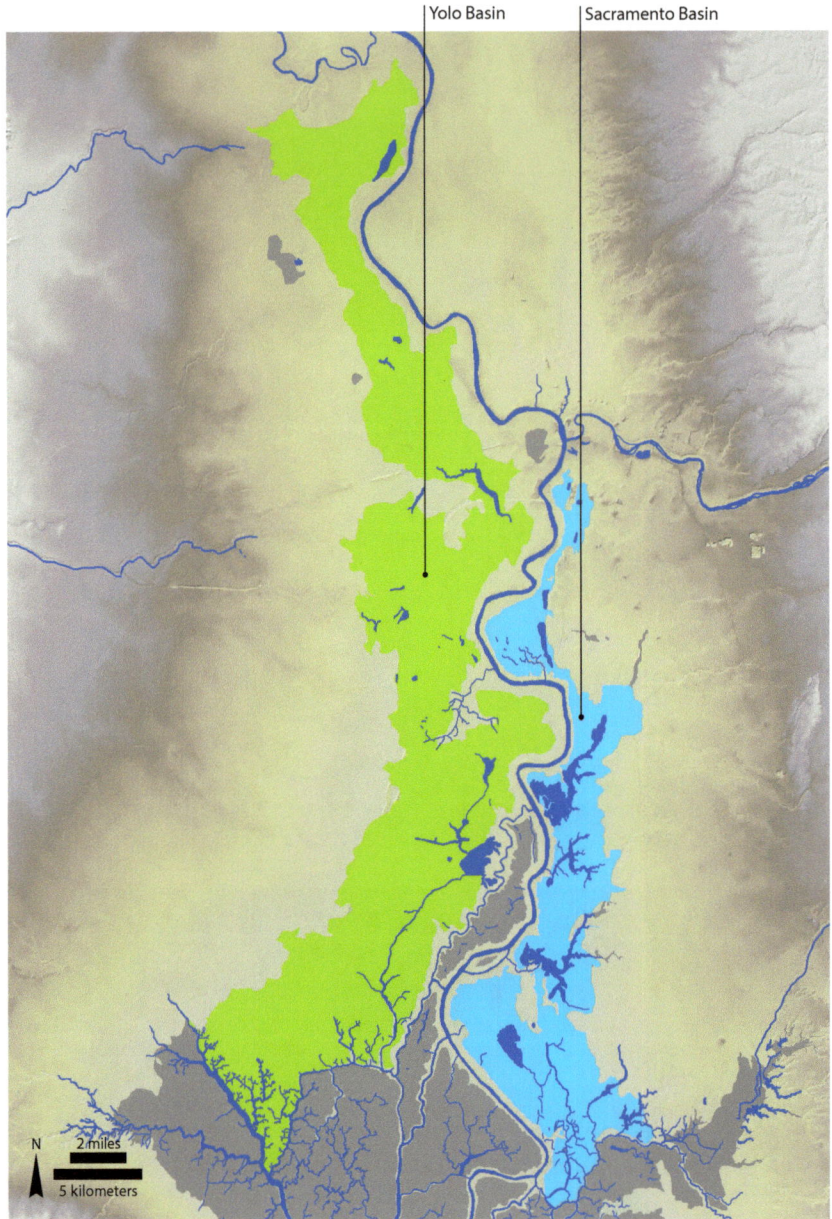

Yolo Basin   Sacramento Basin

Figure 5.7. Comparing the Yolo and Sacramento basins. The Yolo Basin (green) was over twice as large as the Sacramento Basin (blue), with a much lower proportion of area in ponds or lakes. The Sacramento Basin was characterized by greater topographic variability and consequently supported larger and more frequent lakes that were trapped by higher ridges of land preventing their drainage. At the southerly ends, each basin drained into and was fed by a large slough that brought tides into the lower regions of the basins and provided passage for floodwaters into the central Delta.

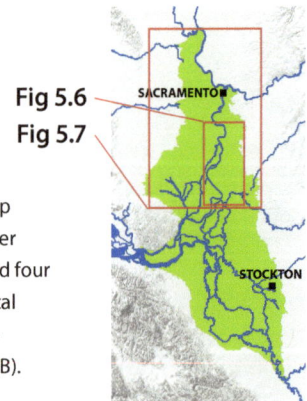

Fig 5.6
Fig 5.7

Figure 5.6. The elevation of the margin of tule as indicated by General Land Office (GLO) surveys. This map (A) of the region around the Pearson District shows the nineteenth century GLO points where the surveyor either entered or exited tule. It illustrates that the tule margin largely marks the extent of tidal reach (between two and four feet) in the lower compartment of the basin. The points are symbolized by their associated elevation in historical USGS topographic maps (USGS 1909-1918). In the non-tidal portion of the basin, tule is found well above these elevations. The distribution of these and other GLO points that mark the edge of tule is shown in the bar chart (B). The points fall most frequently within the range of high tide elevations (2-4 feet / 0.6-1.2 m).

view of "dry land on both banks, with groves of trees" along the Sacramento River to the north (Durán and Cook 1960). The basins followed a basic pattern of higher-elevation riparian forest along the rivers and valley plain at the upland edge, tule-dominated emergent wetland within the extent that was regularly flooded, and lakes and ponds occupying the lowest elevation positions (Fig. 5.9; Browning 1851). These landscape components were much larger than the tidal islands of the central Delta and were topographically defined by the bounding natural levees and upland alluvial fans rather than by wide tidal channels.

Soil properties

Historical soil patterns reflected the flooding patterns. In times of river overflow, the velocity of sediment-laden flood waters decreased first across the natural levees and then across the plain of the flood basins, dropping larger particles from the water column closer to the river and forming the loamy soils of the natural levees. The remaining finer sediments deposited within the basins and formed soils with a greater clay fraction (Bryan 1923, Etcheverry 1924). The resulting soil patterns were captured by historical soil survey maps and descriptions (Holmes and Nelson 1915, Carpenter and Cosby 1930). The natural levee soils were defined by "unweathered recent

Figure 5.8. The vast, flat, and flooded Yolo Basin is seen in this photograph of what appears to be a predessor road to I-80. Broad shallow flooding can be seen in the background. Floods previously supported the emergent wetlands dominated by tule, though these had been cleared by the time of the photograph. (photo ca. 1900, courtesy of the Center for Sacramento History, Yolo Basin Collection, 1981/001/065)

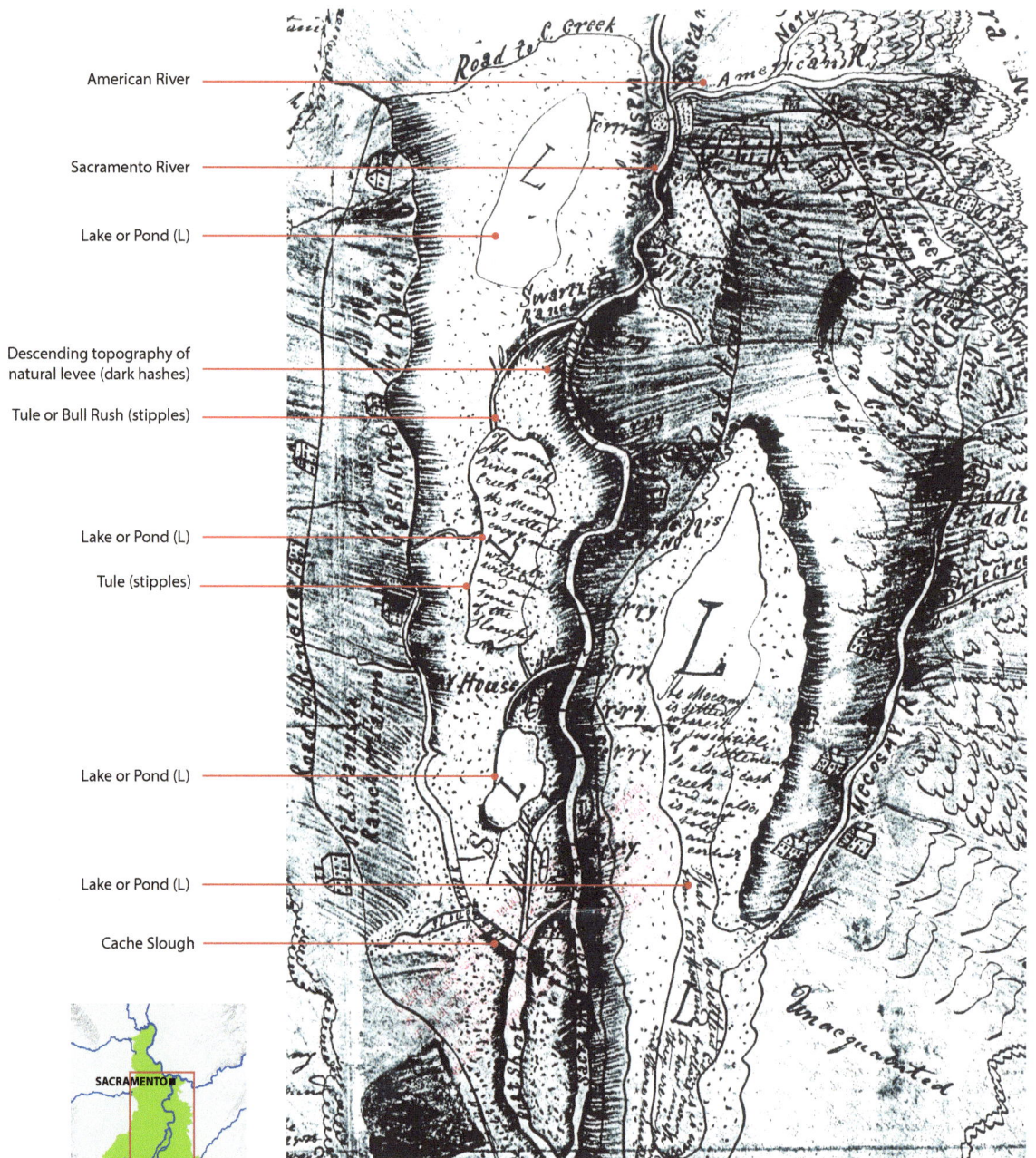

American River

Sacramento River

Lake or Pond (L)

Descending topography of
natural levee (dark hashes)

Tule or Bull Rush (stipples)

Lake or Pond (L)

Tule (stipples)

Lake or Pond (L)

Lake or Pond (L)

Cache Slough

SACRAMENTO

STOCKTON

Figure 5.9. An 1851 sketch of the basin landscape pattern. Though the author of the letter
and this map got few of the facts wrong (e.g., Cache Creek did not connect directly to Cache
Slough), he does convey the character of the basin landscape. To describe the pattern and his
map, George Browning (1851) stated that the land near the river descends (the dark pencil
hashes), "then comes the Tola or Bull Rush [stipples]…and then comes the Lake or Pond
[marked with 'L']." (Browning 1851, courtesy of The Bancroft Library, UC Berkeley)

alluvial stream deposits" (Carpenter and Cosby 1930) and referred to as
"heavy sediment land" or in some places "light sandy loam" (Tucker 1879c).
These higher, well drained levee lands provided ideal conditions for the
orchards planted along the river beginning in the 1850s (as well as towns;
Fig. 5.10; Tucker 1879d, Holttum 1879 in Paterson et al. 1978). Within the
basins, the soil surveys mapped various clay loams and mucky clay loams,

noting "heavy texture," "partly decayed organic matter," and "poorly drained" soils (Carpenter and Cosby 1930, Cosby and Carpenter 1931). In the basins, organic matter content was relatively high and formed characteristic mottling patterns.

While the basin soils had greater organic content than the adjacent natural levees and alluvial fans, they were also less uniformly organic than the soils of the central Delta landscape, reflecting the greater influence of fluvial as opposed to tidal processes. An inspection of soils a mile from the river on Grand Island reported on "a fine yellowish sediment of tule deposit" about a foot deep with a loam and clay soil "checkered with streaks of decayed vegetable matter" for four feet below that (State Agricultural Society 1872). Shallow peat soils were found at the lower extents of the basins (Carpenter and Cosby 1930), where organic matter had only recently begun to accumulate due to the recent estuarine transgression as sea levels rose. For example, reclamation documents reported that the Pearson District (Reclamation District 551) at the lower end of the Sacramento Basin was covered by a three to four foot deep peat layer over clay (Tucker 1879d).

Though the processes that influenced the soil properties may no longer be functioning, these soil patterns are still reflected in the landscape today (see Fig. 5.10). Natural levees are still the best places for orchards, and clayey soils are still found throughout the lower basin lands. The thin layers of peat that once existed at the tidal margins have largely disappeared through oxidation (Carpenter and Cosby 1930, Cosby 1941).

Relationship to emergent wetland characteristics

Hydrology, topography, soils, and disturbance regimes affected emergent wetland characteristics in the basins. As one newspaper article summarized, "the heavier the soil, dampness being equal, the heavier the growth of trees and vegetation" (*Pacific Rural Press* 1871). The presence of water in the basins late in the growing season translated to vegetation communities that contrasted sharply against the rest of the valley, as noted by Jepson (1893) of Delta vegetation: "The herbaceous vegetation is, therefore, late aestival and

The amount of mineral matter in the soil becomes progressively less with distance from the sloughs.

-CARPENTER AND COSBY 1930

Figure 5.10. Orchards occupying the natural levee lands. Today, orchards almost completely cover the historical extent of riparian forest (outlined in orange). The rich alluvial soils of the natural levees that once supported the lush gallery forests of the Central Valley are excellent for the pears and other orchard crops that can be seen extending back from the levee roads today. Their extent is limited by the soils of heavier texture in the interior. (USDA 2009)

autumnal. It succeeds the dry season, as that of the plains adjacent precedes it." The combination of water, temperature, and long growing days yielded highly productive and lush plant growth in the otherwise dry summer and fall.

The basins were defined by the extent of tule-dominated wetland. The fine clay soils supported dense, tall (usually over 10 feet/3 m) stands of tule (McGowan 1961). A historical soil survey that encompasses the upper Yolo Basin offers further detail, noting that the "native vegetation consists of a thick growth of tules, smartweed, mint, and other aquatic or semiaquatic plants" (Mann et al. 1911). Numerous accounts remark on the dense tules and several rare photographs of land reclamation activities in the Yolo Basin lend insight into high levels of biomass productivity within the north Delta basins (Fig. 5.11). One traveler described "an immence groth of weeds and rushes so high and strong that a horse is unable to breake through" (Clyman and Camp 1928[1848], spelling as original). Another similar account describes "simply immense rushes, which cover the ground with an almost impenetrable thicket...frequently attaining the height of sixteen to eighteen feet" (Sprague and Atwell 1870). The abundance of plant matter in the Yolo Basin is remarked upon in Swamp and Overflow Land records: "a rank growth of new tule and masses of drifting tule of former seasons" (Box 5.1; Board of Swamp Land Commissioners 1864-5).

These and other historical accounts confirm the presence of taller, more homogeneous (i.e., with less willow and other species intermixed) stands of tule within large portions of the north Delta basins compared to the central Delta. Wind and wave action and climatic controls (e.g., maritime influences) likely affected these patterns, where the more structually sound *Schoenoplectus californicus* would have been more prevalent in the windy western Delta than the tall *S. acutus* that prefers more protected areas (Keeler-Wolf pers. comm). It may also be that the difference reflects the greater dominance of fluvial processes within the upper reaches, or upper deltaic plain, of the Delta (Pasternack and Brown 2006). The more frequent, higher magnitude, and higher sediment-load flood events may have helped maintain the flood basin wetlands at a younger successional stage, with fewer willows than in the central Delta. Mason (n.d.) raises the issue of different successional marsh and swamp stages within the Delta as an interpretation of the communities he found in the mid-1900s.

It is important to consider that the large scale of the basins translated to similarly large-scale habitat type patterns. In contrast to central Delta wetlands, basin wetlands occurred within a mosaic of other floodplain habitats, including riparian forest, willow thickets, lakes and ponds, and a variety of seaosnal wetland types, arranged according to the major basin landforms (Brown and Pasternack 2005). This is also different from the south Delta floodplain landscape, which was comprised of patchworks of habitat types at a more local scale (see page 351) the north Delta basins contained larger areas of single types. For instance, small patches of seasonal wetlands or willow thickets were less frequent within the matrix of emergent wetland in the north Delta basins.

Figure 5.11. Dense stands of tule, well over head height, are seen in this photograph of land reclamation in the Yolo Basin. The image lends credence to the many narrative accounts of tule being over ten feet high in the basins. (Tule 1916, Holland Land Co., D-118, courtesy of Special Collections, University of California Library, Davis)

Natural levees

Natural levees and associated crevasse splays along the river are important geomorphic features defining the flood basins. Natural levees are formed through depositional processes and are commonly found along valley reaches of larger rivers. They form through gradual accumulation of coarser sediments deposited from of the water column when floodwaters overflow the river channel. They are thought to develop particularly in situations where fast moving water within a channel meets the slow moving water of a floodplain (Atwater and Belknap 1980). The majority of the sediments deposit relatively close to the channel, building elevated rims with lower floodplains behind (Bryan 1923). Until dams altered flood regimes and artificial levees prevented most overflows, levees were active geomorphic features of the north Delta landscape.

Levee height was calibrated to long-term flood heights as their presence depended upon the sediment deposited during overflow events (Fig. 5.12; Gilbert 1917). Natural levee height consequently steadily decreased toward the Delta due to spreading of floodwaters across the wetlands. Natural levees rose over ten feet (3.0 m) above mean water surface, ranging between 5 and 20 feet high (1.5-6.1 m; Bartell 1912, Bryan 1923, Thompson 1957, Atwater et al. 1979). Many early travelers en route to Sacramento remarked on the height of the banks they passed; some stating they were about 10 feet (3 m) high and others 20 (6 m; Wright ca. 1850b, USDA 1874, Hodgdon 1881, Hoag 1882). The banks were between 20 and 25 feet (6-7.5 m) above mean water surface near the Feather River and close to 20 feet (6 m) at Sacramento (Wilkes 1845). Near the head of Grand Island, banks were lower, reportedly 14 feet (4.3 m) above low water (as referenced in Suisun Bay) (Rose et al. 1895) or about 9.5 feet (2.9 m) above mean sea level (Unknown ca. 1900). Natural levee height continued to diminish gradually until reaching tide levels at about Rio Vista (*Sacramento Daily Union* 1862a, Thompson 1957). Slightly elevated banks could be detected down the Sacramento to the Delta mouth, suggested by scrub vegetation illustrated in maps and low banks of sediment reported in reclamation and engineering reports (Ringgold 1852, Tucker 1879e, Rose et al. 1895, Thompson 1957).

BOX 5.1. EARLY CHANGE IN THE FLOOD BASINS

Although large portions of the flood basins were not reclaimed until after 1910, some areas were modified substantially before 1860 by ditching and leveeing, as well as grazing and even full-scale farming. Large floods remained relatively unhindered, however. Testimony taken by the Swamp Land Committee in 1861 provides some insight into the early changes that took place within the north Delta wetlands.

Conveniently, the exact point in question was whether early reclamation efforts had caused incorrect designation of lands that were once worthy of "swamp and overflowed" definition (an important point as swamp and overflowed land was property of the state and all other unclaimed land that of the federal government; see Box 2.3). One witness stated that "at least one-half of the land in said townships [T5-7N R4E], now reclaimed and in cultivation, had in eighteen hundred and fifty, a growth of tule upon it" (Denn in CA Swamp Land Committee 1861). Another explained that by the time the General Land Office surveyed the land, "all of the land returned by him in said survey had been reclaimed and laid dry for a long time by the erection of levees and the closing of inlets from said river. At the time said survey was made…it was impossible for any one to tell what the character of said land was previous to its reclamation" (CA Swamp Land Committee 1861). Testimony from the Sutter land case trial provides additional evidence of early change. In 1860, a witness stated that the lake below Sutterville had already been dried completely through the blocking of inlets to prevent flooding, and concluded that "the bed of that lake proves to be the best on the river for cultivation, and various settlers are now cultivating it in preference to any other land" (Sanford 1860). Taken together, this testimony suggests that some areas of the Sacramento Basin had been cleared of tule by the time most surveys of the area were conducted.

The tule lands were lauded by many as offering good grazing when the rest of the valley lands were dry. During the summer and fall, cattle, sheep, and horses were pastured on flood basin lands (Sprague and Atwell 1870, McConnell 1887, Dunn 1915). This caused an unknown extent of tule to be "partially destroyed – tramped out by cattle" (Cleal 1861). The tule lands were especially intensely grazed during times of drought (Sacramento Valley Reclamation Co. 1872). Although it describes conditions in the Sacramento Valley north of the study area, an 1862 account suggests the potential magnitude of this impact:

> Previous to eighteen hundred and sixty-one, the tule lands were almost sole pasture of the immense herds of cattle then
> in the county; and they had, within the knowledge of residents, receded from earlier limits to the extent of more than a
> mile. (Mathews in Houghton 1862)

Another account of the effects of tule grazing is given by an 1851 newspaper article, which reported that "where the tules have been destroyed, the finest grass, intermixed with clover, has taken its place" (*Daily Alta California* 1851). It also notes that the destruction occurred rapidly.

These accounts remind us that caution is warranted when interpreting pre-1850s conditions from late 1800s sources. In response to this issue, our mapping efforts minimized uncertainty by involving extensive source intercalibration. Where possible, pre-1850 accounts were combined with more spatially detailed, but later topographic and other cartographic sources. Consequently, no single source provides the information necessary to interpret spatially explicit early 1800s habitat patterns. However, the extent of mapped freshwater emergent wetland should be assumed to represent a minimum early 1800s extent.

Figure 5.12. Natural and artificial levees along the Sacramento River. Young orchards are seen at left at the general elevation of the natural levee running along the Sacramento River. The new artificial levee, preventing overflow from the high stages of the river, is superimposed upon the broader, lower natural levee. The title of this photograph reads: "Front levee of Pearson District at Vorden. At right Sacramento River. Sacramento County, California. April 3, 1906." (photo by Gilbert 1906, courtesy of the USGS Photographic Library)

The natural levees of the Mokelumne River's distributaries were about seven feet (2.1 m) lower than the Sacramento River levees at comparable points upstream (*Sacramento Daily Union* 1862a, Thompson 2006). They were about seven to eight feet (2.1-2.4 m) high at the head of Staten Island and descended to tide levels at the lower ends of the island (CA Swampland Commissioners 1861, Thompson and West 1880). The Mokelumne River's north fork along Staten Island was thus described as "low and nearly level, elevated above high tide only an average of about two feet" (*Sacramento Daily Union* 1862c).

Width tracked a similar pattern upstream to downstream, decreasing from close to a mile (1.6 km) wide in upper reaches to less than a tenth of a mile (0.16 km) at the foot of Grand Island (USDA 1874, Hoagland 1881, USGS 1909-1918). Historical accounts describe the basin wetlands as generally set back about half a mile from the river due to the intervening supra-tidal natural levees (Browning 1851, Clyman and Camp [1848]1928). Crevasse splays locally extended the reach of riparian forest into the wetlands (Bryan 1923, Atwater 1982). Babel Slough and the high land known as Dodson's Mound extending east from Randall Island near Stone Lake both extended around three miles into the wetland interior, with width and height dimension similar to the natural levees (*Sacramento Daily Union* 1862a, Bryan 1923).

Many crevasses remain open and discharge water into flood basins whenever the river is in flood. Levee building takes place along these currents of water just as it does along the main river. In consequence sinuous double-crested ridges are built out into the flood basins.

-BRYAN 1923

The banks of the natural levees were apparently quite steep next to the river and fell gradually on the back side until it reached the elevation of the wetlands (CA Swampland Commissioners 1861, *Sacramento Daily Union* 1862a, Hoagland 1881, Hoag 1882). The "high bluff banks" (Green 1881) were blamed for the fact that "horses could not reach the water to drink" (Bidwell [1884]1904) and boats could only be hauled with difficulty to the top of the bank (Wright ca. 1850b). Profiles of the river and levees made by the California Debris Commission in the early 1900s illustrate the variation in form of the natural levees as they descended toward the central Delta (Fig. 5.13).

Natural levees in the Delta were apparently more stable than those upstream of tidal influence, above the mouth of the Feather River. Above that point, the river moved more actively within its floodplain (Unknown 1891). Downstream, recent cutoff meanders and other features common to floodplain landscapes, such as oxbow lakes and meander scroll topography, were absent (USGS 1909-1918, Bryan 1923). Perhaps most importantly, extensive natural levee deposits are found only along the present channels, which to soil scientist Cosby (1941) indicated "that all the major streams and most of the minor ones have occupied essentially their present positions during the entire period of organic accumulation." This large-scale stability of physical features contrasts with the higher-gradient parts of the rivers farther upstream.

Tidal influence

Tides accessed the lower Yolo and Sacramento basins through Cache Slough and Snodgrass Slough, respectively. Due to the relatively flat topography, tidal influence likely extended up past Babel Slough in the Yolo Basin and to Stone Lake in the Sacramento Basin, about 38 miles (61 km) and 40 miles (64 km) from the foot of Sherman Island, respectively. We estimate from the historical habitat type mapping that about 38,000 acres (15,400 ha) in Yolo Basin (contiguous wetland north of Cache Slough) and about 14,500 acres (5,900 ha) in the Sacramento Basin (north of Snodgrass Slough mouth) were wetted by spring tides at low river stages (see Fig. 5.3). Mapping was based on the assumption that the extent of tidal influence falls within 3.5 feet (1.1 m) of mean sea level (see page 66; Atwater 1982). This matched the elevation extent of Yolo Basin tidal wetlands mapped by USGS in the early 1900s.

Early sources relating to efforts to ditch and drain the basins also provide supporting evidence for the mapped areas of tidal influence (Box 5.2). An 1864 reclamation map marked tide gates on ditches connected to Stone Lake, and "tide water" was reported at Little Snodgrass Slough (Reece 1864, Board of Swamp Land Commissioners 1864-5). Ditches up the Sacramento Basin are tidal today nearly to Freeport (Van Löben Sels pers. comm.). Before the first ditches of the Sacramento Drainage Canal were built in the 1860s, it is unlikely that tides extended as far up the basin. The wetlands above Stone Lake were likely primarily non-tidal due to their higher elevations, though a few larger channels off of the Sacramento may have

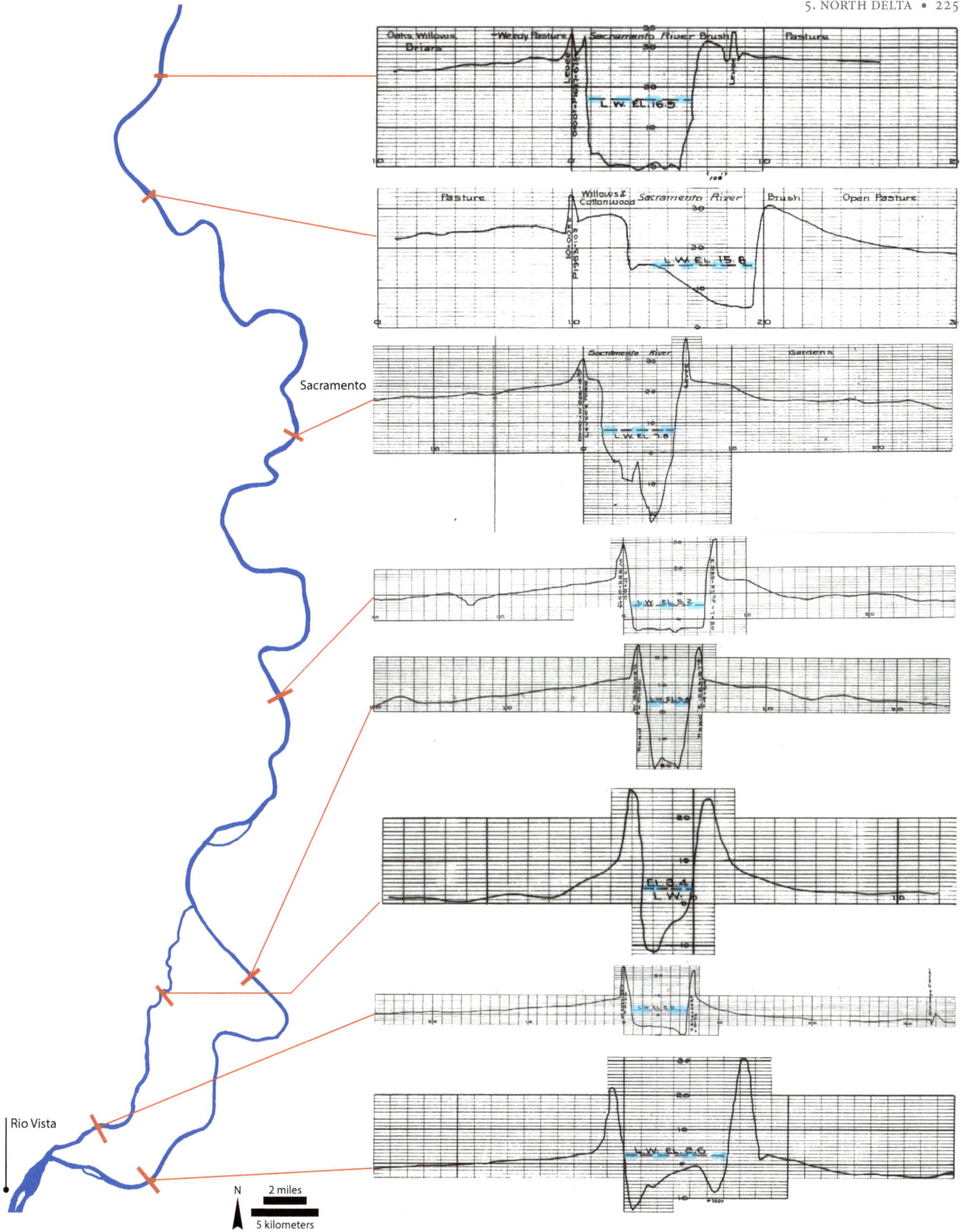

Figure 5.13. Cross sections of the Sacramento River in 1908. Eight cross sections spaced relatively evenly along the Sacramento River (including Steamboat Slough) from just downstream of the Feather River confluence to the foot of Grand Island were selected from many profiles made by the California Debris Commission surveys to illustrate the gradual change in the cross-sectional profile of the river moving downstream. Generally, natural levee height and breadth decreases descending the river. This is most clearly seen when comparing the elevation of low water ("L.W.", blue dashed line) to that of the land surface behind the artificial levees, which are seen as high peaks directly adjacent to the river. Lines across the channel in the map are relative to the total length of the profile. The profiles are not shown at the same scale. (Wadsworth 1908a, courtesy of the California State Lands Commission)

BOX 5.2. EFFECTS OF HYDRAULIC MINING DEBRIS ON NORTH DELTA BASINS

When using post-1860 historical sources to gain understanding of the early 1800s landscape, it is important to consider the potential impact of hydraulic mining debris, particularly in the Sacramento system. The impact on channel geometry in the form of drastically raised bed levels in the Sacramento River is well known and documented (see Box 1.2, pages 22 and 237; Gilbert 1917). The raised bed levels reduced tidal range and extent up the Sacramento River: at the peak of debris around 1913, tides were barely perceptible nine miles below Sacramento, while historically the tide range had been around two feet at Sacramento (Hall 1880, Young 1880, Mendell 1881, Taylor 1913, Gilbert 1917). The raised bed elevations and restricted access to former wetland areas due to leveeing led to increasing floodplain elevations (Taylor 1913).

Less well known is the extent of effects on channel planform and wetland elevations. In his treatise on hydraulic mining debris, USGS scientist Grove Karl Gilbert (1917), thought that "probably that the principal portion [of the suspended load] was received by the inundated lands [wetlands]." Regarding its distribution, however, a national report on freshwater tidelands included a summary stating that though the mining debris had degraded land near Marysville, there was not "much injury below Sacramento City" (Williams in Nesbit 1885).

In the American Basin, there are a few reports of the basin actually staying wetter later in the season as a result of mining debris. This is attributed to the fact that with the bed level increases in the channel, less water was able to drain back into the river from the basin, keeping Bush Lake at the base near the American River mouth much wetter (Rose et al. 1895, Bryan 1923). Other reports suggest significant localized changes along the river channels and at point bars, where additional sediment likely allowed rapid growth of willows and other riparian vegetation (Brewer 1974).

Aside from the changes of the mainstem channels, it appears that hydraulic mining debris, especially when considered against the other dramatic reclamation and levees efforts occurring during that time frame, had comparatively little large-scale impact on the form and function of the wetlands occupying the north Delta basins. Additional inorganic sediments were certainly added to the marshes at rates greater than before mining began. A general statement for all of the Sacramento Valley is made to this effect in an 1878 newspaper article:

> Wherever the tules have been covered up with sediment by deposit from the rivers in times of overflow, so that the
> fall fires have nothing to feed upon, there generally soon appear great numbers of young willows of different kinds,
> cottonwoods and other kinds of soft timber trees…There are at this time thousands of acres of such low land forests
> where, ten years ago, grew nothing but high tules. (*Sacramento Daily Union* 1878)

This may have been more the case north of the study area. We generally found that historical accounts of basin characteristics (e.g., tall dense tule, large lakes, and annual overflow) within the study area were generally consistent before and after hydraulic mining began, where spatially explicit data were available. It seems likely that before hydraulic mining tidal wetlands were of high elevation; they appear to have maintained elevations approximating high tide levels historically (Gilbert 1917, Atwater and Belknap 1980). This supports the hypothesis that Delta wetlands were able to keep pace with slowly rising sea levels without substantial sediment input (Windham-Myers pers. comm.). Instead of marsh accretion related to deposition of inorganic sediments typical of salt marshes, the Delta's freshwater wetland elevations may have primarily related to balancing peat oxidation rates, where above certain inundation frequencies, peat was accreted instead of oxidized (Atwater and Belknap 1980).

brought tides to the lakes and ponds occupying lower positions to the north (Lienhard and Wilbur 1941). The effect of ditching on tides was discussed by one advocating draining the Yolo Basin, who recognized that drainage efforts might be counterproductive if "the fall through the canal should be found so light as to permit the tide to ebb and flow through it" (*Sacramento Daily Union* 1853).

Aside from the main tidal connections to large sloughs at the lower ends of the basins, textual accounts and inference from maps suggest that there may have been additional points of entry along the levees of the Sacramento. These may have been tidally connected only during certain times of the year. A detailed account of an early trip to Sutter's Fort along the banks of the Sacramento includes such points: "We also found lagoons of varying breadth and width formed by marshy areas whose waters flowed back swiftly into the Sacramento with the ebbing tides" (Lienhard and Wilbur 1941).

It is difficult to determine the relative influences of tides spatially; tidal influence very gradually lessened over the span of tens of miles along the basins. Only a portion of the mapped tidal wetland area was actually wetted by twice daily tides, though it is assumed to have been wetted at least by spring tides in times of low water. This lessoning of tidal influence across the wetland is illustrated in the Mokelumne land grant map by a zone of tule outside of a line denoting the extent of daily high tides (Fig. 5.14). In the Yolo Basin, the most southern extent was similarly influenced, representing perhaps 25% of the tidal area in the basin. At an even more local scale within that area, hydroperiods also varied. For instance, a Prospect Island reclamation document states that "high tides flooded the lower lying portions of the land" (Mellin 1918).

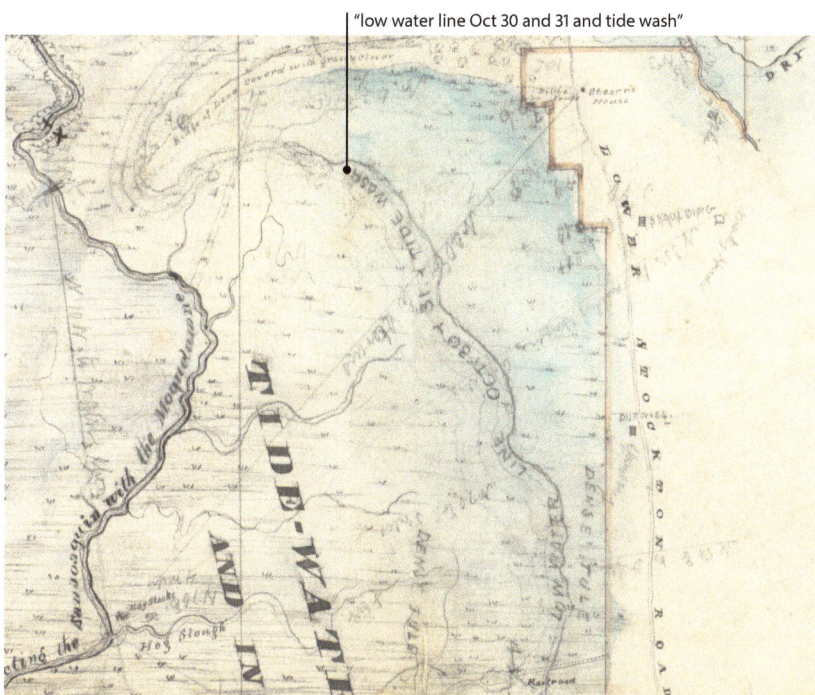

"low water line Oct 30 and 31 and tide wash"

Figure 5.14. Lessening tidal influence at the upland edge is illustrated in one of the main exhibits of the Mokelumne land grant court case. The line of regularly inundated tule during times of low water is labeled as "low water line Oct 30 and 31 and tide wash," with the wetland to the east depicted with a lighter colored blue to indicate the lesser degree of tidal influence. Spring tides would have extended beyond this line. For example, other testimony attests that standing water was found along the less tidally-influenced margins (light blue) during the dry season. (Von Schmidt 1859, courtesy of The Bancroft Library, UC Berkeley)

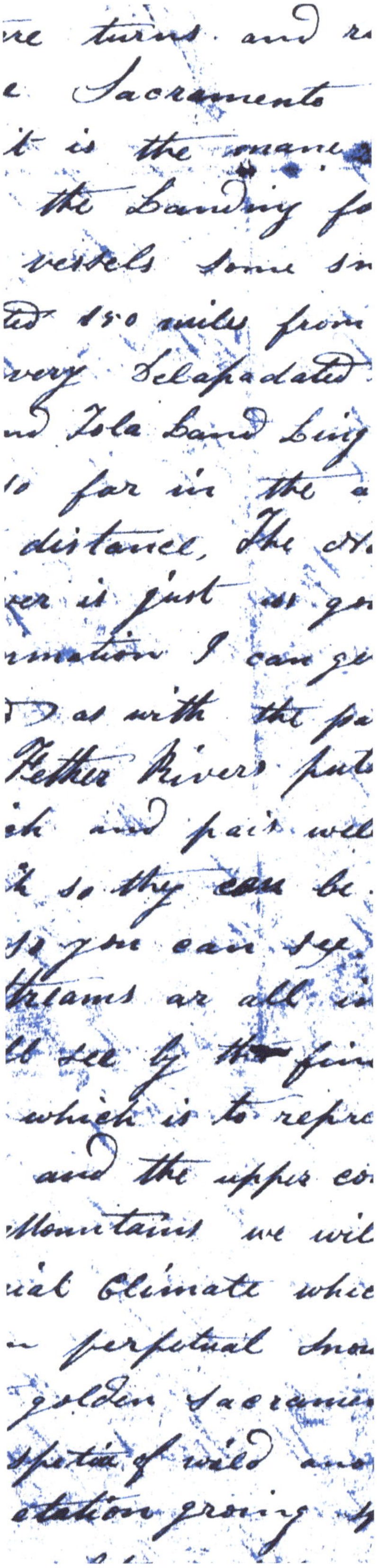

The outer limits of tidal influence up the basin are particularly challenging to define. This transition occurred apparently without a pronounced shift in vegetation – tule maintained dominance throughout. Furthermore, because most tidal connections lay at the very southern extents of the basins, tides had to travel great distances over emergent vegetation to reach full protential related to tidal elevation. Due to frictional effects in the uppermost parts of the basin, tidal influence may not have been very significant, even if elevations were well within tidal range. It is also possible that an even greater area than the mapped tidal extent was affected by extreme tide events or simply by water levels maintained by tides. For these reasons, the mapped extent of tidal wetlands in the basins should be understood to experience a spectrum of hydroperiods depending on spatial and temporal factors.

The extent of tidal influence also shifted seasonally. Tides probably had little influence during times of flood and were likely also limited during the lowest river stages due to the restrictions from natural levees as previously described. Additionally, some islands, such as Randall, Merritt, and Sutter, were completely enclosed by natural levees and were likely isolated from tidal access during periods of low water (Sprague and Atwell 1870, Atwater 1982). In the mapping process, we assigned these areas a tidal classification because land surface fell within tide range, but we included notes that they may have been isolated from tides.

Landscape character in first-hand accounts

First-hand narratives of the Delta can provide a sense of place that is difficult to acquire from maps or aerial photography. These descriptions often convey how individual habitat types fit together to form the larger landscape. Several accounts deserve special mention here as they capture the landscape character of the north Delta.

The unfamiliar territory of the Delta was crossed by numerous gold seekers beginning in 1849, many of whom were writing home to family and attempting to describe the landscape they found themselves in. One of the surviving letters is that of a man named George Browning (1851), who described the landscape pattern of the basins in a letter to his father. The letter refers to an accompanying map (see Fig. 5.9), and it is excerpted here. (The numerous spelling errors are from the original; Fig. 5.15).

> Now for a discription of the vally of the Sacramento, you will see by the map (as that is the only way that I can begin to explane if so you can under stand it) that there is but very little good Land at most along cloast to the river and on some of the Sloughs the Land is represented by being painted with a pencil and in all cases it is the highest next to the River and desends back as the pencil marks does. In an average it is about ½ mile in width then comes the Tola or Bull Rush all it differs from the Rush in the States it grows abought 10 ft in hight and is about 1 inch thick or more at the but and grows even all over the ground and not in bunches or on tusick as it does in the States and grows as thick as it can stand. This is represented by dots with a pen this runs of gradully untill it gets to deep

Figure 5.15. Text from George Browning's letter home to his father. (Browning 1851, courtesy of The Bancroft Library, UC Berkeley)

for Tola and then comes the Lake or Pond. When the River raises in the spring it raises the Tola as we call it till it comes to the high ground and last spring the River flowed its Banks and then it was one purfect Lake from one Mountain to the other or to the bench Land of the Mountain.

A second, pre-Gold Rush, detailed account from a traveler to Sutter's Fort in 1846 (time of year unknown) gives a sense of what the north Delta landscape looked like on the ground. Opting to make the trip on foot along the banks of the Sacramento instead of by boat, pioneer Heinrich Lienhard describes the many scenes and obstacles he passed along the way. Excerpts from his narrative illustrate the variety and complexity of the landscape at the local scale and provide some perspective from which to synthesize information in the rest of the chapters.

Lienhard began his trip up-river approximately at the head of Grand Island, having passed by boat along the island "stretches of swamps, thick with tules and reeds, that extended far off into the back country" as well river banks covered with trees (Lienhard and Wilbur 1941). Once on foot, he recounts:

> Having gone only a few yards we reached a slough; here we were forced to wade through water up to our knees. A frightened elk suddenly bounded out of the water with great leaps and soon vanished in the brush...We soon discovered that we could travel faster if we followed the grassy area that lay between the river and the forest, and avoided the swamps further inland...Unfortunately, the extensive tule-covered marshes that often reached far back in the forests were occasionally submerged in deep water. We also found lagoons of varying breadth and width formed by marshy areas whose waters flowed back swiftly into the Sacramento with the ebbing tides, and at such places there was always unavoidable delay, for we had to find out whether we could get through without running into deep currents. Several lagoons were crossed where the cold water reached to our hips and in places even up to our shoulders; these icy baths I did not enjoy...

> No wild animals except some coyotes, and a few wildcats had been seen thus far, although frequently we came across deer, especially bucks; once, in a grass-covered area between the forest and swamp, were seen grazing. Our route continued through less open country now and we were forced at times to wade through marshy stretches. Finally, a place was reached that was too deep to cross on foot. At such times we usually tried to find trees with branches broad enough to reach to the opposite side of the slough. Many such trees were available; we succeeded in our attempt; and by clinging to one large branch after another were able to approach so close to the opposite bank that one short jump landed us on dry ground.

> Noon found us in a dense forest near a deep, broad arm of the river...On the opposite bank was a broad-limbed juniper tree [species unknown] against which a large sycamore that formed a kind of natural path leading over to the other trees had fallen. Without stopping to think, I used this as a bridge and, by balancing myself on one of its strongest branches at a height of ten or twelve feet above ground, reached the opposite bank. While crossing at this elevation, I had seen large numbers of vultures, turkey buzzards, ravens, crows, and magpies perched on a sycamore tree nearby, and knew there must be a carcass somewhere in the vicinity. I had not looked around to see where it was, for I did not suspect any danger,

and was busy lowering myself from limb to limb to the ground. I called to my two comrades to hurry across so we could continue without too much delay…Whenever he had to climb anywhere McDowell was always slow…When he released them [branches], they swung back with a loud cracking sound, and as McDowell let go and landed on the ground with a heavy thud, the branches crashed more noisily than before. Then came a shrill whistle from the sycamore tree, and the birds flew off together as if they had been shot from a cannon. Thick underbrush that cut off the view grew between us and this particular tree. The noises and sounds of nature were far more obvious to the Indian than to we two American green horns traveling through the forest for the first time, and he seemed frightened and excited by the loud whistling sounds; he stood on his tiptoes, looked in the direction the sounds had come from, then all around without stirring.

Observing his agitation I asked, "Is it a wolf?"

"No, no," he said.

"Is it an elk?"

"No, no," he replied again.

"Is it a grizzly bear?"

"Yes," he whispered quickly…Why the gray rascal allowed us to escape unmolested is a mystery; perhaps it had watched us make the strange trip over the bridge, heard McDowell call, and seen the branches shake…

For the first time in many days the sky was clear, and the sun shone with a welcome warmth; as I started out alone the path was dry and I hoped that I would not be forced to wade again through deep water. But I recognized my error too late when I reached another wet area that seemed almost too deep for wading. It was not long before I stumbled on a place where several trees had been interlaced with wild grapevines that formed a kind of net or hammock. This seemed to be a favorable point to cross, and so I climbed up through these bushes, trees, and vines, hoping in this way to keep above water; but now I discovered that, like a fly in a spider's web, I had difficulty in getting out, and once narrowly escaped falling into the water.

Eventually I reached the other side safely. Within the next few miles no obstacles were encountered, but later I found several places where I was compelled to wade through water so deep that it came up to my shoulders." (Lienhard and Wilbur 1941)

Parts of this passage are discussed later in the chapter within the context of related historical data to synthesize particular habitat characteristics.

SEASONALITY AND FLOW

The landforms of the north Delta and the vegetation communities occupying them were adjusted to extremes of flood and drought in the Sacramento Valley. The landscape looked very different and served different functions depending on the time of year. The "inland sea" of the Sacramento Valley takes its name from the estensive flooding that occurred within the basins (Fig. 5.16; Belcher 1843, Grunsky 1896, Bryan 1923, Kelley 1989). Water often remained on the surface for a significant portion

Figure 5.16. View of flooding along the Sacramento River. This 1927 photograph of North Sacramento shows oaks and willows lining the river and in some cases within the flooded area, as would have been the pattern historically within the flood basins. (photo by McCurry, courtesy of the California History Room, California State Library, Sacramento)

of the year, perhaps about six months, until the land became dry by late summer. These inland seas hydrologically connected the river to its surrounding floodplain environment. Habitat conditions and species responded to the seasonal patterns of flooding and drying, as aquatic species found access in the wet season and terrestrial in the dry. The tribes of the Delta would have also responded to the seasonality: it was known that some moved from the river banks to "the ridges," or upland margins of the basins and foothills, during the floods (Robinson 1860).

The series of basins filled during floods historically, with the majority of water entering at their heads (e.g., at the present Fremont Weir in the Yolo Basin) and draining to the south (Young 1880). While considerable flood volumes contained within the river channel moved quickly to the Delta mouth, often an even greater volume of overflow passed more slowly through the basins to gradually drain as the river stage fell (Gilbert 1917). As relief "retention basins" (Thompson 1957) or "regulating reservoirs" (Heuer 1900), they served to reduce the peak flood flows of the Sacramento

The great basins…act as enormous regulating reservoirs…Their effect at all times – though in a less degree when full than when empty – is to cut down the crest of the great flood waves passing through them, and to distribute their discharge over longer periods than if the river were confined to its channel; so that, on the whole, the discharge of the river below the reservoir will never be either as high or as low as if the reservoir were not there.

—DABNEY 1905

River (Dabney 1905, TBI 1998). The timing of water levels in the basins and river was summarized by Sanford, a witness to the Sutter land case: "When the river first rises the river is always higher than the lake; when the river falls the lake is the highest; at the highest rise it might have been a foot lower in the lake" (Sanford 1860). The presence of tall emergent vegetation served to further slow the passage of water (Young 1880). The water stored in the basins contributed freshwater to the central Delta through the summer months (Gilbert 1917, Thompson 1957). Once the water entered the basins there were few opportunities to exit; water was forced to travel the length of the basins, from north to south, to drain through the few outlets at the base.

Flooding extent

The regularly flooded extent of the north Delta landscape can be interpreted as the boundary of the flood basins (Bryan 1923), which aligns with the extent of mapped perennial emergent wetland. Approximately 80,000 acres (32,370 ha) in the Yolo Basin (north of Cache Slough), or 49% of the land surface within the 25-foot contour (164,400 ac/66,530 ha), was annually overflowed and occupied by perennial wetlands. The actual extent overflowed in extreme floods was often much greater than that of the mapped perennial emergent wetland – seasonal wetlands were often temporarily flooded during the rainy season. For example, an additional approximately 17,000 acres (6,880 ha) were flooded in the 1878 floods (Hall ca. 1880c; Fig. 5.17). This area of inundated seasonal wetlands often extended as broad swaths bordering the perennial wetlands. In the Yolo Basin region, aside from the riparian forest along the major channels, seasonal wetlands and wet meadows occupied the remaining area within the 25-foot contour. These lands were temporarily or occasionally saturated during the wet season, primarily from the smaller streams emerging from the foothills of the Coast Range.

The upland edge of the flood basins was a broad area where the degree and timing of overflow was intermediate between the wetter flood basins and drier alluvial fan slopes. For example, the eastern Delta margin just south of the Mokelumne River was characterized by variable flooding extents, related for the most part to Mokelumne River overflow. The edge of the regularly flooded area did not exist as a clear line on the ground. The western edge of the Mokelumne land grant boundary was conveniently located along the line of tule (Von Schmidt 1859), but the accompanying case testimony adds more detail, suggesting that the flooded extent was variable depending on the season and year. For example, the Mokelumne land case testimony includes statements that the grant line and land to the east was overflowed "during freshet" (e.g., high river stages; Beaumont 1859a) and sometimes overflowed "more than half a mile over the west line" (Gray 1859). This transitional edge was mapped as seasonal wetlands. To the west of the line, the same witness states that for two months of the year, the land could be traversed by buggy for a half-mile to mile into the tule before it got too wet (Beaumont 1859a).

The basins as they exist today are more limited. Much of what is now thought of as the Yolo Basin consists of the seasonal floodplain lying west of the Yolo Bypass – an area occupied primarily by seasonal wetland historically as well. The core of the historical Yolo Basin, however, lay to the east of the Bypass, which is mostly farmland today. South of Sacramento within the Yolo Basin (both east and west of the Bypass) today, roughly 11,800 acres (4,775 ha) of seasonal and perennial wetlands exist (Hickson and Keeler-Wolf 2007). Of these, 5,700 acres (2,300 ha) are likely perennial and thus comparable to the 67,700 acres (27,400 ha) of perennial wetlands present in that same area historically, representing a 92% decline in perennial wetland area and a shift toward seasonal wetland from perennial wetland within the Yolo Basin.

Origin, timing, and frequency of flow

Although the hydrographs of both the Sacramento and San Joaquin rivers reflected snowmelt in the Sierra Nevada, a greater proportion of the contributing watershed on the Sacramento River were at elevations lower than the San Joaquin, including many rain-fed Coast Range streams. This meant that flooding generally occurred earlier in the season, often with higher peak flow events (see page 10; USDA 1874). Floods tended to occur between December and April, with high flows sustained by snowmelt for a number of months (Kybruz 1854, Gilbert 1917). This contrasts with the San Joaquin River, where floods occurred most frequently in the later spring and early summer as rainfall-driven floods were less common.

The Sacramento River frequently passed much of its flows into the basins, though not necessarily every year (Fig. 5.18; Mathews in Houghton 1862, State Agricultural Society 1872). In reference to the American Basin, a witness to the Sutter land case stated that the Sacramento River flooded the basin every third year (Kybruz 1854). The river was also known to overflow multiple times in a single year (Hall 1856). One witness attempted to reconstruct the major floods within the southern part of the Sacramento Basin near the head of Snodgrass Slough, reporting that during the 1850s the land was inundated for several months about every other year (Fig. 5.19; Greene in CA Swampland Commissioners 1861). Sustained late-season inundation would have been supported by a combination of high flows from snowmelt and rainfall runoff.

In the Sacramento Basin, floodwaters from the American and Sacramento rivers flowed south to meet the floods of the Cosumnes and Mokelumne in the present-day Pearson District and McCormack-Williamson Tract area (Green 1882, Payson 1885). Inundation within the Pearson District was historically attributed primarily to the Sacramento River; to the east overflow was more directly attributable to flooding on the Mokelumne and Cosumnes rivers (Wallace 1869). Floodwater would be trapped until water levels fell, permitting waters to drain through Snodgrass Slough and into the Mokelumne River (*Sacramento Daily Union* 1862a). For example, General Land Office surveyor William Lewis (1859a) reported a foot of

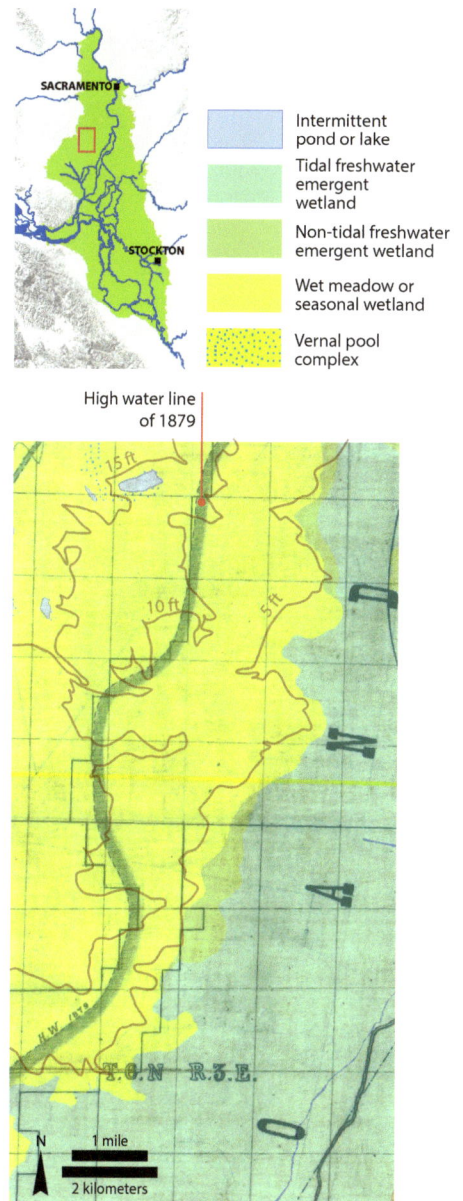

Intermittent pond or lake

Tidal freshwater emergent wetland

Non-tidal freshwater emergent wetland

Wet meadow or seasonal wetland

Vernal pool complex

Figure 5.17. The line of high water during 1879 is shown on a State Engineering map. The historical habitat mapping is overlaid to illustrate the generally mile-wide zone beyond the edge of tule (darker green) that this flood extended. The seasonal wetland complexes (light green) at the basin margins were overflowed in this manner during the larger floods. Generalized elevation contours from the early 1900s USGS topographic maps are included to aid interpretation. (Hall ca. 1880c, courtesy of the California State Archives)

234

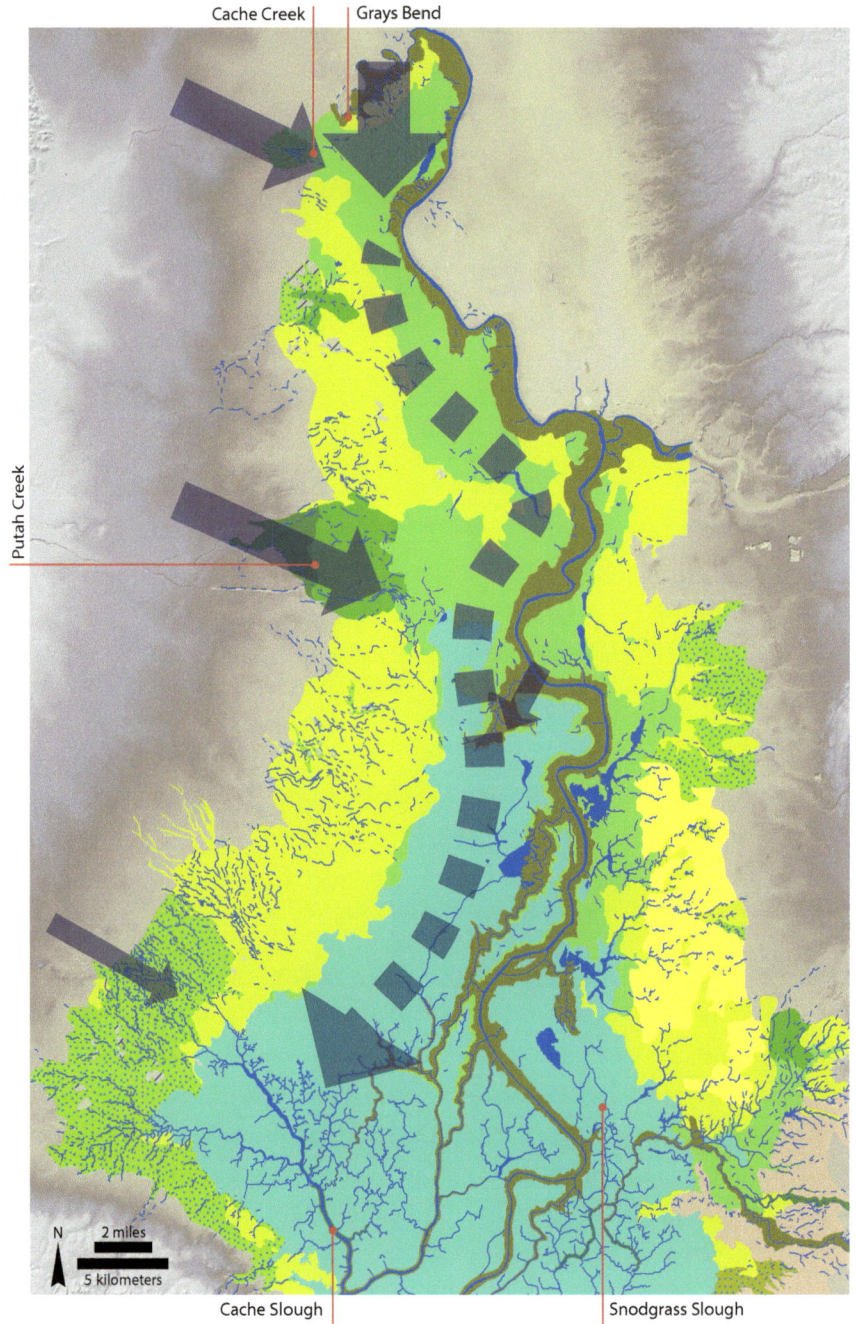

Figure 5.18. The pattern of flood flows is illustrated conceptually for the Yolo Basin. The solid-fill blue arrows indicate direct inputs from river and stream channels (e.g., the Sacramento River near Gray's Bend; Putah Creek). The dashed blue arrow indicates the slower-moving basin flows from north to south that moved as a broad sheet through the wetlands toward its exodus at the mouth of Cache Slough.

water in the tules on January 4 along his survey line in the middle of the McCormack-Williamson Tract, which became deeper southward.

Inundation could also occur in the early summer. On June 13, 1833, trapper John Work passed near the Mokelumne River (Maloney and Work 1943). He wrote in his diary that "the river had overflowed its banks so that we cannot encamp on them nor indeed except in some places approach the river. The lake where we encamped yesterday continues on to the river." He also noted the differences in water temperature, complaining that the shallow water within the flooded basin he was traveling along was "very warm and we cannot get to the river where it might be a little colder" (Maloney and Work 1943).

The many small watersheds entering the valley from the Coast Ranges and eastern foothills were an important annual source of water. These streams spread into distributaries across their alluvial fans, discharging all of their flow into the basins. A general account describes that the streams "loose themselves in the vally [sic] and spreading in all directions form extensive lakes of water" (Clyman and Camp 1928[1848]). Though dry in the summer (USDA 1874, USGS 1909-1918, Moerenhout [1849]1935), these small systems carried substantial flows in the winter (Hilgard 1884, Vaught 2006). During storms, flows from the Coast Range streams, particularly Putah and Cache creeks, could be as great as low water flows in the Sacramento (Sacramento Daily Record-Union 1892). Flooding was sometimes solely attributable to these streams. On November 3, 1849, for example, a group traveling west from Sacramento "crossed the Tule safely, but found the road beyond extremely heavy and covered in some places with water" (Derby and Farquhar 1932). They soon found that this was due to flooding in Putah Creek. Thus, while Sacramento floods were important to the basins, so too was the influence of the entire annual flows of these smaller systems in terms of flood timing and inundation depth and frequency.

Flood magnitude

Flood flows through the basins were considerable and could bear more than the volume of the river channels during floods of significant magnitude (Heuer 1900, Gilbert 1917). The Yolo Basin in particular had a large flood capacity, as documented by numerous early accounts. Rose et al. (1895) reported that the flood of 1879 brought 66,000 cfs (1,870 cms) through the basin and the 1881 flood, 185,000 cfs (5,240 cms). For comparison, the Sacramento River's average discharge was 30,000 cfs (850 cms) with flows at low water averaging between 7,000 and 8,000 cfs (200-227 cms; California Debris Commission 1910). The channel had a maximum capacity of 110,000 cfs (3,115 cms; McClatchy 1916). Early 1900s estimates reported that the maximum discharge for the valley was 660,000 cfs (18,690 cms; McClatchy 1916). A 1906 engineering document reported flood capacity in the Yolo Basin was 1.15 million acre-feet (1.42 km³; Newell 1907), which is supported by data showing that the capacity of the Yolo Basin during the

Figure 5.19. Periods of overflow. This graphic shows the months in selected years from 1850 and 1860 when the region near Snodgrass Slough was overflowed. This information was summarized from recollections of Josiah B. Greene, who testified before the Swamp Land Committee (Greene 1861). The lighter blue represents flooding of lesser extent – periods when the whole area in question was not overflowed. These spring season inundations were a signature of high flows from the season's storms augmented by snowmelt.

Question 29. In the overflow of which you have spoken, covering Sacramento city, about what season of the year did the waters generally recede, and how long did they remain up?

Answer 29. In 1849-'50 the water receded sometime in the month of February and rose again, then receded again sometime in June. The winter of 1850-'51 there was no overflow.

—SANFORD 1860, TESTIMONY FROM THE MOKELUMNE LAND GRANT COURT CASE

When not in very large volume, they [waters] are held back by the growth of tules, and do not find their way rapidly down the steep grade of the basin; but, after filling the deeper depressions thereof, they are delivered gradually through Cache Slough… When, on the contrary, after the basin has been partially filled, there is a large accession of water from the creeks or the river suddenly precipitated therein, it delivers at its lower end through Cache Slough, and over its rim into Steamboat Slough, a large flood volume in advance of the rise which comes regularly down the river, and thus temporarily gorging the river below Grand Island, creates a perfect water-dam in the Steamboat Slough channel, and causes an elevation of the flood up-stream as far as Sacramento.

—YOUNG 1880

In one instance, the counter current carried a barn two miles up the river, and deposited it on the opposite bank, where it now stands.

—MATHEWS, IN HOUGHTON 1862

1907 and 1909 floods was 1.126 million acre-feet (1.39 km³; California Debris Commission 1910).

In larger floods, flow through Cache Slough could be so great as to hydraulically dam the Sacramento River, occasionally causing the river to flow upstream (*Sacramento Daily Union* 1862a, Tucker 1879c, Young 1880, Thompson 1957, Atwater et al. 1979). This phenomenon was reported in the *Pacific Rural Press* (Ryer 1884): "it has been within the observation of every river pilot that the current carrying driftwood and other floating bodies runs for several miles up Old [Sacramento] River and Steamboat slough towards Sacramento." An engineering report stated that the stage of the river was level from the mouth of Cache Slough to Walnut Grove during these periods of high floods (U.S. Congress 1916). Today, the Yolo Bypass has the capacity of 80% of the flood volume of the Sacramento (Sommer et al. 2001).

Basin lands could often be flooded over five feet deep (Thompson 1957), up to 15 feet (1.5-4.6 m) in its deepest portions (Fig. 5.20; Mathews in Houghton 1862, Rose et al. 1895). Inundation during floods was sometimes so deep as to require a boat to cross (*Sacramento Daily Bee* 1881). In the Sacramento Basin, just south of Stone Lake, one man testified that he needed a boat to cross in the spring of 1850 (Hazen in California Swampland Committee 1861). A witness to the hearings determining swamp and overflowed land boundaries, who lived on the Sacramento River bank in the Pearson District, claimed:

> Boats have started from my house and gone to the city of Sacramento for provisions, passing over the lands described in said townships, situated on the east side of the Sacramento River, without ever touching said river." (Summers in California Swampland Committee 1861)

The depth of extreme floods over the banks before significant artificial levees were built was reported to be about two feet (0.6 m) at the City of Sacramento, about three feet (0.9 m) at Freeport, and eight feet (2.4 m) at Rio Vista (Mathews in Houghton 1862).

Receding waters

Once rains ceased to fall and the majority of snow had melted, flows in the rivers and streams diminished dramatically. The Sacramento River's low water flows were around 7,000 to 8,000 cfs (198-227 cms), or about 1% of maximum flood flows. The American River was easily forded at its mouth, which was only a few feet deep. The Cosumnes and Mokelumne rivers, the only other two substantial Sierra Nevada rivers entering the north Delta, were reduced to very low stages. Within the channels of the Cosumnes Sink near its confluence with the Mokelumne, it was reported that "there are places where you cannot distinguish a current, though there is water in those places" (Gray 1859). Putah and Cache creeks were generally dry by the end of the summer, though they maintained pools in places and kept their "sinks" at the edge of the Yolo Basin wet.

Though the majority of floodwater occupying the basins from a single flood exited within days, a substantial portion remained behind. Large expanses of overflowed area could remain for several months (Prentice 1856, Gilbert 1917). This pattern was described in testimony of the Sutter land case: "waters run out the sloughs for a certain length of time until the sloughs fail to drain, when they evaporate all summer, until the next rains" (Sanford 1860). In general, water was found on the surface for five to six months of the year, with some parts remaining overflowed year-round (Buchannan 1853, Hatch in *Sacramento Daily Union* 1854b, Holttum 1879 in Paterson et al. 1978). At their lower tidal ends, the north Delta flood basins were saturated at the surface year-round, wetted by the tides. At their upper ends, the basins could become dry late in the season, though retaining enough near-surface moisture to support the dense growth of tule. Localized depressions, ponds, and lakes would stay wet through the year, filled by overland flow from floods and high water tables, but disconnected from means of drainage (channels; Rose et al. 1895).

Figure 5.20 Depth of inundation within the basins. It was standard protocol for the GLO surveyors to record the average known or estimated typical depth of inundation for areas "subject to inundation." Surveyor William Lewis (1858a, 1858-1859, 1859a) was particularly thorough in recording the depth of flooding the land within the lower Sacramento Basin could be subject to, as summarized in this map.

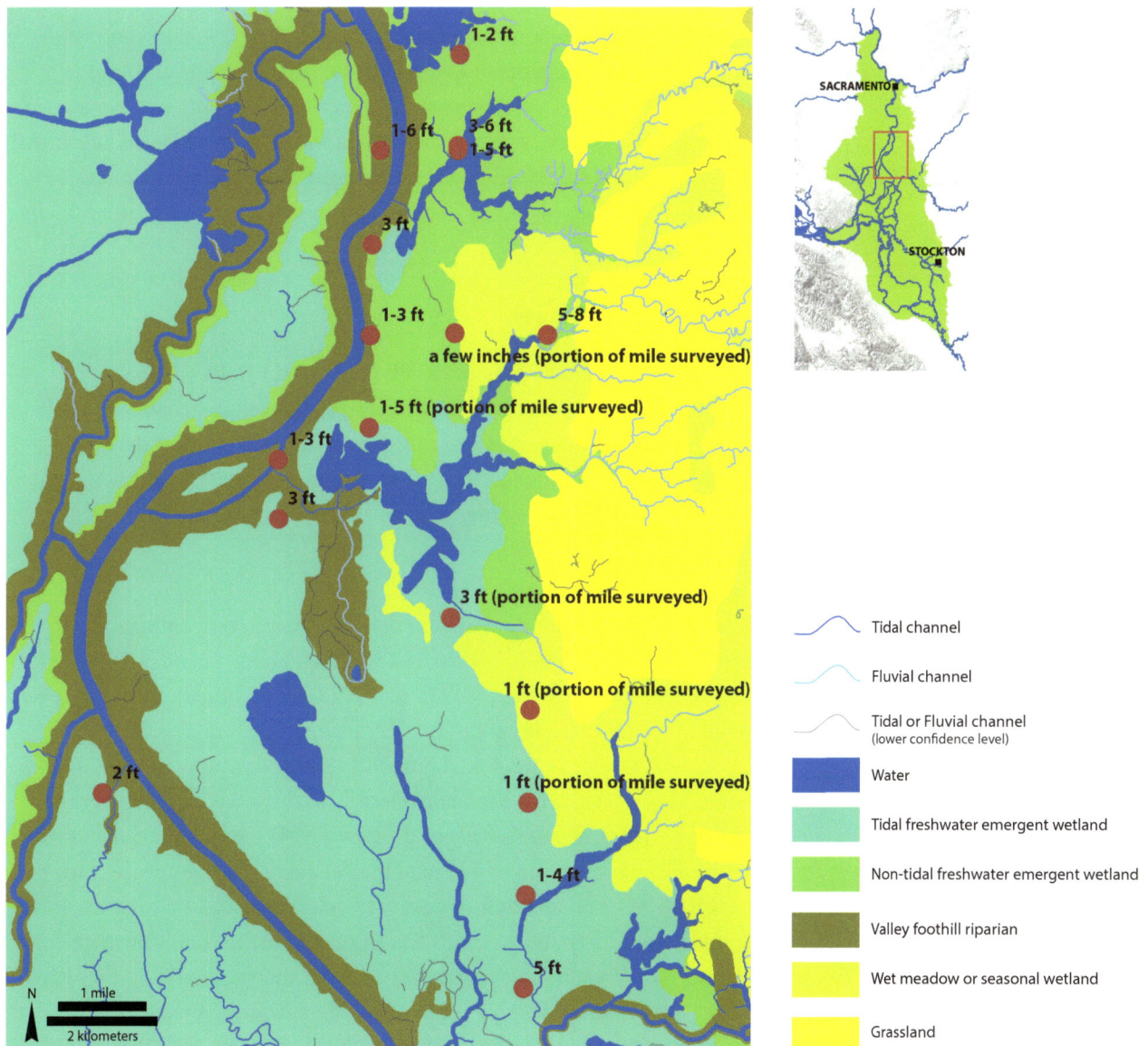

Tidal channel

Fluvial channel

Tidal or Fluvial channel
(lower confidence level)

Water

Tidal freshwater emergent wetland

Non-tidal freshwater emergent wetland

Valley foothill riparian

Wet meadow or seasonal wetland

Grassland

General descriptions of the basins state that water remained within the tules until sometime between July and September (McClatchey 1860, *Sacramento Daily Union* 1860, Algier 1863, Hall in Board of Swamp Land Commissioners 1864-5, McGowan 1961). Others state that the basins were impassable half the year (Derby and Farguhar 1932) or became passable in the fall (*Sacramento Daily Union* 1860). Geomorphologist Kirk Bryan specified that even with the drainage provided by the Tule Canal (see Box 5.3), the Yolo Basin could sustain water through the dry season up to Sacramento and would stay wet enough to support tule to its northern limit near Cache Creek (Bryan 1923). In the Yolo Basin near Putah Creek, a newspaper article reported that the water in the tule was two feet on May 15, 1851, a lower than average rainfall year (*Sacramento Transcript* 1851b).

CHANNELS AT THE FLUVIAL-TIDAL INTERFACE

The mapping and supporting material demonstrates that channel density and morphology varied widely yet predictably across the north Delta landscape – depending on relative fluvial and tidal influence and degree of connectivity to primary fluvial (and sediment) sources. The main river channels control the landscape patterns of the north Delta. In contrast to tidal wetlands and floodplains of the central and south Delta, the non-tidal upper basins of the north Delta were crossed by relatively few defined channels. At the edge of the basins, small crevasse splay channels cut across the large natural levees, and upland drainages spread along their alluvial fans. As the landscape transitioned toward the tidal Delta, low order tidally-influenced channels became more numerous within the basin wetlands, while distributary channels such as Elk, Sutter, and Steamboat Slough branched off the main river (CDFG and YBF 2008).

The following sections provide details about north Delta channel characteristics. The discussion is necessarily not comprehensive. The first sections cover characteristics of the larger mainstem rivers of the north Delta, with later discussion focused on defining characteristics and large-scale patterns of the low order tidal and overflow channels.

Sacramento River morphology

The channel planform downstream of the Feather River confluence was characterized by a wide meander belt width and long meander length (Fig. 5.21; Belcher 1843, Wilkes 1849, Bryan 1923). In addition to being the entry point of the largest tributary to the Sacramento River, this confluence coincided with the limit of tidal influence. Below this point, the natural levees also appear to have remained largely in the same location over their period of development (see page 221; Cosby 1941). Features characteristic of the meandering river upstream of the Feather River confluence, such as oxbow lakes, were not common. Early travelers noted that navigation was comparatively easy and generally free of major obstructions downstream of the Feather River. The first major impediment was met at the Feather River mouth, where several early explorers in different years noted a sand bar "extending the whole distance across it [the Sacramento]" (Wilkes 1845).

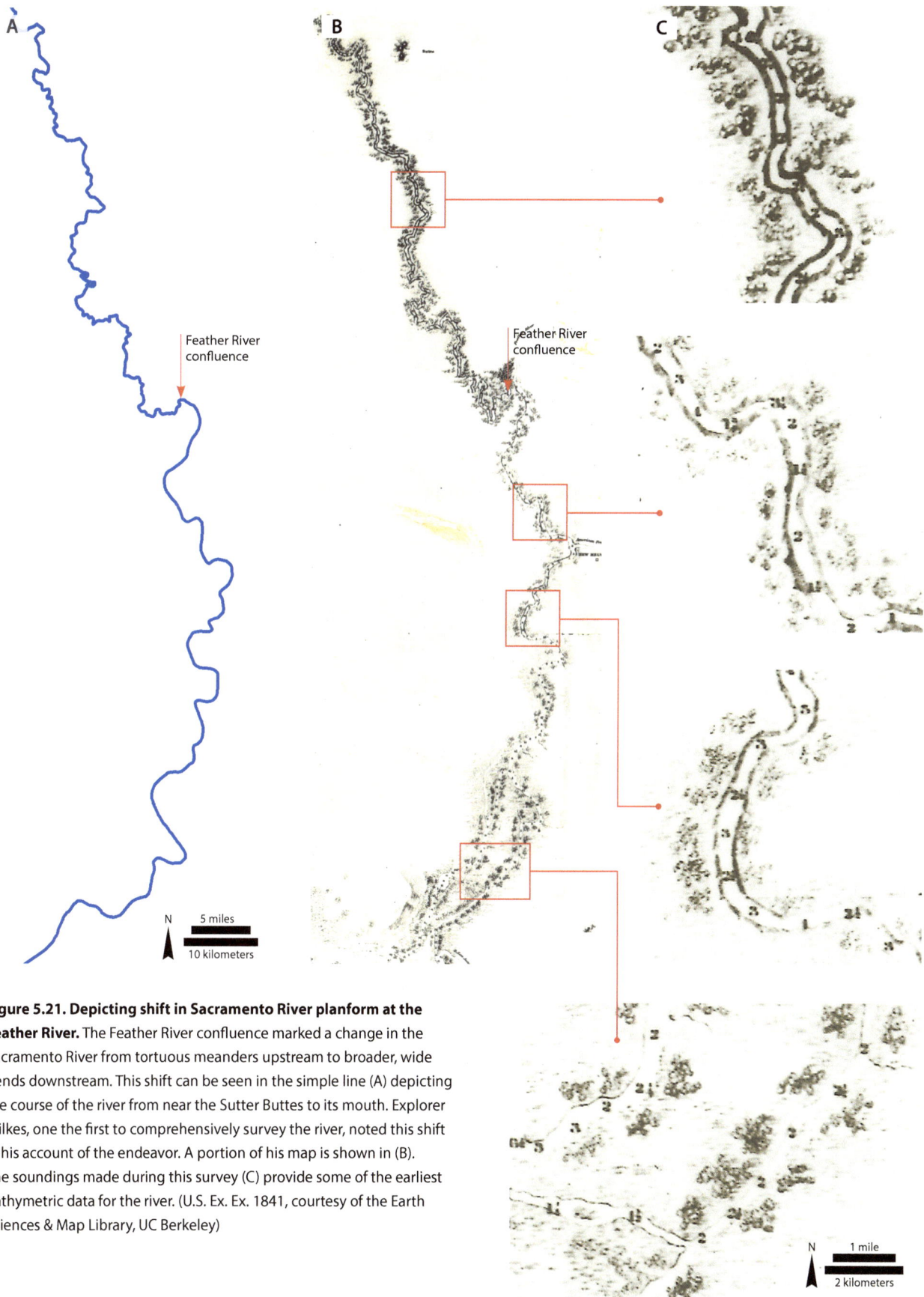

Figure 5.21. Depicting shift in Sacramento River planform at the Feather River. The Feather River confluence marked a change in the Sacramento River from tortuous meanders upstream to broader, wide bends downstream. This shift can be seen in the simple line (A) depicting the course of the river from near the Sutter Buttes to its mouth. Explorer Wilkes, one the first to comprehensively survey the river, noted this shift in his account of the endeavor. A portion of his map is shown in (B). The soundings made during this survey (C) provide some of the earliest bathymetric data for the river. (U.S. Ex. Ex. 1841, courtesy of the Earth Sciences & Map Library, UC Berkeley)

With a large flood volume spreading into the broad flood basins adjacent to the river, the river channel capacity was adjusted to discharges that were a fraction of the total flood volume (see page 235; Gilbert 1917). Like many low gradient rivers, it overflowed its banks regularly (*Californian* 1848, Young 1880, Leopold 1994). The Sacramento River can be thought of as a low flow channel, with a capacity of 110,000 cfs (3,110 cms), just 16% of the maximum estimated flood discharge of 660,000 cfs (18,690 cms; McClatchy 1916). Instead of increasing in flow capacity downstream like most rivers, the Sacramento River historically decreased in size between Colusa and the Feather River confluence, to a point where, as USGS scientist Grove Karl Gilbert (1917) estimated, the channel's capacity was just 10% of total flood flows. Channel size increased downstream of the Feather, but only to a capacity of about 27% of what was required to contain most flood flows (Fig. 5.22; California Debris Commission 1910)

The points where each of the major rivers entered the channel, and where the main distributaries left the channel, marked distinct shifts in the character of the river. For the purposes of discussing differences in channel geometry, a late 1800s engineering report divided the Sacramento River into distinct reaches: Feather River to American River, American River to the head of Steamboat Slough, head of Steamboat Slough to Cache Slough, and the tidally-dominated reach between Cache Slough and the Delta mouth (Young 1880).

Just above the mouth of the Feather River, the Sacramento River was only around 330 feet (70 m) wide (Wilkes 1845, USGS 1909-1918). Below this it widened substantially to over 490 feet (150 m). From there to the American River we estimate the river was on average about 575 feet (175 m) wide (estimated from width measurements taken every mile). This agrees

Figure 5.22. Minimum Sacramento River channel capacity was recorded at the Feather River confluence (blue line, Knights Landing) in a 1910 report by State Engineer William Hammond Hall. The channel increased in capacity downstream from that point (blue line), but was only a fraction of the estimated capacity required to carry most flood flows (purple line). The corresponding channel widths for these locations, as listed in the same report, are shown as red dots and correspond with the right-hand axis. (California Debris Commission 1910)

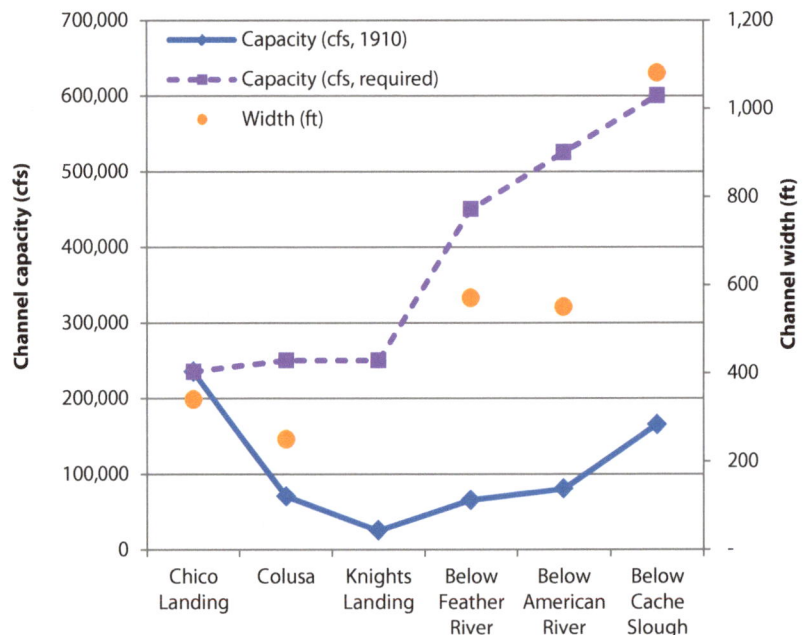

with the reported width of 571 feet (174 m) by the State Engineer in 1910 (California Debris Commission 1910). Between the American River and Steamboat Slough, about 27 river-miles (43 km) downstream, the mapping shows the historical channel was on average 550 feet (168 m) wide, the same as reported by the State Engineer (California Debris Commission 1910). Below Steamboat Slough the channel became significantly narrower, due to flows through that distributary. The average width for that reach down to Cache Slough was about 380 feet (115 m; Fig. 5.23).

We found that channel widths based on maps such as the early USGS topographic maps and the California Debris Commission mapping generally agreed with pre-hydraulic mining debris and reclamation accounts of channel width (Randall 1882). Generally, accounts describe the river as being from 600-900 feet wide (184-274 m; Taylor 1854). Just above the mouth of the American River, the 1841 U.S. Exploring Expedition reported that the channel was 800 feet (244 m) wide. Travelers to Sutter's Fort found the river around 600 feet wide (184 m; Clyman and Camp 1928[1848], Kerr 1850). For comparison, mapped channel width varies between 525 and 920 feet (160-280 m) for the mile above the American River confluence.

It was commonly reported in early histories that before hydraulic mining debris impacted river depth, boats traveling to Sacramento could rely on seven feet of water in the channel and 10 to 11 feet (3-3.4 m) at high tide (Upham 1878, U.S. War Department 1898, U.S. Congress 1916). Two surveys were conducted of the river before 1850, in which soundings no less than 2 fathoms (12 ft/3.7 m) were recorded up to Sacramento (see Fig. 5.21; U.S. Ex. Ex. 1841, Ringgold 1850a). In the written report accompanying his map, Wilkes concluded that vessels with a 12 foot draft could reach Sacramento (Wilkes 1849). Notable bars were found along certain reaches, but usually deeper channels ran alongside; the bars posed trouble for the larger boats during low tide (Abella and Cook 1960).

The mouth of the American River at Sacramento was deemed the "head of navigation during the dry season, or the stage of low water" (Wilkes 1845) because of bars in the reach between the Feather and American (Upham 1878). These were more or less permanent features in the channel, with distinct names such as Sixmile bar, Tenmile shoals, and Twelvemile bar (Young 1880, USGS 1909-1918). Individual flood events likely removed the bars temporarily, as was reported for the bar at the Feather River mouth in 1849 (Lewis Publishing Co. 1891), but they appear to have readily re-established. Bars and shoaling seem to have been less common downstream of Sacramento (Bryan 1923). Notable exceptions were Haycock Shoals near Babel Slough and Hogsback Shoals on Steamboat Slough (Ringgold 1852, Young 1880).

Leveeing, channel modifications, and closure of small overflow channels across the natural levees occurred along the Sacramento River beginning in the 1850s (Box 5.3). The rising floodplain elevations were attributed to these efforts, as well as to hydraulic mining debris (Taylor 1913). Also significant

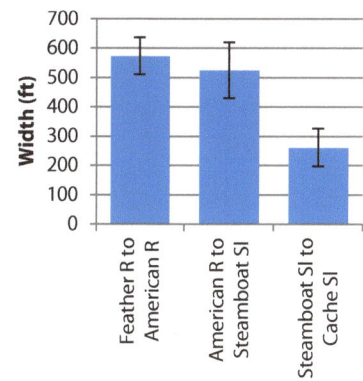

Figure 5.23. Average channel widths for three Sacramento River reaches is shown based on the GIS. Channel width decreased somewhat between the upper and middle reach (though variation was great) and then more notably along Grand Island. This last drop in width occurred primarily because the mainstem channel flow was relieved by several large distributaries, including Steamboat Slough, which was almost as wide as the main river. As its name suggests, this slough was the route often taken by boats because it was more direct. We averaged measurements taken about every mile along the river. Error bars represent one standard deviation. (primary mapping sources include: Wadsworth 1908a, USGS 1909-1918, Atwater 1982)

The river spreads out considerably and in two places the boats ran aground because the tide was very low. However there is a [deeper] channel along the banks.

—ABELLA AND COOK 1960, OBSERVED ON OCTOBER 26, 1811

BOX 5.3. THE REALIGNMENT OF THE AMERICAN RIVER AT ITS MOUTH

The American River historically entered the Sacramento River about a half a mile (0.8 km) downstream from its present junction. It made a sharp bend southward through a thickly forested floodplain before joining the Sacramento River just north of the present-day I Street Bridge (Fig. 5.24). Along this lower reach, the banks were only between 4 and 10 feet (1.2-3 m) high (Rose et al. 1895). Historically, a sand bar occupied the mouth, which according to travelers from Sutter's Fort in 1849, "at extreme low water is exposed, forming a small island in the middle of the river" (Derby and Farquhar 1932). The sandy substrate of the channel bed continued upstream past the Central Pacific Railroad Bridge (Rose et al. 1895). In the fall, during low water, the water was only a few feet deep (Derby and Farquhar 1932). Floodwaters from the American contributed significantly to the rising stages in the Sacramento River, causing floods in Sacramento and southward (Rose et al. 1895). In fact, the American River was blamed for some of the worst floods in the early decades of the city's existence.

Initial efforts to move the channel were reported in the *Sacramento Daily Union* in 1862. Substantial work to clear the channel of vegetation had been conducted and the channel was at that time being straightened by establishing "a channel three hundred and fifty feet wide through the bluff of the Sacramento river" (*Sacramento Daily Union* 1862b). This new straight canal extended from just below the American River railroad bridge to the mouth. The work was completed in 1868 and reportedly substantially reduced the risk of flood in Sacramento (Thompson and West 1880).

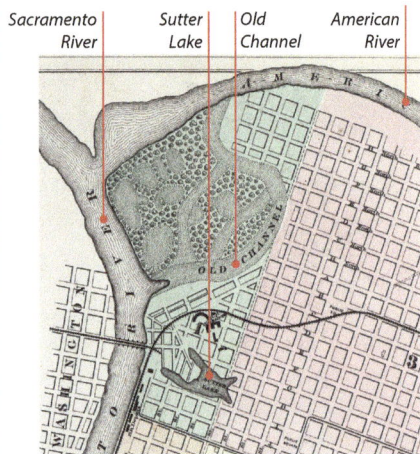

Figure 5.24. Old route of the American River. Before 1862, the American River's confluence with the Sacramento River was about a half mile downstream of its present confluence. It was rerouted to a more direct route to relieve flooding pressures on the city. (Ray 1873, courtesy of the David Rumsey Map Collection, Cartography Associates)

was the closure of several distributary channels that were historically connected to the Sacramento River during low flows. By 1880, Elkhorn Slough was dammed at its head at Clarksburg (Hall ca. 1880c, Bryan 1923), Hensley Slough (which formed Randall Island) was closed and filled in by 1879 (Tucker 1879d), and Tyler Slough (near Walnut Grove) was eliminated by 1884. The closure of Tyler Slough meant the elimination of "a large escape" of floodwater into the Mokelumne River and central Delta islands from the Sacramento River (Ryer 1884). Today, the Delta Cross Channel positioned north of Walnut Grove provides a similar, though regulated, connection. Aside from the dramatic changes in bed level associated with hydraulic mining debris in the late 1880s (see Fig. 1.14), there do not appear to have been other significant trends of widening or narrowing of the Sacramento's channel. This is likely because artificial levees were built on top of natural levees, largely fixing the channel in place. However, several significant local changes were found, particularly for the lower reaches.

STEAMBOAT VERSUS OLD RIVER Steamboat Slough is the largest distributary of the Sacramento River within the Delta (Young 1880). It exits the river just south of Courtland and rejoins it at the mouth of Cache Slough. Grand Island, 17,000 acres (6,880 ha), is formed by it and the main channel of the Sacramento. Steamboat Slough is significantly shorter, just 11.7 miles (18.8 km) compared to the main river's 18.2 miles (29.3 km). Though most early navigation prior to hydraulic mining occurred along Steamboat Slough for this reason, the eastern Sacramento River branch was considered to be the main river channel (Ringgold 1852, Kerr and Camp 1928, Lienhard and Wilbur 1941). This is noted as early as 1817, when Spanish explorer Fray Narciso Durán wrote, "we came to a stream to starboard leading to the east, and they say that this is the turn which the principal river makes," before choosing to head north up Steamboat Slough (Durán and Chapman 1911). The main branch of the Sacramento was also referred to as Old River (Hutchings 1859, Tucker 1979c).

Though the longer Sacramento River channel was deemed the main river channel, the minimum depths within the channels do not appear to have differed substantially from that of Steamboat Slough. An 1850 survey recorded soundings as low as seven feet (2.1 m) at low water in both channels (Ringgold 1850a). For much of its length, the main Sacramento River channel soundings were above 12 feet (3.7 m), save for a mile or so above Ida Island, at the foot of Grand Island. On Steamboat Slough, the shallow 7 foot (2.1 m) sounding was located at the well known Hogsback Shoal midway up the channel. The U.S. Exploring Expedition map of 1841 also shows these two locations to be the shallow points in the channels, with soundings of 1.5 fathoms (9 ft/2.7 m). Hogsback Shoal was the primary point along Steamboat Slough where navigation was significantly affected. Boats had to carefully navigate its shallow waters; often steamboats waited for high tide before crossing (Hutchings 1859, McGowan 1939). Though the early navigational charts do not show a substantial difference in depth between the two channels, later State engineering reports show that Steamboat Slough was the deeper channel before hydraulic mining debris raised its bed (Young 1880). This was attributed to the greater channel slope in comparison to the main river.

Hydraulic mining debris raised the streambed to the point where it had to be closed to larger vessels (see Box 1.2). The depth in Steamboat Slough decreased from a reported average of 12 feet (3.7 m) in 1853 to just 5 feet (1.5 m) in 1879 (Jacobs 1993). As a remedy to the flood and navigation problem, State engineers proposed drastic modifications in an 1880 report that would have closed off the main branch of the Sacramento River and made Steamboat Slough the only channel (Young 1880).

SNAGS Early accounts of the Sacramento River note that the channel was generally free of large wood obstructions between its mouth and the City of Sacramento (Bryant 1848, Revere 1849, Gerstäcker 1853). Whereas two of the three branches of the San Joaquin River accumulated woody debris, or rafts (Gibbes 1850b, see page 366), evidence of similar large accumulations of woody debris was not found for the Sacramento River. However, individual fallen trees near the banks and submerged logs were documented (Johnson 1851, Ringgold 1852). An expedition, likely in the Old Sacramento River branch along Grand Island, reported "many logs" making their travel difficult (Durán and Chapman 1911). Once travelers departed from the main river channel for one of the smaller distributaries such as Sutter Slough, however, wood in the channel was more of an impediment to travel, which for some vessels made the channels unnavigable (Sprague and Atwell 1870). Unlike the Sacramento River mainstem, flood flows were of insufficient force to remove the accumulation of debris in these side channels.

Mokelumne and Cosumnes rivers meeting the tidal Delta

The Mokelumne and Cosumnes rivers imparted their signatures on the Delta, distinct from those of the San Joaquin and Sacramento rivers. This transitional area in the vicinity of New Hope Tract, McCormack-Williamson Tract, and the upper portions of Staten and Tyler islands was profoundly affected by the fluvial influence of these rivers, a part of the Delta's upper deltaic plain (Coleman 1976, Brown and Pasternack 2004). Given its position at the Delta margin, the associated tidal wetlands had only recently (in geologic time), been transgressed by tides as sea levels rose. While tides did affect the area, the landforms and function of the landscape were largely driven by fluvial processes (Florsheim and Mount 2002).

Upon exiting the foothills, the Mokelumne River meandered as a single-thread channel along a relatively narrow and densely wooded floodplain (Norris 1853a, Thompson 1862). Tidal influence was perceptible upstream of the Dry Creek confluence (Gray 1859, Thayer 1859, Van Scoyk 1859, Watson 1859b, Rhodes in Mendell 1881). The Cosumnes River spread into numerous distributaries close to nine river-miles (14.5 km) above its mouth. These then coalesced into a single primary tidal channel several miles before the river's confluence with the Mokelumne River just upstream of Benson's Ferry (now the Thornton Road crossing).

At this point, the Mokelumne River mapping showed an approximately 140 feet (43 m) wide channel (based primarily on the 1914 Debris Commission survey of the river). Boats often traveled to Benson's Ferry, though the official head of steamboat navigation was downstream at New Hope Landing (the head of Staten Island; Payson 1885). The river largely maintained a relatively narrow (about 110 ft/34 m wide) channel downstream to where it branched around Staten Island. A late 1800s U.S. Army report figure was over 30% narrower (70-80 ft/21-24 m; Payson 1885).

Downstream of the island, both forks widened substantially. Within the first three miles, the North Mokelumne channel widened from about 100 to 300 feet (30-90 m). The South Fork was narrower on average, maintaining a channel between 130 and 165 feet (40-50 m) for most of the first three miles downstream from the head of the island. Mokelumne land case testimony includes estimations and point measurements of channel width (Fig. 5.25). Unfortunately, however, several witnesses contradicted each other and some measurements seemed improbably wide (often around twice as wide as the mapped channel; Davis 1859, Watson 1859b, Gray 1859). In spite of this, testimony illustrates that within a mile downstream from the head of Staten Island, the two branches of the Mokelumne almost doubled in width. This coincides with the rapidly diminishing levee height and increasing tidal influence. This testimony also confirms that the North Mokelumne was by far the wider and deeper channel of the two branches (Davis 1859).

Early channel depth measurements on the Mokelumne at Benson's Ferry and several miles below the head of the island also are available from the Mokelumne land case testimony. At Benson's Ferry during low tide, low water, one witness reported depths of 1.5 feet (0.5 m; Watson 1859b). Along the North Mokelumne, witness George Gray (1859) reported that the channel was on average 12 feet (3.7 m) deep. Along the South Mokelumne, he stated that most of the reach was 20 feet (6.1 m) deep, but that the upper three-quarters of a mile was less than 10 feet (3.0 m). This is in general agreement with another witness, William Watson (1859b), who took 10 soundings spaced within the first mile downstream of the head of Staten Island along both forks. In the North Mokelumne, his soundings were from 10.5 to 15 feet (3.2-4.6 m); in the South Mokelumne, the soundings ranged between 5 and 12 feet (1.5-3.7 m). These surveys were conducted before large artificial levees were in place, though some reclamation efforts had begun. More significantly, witnesses noted that they had seen a recent rise in bed level and decrease in summer freshwater inflows as a result of mining efforts upstream. For instance, one witness stated that at Benson's

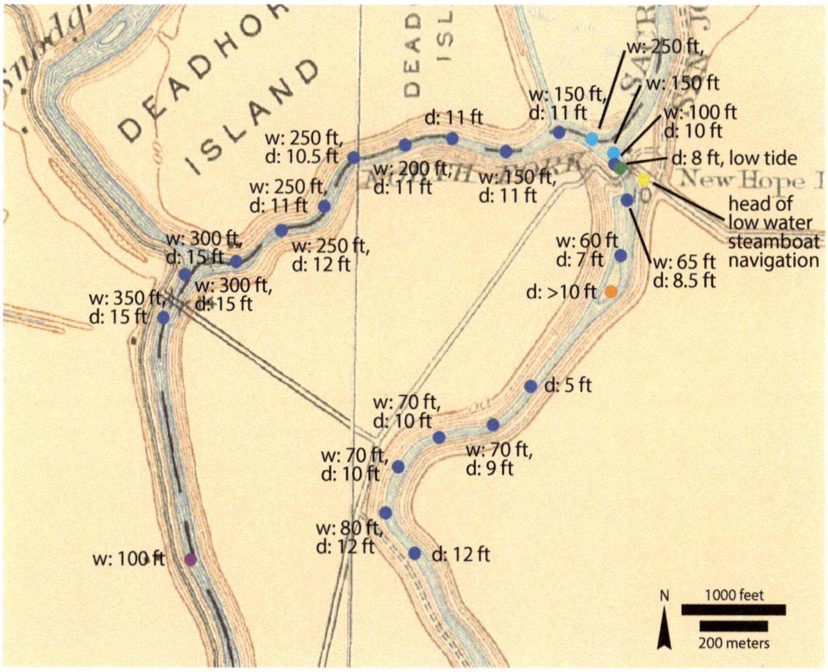

Figure 5.25. Mokelumne River width and depth from 1859 testimony. These data illustrate both the wider and deeper channel of the North Mokelumne River and the rapidly increasing width as one traveled downstream. Measurements taken by witnesses testifying in the Mokelumne land case trial were given in depositions to the case. These were georeferenced as accurately as possible in order to illustrate the trends in width (w) and depth (d) of the two forks of the river within the first mile or so of the head of Staten Island. (USGS 1909-1918)

Ferry "there is about 2.5 feet [0.8 m] less" water at low tide than when he first observed the channel, over two and a half years prior (Davis 1859).

Soundings from the early 1900s are available from Debris Commission surveys (California Debris Commission 1914). These data are suggestive of a deeper channel in comparison to that stated in 1859 testimony, though most measurements fall within the range of earlier evidence. The Debris Commission soundings from the deepest part of the channel are within 8 to 23 feet (2.4-7.0 m) at low water between Benson's Ferry and the head of Staten Island. Along the first three miles of the North Mokelumne, soundings range between 6 and 29 feet (1.8-8.8 m). On the South Mokelumne for the first three miles, soundings range between 8 and 27 feet (2.4-8.2 m; California Debris Commission 1914).

SNAGS Historically, the Mokelumne River was marred by substantial amounts of woody debris in its upper tidal reaches. The channel was narrow and lined with dense forest particularly above the head of Staten Island. Individual fallen trees and accumulated masses of debris, or rafts, were present (Gibbes 1850a, Matthewson 1859, Payson 1885). This material would have affected flows, potentially encouraging increased floodplain inundation and the development of backwater habitat (see page 366). Snag-boats were funded in the later 1800s by the federal government to clean out navigable channels. An 1881 report created in preparation for such removal for the Mokelumne River stated that from the Galt Ferry at New Hope to New Hope landing, there were "79 snags, 21 overhanging trees, 7 rafts" (Rhodes in Mendell 1881). Below the head of Staten Island, the North Mokelumne channel was apparently "excellent," that is, generally free of obstructions, only three snags being reported. In contrast, on the South Mokelumne, the report states that within the upper one and a half miles there "are 20 snags, 12 overhanging trees, and one raft." Later, a reported "160 snags and 314 overhanging trees" were removed along the river between Snodgrass Slough and Benson's Ferry (Payson 1885).

Low order channel characteristics

In addition to the rivers and associated distributary channels (e.g., Elk Slough), numerous small tidal and non-tidal channels were found in some parts of the north Delta landscape; in other locations, particularly in the upper non-tidal portions of the basins, defined channels within the wetlands were sparse or absent. In the GIS, we assembled a total of about 460 miles (740 km) of low order channel within about 190,000 acres (76,890 ha) of perennial wetland (including Grand Island and upper Andrus, Tyler, and Staten islands). Over half of that length occurred below the head of Grand Island. The historical habitat type mapping suggests that close to 60% of these channels were tidal. Of the mapped channels, we estimate that 62% were definitely present in the early 1800s ("high" confidence level), 30% were probably present ("medium" confidence level) and 8% were possibly present ("low" confidence level). Overall, this suggests average channel density within the range of 9 to 15 feet per acre (0.68-1.13 km/

km²). However, density was highly variable across the area. Channels were concentrated in the lower, most tidally-influenced portions and where streams spread into numerous distributary channels that intersected the basins. Outside of perennial wetlands, the mapping includes an additional 550 miles (885 km) of low order fluvial channel within the 25 foot contour, many of which were ephemeral streams to the west of Cache Slough.

The north Delta's low order channels can be identified and characterized by their landscape position: tidal channels at the lower ends of the basins (e.g., Cache Slough), channels formed by fluvial processes that crossed natural levees and dissipated into adjacent basins, small ephemeral systems that lost definition at the upland margins, and branching distributary networks that occupied the wetland "sinks" of the larger upland systems such as Putah Creek and the Cosumnes River. The first two types are discussed in this section; wetland sinks are addressed on pages 294-300.

TIDAL CHANNEL NETWORKS AT THE BASIN OUTLETS At the outlets of the Yolo and Sacramento basins, Cache Slough and Snodgrass Slough, respectively, conveyed tides into the lower portions of the basins. These points of access were made possible in part by the low natural levees found at the southern extents of these basins. The tidal channels of the lower north Delta functioned much like those of the central Delta, with year-round direct connection to the twice-daily ebb and flow of tides through their mouths. However, these tidal channels experienced relatively greater flood disturbance given their positions at the downstream end of the basins' flood flows.

Of the three branches encountered by sailors just north of Rio Vista as they traveled upstream on the Sacramento, Cache Slough was the least important for travel, although it was known to be easily mistaken as the primary route to Sacramento (Palmer et al. 1881). One of the first surveys of the area concluded:

> The West Fork [Cache Slough], and the sloughs connecting with it, are not navigable except for small boats; originally, they were successfully frequented by trappers, for otter and beaver. On the west, the waters terminate and waste themselves in the swamps and mud flats.
> (Ringgold 1852)

However, the slough apparently maintained sufficient depth for navigation of small craft to Maine Prairie, for a time an important shipping point for Solano County (Munro-Fraser 1879). Though the slough was connected to the Sacramento River at several points, only the main slough provided substantial tidal access. The Cache Slough network can therefore be considered a large blind tidal channel network lying at the southernmost part of Yolo Basin.

The slough had significant tidal capacity, which also served the Yolo Basin during flood (Young 1880). Tidal range was reported to be "from nearly six feet at low water to about one foot at extreme flood stages" (Rose et al. 1895). The slough proper was generally between 330 and 660 feet (100-200

m) wide for much of its length. Lindsay Slough, the largest branch of Cache Slough, was around 330 feet (100 m) wide and extended northwest into the vernal pool complex of today's Jepson Prairie Reserve. Most of the channel length associated with Cache Slough lay along the northeast bank of the slough, where Shag, Prospect, Miner and Elkhorn sloughs branch into the lower Yolo Basin. Both Miner and Prospect sloughs were flow-through, with Miner connecting to Steamboat Slough and Prospect connecting to Miner Slough. Shag and Elkhorn sloughs may not have been strictly blind tidal channels either, though their connections were less significant. Shag Slough did not appear to have connected to Prospect Slough, however, suggesting that today's Liberty Island was not an island historically.

The Cache Slough network evidently had some of the highest channel density in the Delta. As mapped, the network consisted of about 150 miles (240 km) of tidal channel influenced primarily by tides, 76% of which were mapped with high interpretation certainty. This channel length has an associated density of approximately 25 feet per acre (1.87 km/km^2), if the estimated contributing area basically bounds the extent of channels. By comparison, densities around only five feet per acre (0.37 km/km^2) were found within the rest of the tidal wetland area extending north in the Yolo Basin (with most of the length coming from a single channel, Duck Slough). However, since reclamation occurred later in the lower Yolo Basin, we may have been able to map historical channels at a more detailed level for this area. For example, a highly detailed map made in 1920 still shows the historical channel network, though most had been dammed by that time (Fig. 5.26; Wheeler 1920). This should be considered when compared to channel densities of the central Delta. However, numerous visual inspections suggest that the density differences were significant: in the Cache Slough area compared to elsewhere in the central Delta, channels in unreclaimed portions shown in the early USGS topographic maps appear denser, signatures in historical aerial photography (taken when virtually the entire Delta had been reclaimed) also seem relatively denser, and the frequency with which tidal channels branch off from the mainstem is greater.

The Cache Slough channel network is also distinct from other areas of the Delta. Compared to the nearly 10 mile (16 km) long tidal channels at the eastern Delta margin (which were mapped as 4th order or lower), Cache Slough tidal channels were more truncated in form, with even large branches such as Shag Slough (a 5th order channel) only extending a few miles before terminating within the wetlands. This pattern was more of a classic dendritic planform than characterized the sinuous tidal channels of the central Delta. Such patterns may also reflect a more recently developed tidal network (Fig. 5.27; Mount pers. comm.). The more recent estuarine transgression towards the Delta margins and frequent flood disturbance in the Yolo Basin supports this latter hypothesis. It may also relate to the relatively higher channel density found within this area.

In time of flood the navigation of the North Fork is made difficult on account of the free discharge of the flood-water, being obstructed at the mouth of Snodgrass Slough. The Cosumne [sic] River overflows near its confluence with the Mokelumne River, and discharges this body of water, under the most unfavorable conditions, through Snodgrass Slough.

—PAYSON 1885

Additionally, the limited spatial extent of the Cache Slough tidal channel network suggests that it may have been difficult for tides to regularly flood all the land within the high-tide elevation level in the channels. Without tidal channels to carry the tides, they would have had to spread great distances through dense vegetation which would have retarded passage and may have prevented tides from reaching their full potential extent before the turn of the tides. This issue is hinted at by an early newspaper article expressing the possibility that a drainage canal within the basin might actually facilitate the propagation of tide water up the basin (*Sacramento Daily Union* 1873c).

The only other major tidal network of the north Delta basins was associated with Snodgrass Slough. It was the primary conduit transporting tidal flows into the lower Sacramento Basin; one early account identifies "a tule drained by Snodgrass slough" (*Sacramento Daily Union* 1862a). As it was not directly connected to a major upstream sediment source, large natural levees did not form along the channel and restrict tidal communication with the surrounding wetland. While tidal exchange from the Sacramento and Mokelumne rivers was limited by natural levees, waters flowed easily from Snodgrass and through the many smaller tidal channels within its network. This pattern is illustrated by historical maps of the area. For instance, one of the earliest maps shows many trunks of tidal channels branching off of Snodgrass Slough into the wetland plain. By comparison, only few are seen along the Sacramento and Mokelumne channels (Fig. 5.28; Reece 1864). Most tidal channels branched off of the east bank of Snodgrass Slough. The absence of large tidal channels to the west is

Figure 5.26. Channel detail in the Cache Slough area. A detailed map (A) shows small tidal channels in the vicinity of Shag Slough, many of which were dammed by this time. This map suggests relatively high channel densities for the Delta – around 25 ft/ac (1.87 km/km^2). A number of the channels can be seen as remnant signatures in the historical aerial photography (B). Within the agricultural fields, these show up as lighter-toned signatures in the soil, reflecting the more inorganic sediment banks of the sloughs (A: Wheeler 1924, courtesy of the Solano County Surveyor; B: USDA 1937-1939)

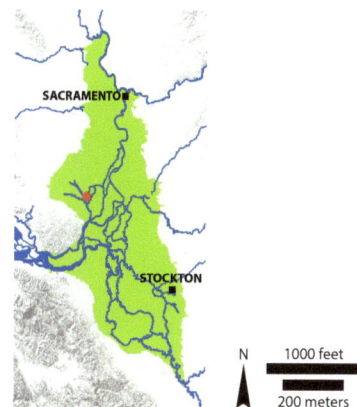

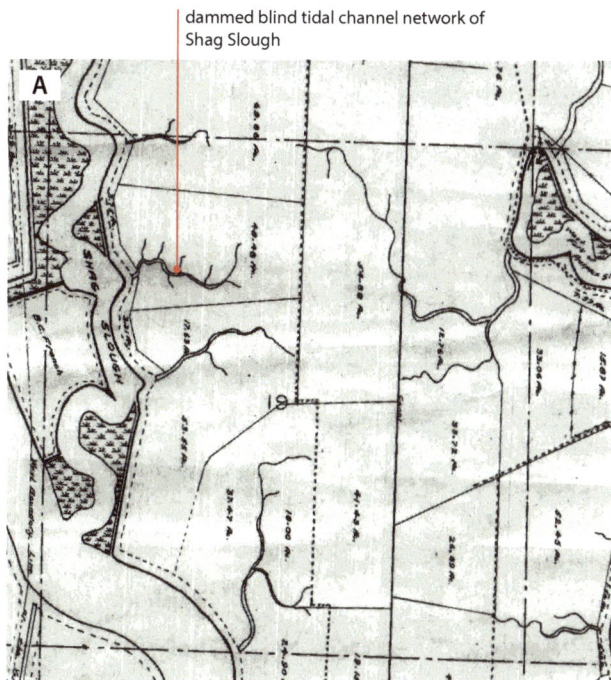

dammed blind tidal channel network of Shag Slough

channel signature seen in the lighter-toned soil

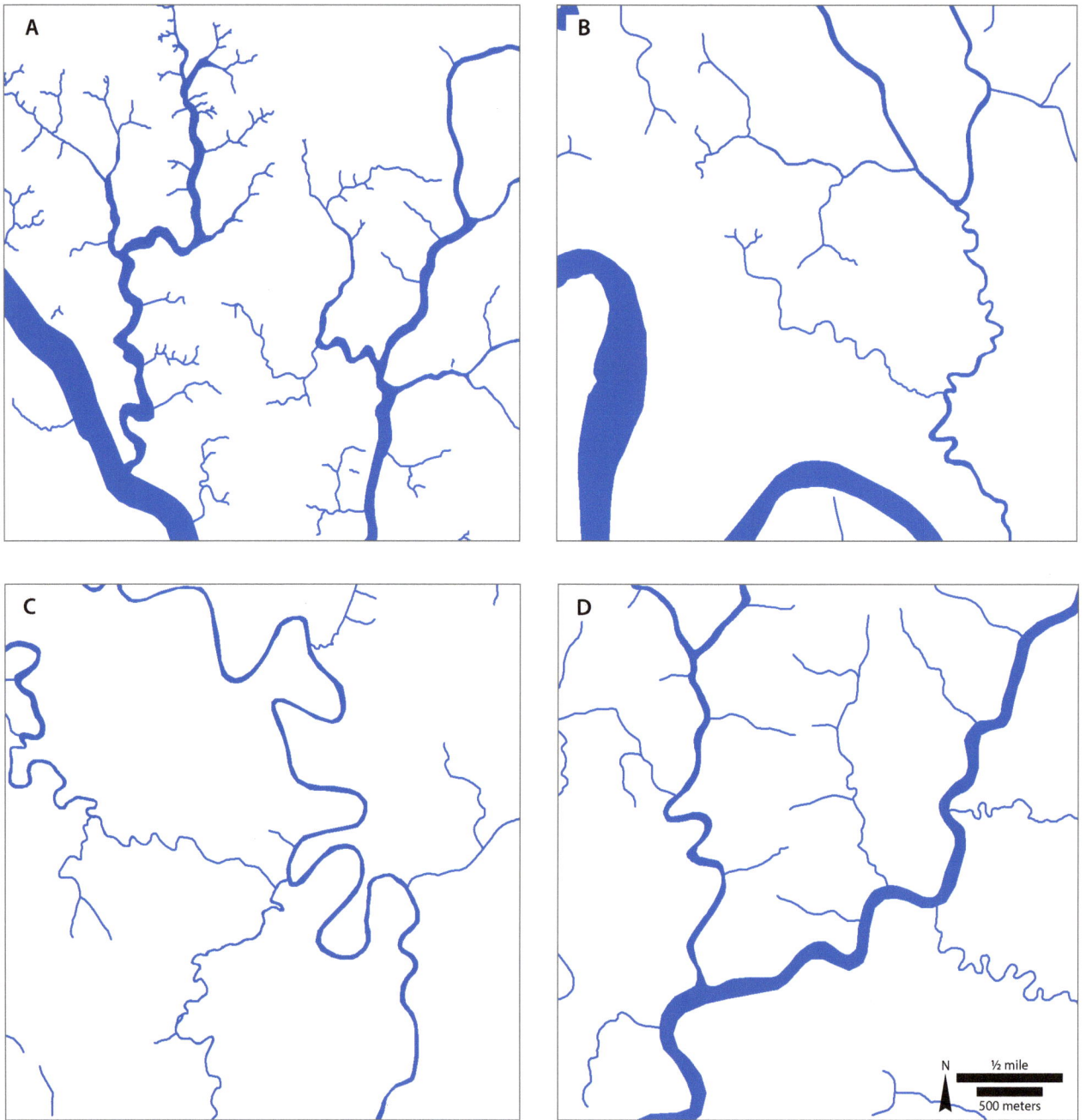

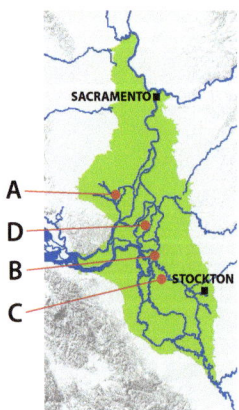

Figure 5.27. Comparison of channel planform. Cache Slough (A) had a more truncated network than sloughs found elsewhere in the Delta. Shown here is part of Venice Island (B), Whiskey Slough (C), and part of Tyler and Staten islands (D). This may relate to the frequent floods that passed through Cache Slough from the Yolo Basin, perhaps keeping the network in a younger developmental state.

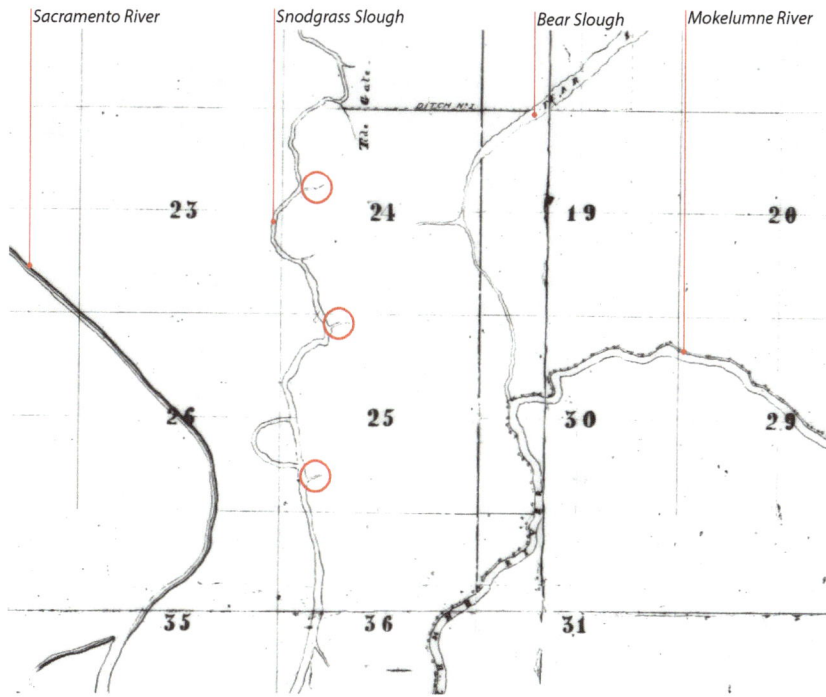

Figure 5.28. The trunks of tidal channels branching off of Snodgrass Slough into the tidal wetland plain are shown in this 1864 map (examples circled in red). Snodgrass Slough likely carried the majority of the tidewater into the wetlands north of the Mokelumne River as very few similar trunks of channels are seen along the Sacramento or Mokelumne river channels, likely related to obstruction from the natural levees. (Reece 1864, courtesy of the California State Lands Commission)

supported by a reclamation summary for the Runyon District, (Pearson District), which stated "there are no wide sloughs to dam" (Tucker 1879d). Mapping shows a total of over 35 miles (56 km) of tidal channel (including flowpaths through wide sloughs, ponds, and lakes) within the Pearson District and McCormack-Williamson Tract area, about 17.5 miles (28.2 km) of which appear to have been directly connected to Snodgrass Slough. Using several different interpretations of contributing area, the mapping suggests an associated channel density in the range of 16 to 21 feet per acre (1.2-1.6 km/km²).

Historically, Snodgrass Slough terminated within the wetlands just south of where present-day Russell Road bends north (Fig. 5.29; Tucker 1879d). Today, Snodgrass Slough is leveed and continues northerly to the Sacramento River levee in a canal. Sources suggest that Snodgrass Slough historically had no substantial tidal connection to any other major channel or lake. Substantial ditching occurred in the 1860s to create connections and drain water through to Snodgrass Slough from the lakes within the Sacramento Basin (Hall in Board of Swamp Land Commissioners 1864-5, *Sacramento Daily Union* 1873b). Connections at high flows may have been present, however. Land case testimony includes a description that "at high water there is a slough the course of which I have marked down…emptying into…Snodgrass slough" (Lambeth 1859). This speaks to the additional function of Snodgrass Slough as a conduit of floodwaters during the wet season, just as Cache Slough functioned for the Yolo Basin (Payson 1885).

CHANNELS THAT CROSS NATURAL LEVEES In addition to the tidal channel networks positioned at the foot of the basins, secondary channels bisected natural levees throughout the north Delta, serving to connect the river to the basins and deliever flood water. They were referred to as "sloughs," "small sloughs,"

These basins become partly filled by the first freshets of the season, whose waters escape through deep crevasses and sloughs into them.

—YOUNG 1880

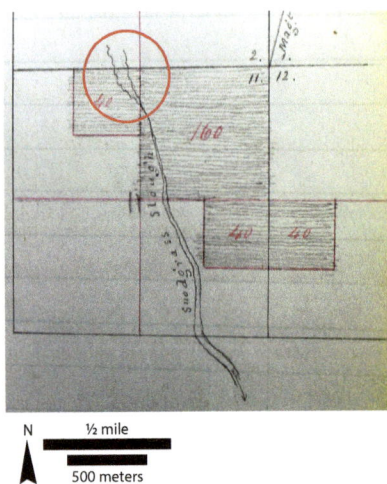

Figure 5.29. Snodgrass Slough terminating in the wetlands. According to this swampland survey sketch, Snodgrass Slough branched and ended just south of where Russell Road today bends to the north. (circled in red, Cleal 1855, courtesy of the Center for Sacramento History, Swamp and Overflowed Land Surveys, No. 129)

"crevasses," or "lagoons" (Flint 1860, Young 1880, Rose et al. 1895, Bartell 1912, Lienhard and Wilbur 1941). Some were deep enough to sustain tidal flows, at least during higher river stages. Most, however, served solely as conduits for flood water. All were primarily formed and maintained by fluvial processes. As one observer noted, the channels "do not run *into* but *from* the river" (Logan 1865). To describe how the basins became flooded, one person testified that the "tula land is annually overflowed by means of sloughs through the timber land [riparian forest], through which the waters run from the river during the wet season" (Fowler 1853). Some also served to drain the basins for a period after large floods: "the waters run out of the sloughs for a certain length of time until the sloughs fail to drain, when they evaporate all summer" (Sanford 1860).

These small secondary channels, or crevasses, were found along the natural levees. They lost definition shortly after reaching the wetlands beyond. Many of these features likely formed in single events, only serving as overflow channels for brief periods of time (Bryan 1923). Their signatures are evident in the historical aerial photography and in the early USGS topographic maps as short narrow depressions (Fig. 5.30; USGS 1909-1918, USDA 1937-1939). This ephemeral and shifting quality, along with associated sediment deposition and disturbance of vegetation communities, affected the dynamics and complexity of the riparian forest. This complexity is conveyed in early textual descriptions, such as this account of the river banks near the City of Sacramento: "deep sloughs creased its site whose beds were a bramble of grapevines, blackberry bushes and other undergrowth, while big white oak trees…dotted the space between these depressions" (Fairchild 1934).

These secondary channels were less well defined than the tidal channels of the central Delta, which tended to be fairly deep, commonly navigable channels where they branched from the mainstem river channels. This can be seen in the Mokelumne land case testimony, where, when asked about the number of sloughs leaving the river on its left bank between Benson's Ferry and the head of Staten Island, a witness stated that he found "no slough, but two or three small inlets, the smallest not being over two feet wide, and the largest not more than four" (Sherman 1859). However, another witness, responding to the same question, stated "there must be five or six small sloughs" (Van Scoyk 1859). This disagreement likely lies in the fact that Sherman's definition of slough is more akin to the tidal sloughs of the central Delta and therefore he did not count the secondary channels to which Van Scoyk referred. Earlier in his testimony Van Scoyk described the general character of these banks as "cut up with little sloughs here and there," which relays a character very different from the tidal channels found farther downstream, such as Sycamore Slough.

While the majority of these secondary channels bisecting natural levees flowed only during the highest river stages, a few were deep enough to maintain tidal connectivity for much, if not all, of the year (Board of Swamp Land Commissioners 1867). Most were found along the lower reaches of the Sacramento River and many appear to be associated with some of the

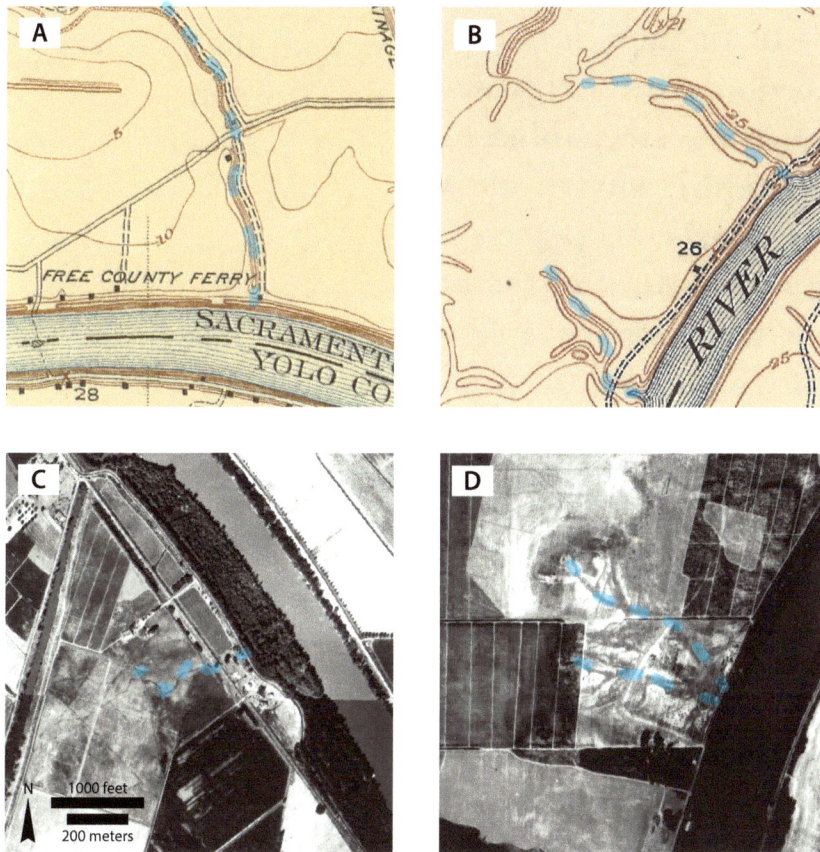

Figure 5.30. Examples of secondary channels crossing natural levees of the Sacramento River. These overflow channels (dashed blue lines), or crevasses, carried floodwater unidirectionally into the basins and flowed for only brief periods during the year. They were likely created by single events and may have been active for relatively short periods of time. (A, B: USGS 1909-1918; C, D: USDA 1937-1939)

larger crevasse splays, such as Babel Slough. The slough, referred to as a "large distributary channel" by geomorphologist Kirk Bryan (1923) and as the slough that "connects the tule with the Sacramento River" (Sprague and Atwell 1870), likely maintained tidal flows through its channel year-round.

In addition to Babel Slough and the larger Sutter and Elkhorn sloughs, we mapped several additional channels that were connected to the Sacramento River north of the head of Steamboat Slough as possibly tidal. One of these was Beaver Slough (one of several in the Delta). It "put out from Hensley's slough [the slough forming Randall Island], and emptied into the tule" (Green 1882). An early 1863 reclamation map shows it as one of only two other channels on the east side of the Sacramento between Steamboat Slough and the City of Sacramento, suggesting it was historically one of the more significant channels on the east side of the river. However, early modification, particularly the closure of Hensley Slough in 1865, makes interpretation of its tidal status difficult; the slough may very well have experienced tidal influence only at the highest river stages. One account that lends support to this interpretation is an 1858 GLO survey that notes a "dry slough 60 links [40 ft/12.2 m] wide" near where the Beaver Slough network nears Stone Lake.

Early maps showing prominent channels and revealing topographic evidence suggesting appropriate elevations for tidal connections helped in the mapping process (Gibbes 1869, Secretary of State 1866-1877, Wadsworth 1908a, USGS 1909-1918). Additionally, though only a few textual descriptions of these channels' tidal character are available and they are rarely spatially explicit, several offer useful information. The account of Heinrich Lienhard's 1846 trek (season unknown) on foot along the west bank of the Sacramento River includes the crossing of multiple "lagoons." Between the head of Grand Island and just upstream of Babel Slough, his narrative mentions two channels that were too deep to cross on foot, which were a combination of Sutter, Elk, and Babel sloughs. In addition to these, Lienhard also described "lagoons of varying breadth and width formed by marshy areas whose waters flowed back swiftly into the Sacramento with the ebbing tides…several lagoons were crossed where the cold water reached to our hips and in places even up to our shoulders" (Lienhard and Wilbur 1941). This description lends support to the idea that there were more channels than just the major distributaries of the Sacramento that maintained tidal connection to the river.

Well before the basin wetlands were farmed, these secondary channels off the Sacramento River had been transformed from their historical conditions. Since they threatened the newly constructed homes and blossoming cultivation along the Sacramento River's natural levee lands in the 1850s, they were dammed and filled early. An 1860 agricultural report attests to this, stating that with "farms being opened all along its banks, the small sloughs, which at high water discharged a portion of the surplus into the tule, have been closed up, so that none of its waters now go upon the tule" (Flint 1860). The closure of these overflow outlets had an early impact on how floodwaters were routed to the central Delta and on the hydrologic and ecological connections between river and flood basin. The subsequent infamous and frequent levee breaks occurred as a result of the natural tendency of the river to spill into its adjacent flood basins.

It should also be considered that surface water connectivity was not the sole means through which the river communicated with its floodplain. Though surface water connectivity between river and flood basin is the most important ecologically, the rivers also communicated with the surrounding lowlands via their connected water tables. Water levels in the basins' sloughs were affected by river levels. A witness for the Sutter land case explained that "all or nearly all of the sloughs are supplied, more or less, by percolation or seepage from the river" (Denver 1860). These connections supported groundwater elevations near the surface and helped maintain the wetlands and sustained the seepage of freshwater into the central Delta through the summer months.

CONTRASTING CHANNEL MORPHOLOGY OF ELK AND DUCK SLOUGHS Channel morphology varied substantially depending on the relative influence of fluvial processes. Whether or not a channel was directly connected to riverine inputs and the associated sediment supply affected its shape. Most apparent was the impact on the relative height of natural levees. This can be seen by comparing two channels of the Yolo Basin: Elk Slough and Duck Slough (Fig. 5.31a).

Elk Slough is a distributary of the Sacramento bounding Merritt Island. Sediment-laden flood flows were directed through this channel and built natural levees like those of the Sacramento River. These levees were well over five feet (1.5 m) above sea level and consequently supported a dense riparian forest (Fig. 5.31b). Duck Slough, on the other hand, received floodwaters only indirectly into its head in Big Lake. Since the floodwaters had already spread into the Yolo Basin before entering the channel, most sediment had already dropped out of the water column and thus did not build natural levees (Fig. 5.31c). Duck Slough did, however, receive some direct flood flows, but through its lower end from Miner Slough, which was directly connected via Sutter Slough to the Sacramento River. This is evidenced by the fact that natural levees extended along this lower reach of Duck Slough for about two and a half miles (4 km).

THE ABSENCE OF CHANNEL The broad natural levees of the Sacramento River largely prevented the establishment of extensive secondary or overflow channels extending into the lowlands, as was common in the more typical floodplain environment of the south Delta. This effect of natural levees on channel planform is particularly striking for the channels of Tyler Island, which is bordered on the west by the relatively high natural levees of Georgiana Slough and on the east by the comparatively low levees of the North Mokelumne River. Virtually all of the tidal channels extending into the island originate from the North Mokelumne River. Notes on reclamation of nearby Andrus and Brannan islands point out that the first levees, built in 1858, were constructed to "keep out the tide water from the San Joaquin River," suggesting that these islands were primarily wetted through tidal channels connected to the San Joaquin and that there were few, if any, tidal channels exiting the Sacramento River (Tucker 1879f).

Channels were sparse, particularly within the upper portions of the basins (see Fig. 5.3). As stated by Jepson (1893), the "tule lands northward from Cache Slough…extend untraversed by any water course to and beyond Putah Creek." Most overflow channels along the natural levees lost definition upon entering the wetlands. Water flowed as broad sheets through the emergent vegetation of the basins, lacking the scouring energy to create defined channels (CDFG and YBF 2008). The accumulated winter flood waters were described as "creeping slowly along toward tide water, not in a direct or free channel" (Board of Swamp Land Commissioners 1864-5). The lower depressions filled within the basins during floods and, without channels to provide sufficient drainage, formed the lakes and ponds common to the tidal margins and upper basins. In response to this general absence of channels, both the Yolo and Sacramento basins underwent early and extensive coordinated efforts to establish systems of drainage canals that ran the full extent of the basins (Box 5.4).

LAKES AND PONDS OF THE WETLANDS

Lakes and ponds, located in the lowest parts of the basins, were prominent features of the north Delta landscape. They were filled primarily by annual flooding and gradually drained and evaporated over the course of the dry season. During the dry season, many became hydrologically disconnected from the rivers. Many of the larger and deeper features were maintained

The plan of reclamation contemplates first, to facilitate the drainage of waters which almost annually accumulate in the basins of Cache and Putah Creeks, thence spreading over the entire District, creeping slowly along toward tide water, not in a direct or free channel, but across an uneven surface of miles in width, obstructed by a rank growth of new tule and masses of drifting tule of former seasons.

—HALL IN BOARD OF SWAMPLAND COMMISSIONERS 1864

Duck Slough | natural levees | Elkhorn (Elk) Slough

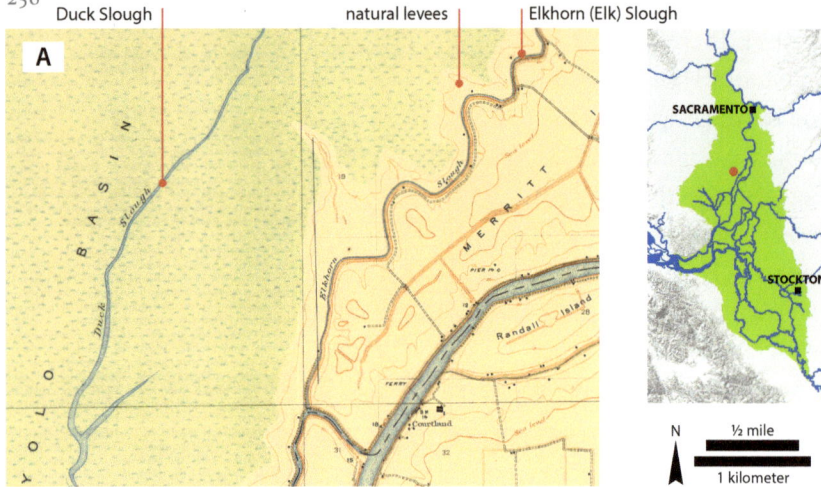

Figure 5.31. Comparing natural levees along Elk Slough to low banks of Duck Slough. The map (A) shows the supra-tidal natural levees along Elk (then Elkhorn) Slough, which contrast with the low wetland shown lining Duck Slough. Elk Slough was a distributary of the Sacramento River, receiving direct flood flows with the sediment to build natural levees, whereas Duck Slough received non-channelized flood flow that had already released most of its suspended sediment after passing slowly through the Yolo Basin. This resulted in different vegetation communities along the banks: gallery riparian forest along Elk Slough (B) and emergent vegetation and other wetland associated species along Duck Slough (C). (A: USGS 1909-1918; B: Unknown ca. 1890, from the collection of Bernice Krull, used with permission from the Yolo County Historical Society; C: Duck Slough, Holland Land Co., D-118, courtesy of the Map Collection of the Library of UC Davis)

BOX 5.4. TULE CANAL

Talk of constructing extensive canals to drain the north Delta and Sacramento Valley basins can be found in newspapers as early as 1847 (*Californian* 1847). Though it was recognized that the system of canals would do little to prevent flooding during the wet season, it was thought that the canals would hasten the removal of lingering floodwaters that otherwise remained in the basins through the dry season (*Sacramento Daily Union* 1853). Proposals to drain the tule lands of the Yolo Basin were found in some of the first bills to pass before the California legislature (*Sacramento Daily Union* 1853). Various options were still being considered in 1860, but by November 1864, the Tule Canal was complete (Flint 1860, Bailey [1918]1927). The *Sacramento Daily Union* (1864) reported at its completion that the canal was "about twenty-four miles in length, has twenty-one feet fall, is five feet deep on the average." Swampland District 18, which encompassed most of the Yolo Basin and was the largest district to be organized, had orchestrated this reclamation feat. The canal was positioned along the lowest part of the basin and was connected to the sinks of Cache and Putah creeks, in order to "facilitate the drainage of waters which almost annually accumulate in the basins of Cache and Putah Creeks" (Hall in Board of Swampland Commissioners 1864).

The Sacramento Basin also faced early ditching efforts to connect and drain the many lakes that lay along the extent of the basin. Work began on the Sacramento Drainage Canal in 1868, only a few years after the Tule Canal (*Sacramento Daily Union* 1873b). While perhaps not accomplishing all the reclamation and drainage expected, these early systems of canals did much to alter the natural hydrology of the basins, increasing connectivity between different parts of the basins during low flow periods (Fig. 5.32). This decreased residence time and with it the time for exchange between water and wetland as well as use for aquatic species. Many of the ditches are maintained today.

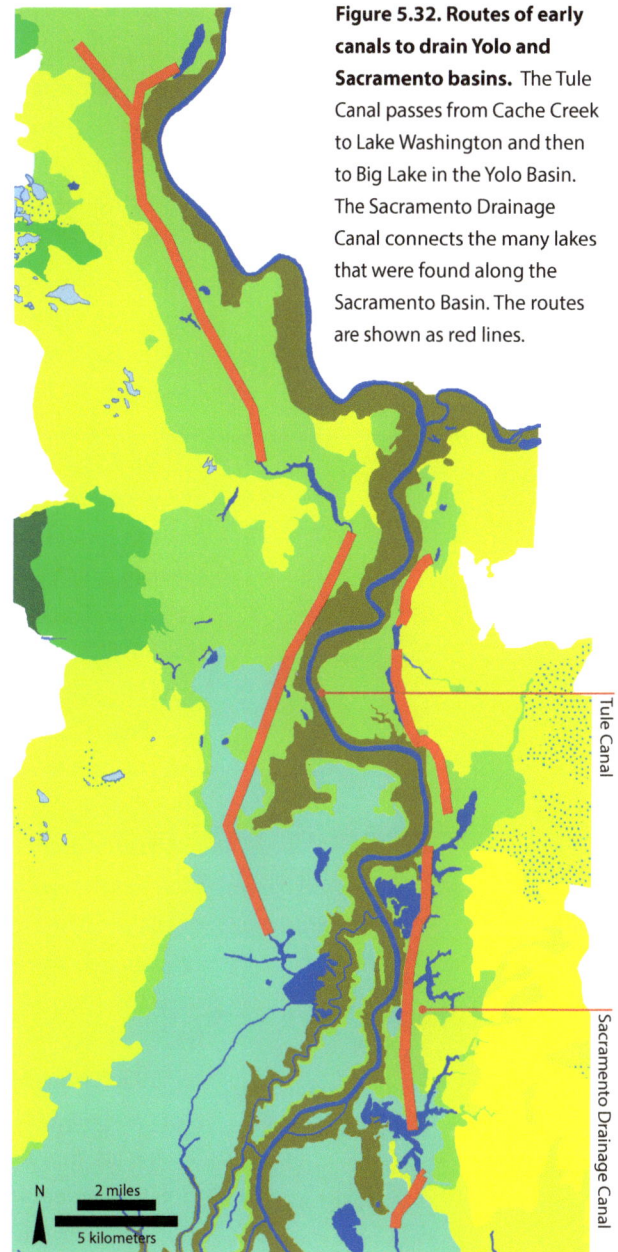

Figure 5.32. Routes of early canals to drain Yolo and Sacramento basins. The Tule Canal passes from Cache Creek to Lake Washington and then to Big Lake in the Yolo Basin. The Sacramento Drainage Canal connects the many lakes that were found along the Sacramento Basin. The routes are shown as red lines.

Tule Canal

Sacramento Drainage Canal

N

2 miles
5 kilometers

Tidal channel

Fluvial channel

Tidal or Fluvial channel (lower confidence level)

Water

Intermittent pond or lake

Tidal freshwater emergent wetland

Non-tidal freshwater emergent wetland

Willow thicket

Willow riparian scrub or shrub

Valley foothill riparian

Wet meadow or seasonal wetland

Vernal pool complex

Grassland

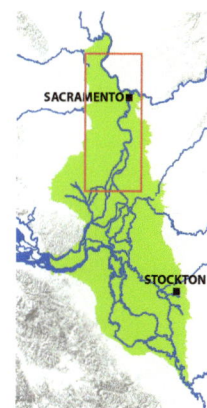

258

Question 47. What do you mean by the term lake?

Answer. I mean by the term lake a place that is generally overflowed; the place I am speaking of was always considered a permanent lake; there may have been dry places in it at some seasons; I never saw any, but have heard so...

Question 48. Over what portion of the lake did tule grow?

Answer. Over the greater portion—more than half; tules will not grow in places where the water at its lowest stage is over four or five feet.

—SANFORD 1860

For aught we knew we might be attempting a lake half a mile in width and twenty feet deep in the middle. Luckily we struck no places in which the water came above our breasts.

—WRIGHT CA. 1850A, IN THE PEARSON DISTRICT

through the dry season by high groundwater levels. Conditions varied substantially depending on the time of year. The lakes and ponds were bordered by tules and communicated directly with the rivers during the wet season. Some partially dried out, such that their size fluctuated dramatically over the course of the year. Even those positioned within tidal elevation ranges were somewhat, if not completely, isolated from tidal influence either through the lack of direct channel connections or simply due to the great distance from tidal sources. Though environmental conditions (e.g., nutrients, temperature, hydrologic connectivity) fluctuated depending on the season and year, the lakes and ponds were relatively stable features within the landscape. That is, they do not appear to have been ephemeral features that would appear one year in a flood and be gone the next.

The lakes and ponds of the north Delta were historically more abundant and on average larger than those elsewhere in the Delta. Early maps and textual descriptions, as well as early landscape photography, convey the character of the habitat (Fig. 5.33). Within the north Delta freshwater emergent wetlands, we mapped 48 lakes and ponds greater than five acres in size. Together, they cover 4,572 acres (1,850 ha), representing 84% of the total area of lakes and ponds mapped within the entire study area's perennial wetlands. The largest lake, Beach Lake, was over 1,000 acres (404 ha). Four lakes in the Yolo Basin and five in the Sacramento Basin were over 100 acres (40 ha). Thirteen were over 80 acres (32 ha), while the majority (26) covered less than 20 acres (8 ha; Fig. 5.34). Confidence was higher for larger features: while 91% of the area was classified with a high interpretation certainty level, 65% of the features were classified with a high level of interpretation certainty. There were a few cases as well where water bodies mapped in early twentieth century sources were absent in earlier sources and were therefore not mapped as open water (Box 5.5). We mapped an additional 44 lakes and ponds (a total of 1,507 ac/610 ha) outside of the emergent wetlands, most of which were seasonal and associated with vernal pool complexes (e.g., Jepson Prairie Reserve). We did not attempt to map features under five acres in a comprehensive manner, though we identified an additional 36 ponds of under five acres (amounting to 72 ac/29 ha total) in the early 1900s USGS topographic maps and other post-1900 sources.

Water depth in the lakes and ponds was variable. Even the beds of larger lakes may have been on average only a few feet below the general elevations of the basins (Etcheverry 1924). Early travelers often found that they could wade across, and that hydrophytic plants such as water lilies covered portions of the surface (Wright ca. 1850a,b, Lienhard and Wilbur 1941). Others were apparently deeper; Beaver Lake on Grand Island was reportedly "two to thirty feet deep" (Board of Swamp Land Commissioners 1867) and photographs from the early 1900s show people diving into Lake Washington. For the larger lakes within tide range, depths apparently reached well below mean sea level. An engineer's report stated that "a shallow lake bed" (likely of Beaver Lake) in Grand Island was "10 to 15 feet

Figure 5.33. "**Big Lake before draining.**" Though the edges have been cleared of native vegetation cover (formerly tule) by this time, the expanse is striking in this early 1900s photograph. (Big Lake before draining, Holland Land Co., D-118, courtesy of the Map Collection of the Library of UC Davis)

lower than low water in Suisun Bay" (Rose et al. 1895). Secret Lake was apparently over 30 feet (9.1 m) deep before it was drained (Van Löben Sels n.d.), and engineers later found the former lake bed "9.7 feet below mean sea level" (Unknown ca. 1900). A Merritt Island lake was also reported below sea level (Russell 1940). Unfortunately, the counteracting effects of land subsidence and filling in of depressions over the course of reclamation make it difficult to interpret these post-reclamation elevations. However, this information can be used to bracket our understanding of lake bed elevations rather than give definitive historical depths.

The following section discusses the landscape position, seasonality, hydrologic connectivity, and associated biota of the north Delta's lakes and ponds.

Landscape position

In the north Delta, the lowest elevations of the basins were typically occupied by relatively large bodies of water (see Figures 5.3 and 5.9). As one young man described, the tule "runs of gradualy untill it gets to deep for Tola and then comes the Lake or Pond" (Browning 1851; spelling as in original). The saucer-like shape of the basins caused water to pool at their lowest point, making drainage difficult (Van Löben Sels 1902). These low spots

Figure 5.34. **Distribution of pond and lake size.** The size distribution of the 48 lakes and ponds mapped in the north Delta perennial wetlands. Ponds below 10 acres (4 ha) in size were most common, but there were 13 lakes over 80 acres (32 ha) in size.

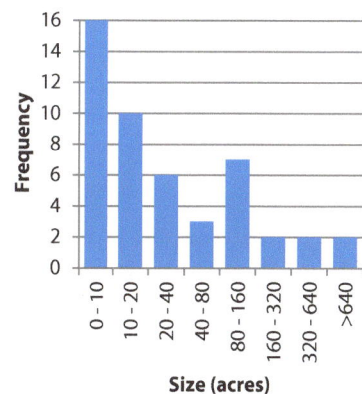

BOX 5.5. LAKE OR WETTER WETLAND?

One complication of the mapping process was that some lakes shown by turn of the twentieth century sources may have been only ponds or depressions occupied by tule in the early 1800s. Hydrologic modifications and land subsidence during the late 1800s could have affected drainage patterns and allowed lakes to become more established features. The construction of dams on sloughs, roads, levees, and ditches as well as farming all changed drainage patterns in significant ways. In some cases, this could have prevented water from draining during times of flood and kept the tides from transmitting up through the marsh plain vegetation. One example of this possible transformation is a lake on the edge of the Yolo Basin that is mapped by the early USGS topographic maps (Fig. 5.35).

Though earlier maps of equivalent detail are not available, maps from the nineteenth century for this area that do show other lakes do not show this lake. The most significant evidence suggesting that this lake may not have been a persistent feature in the landscape in the early 1800s comes from the GLO field notes. A survey line goes directly through the lake mapped in the USGS maps and the surveyor, William Lewis (1859c), followed the line directly through the area in late January, noting only that he was on the "margin of swamp and overflowed land" and that the "line follows margin of swamp and overflowed land to 50 chains." Consequently, we did not map this feature as a lake, but instead as perennial emergent wetland. However, the interpretation certainty level for this area was reduced because of the apparent conflict in historical sources.

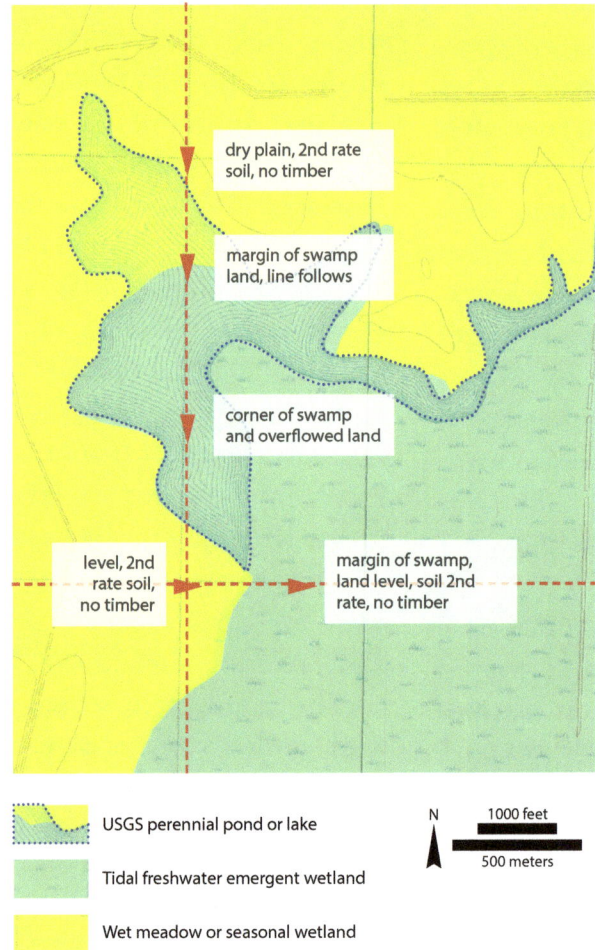

dry plain, 2nd rate soil, no timber

margin of swamp land, line follows

corner of swamp and overflowed land

level, 2nd rate soil, no timber

margin of swamp, land level, soil 2nd rate, no timber

USGS perennial pond or lake

Tidal freshwater emergent wetland

Wet meadow or seasonal wetland

N   1000 feet
     500 meters

Figure 5.35. Lake or depression?
GLO surveyor, William Lewis, passed through this area on the western edge of the Yolo Basin on January 25, 1859 (red arrows). Had there been a large perennial pond then as there was in 1916, when the USGS topographic map (base layer) was made, Lewis should have noted the lake and meandered around its boundary. Instead, he passed right through the bounds of the 1916 lake (outlined with dark blue dots). The fact that this survey was in the winter makes it unlikely that this was a perennial lake in the early 1800s. A likely scenario is that later modifications helped transform a natural depression into the lake. It had been drained by 1937, as revealed in historical aerial photography. (Lewis 1859; USGS 1909-1918)

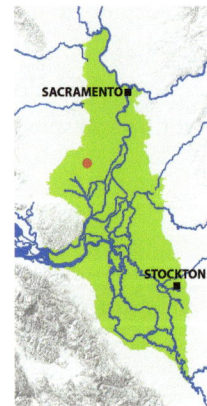

SACRAMENTO

STOCKTON

could occur in the center of the basin (e.g., Secret Lake within the Pearson District; Fig. 5.36a) or closer to the wetland edge (e.g., Big Lake within the Yolo Basin; Fig. 5.36b). The pattern was also exhibited in the north Delta islands that were surrounded by natural levees, such as Sutter and Grand islands. This landscape position contrasts with that of the marsh pannes or ponds common to tidal marsh landscape downstream in San Francisco Bay, which lay at the highest elevations of the marsh plain (Leopold et al. 1993, Collins and Grossinger 2004).

The basin landforms affected position at the broad level: in the Yolo Basin, some lakes may have been former channels constricted by the western alluvial fans and the Sacramento River (e.g., Washington Lake). In the Sacramento Basin, the river's natural levees appears to have restricted the drainage of the small eastern distributaries (Atwater pers. comm.). Lake position also related to how floodwaters were routed through the basin; lake sites were the areas most deprived of inorganic sediment supply (CDFG and YBF 2008). For example, an explanation for Big Lake's position against the back of Elkhorn Slough's natural levee may be that the predominant floodway was directly south down the center of the basin. The fine sediments in the water column would have had greater opportunity to settle within the center of the basin instead of along the edge where Big Lake was. The wetland plain would have therefore aggraded at a slightly greater rate in the center relative to the edge.

Smaller bodies of open water also occupied depressions formed by more localized topography throughout the basins. For instance, a map originally created in 1841 showing the upper part of the Sacramento Basin illustrates the position of "lagunas" within the wetlands in the southern portion of the present City of Sacramento (Fig. 5.37). These were the first of the series of lakes and ponds within the basin to receive the waters from the Sacramento and American rivers during floods. Topographic heterogeneity was influenced at the upland edge by the alluvial fans of streams draining from the foothills and, on the natural levee edge, by the meanders, secondary channels, and crevasse splays of the Sacramento River.

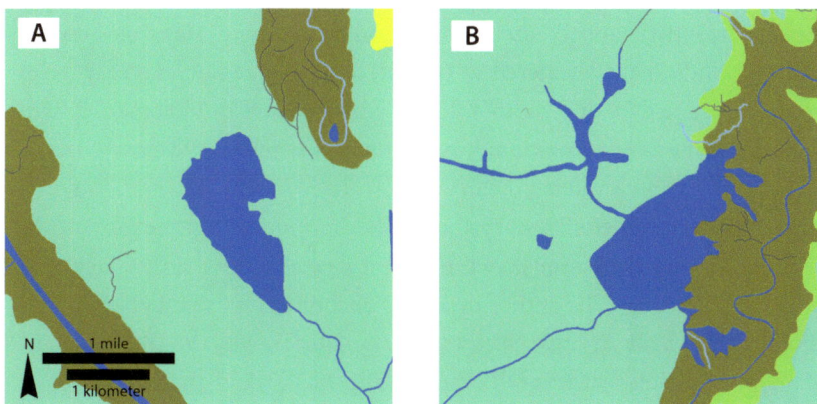

After two hours hard work we got the boat up the bank. To drag it over the level ground on its sharp keel was comparatively easy. At last we got it to the lake.

—WRIGHT CA. 1850B

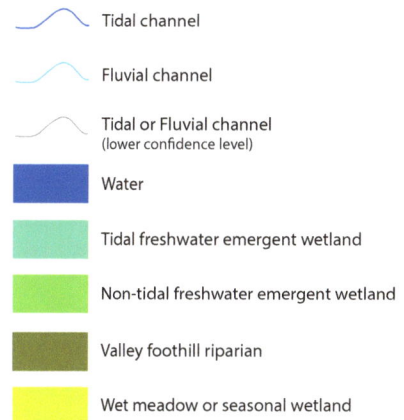

Figure 5.36. Comparing landscape position of two lakes. Lakes and ponds occupied the low elevation positions of the basins, where drainage was limited. Depending on flow and depositional patterns, the lakes were sometimes positioned at the center of the basin, like Secret Lake (A), and sometimes at the back of natural levees, like Big Lake (B).

Figure 5.37. "Lagunas" of the upper Sacramento Basin. These "lagunas," shown just south of the boundary of tule, were relatively small depressions at the north end of the flood basin. In addition to the several large lakes of the basins, numerous ponds occupied localized isolated low spots in the north Delta basins. The map was origianlly created in 1841, later recreated in 1854. (Vioget 1854, courtesy of The Bancroft Library, UC Berkeley)

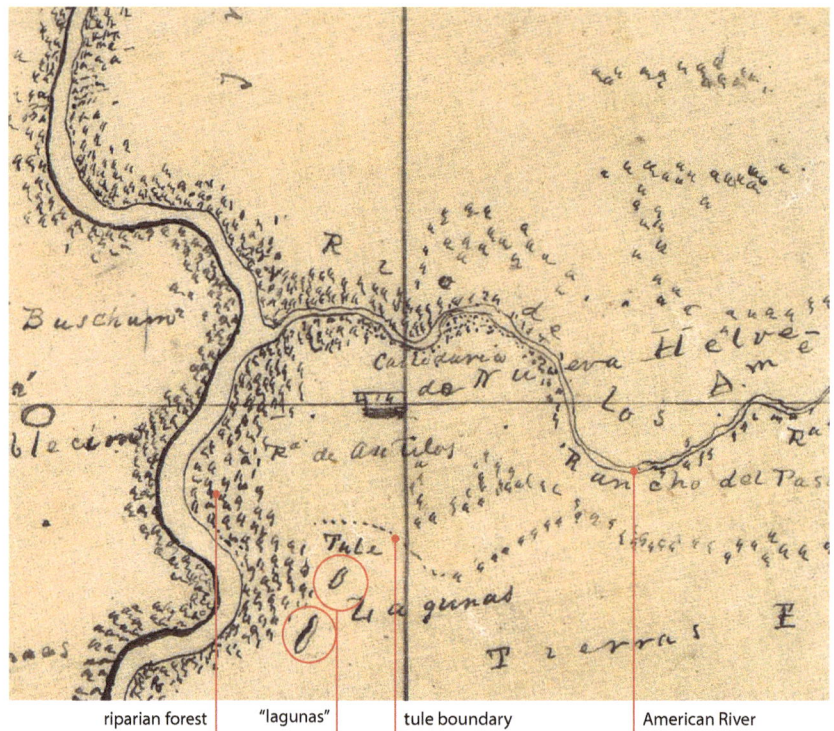

riparian forest　　"lagunas"　　tule boundary　　American River

The back land, which is now entirely cultivated, was then all tule and small lakes.

—LEALE 1939, REFERRING TO CONDITIONS IN THE 1860S

The shapes of the lakes and ponds appears to have been quite complex. In relating his experiences as a duck hunter in the north Delta, author William Wright (ca. 1850b) mentions that a lake's edge has "many coves and slough-like branches." He used a similar description for Secret Lake, describing how the lake's "numerous ponds and creek-like branches" provided rich hunting grounds (Wright ca. 1850a). The features' local-scale complexity resulted in high edge to area ratios and substantially increased the capacity of exchange between the marsh and aquatic environment. Crenulated borders of this nature for the majority of the lakes are also depicted in many of the more detailed maps (e.g., USGS 1909-1918). Using the mapping of Big Lake, we calculated an edge to area ratio about three times greater than that of a circle (a circle has the lowest possible ratio).

Beyond the scope of our mapping effort was a level of small (generally <5 ac/8 ha) shallow depressions. These features should be considered as part of the matrix of the mapped emergent wetland habitat type. Many of these shallow ponds likely dewatered by the end of the dry season and may have become occupied by tule without flooding disturbance or biological activity to keep them free of vegetation. It is unknown whether they were recognized as ponds during the dry season. Though maps do not always depict this level of detail, a story about duck hunting within the present-day Pearson District gives a sense of the finer scale complexity of this basin landscape in winter. The author describes encountering "small pools abounding in mallard" near Secret Lake and after leaving the lake, coming to another "region that was full of ponds and small lakes" that were from "one hundred to three hundred yards in width [91-274 m]" (Wright ca.

1850a). This is quite an adventure the writer relates, involving wading "straight through" the ponds at night with "a load of twenty-five mallard, half a dozen geese and a number of small ducks," eating raw goose for dinner, and sleeping on a bed of wings.

In the wetlands formed by the Cosumnes and Mokelumne rivers, a series of lakes and ponds was arranged in a slightly different landscape pattern associated with the region's topography and geomorphology. They generally occupied small, short upland drainages that fed into the floodplain. They can be seen in early maps of the area, including *diseños* and USGS topographic maps from the early 1900s (USDC ca. 1840d, USGS 1909-1918). These features have the appearance of small drowned valleys that were too small to have significant sediment sources of their own and may have been partially blocked through the process of floodplain development (Florsheim and Mount 2002). These were the more persistently inundated features of the Mokelumne-Cosumnes floodplain. Because of their elongated shape that followed drainages, the features were sometimes referred to as sloughs rather than lakes: GLO surveyor Thompson (1862) recorded "large tule slough," "tule and water," and "timbered slough." Also, early maps refer to Beaver Lake (which drained into the present-day McCormack-Williamson Tract) both as "Beaver Lake" and "Beaver Slough" (Reece 1864, USGS 1909-1918). A general account of the area is found in the diary of trapper John Work, who found it difficult to reach the Mokelumne River in January 1828 because "for a considerable distance up it is so surrounded with swamp and deep gullies full of water that it cannot be approached but at one or two places" (Maloney and Work 1943).

The lakes and ponds were bordered by emergent vegetation, which sometimes extended for miles beyond the edge of the lake. Secret Lake, for example, was located "far out in an impenetrable tule swamp of immense extent" (Wright ca. 1850b). As these areas remained the wettest through the season, emergent vegetation was densest close to the lakes' edges (Lewis 1858a). Some features were closer to or partially surrounded by forest or dense underbrush, particularly those positioned close to natural levees (e.g., Stone Lake) or near upland drainages (Fig. 5.38). In some cases, such as Lake Washington, we did not map the partially forested edges due to lack of spatially explicit evidence concerning location and extent. Instead, this pattern should be considered as one type of ecotone sometimes found along lakes in the non-tidal regions of the basins. Willows, wetland-associated species such as button bush (*Cephalanthus occidentalis*), and (less commonly) oaks were found in this community (Fig. 5.39; State Journal Office 1854, Sanford 1860). Only a few ponds were mapped completely within the riparian forest zone, two of which were found at the site of the present day City of Sacramento; the third was found at the tip of Dodson's Mound, the finger of land that extended south from Randall Island into the Pearson District.

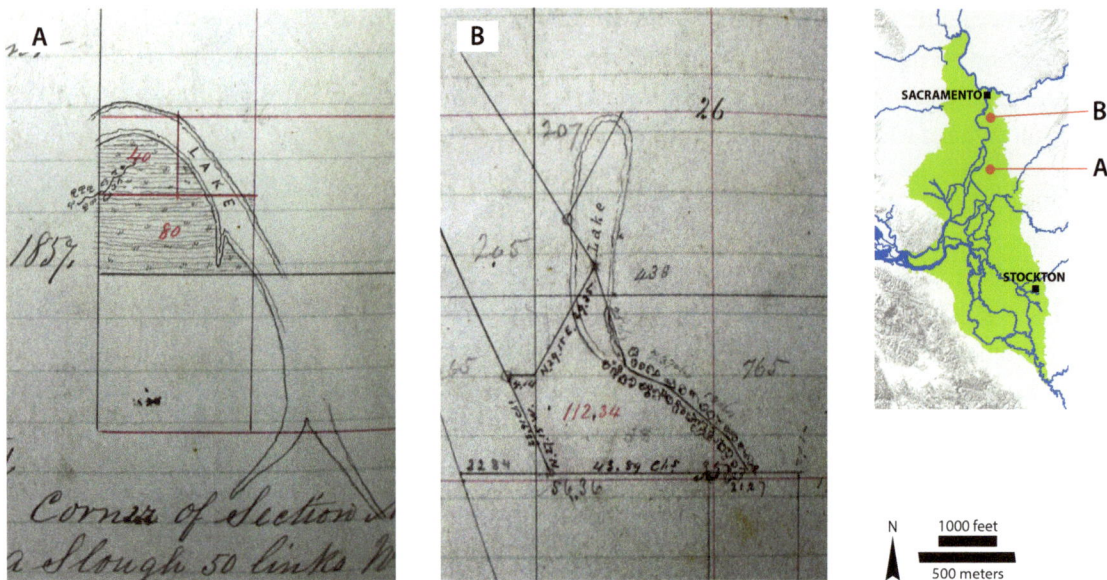

Figure 5.38. Depictions of trees near lakes in the north Delta. These two maps from swampland surveys illustrate how secondary channels across natural levees may have brought forest close to the lake margins (A), and in other cases, how upland drainages feeding directly into the lakes were bordered by riparian forest, which continued along the lake (B). (A: Cleal 1859, courtesy of the Center for Sacramento History, Swamp and Overflowed Land Surveys, No. 134; B: Cleal 1858, courtesy of the Center for Sacramento History, Swamp and Overflowed Land Surveys, No. 158)

Figure 5.39. A mix of species at the lake edge can be seen in these two photographs of Lake Washington. Dense willows mixed with other species are shown in A, while tule bordered the lake margin elsewhere (B). (photos by McCurry ca. 1910, courtesy of the California History Room, California State Library, Sacramento)

The extreme flooding regime and the extensive wetlands made reclamation difficult, and as a result the basins were some of the last areas in the Delta to be reclaimed. The lakes were most difficult to drain. A system of pumping plants employed for continuous drainage of the area that was once Big Lake was referred to in 1923 Reclamation District 999 records (Board of Trustees 1923). Despite these drainage efforts and decades of farming, former "lake beds" were referenced in the historical record. For instance, the roughly 395-acre Secret Lake shown on an early Reclamation District map (Reece 1864) was described decades later in early twentieth century descriptions of "an old lake bed of about 350 acres" (Etcheverry 1924) and "300 acres... known as 'lake bed'" (Van Löben Sels 1902). Portions of several of the larger north Delta lakes persist today, such as Stone Lake, Beach Lake, Lake Washington, and Beaver Lake. However, they are no longer integrally connected to the surrounding landscape, including seasonal dynamics of the Sacramento River and the larger basin processes. The former lakes often remain as distinct topographic depressions, often detectable in modern LiDAR imagery (Fig. 5.40).

Landscape position greatly affected the functions of these features. Open water aquatic habitat today is far more connected to the main tidal channels of the Delta and generally deeper than lakes and ponds were historically. Instead of being surrounded by deep tidal channels, bodies of open water were often historically surrounded by extensive wetlands fed by water that had passed through many miles of these wetlands, often without connecting channels. Furthermore, the seasonally flooded lands of the Yolo Basin have been shown to provide valuable foraging habitat for juvenile fish (Sommer et al. 2001), inspiring questions about the potentially significant functions served by the lakes and ponds (and their surrounding perennial wetlands) of the recent past.

Seasonality and hydrologic connections

The lakes and ponds of the north Delta experienced substantial changes in hydrology over the course of a year. Floodwaters passing southward through the basins filled them annually and would gradually evaporate through the summer months. The seasonal pattern of being connected to the river during the winter and disconnecting later in the season is

Figure 5.40. Big Lake after drainage. Early 1900s USGS topographic map (A) shows Big Lake in its historical form. Its former locations is still discernable in the 1937 aerial photography (B) through the pattern of drainage ditches. Modern LiDAR (C) picks up the topographic depression that still clearly defines the former boundary of Big Lake. (A: USGS 1909-1918; B: USDA 1937; C: DWR 2008)

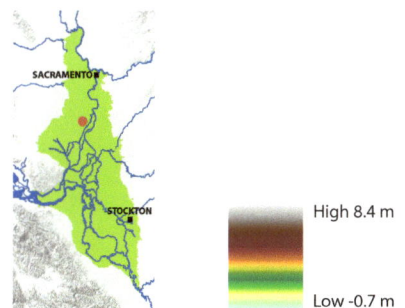

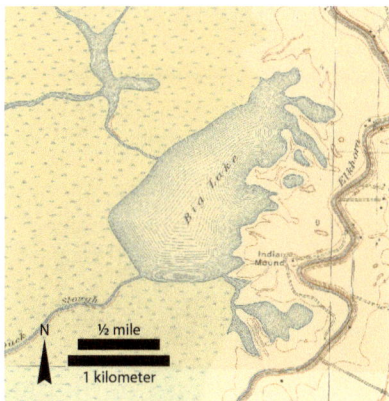

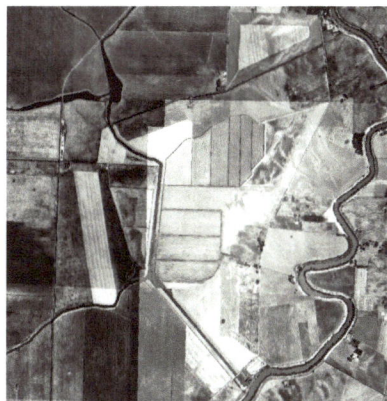

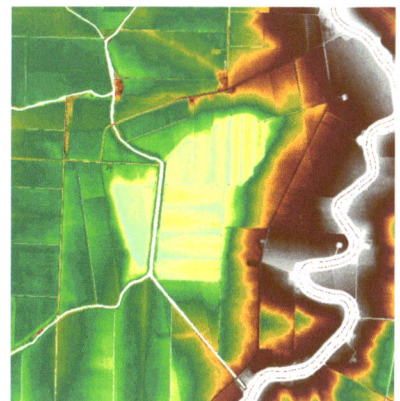

described by many historical accounts, such as explorer William Wilkes' (1845) "small lakes or bayous" that "filled at high water, but become stagnant during the dry season." To some, the lakes and ponds were simply "holes where the water would stand some time" after the river stage had fallen (Hardenburgh 1860). Many were also fed directly through channels connected to upland drainages, particularly those in the Sacramento Basin.

The size of the north Delta lakes and ponds expanded and contracted greatly. A witness in the Sutter land case also had trouble estimating lake area, stating that the size "depends on the stage of the water" (Hall 1856). Because many of the lakes appear to have been more like broad, shallow, flooded areas than deep features with well defined perimeters, the long drying period of the summer would have impacted size. Some features referred to as lakes during the wet season may have shrunk to small ponds by the end of the dry season.

At last we got it to the lake and dragged it into the water, but it did not float. A man uncoiled the long rope at the bow and went ahead to tow the craft, but when the rope was all out the water had not yet reached his knees. "No use trying to float her," cried he – "Why I can wade all over the blamed lake!"

—WRIGHT CA. 1850B

An example of such a feature is found in Wright's (ca. 1850b) account. After laboriously dragging a boat over the natural levee to a lake on the other side, he and his hunting companions discover (much to their astonishment) that the lake was so shallow that the boat was grounded and that the lake "could be waded in all parts, except a small streak in the middle." In the process of reconciling this wet season description of size and shallowness with the lake's stated location on Randall Island, the feature was not mapped as a permanent lake. Support for this interpretation is found in the absence of a mapped lake in early maps that do include other lakes we mapped in the vicinity (Reece 1864). Seasonal conditions of this nature were likely found throughout the north Delta landscape, particularly where topographic features (e.g., the natural levees surrounding Randall Island in this case) hindered drainage.

Seasonal variability also influenced hydrologic connectivity. During the wet season water depths were usually high enough to provide hydrologic connection across the majority of a basin. The dry season, however, revealed a subtle topographic variability that nevertheless caused large portions of the basin to become comparatively hydrologically isolated. According to testimony concerning the upper Sacramento Basin lakes, when the slough in question "gets full it communicates with the whole chain of lakes to Sacramento. These lakes are divided in summer by banks of sloughs, when the water is at its lowest" (Sanford 1860). The greater proportion of area mapped as lakes relative to emergent wetland area in the Sacramento Basin likely relates to the basin's more constricted size and topography. Many of the lakes were so close together that they appeared as a chain of lakes progressing down the basin (Reece 1864). This chain was not directly connected together until development of the Sacramento Drainage Canal in the late 1860s. One newspaper article listed eight lakes that were linked by the canal, which extended from Sacramento to Snodgrass Slough (see Box 5.3; *Sacramento Daily Union* 1873b).

Many lakes and ponds were hydrologically connected to the river by channels. Some were connected via tidal channels, as was the case of Big Lake. Its connection to Cache Slough was over 12 miles (19.3 km) long through the tidally influenced Duck Slough. These tidally influenced connecting channels were typically fairly long. Other lakes and ponds were connected to overflow channels or intermittent upland streams. Sutter Lake provides an example of a seasonally active connection to the river. This lake was once connected to the river at high stages by an overflow channel across the natural levee, one of the primary routes through which the City of Sacramento was likely flooded (*Daily Alta California* 1852a). Other seasonal direct hydrologic inputs came from upland drainages. Numerous examples of this were found, particularly along the upland edge of the Sacramento Basin. Still other lakes appear to have had no substantial connecting channels, such as Secret Lake. In addition to cartographic evidence, the absence of natural channels leading to it is documented by Wright's (ca. 1850a) duck hunting account, in which he remarked that the hunters on the lake had "constructed a ditch or small canal navigable for duck-boats." Though this area extending up to Stone Lake likely experienced the tides (Reece 1864, Tucker 1879d), the lack of extensive channel networks suggests a muted tidal influence. Decades later, extensive ditching, pumping, and the installation of a tide gate reclaimed Secret Lake and also drained the surrounding area (Van Löben Sels 1902).

An important consideration in lake function is the role of high groundwater in maintaining surface water levels. Groundwater was at or near the surface for most of the area historically occupied by emergent vegetation (Bryan 1923, Fox 1987a, TBI 1998). In his map of the Sacramento Valley, geomorphologist Kirk Bryan (1923) specified that water depths for much of the American, Sutter and Yolo basins "ranges from a maximum of 20 feet along the river bank to only a few inches in parts of the basins." (Bryan 1923, Fox 1987a). It can be expected that most lakes and ponds intersected the historically high groundwater tables of the basins. This is suggested by reports for Reclamation District 999 (in the Yolo Basin) stating that water levels were at the surface for 2,700 acres, within a foot of the surface for another 2,700 acres, and that the area of Big Lake had standing water upon it (Unknown 1919).

The impact on water quality is another point to consider in the context of hydrologic connectivity and water retention within the basins and their lakes. Water passing through the basins had long residence times. The time water spent traveling through wetland vegetation and being retained in lakes and ponds would have allowed for chemical transformations and nutrient exchange between these environments, an attribute largely missing from the landscape today. Warning of stagnant warm water promoting disease, an early medical journal article noted that "during its journey towards the south," Sacramento Valley water would "under the influence of a hot sun, undergo great modifications" (Logan 1865). There would have been significant opportunity for wetland organic matter to be released into

The surface of the country being more or less irregular, when this low or tule land south of R street is over flowed entirely, when the water recedes there are many ponds left where water remains. In some of those ponds the water stands nearly if not quite all the year round; consequently this water gives moisture to the tule.

—DENVER 1860

the aquatic environment, impacting the nutrient cycling of the marsh. This interaction is expressed in an early observation that the "water of the tule marshes" was "so thoroughly impregnated with decaying vegetable matter that it looked more like sherry than water" (Wright ca. 1850a). High marsh productivity is suggested in the text that follows:

> In order to see the strange creatures in the water no microscope was required; they were visible to the naked eye and in size ranged from an inch in length down to mere points...which would not have been suspected had they not been gifted with powers of locomotion. In lying down to drink from the edge of a pool we had before us for study a whole universe of animalcules. Though we steered clear of such creatures as were above half an inch in length we paid no attention to the little fellows. (Wright ca. 1850a)

Water temperature in the lakes would also have generally been higher than in the shaded, deep, fast moving rivers, particularly in the summer months. One account explicitly addressed this factor, with the observation that at a lake near the Mokelumne River "the water is very warm and we cannot get to the river where it might be a little colder" (Maloney and Work 1943). The overall slow movement of water as a result of the basin morphology through the broad freshwater wetlands likely had significant water quality ramifications.

Evidence for selected species

This section considers a few points about the potential impacts of certain biological factors on the historical landscape. These factors relate most closely to the lakes and ponds landscape component and are consequently discussed here, but should also be considered more broadly for their role in the overall landscape function.

Lakes and ponds of the Delta were historically occupied by aquatic plants including pondweed (*Potamogeton* spp.and *Suckenia* spp.), yellow pond lily (*Nuphar polysepala*), floating water primrose (*Ludwigia peploides)*, knotweed (*Polygonum* spp.), and wapato (*Sagittaria* spp.; Fig. 5.41; Brewer et al. 1880, Jepson 1901, Jepson 1904, West 1977, Mason n.d.). As evidence of the prevalence of pondweed (predominantly freshwater species), seeds were present for most of a peat core taken in the vicinity of Suisun Bay (Peyton Hill; Goman and Wells 2000). The presence of these aquatic species in the historical Delta supports that shallow, slow moving water associated with lacustrine environments and low-energy tidal sloughs was common.

The historical presence of yellow pond lily is of note given its general absence in the Delta today. Botanist Willis Jepson (1901) recorded that it was found in the vicinity of Stockton in lakes and sloughs and botanist Herbert Mason (n.d.) included it as part of the "epihydrous mosaics" vegetation community. Its historical presence has been confirmed, including for Stone Lake, through sediment coring and pollen analysis (West 1977). One historical account that explicitly mentioned "lily pads" expressed surprise upon the discovery that a

Figure 5.41. Floating aquatic vegetation (likely *Ludwigia peploides*) on a lake in 1905 is seen in this photograph of USGS surveyors at work. Minority species (perhaps speedwell, *Veronica* spp., and smartweed, *Persicaria* spp.) may be present in the foreground (Baye pers. comm.; photo by Rogers 1905, courtesy of the Center for Sacramento History, Hubert F. Rogers Collection, 2006/028/112)

lake that hunters had thought (at night) was covered in ducks, was in fact "for a distance of one hundred yards out thickly covered with lily pads" (Wright ca. 1850b). The plant's rooting depth of up to six feet (1.8 m) indicates relatively shallow waters. Yellow pond lily tubers and seeds were eaten by animals and the seeds may have also been harvested as a food source by indigenous tribes. The use of yellow pond lily in traditional cultures has been well documented for the Klamath Lakes region, where tribes referred to the plant as "wocas" or "wocus" (Gatschet 1890, Deur 2009). It is relevant to consider that indigenous management for this food source may have affected Delta vegetation patterns.

The millions of migrating waterfowl along the Pacific Flyway that seasonally blanketed the wetlands of the Sacramento Valley and Delta were, and still are to a lesser extent, important actors in the Delta ecosystem (Fig. 5.42). They depended on the primary production of floating and submerged aquatic vegetation in the wetland complexes (Garone 2011). While the origin of the larger lakes was likely related to physical processes, the clearing of submerged aquatic vegetation by the activity of feeding waterfowl may have been an important factor in maintaining smaller ponds. A field entry by botanist Willis Jepson (1904) for Suisun Marsh gives an indication of how effective geese and other waterfowl were in consuming submerged aquatic vegetation. In discussing ponds that had previously been filled with sago pondweed (*Stuckenia pectinata*), he stated:

> Now the ducks have cleaned it out so well that we had no trouble in going anywhere. The effect of disturbed areas – where ducks have been feeding – is something like the rooting over of a new field by hogs when the vegetation is young and shining white roots are exposed. (Jepson 1904)

He concludes that geese can "clean out areas of 5, 10 and 20 acres or even more. Many of the duck ponds, now used by the hunters have been made in this very way." This suggests one possible mechanism by which the smaller open water features of the Delta may have been maintained.

Beaver likely influenced local-scale habitat complexity through their consumption and harvesting of tule and willow, creation of "beaver cuts" several feet deep across the landscape, and dam and dwelling construction (see Box 4.2). The apparent prevalence of "beaver cuts," as documented in several accounts, is particularly intriguing (Beaumont 1861b, Board of Swamp Land Commissioners 1867, Tucker 1879e, Soares pers. comm.). One description suggests the beaver cuts may have been significant in promoting hydrologic connectivity:

> The ground forming the basins of the lakes was full of beaver holes and when we broke through into one of these down we went over head and ears in the water. Luckily we struck no places in which the water came above our breasts but as the break through into the subterranean excavations of the beaver always gave us a perpendicular drop of about two feet we were very frequently in over our heads. At first these plunges caused a halt and some talk, but presently such mishaps became so frequent that the man who was still on top merely halted until the sound of gasping and sputtering informed him that his companion's head was again above the surface. (Wright ca. 1850a)

This account suggests that beaver burrows or paths many have introduced significant local topographic variability, which may have offered pathways of water between ponds.

In understanding the ecological functions served by the Delta, there are numerous questions concerning how native fish utilized and moved through the Delta. The lakes and ponds, as well as perennial and seasonal wetlands of the north Delta, were important habitat for many of the native fish species (Schulz 1979, Moyle pers. comm.). Historically, there were large populations of fish species associated with slower moving and shallow waters and floodplains. These species included the Sacramento perch (*Archoplites interruptus*), hitch (*Lavinia exilicauda*), Thicktail chub (*Gila crassicuada*, now extinct), Sacramento blackfish (*Orthodon microlepidotus*), and splittail (*Pogonichthys macrolepidotus*; Turner 1966, Moyle pers. comm.). One account refers to previously catching perch in the two larger lakes within the present limits of the City of Sacramento (McClatchey 1860). The native fish were a primary source of food for the indigenous tribes of the Delta, as revealed in archaeology studies (Schulz and Simmons

The geese keep feeding as long as they can get at the roots and they can get at the roots as long as they have something to pull against, that is as long as they can pull against the bottom. Sometimes these "geese wallows" become 4 or 5 ft deep, as the waters recede the geese work down. They will clean the roots out completely or the next year they get at the tender shoots and complete the job…2000 Canvasback will clean the tubers out of a pond in a night; the sound of their eating (for they are voracious eaters) is like the guzzling of hogs!

—JEPSON 1904 ON SUISUN MARSH

When the water had receded sufficiently hundreds of crane and storks ate them [thousands of fish] up.

—VAN LÖBEN SELS N.D.

Figure 5.42. Waterfowl seen wintering on a lake of the north Delta. Along the Pacific Flyway, the Delta annually received thousands of migrating waterfowl. This had a significant impact on the function of the marsh, as well as habitat modification through, for example, the consumption of aquatic vegetation. (courtesy of William G. Miller, Cole~Miller Photography, December 31, 2011)

1973, Fagan 2003). From the analysis of remains in and near the Delta, one researcher concluded that "the smaller fish were evidently native to the sluggish, semistagnant marshes and sloughs of the level plain through which coursed the lower Cosumnes and Mokelumne rivers" (Cook and Heizer 1951). The historic prevalence and the life history traits of these fish point to a landscape dominated by slow-moving water and marsh environments. Transformations away from such conditions in the Delta over the past 160 years have contributed significantly to species population declines (Alley et al. 1977, Moyle 2002).

Hydrologic connectivity within the floodplain is a necessary element for fish to be able to pass back into the river channels once floodplains begin to dry. This is not only important for migrating species such as salmon but for resident species such as splittail and Sacramento perch, which show strong adaptations for floodplain spawning and rearing (Moyle et al. 2004, Moyle et al. 2007, Crain and Moyle 2011). Necessary connectivity was generally available within the basins historically: inundations of several feet would occur for durations of weeks or months, and water generally remained on much of the wetland surface through the spring and early summer. This would have likely provided sufficient time for fish to move through the flooded basins, where species were cued to signals of depth, temperature,

There is at present a lake about three miles long and from one-eighth to seven-eighths of a mile wide, and from two to thirty feet deep in the center of the Island [Grand], from which most excellent fish are taken.

—BOARD OF SWAMPLAND COMMISSIONERS 1867

and water clarity. This idea is conveyed in an early newspaper article on the "Sacramento fisheries":

> The small fish run into the sloughs and lakes as soon as the water gets sufficiently high, and return to the river when it begins to get low, at which times they are taken in unusually large numbers...During the high stage of water these lakes all communicate with the Sacramento (*Sacramento Daily Union* 1854a).

This narrative both links the small fish species (e.g., "perch, chub, suckers, hard-heads, narrow-tails, etc.") to the lakes and ponds and describes how they moved from floodplain to river (Fig. 5.43; *Sacramento Daily Union* 1854a). Migratory salmon accessed excellent floodplain rearing habitat through these connections. At a time when flooding within the Yolo Basin still occurred regularly, people reportedly fished for salmon in the man-made Tule Canal that ran the length of the basin (see Box 5.3), indicating that the north Delta wetlands were naturally accessed by these fish (*Sacramento Daily Union* 1889). Salmon would have benefited from the floodplain's capacity to support fish with connections to the main river via channels as important migration pathways (Jeffres et al. 2008, Sommer et al. 2001).

Scientists today are often limited in their study of native fish to those in modified habitats and whose populations have been declining since well

Figure 5.43. Seine fishing in the Delta. A group harvest fish along the banks of an unknown waterway. In addition to the fish in the net, several individuals hold large fish. (photo ca. 1900, courtesy of the Center for Sacramento History, Ralph Shaw Collection, 1998/726/0776)

BOX 5.6. A NOTABLE DECREASE IN SALMON BEFORE 1900

Salmon and other native fish populations of the early 1900s and within the time period of modern record-keeping were already highly impacted by decades of fishing, water diversions, damming, mining, wetland reclamation, and logging. A few historical notes are included here for the purpose of fostering a longer-term perspective on population declines, recognized as the challenge of "shifting baselines" (Jackson et al. 2001).

During the second half of the nineteenth century, salmon fishing accounts record a noticeable decline in the number of fish harvested. As early as 1860, a popular magazine of the day included an article on the Sacramento River salmon fishery, beginning with the statement: "Salmon fish are fast disappearing from our waters" (Kirkpatrick 1860). The Stockton Independent in 1874 reported that salmon "had become almost extinct from these streams [San Joaquin tributaries]" (Crow 2006). By 1889, salmon had reportedly become "scarce" in the upper Sacramento River and the Delta, though that spring was reportedly a good season (*Sacramento Daily Union* 1889). A 1915 Solano County history discussed the importance of the salmon fishery for the local economy, but referred to salmon fishing as a passing industry: "for some reason salmon seem to be disappearing from the waters of the Sacramento" (Dunn 1915).

Perhaps the most significant early impacts were water diversions, sedimentation, and channel modifications associated with gold mining in the Sierra Nevada. Additionally, reclamation eliminated floodplain area, which had likely provided important rearing habitat for young smolts (Sommer et al. 2001, Moyle pers. comm.). Direct catch of salmon had a significant impact on population numbers as well. Fish were sold in San Francisco (then Yerba Buena) as early as 1840 (Davis 1889). In 1853, newspapers reported that the fishing fleet consisted of 60 boats that together caught between 300 and 600 fish a day (*Daily Alta California* 1853). In 1860, it was reported that the best place for fishing on the Sacramento was near Rio Vista, a place "so far from the mining region, that there is a clearer and larger body of water than can be found anywhere else on the river" (Kirkpatrick 1860). In that year 337,400 pounds of fish were sold. By 1864, canneries began springing up along the lower Sacramento River (Ogden 1988, Jacobs 1993). In 1880, 10.8 million pounds of salmon were caught, and the peak catch on record was in 1909 (12 million pounds; Yoshiyama et al. 1998). At its height, commercial fishing consisted of 20 canneries in 1882. The last cannery closed in 1919 (Rensch et al. 1966, Jacobs 1993).

before modern record keeping (Box 5.6; Jackson et al. 2001). The challenges associated with drawing conclusions regarding habitat and species connections can be addressed in part by the historical perspective. Historical studies are important for explaining physiological and behavioral adaptations of fishes that seem divorced from present distribution patterns. For example, Sacramento perch have distinctive larval morphology and behavior that appear to be adaptations for floodplain rearing, though they no longer have access to this habitat (Crain and Moyle 2011). Likewise, Sacramento perch, Sacramento blackfish, and other native fishes have extraordinary physiological adaptations that reflect an ability to survive in isolated, warm, shallow lakes and ponds (Moyle 2002). These are the situations where native people would have had relatively easy access, relating to their abundance in middens.

RIPARIAN FOREST EXTENT AND COMPOSITION

Broad mature riparian forests extended deep into and functioned integrally with the Delta's wetland and aquatic environment, contributing to diversity, productivity, and connectivity. The forests cloaked the natural levees of the major rivers in the Delta and Central Valley and gave the Delta an overall appearance of "an immense timbered swamp" (Bryant [1848]1985). The multi-layered structure of the forest was composed of dense understory and tall canopy intertwined with vines and brambles. It looked, to many, like a jungle and is referred to as gallery forest (forest in an otherwise treeless landscape; Fig. 5.44; Holmes and Nelson 1915, Fairchild 1934, Smith 1977, West 1977). Unlike more saline environments in the rest of the San Francisco Estuary, the freshwater Delta could support trees and other woody vegetation otherwise excluded within the estuary's tidal wetlands.

These forests supported the most ecologically diverse communities in the Central Valley, providing valuable habitat for riparian-associated birds such as the yellow-billed cuckoo (*Coccyzus americanus*), Least Bell's vireo (*Vireo bellii pusillus*), and tricolored blackbird (*Agelalus tricolor*), as well as mammals including the riparian brush rabbit (*Sylvllagus bachmani riparius*; Sands et al. 1977, Vaghti and Greco 2007). Likely due to the plentiful and diverse array of resources located nearby, village sites of Delta tribes were often located on the higher natural levee lands associated with riparian forests (Fig. 5.45).

Early accounts from traveler's diaries provide some of the best descriptive information conveying the complexity of the riparian forest. Sacramento River banks were generally described as "thickly wooded and more interesting in their appearance" in comparison to the scenery downstream in the central Delta (Duvall and Rogers 1957). Vines woven about the branches formed "a matted jungle" (Fairchild 1934), giving the forest "a tangled appearance" from the river (Bryant [1848]1985). The benefits of shade, as well as the perils of snagged rigging and blocked wind from overhanging branches of the dense forest, are discussed in a number of accounts: one expedition waited to proceed "until the high trees should cast

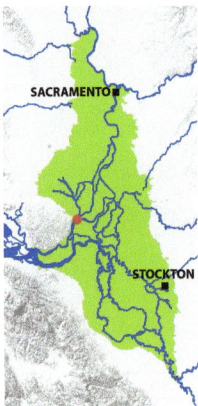

Figure 5.44. Riparian forest at the junction of Steamboat and Sutter slough. Looking upstream, this 1850 sketch portrays the gallery riparian forests of the Central Valley. Downstream, banks decreased in height and sycamores and oaks gradually gave way to willows and other shrubs, then emergent vegetation. (Ringgold 1850a, courtesy of the David Rumsey Map Collection, Cartography Associates)

Sutter Slough   Sutter Island   Steamboat Slough

West Fork   Middle Fork

Mark for entering the second section of the Middle Fork of the Sacramento River

their shadows across the river and afford us more protection" (Phelps 1841 in Dawdy 1989), while another from May 1850 complained that "the trees on each side the river are so high and close that there is scarcely a Breath of Wind" (Kerr and Camp 1928). People were even known to use overhanging trees to cross over some of the smaller waterways (Lienhard and Wilbur 1941). A narrative of a trip downstream along the river via wagon states: "The road is nicely shaded with groves and copses of low oaks and willows, tall sycamores, spreading ash and sprawling buckeyes" (*Sacramento Daily Union* 1860). Other notable general descriptions characterizing the riparian forest are summarized in Table 5.1. Artwork from that era as well as early landscape photography provide other beneficial perspectives. We used early maps showing riparian forest primarily to map forest extent. They rarely provided detailed information as to the character of the forest.

Trees found ideal growing conditions in the natural levees' well drained fertile soils, plentiful available water, and favorable temperatures. The riparian zones along the larger rivers in California were some of the only places where plants were assured ample water supply during the dry summer months (Ornduff et al. 2003). Early travelers to the area usually remarked on the strikingly lush conditions and exceptional size of many of the plants (Abella and Cook 1960, Gregory 1912). Explorers in 1837 noted "oaks of immense size" and measured two exceptionally large trees (27 and 19 feet (8.2 and 5.8 m) in circumference 3 feet (0.9 m) from the base;

Figure 5.45. A north Delta view. The complex and broad riparian forest followed the natural levees of the river, while wetlands lay behind the forest. These natural levee lands were inhabited by the Delta tribes, as well as myriad terrestrial animals. Artwork by Laura Cunningham.

276

Table 5.1. Selected early 1800s narrative accounts of the riparian forest lining the banks of the Sacramento River.

Quotation	Year	Citation
"Each branch [of the river] is covered with trees on both banks, of various kinds and very large. There are many walnut trees and wild grapes but the latter have stems so thick that those who have seen grapes in favorable countries say they have never seen such thick trunks."	1811	Abella and Cook 1960
"All along this river [possibly Steamboat Slough] it is like a park, because of the verdure and luxuriance of its groves of trees."	1817	Durán and Chapman 1911
"thick, dense barriers of trees and shrubs that lined the banks"	1837	Belcher et al. 1979
"The banks of the river, and several large islands which we passed during the day, are timbered with sycamore, oak, and a variety of smaller trees and shrubbery. Numerous grape-vines, climbing over the trees, and loaded down with a small and very acid fruit, give to the forest a tangled appearance."	1846	Bryant [1848]1985
"a narrow ridge of land mostly covered with a growth of oak, cottonwood, willow and sycamore trees, amidst which was a matted jungle of grape and blackberry vines which, with other shrubbery, made it very difficult to penetrate"	1849	Fairchild 1934
"At first, its margin is hedged only by thick underwood, or tule, but higher up both shores are skirted with large trees, chiefly a species of scraggy white oak and sycamore. These are covered with the mistletoe bough and a species of long dry moss, flowing from the branches, with leaves of fairy net work, its light shade of green contrasting beautifully with the dark foliage of the mistletoe and oak."	1851	Johnson 1851
"The banks increase in altitude, gradually, after leaving the mouth of the river, and groves of sycamore and oaks are soon reached, and the soil better adapted for agricultural purposes."	1852	Ringgold 1852

All the trees and roots on the banks afford unequivocal proofs of the power of the flood-streams, the mud line on a tree we measured exhibiting a rise of ten feet above the present level, and that of recent date.

—BELCHER ET AL. 1979, AS OBSERVED IN 1837

One may force his way through the thicket of brambles and underbrush beneath these river trees, and then there is another expanse – not of water, but of the masses of waving tule.

– JEPSON 1893

Belcher et al. 1979). Botanist Willis Jepson (1893) observed that herbaceous vegetation in the riparian forest was able to far exceed usual heights, with annuals "commonly from four to six feet [1.2-1.8 m] in height" and perennials as high as 18 feet (5.5 m).

With positions along the natural levees that extended from 10 to 20 feet (3.0-6.1 m) above the marsh plain, the forest lands were flooded less frequently than the adjacent basins. They were "sufficiently high not to be subject to the usual overflow of the river" (Grant 1853). The levees natually built only as high as the highest waters would deposit sediment. Differences in inundation frequency related to levee height and flood disturbance affected vegetation composition and prevalence of species. With the relative stability of the natural levees, mature stands of riparian forest with sequential successional stages of the forest were able to develop.

The forests are characterized by high tree density of around 124.5 stems/hectare (50.4 stems/ac; Vaghti and Greco 2007). These were true forests: "the grandest hardwood forests in California" (Roberts et al. 1977). Photography of oak groves (Fig. 5.46) helps to convey the sense of oaks "considerably crowded" in some places along the Sacramento River (Williamson 1857).

Patterns emerged at different scales. At the landscape scale, the forests were relatively broad corridors adjacent to the major river channels. On the other side of the forest from the river extended the much broader freshwater emergent wetland zone (Fig. 5.47). This pattern was described in 1867 as "a narrow strip of land, varying from two to eighty rods [33-1,320 ft/10-402 m] in width, bounded on one side by river or slough, and on the other by

Figure 5.46. Riparian forests of the Central Valley. This photograph (A) of an oak grove illustrates the high density the valley oak can achieve. While the location of this photograph is unknown, it conveys the sense of densely packed trees with open understory that is characteristic of some places within the Sacramento River's riparian forest. The modern photograph (B) shows dense riparian forest with grape vines along the Mokelumne River in August 2005. (A: Oak grove, Hindsdale property, Holland Land Co., D-118, courtesy of Special Collections, University of California Library, Davis, B: photo by Daniel Burmester, August 24, 2005)

water and tules" (Board of Swamp Land Commissioners 1867). Along the fluvial-tidal gradient, width narrowed and vegetation community shifted to willows and other riparian scrub species as natural levee height diminished and inundation frequencies increased downstream toward the central Delta. Latitudinally across the natural levees, community structure reflected a similar relationship to inundation frequency and groundwater levels. At the river's edge willow, alder, and other scrub as well as sycamore formed the foreground, while oak was found on the higher parts of the levees, where inundation was least frequent and groundwater levels the lowest (Wells 1909). At the local level, riparian forest was characterized by substantial within-habitat type complexity, where patterns were affected by biological factors as well as localized topographic, hydrologic, and soil characteristics.

Our mapping suggests that around 33,500 acres (13,560 ha) of valley foothill riparian and another 3,000 acres (1,210 ha; 8% of the total) of willow riparian scrub occupied the natural levee lands of the north Delta. This figure represents more than a third of all the remaining riparian forests in the Central Valley today (Katibah et al. 1981, Frayer et al. 1989) and was about 4% of the estimated one million acres of historical riparian forest in the Central Valley (TBI 1998). Most of the historical riparian forest was associated with the Sacramento River and its distributaries. We mapped 3,500 acres (1,420 ha) of riparian forest along the Mokelumne and Cosumnes rivers within the study area. These estimates likely represent a minimum area of historically forested land as there were undoubtedly forested patches within the matrix of mapped emergent wetland habitat type and along lakes, intermittent streams and other low order channels at scales we were unable

Sycamore…grows directly upon the banks of rivers and sloughs, and places which were once the courses of rivers and sloughs…Its roots must go down to the water…The Cephalanthus [buttonbush] would show land that would be overflowed for from four to six months in the year…The willows indicate land that would retain moisture for months, that might be overflowed, and the soil retain that moisture…The oak will grow where it is and where it is not overflowed.

—REDDING 1860

"Along margin of tule [Sycamore bearing trees: 67 m, 73 m, 3 m, and 47 m distant; 46 cm, 61 cm, 101 cm, and 76 cm diameter]"

[Sycamore bearing trees: 6 m and 18 m distant, 61 cm and 91 cm diameter]

"Left bank of Sutter Slough, navigable stream. Slough [65 m] wide"

"Sycamore [76 cm] diameter on right bank of Sutter Slough"

"Low and wet."

"Timber sycamore and oak. Dense undergrown of oak and briars."

1 mile

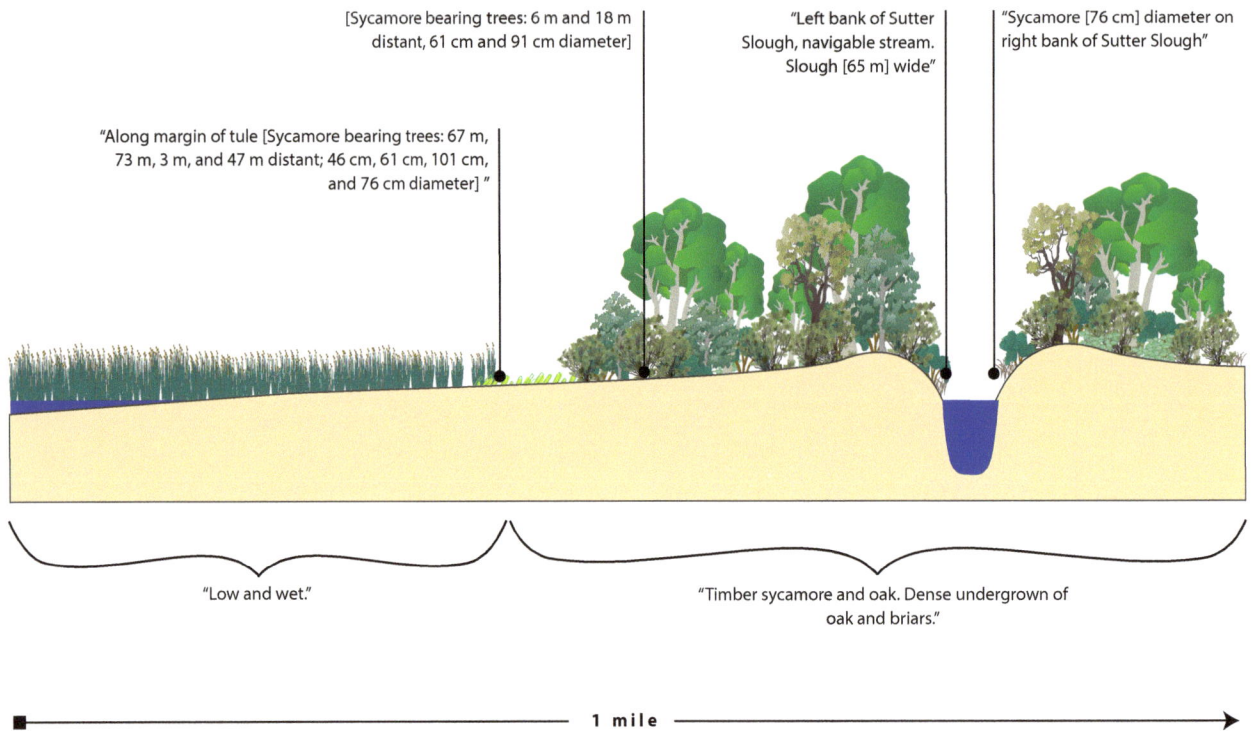

Figure 5.47. Riparian forest, comprised of sycamores and oaks with a dense understory, was positioned along the corridors of natural levees between the river channel and the wide expanse of wetland basin. This profile across Sutter Slough, reconstructed from 1859 GLO field notes and generalized topographic data, illustrates the complexity the riparian forest brought the north Delta basin landscape. (Lewis 1859b)

to map. We also do not discuss the mapped willow thickets in this section as these were generally wetter habitat types occupying positions in the river bottomlands and sinks (see pages 43 and 294).

Today's valley foothill riparian and coastal scrub mapped within the historical extent of these habitat types represents 10% of the area (about 2,700 ac/1,090 ha; Hickson and Keeler-Wolf 2007). Another 4,250 acres (1,720 ha) of modern riparian forest is mapped outside of the historical extent. This compares to an overall estimated 94-98% loss of riparian forest extent in the Central Valley (Vaghti and Greco 2007). Relatively large areas of forests today are found in places such as the Cosumnes River Preserve and the Delta Meadows State Park. There are few places that approximate the complexity and breadth of the native mature riparian forest; most mapped valley foothill riparian forest along the Sacramento River exist as a corridor a few trees wide along the artificial levees. These findings are consistent with the overall 89% decline in riparian forest for the whole Central Valley (Katibah 1984). It is difficult to imagine virtual jungles extending for half a mile beyond the river's edge to meet the tules on the other side. The diminished width and complexity of the remaining forests significantly compromises the forest's capacity to provide essential ecosystem functions (e.g., nesting sites, allochthonous input, woody debris, shading, stabilizing channels, windbreaks, wildlife corridors; Thompson 1977).

Conversion of native mature stands of riparian forest to homesteads and orchards occurred rapidly during the first several decades after the Gold Rush – one report from 1854 described the Sacramento River banks above the American River confluence as "formerly well timbered" (Box 5.7; Fowler 1853). Riparian forests were some of the first areas to be substantially

BOX 5.7. EARLY IMPACTS TO THE RIPARIAN FOREST

The riparian forests lining the Sacramento River were some of the first parts of the Delta to be substantially impacted by rapid settlement in the mid-1800s (Fig. 5.48; McGowan 1939, Thompson 1977). The riparian forest supplied cordwood to steamboats for fuel (Thompson and West 1880). The steamboats' consumption of wood was substantial. For example, on the Willamette River in Oregon, researchers Sedell and Froggatt (1984) found that steamboats used between 10 and 30 cords of wood a day. By the early 1850s, woodcutting was a legitimate enterprise. Cadwalader Ringgold (1852) reported while surveying the Sacramento River: "A lively scene is presented to persons passing up and down the river; at almost every bend and turn, the wood-cutter is seen, and the pleasant sound of his axe heard." A traveler's narrative from the early 1850s describes woodcutting "in the bottom-lands, between Suttersville and Sacramento city," for the sale of oakwood at $15 a cord (Gerstäcker 1853). In another account, a Yolo county history reported that wood was sold to steamboats for $10 a cord in 1850 (Gregory 1913). The effects were noticeable in the 1850s: Several General Land Office surveys of the riparian forest lands include notes such as "large oak timber has been cut down – dense oak bushes" (Lewis 1858c). Although not suitable for construction, the oak wood was also used extensively for fenceposts. Additional pressure was placed on the riparian forests since natural levee lands were ideal locations for homesteads, small vegetable gardens, and orchards. Small farms sprang up to supply miners traveling to the gold fields in the Sierra Nevada, some of which were established by disheartened miners returning to find new beginnings. Unlike the majority of the Delta, preparation of these natural levee lands for cultivation was relatively easy. Drainage was unnecessary, and it was often the case that natural levee lands would be cultivated while the land at the back, the tule land, was largely left alone apart from use as pastures (Hoppe, pers. comm.).

Figure 5.48. Cordwood stacked in an area of cleared riparian forest near where West Sacramento stands today. (ca. 1910, courtesy of the California History Room, California State Library, Sacramento)

impacted. Due to lower flood frequencies, rich, well drained, loamy soils, and easy access to river transportation, the natural levees were some of the first areas to be occupied and farmed by settlers in the Delta. The infrequent flooding of natural levees meant that the land was not classified as "swamp and overflowed land" and was thus available for sale to settlers by the federal government. These banklands were subdivided, bought, and cultivated in the early years while the tule lands behind remained unreclaimed (Fig. 5.49). Early land ownership maps and aerial photography reveal land ownership patterns related to Delta topography that persist today.

Transitions along physical gradients

The transition to valley foothill riparian forest is illustrated by early maps and by the topographic and geologic evidence of gradually increasing breadth and height of natural levee deposits (Fig. 5.50; USGS 1909-1918, Holmes et al. 1913, Atwater 1982). The characteristics of riparian forest tracked the shift from a tidally-dominated to a more fluvially-influenced system, reflected in the elevated depositional landforms of the natural levees. As the natural levees increased in height and breadth ascending toward the City of Sacramento, the forest increased in width and degree of complexity. This pattern of riparian vegetation shifting from emergent vegetation to scrub to trees along the tidal to fluvial gradient was evident along each of the major river channels that entered the Delta.

Upon ascending the Sacramento River, numerous travelers observed the gradual transition of vegetation (Buffum 1850, Ringgold 1852, Belcher et al. 1979, Phelps and Busch 1983). One traveler notes the transition on the Sacramento: "at first, its margin is hedged only by thick underwood, or tule, but higher up both shores are skirted with large trees, chiefly a species of scraggy white oak and sycamore" (Johnson 1851). The oaks and other trees

The marshy land now gave way to firm ground, preserving its level in a most remarkable manner, succeeded by banks well wooded.

—BELCHER ET AL. 1979 AS OBSERVED IN OCTOBER 1837

Figure 5.49. Farmsteads line the natural levee land along the Sacramento River in this county map. This infrequently flooded land was not deemed "swamp and overflowed land" like the wetlands within the basin interior and thus came under the jurisdiction of the federal government, which sold this land to settlers. In the mid- and even late-1800s, levee lands were in high stages of cultivation while the lands at the back remained wetland habitat. (courtesy of the Center for Sacramento History, County Map Book No. 70)

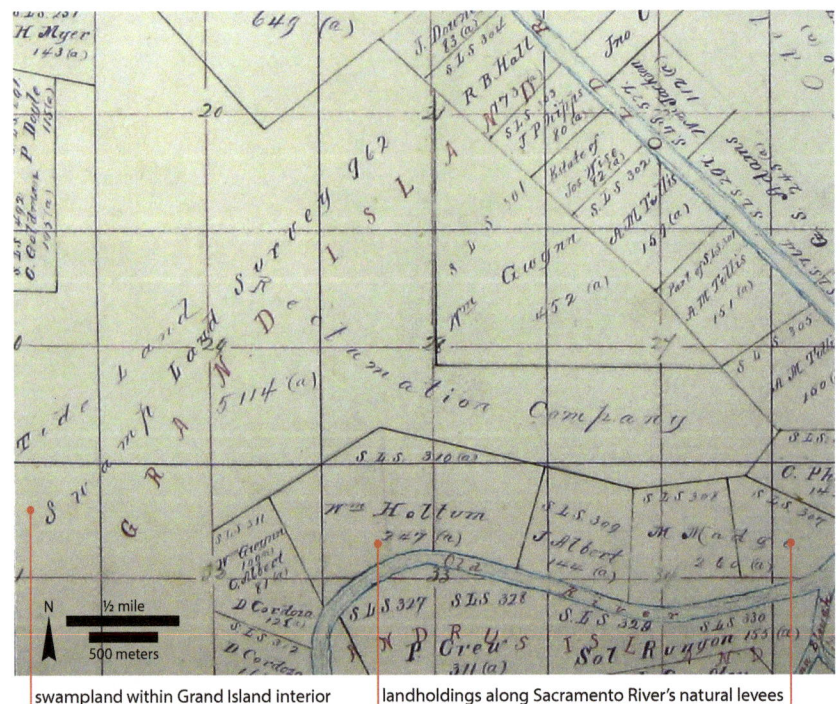

swampland within Grand Island interior   landholdings along Sacramento River's natural levees

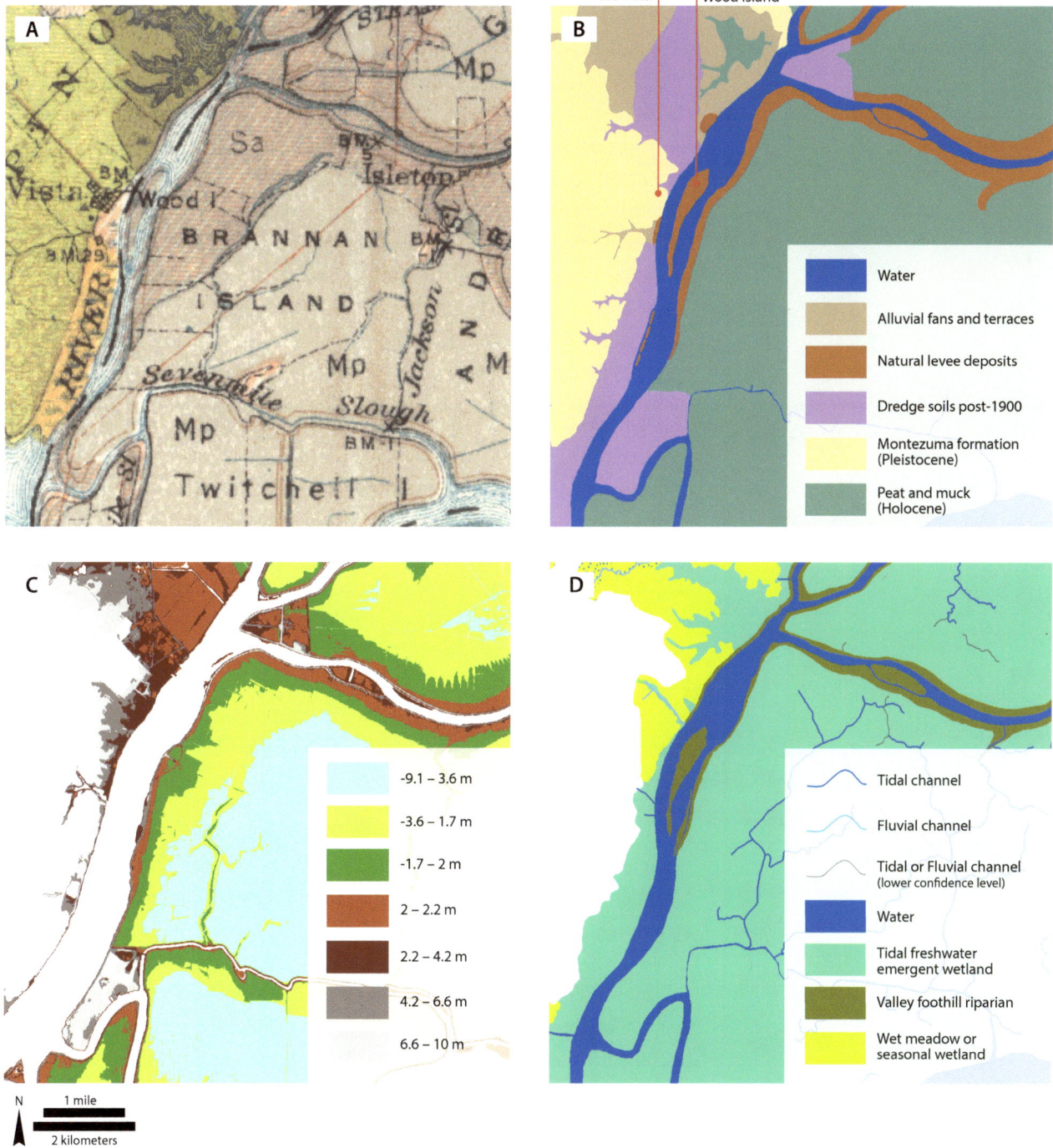

Figure 5.50. Natural levee deposits diminish downstream. Historically, the Sacramento River natural levees did not support riparian forest far beyond the town of Rio Vista. The soil survey (A) shows a shift toward the more organic peat soils (Mp) as the Sacramento clay loam (Sa) diminishes in width along Brannan Island. Geologist Brian Atwater's (1982) mapping (B) shows narrow natural levee deposits (brown) ending just downstream of the historical Wood Island. The recent LiDAR elevation data also illustrates the trend in diminishing natural levees related to the narrower width of higher land. The historical habitat mapping, reflecting this pattern in decreasing riparian forest width, is shown in D. (A: Holmes et al. 1913, B: Atwater 1982, C: CDWR 2008)

associated with the gallery riparian forest that lined the Sacramento River were well established upstream of Rio Vista (Thompson 1957, Abella and Cook 1960, Phelps and Busch 1983). In describing the locality of a new city, Halo Chamo (an early name for Rio Vista), a newspaper reported that "oak, ash, black walnut and many other different kinds of timber are in great abundance" (*Californian* 1847). The transition is also represented in maps and surveys associated with the Los Ulpinos land grant that extended along the west bank of the Sacramento River: vegetation lining the bank is drawn beginning just downstream of Wood Island on the *diseño* (Bidwell ca. 1840); and an 1858 survey for the grant does not use a bearing tree until over a mile upstream of Rio Vista (though some tree cutting may have occurred by this time around Rio Vista; Lewis 1858d).

Though oaks and other large trees were apparently common by this point on the river, the transition from emergent vegetation and scrub dominated banks to the dense, multi-layered, broad forests was gradual, occurring over a number of miles. An early 1900s landscape photograph at Rio Vista illustrates the character of this transition, where low banks with few trees are apparent (Fig. 5.51). The more dense forest characteristic of the Sacramento River was likely not present along the banks until the mouth of Cache Slough (Thompson 1957). Downstream of Rio Vista, sources indicate that trees were found on in-channel islands and that some clumps of willow and other scrub were found within the emergent wetland matrix of the larger islands, such as Sherman Island (Fig. 5.52; U.S. Ex. Ex. 1841, Ringgold 1850a).

Along channels associated with Cache Slough, those that received flood flows most directly from the Sacramento had larger levees and were occupied by larger trees. While present at the mouth and likely along much of Miner Slough, dense riparian forest with large trees such as oaks and sycamore probably was not found along most of the Cache Slough channels. Lower scrub vegetation consisting of willow and alder was likely more common: a GLO surveyor on the bank of Cache Slough just downstream of Miner Slough noted a two-foot (61 cm) diameter alder, stating, "no other bearing tree convenient" (Lewis 1858d). Early evidence of tree or scrub-lined banks is found in the Ringgold 1850 survey of the area, where symbols representing woody vegetation are drawn up Cache Slough to Miner Slough and up Miner Slough (see Fig. 4.22).

The natural levees along the Mokelumne River tended to be about four feet (1.2 m) lower than points on the Sacramento River at the same latitude (Thompson 1957). The transition to forest on the Mokelumne consequently occurred farther upstream. Accounts, particularly those from the Mokelumne land grant testimony, specified that dense forest began approximately a mile below the head of Staten Island (traveling upstream along either of the two forks of the river; Fig. 5.53; Davis 1859, Payson 1885). This is supported by reclamation documents reporting that the only

At the mouth of the river there is very little timber; but in our progress upward we found the oak and the sycamore growing most luxuriantly.

—BUFFUM 1850

Grand
Island

Rio
Vista

Brannan
Island

groupings
of trees

Sherman
Island

Figure 5.51. The low banks at Rio Vista can be seen in this 1916 photograph. Only a few trees are present and the banks are mostly occupied by scrub and wetland species. Though the landscape was heavily modified by this point, the natural topography and basic vegetation patterns remained largely intact. (1916, courtesy of the California History Room, California State Library, Sacramento)

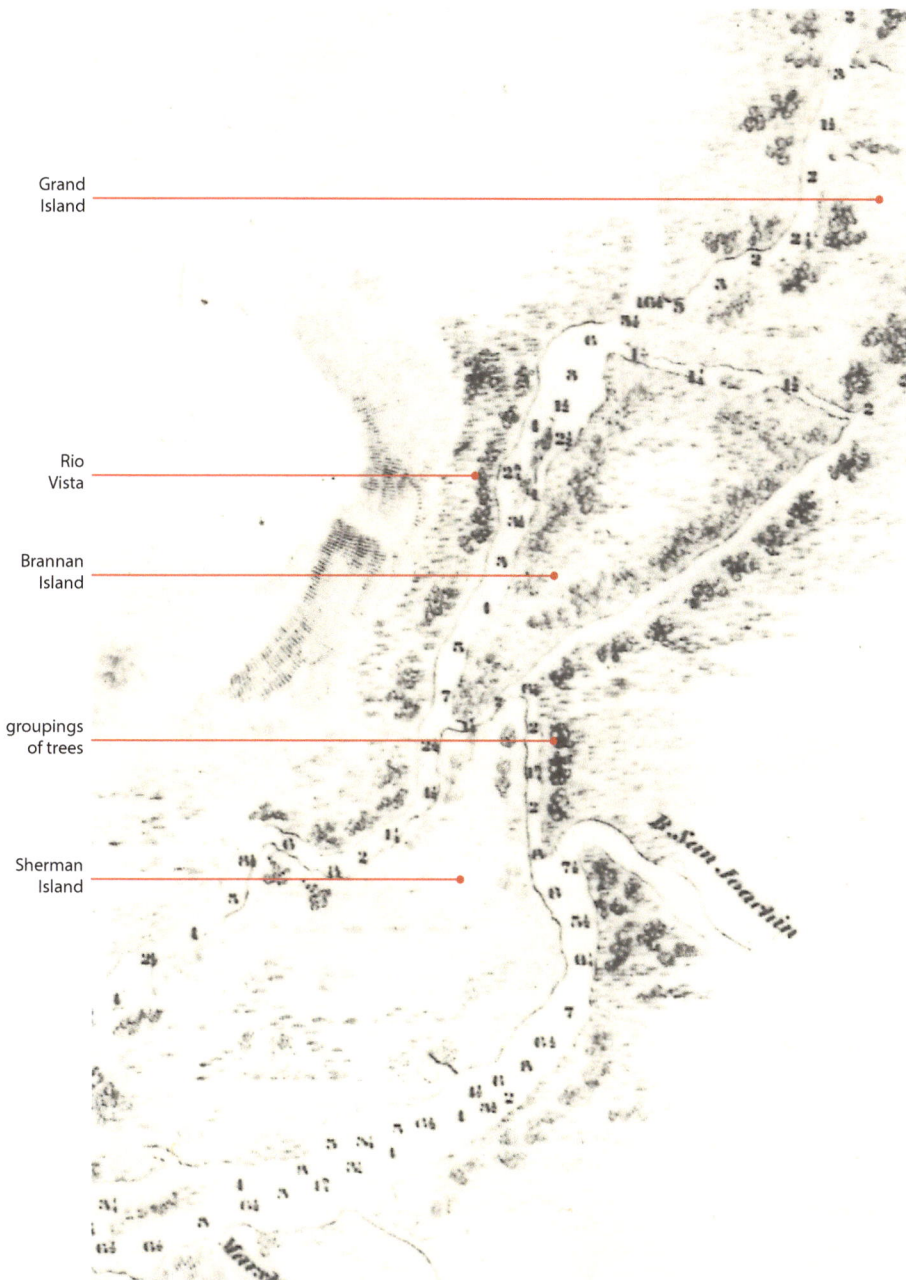

Figure 5.52. Early depiction of riparian forest transition along the lower Sacramento River. The clumps of trees drawn upstream of the head of Sherman Island become bigger and darker. Below Rio Vista, the tree symbols may represent scattered groupings of willows and other scrub or willow-fern swamps, which were found on some central Delta islands (see page 177). (U.S. Ex. Ex. 1841, courtesy of the Earth Sciences & Map Library, UC Berkeley)

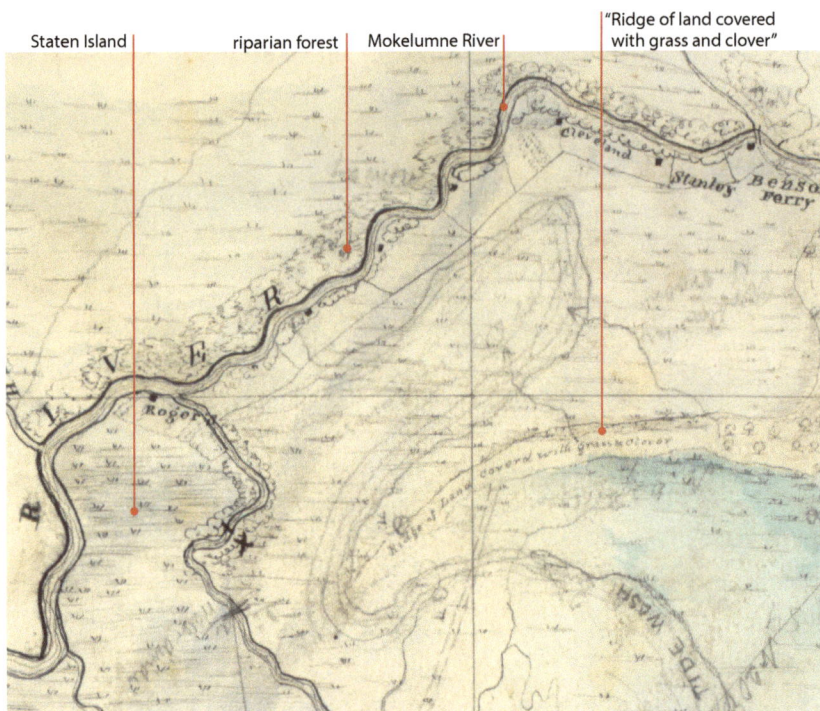

Staten Island | riparian forest | Mokelumne River | "Ridge of land covered with grass and clover"

Figure 5.53. The downstream end of the riparian forest along the Mokelumne River is shown in this 1859 map just downstream of the head of Staten Island. (Von Schmidt 1859, courtesy of The Bancroft Library, UC Berkeley)

extensive area of "bank land" on Staten Island (around 100 ac/40 ha) was found at the head of the island (Tucker 1879a). A surveyor, attempting to survey the river below the head of the island, got about half a mile (0.8 km) downstream before he "found it impracticable and stopped, owing to the dense thicket" (Sherman 1859). The riparian vegetation was quite dense: one witness stated that boats could not get into sloughs above the head of the island "on account of the bush" (Van Scoyk 1859). The transition to denser forest in the upper few miles of the island is also evident in the later War Department surveys of the river conducted to identify obstructions in navigable waters (see page 246; Payson 1885). On the Mokelumne, it was reported that "the main obstructions occur in the upper few miles [before New Hope Landing], and consist principally of snags and overhanging trees" (Mendell 1881). Evidence also suggests that lower willow-dominated scrub continued a distance further downstream tracking the steadily decreasing levee heights. Willows tall enough to obstruct the view of the surrounding landscape were found as far downstream as the mouth of Hog Slough (Sherman 1859).

Mapping the extent and communicating the character of the forest was particularly challenging where forested banks gradually transitioned to primarily emergent wetland along the most southerly reaches. Given the limitations of GIS, we were unable to illustrate the increase in proportion of brush and emergent vegetation and patchiness of the forest as emergent vegetation became more common. Instead, this transition is largely represented by the corridor's steady narrowing and the transition, particularly on the Mokelumne, to a willow riparian scrub or shrub habitat type.

Width of the forest corridor

Riparian forest width varied substantially along the Sacramento River, Mokelumne River, and major distributary channels. At the landscape scale, width increased upstream in relation to increasing natural levee height and breadth (Fig. 5.54; Thompson 1957). It was also related to the size of the stream, with Sacramento River riparian forest wider than on lesser distributaries and the Mokelumne River. According to the historical habitat type reconstruction, the forest width averaged about 390 feet (120 m; for just one side of the river) within the first five miles (8 km) above the foot of Grand Island on the Sacramento River and about 230 feet (70 m) on Steamboat Slough for the same distance. Above the head of Grand Island riparian forest width increased to about a quarter mile (approximately 375 m). Farther upstream, above the mouth of the American, width was greater still, about 0.6 miles (1,000 m). Riparian forest of the upper Sacramento Valley above the Feather River could reach widths over 4 miles (6.4 km; Thompson 1961). Compared to the large areas of wetlands at the back of the forest, the bank lands were comparatively narrow. However, with the perspective of the usual riparian corridor today of only a few trees wide, these forests were broad to an extent that is difficult to imagine in the modern landscape.

Our mapping is the result of synthesis of numerous maps, texts, and surveys that indicate the forest boundaries. The most spatially comprehensive sources were the early 1900s USGS topographic maps and the soil surveys, which were used to infer the extent of forest based on the higher and well drained depositional soils of the natural levees. The inference was facilitated through other more spatially limited and often earlier sources that gave direct evidence of forest extent. The most spatially explicit of those available were the GLO surveys and associated plat maps. Often, we were able to use the elevations associated with notes of entering or leaving "timber" to interpolate the boundary from the USGS topographic maps.

Narratives discussing the riparian forest of the north Delta provide additional valuable early confirmation of width. Describing the forests upon ascending the Sacramento River, explorer Belcher (1843) wrote that they "appeared to form a band on each side, about three hundred yards [274 m] in depth." Most accounts state that the forest lining the river was about a half a mile, with the upper reaches wider that those to the south (Table 5.2; Clyman and Camp 1928[1848], Browning 1851, Bates 1853, Grant 1853, Hilgard 1884, McConnell 1887). Above the mouth of the American to the Feather River, some accounts note wider widths of between a half and three-quarters of a mile wide (0.8-1.2 km; Fowler 1853, Robinson 1854). In describing the extent of the natural levee fertile soils, another account summarized that it "varies in width from one-eighth to one mile" (Sprague and Atwell 1870).

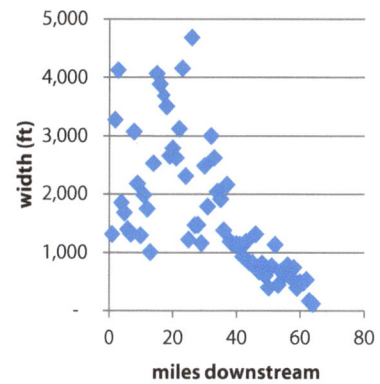

Figure 5.54. Overall decreasing riparian forest width is shown based on measurements of the right bank riparian forest in the historical habitat type mapping. Measurement were taken every mile downstream from the Feather River confluence to the foot of Grand Island.

The Banks generally low are timbered…The Timber on each side of the river is narrow…In the bottoms are Lakes and flags which frequently extend 2 miles from the river.

—SULLIVAN 1934, CONCERNING THE LOWER MOKELUMNE RIVER IN FEBRUARY 1828

Table 5.2. Selected accounts of the width of the "high lands" along the Sacramento River are listed in this table. These lands historically supported broad and diverse riparian forests.

Quotes	Location	Year	Citation
"These appeared to form a band on each side, about three hundred yards in depth"	Sacramento River	1837	Belcher 1843
"There being a large tuly [tule] or rush swamp about half a mile from the river"	near Sutter's Fort	1845	Clyman and Camp 1928 [1848]
"In an average it is about ½ mile in width then comes the Tola or Bull Rush"	Sacramento River	1851	Browning 1851
"The average width of the high land I should judge to befrom one-half to three-fourths of a mile"	American to Feather	1853	Fowler 1853
"There is a strip of timber land about a half mile in width"	American to Feather	1853	Bates 1853
"These strips of land are from a quarter to a half mile in width"	American to Feather	1853	Grant 1853
"There is a narrow strip of land, from half to three-quarters of a mile"	American to Feather	1854	Robinson 1854
"High lands averaging in width about 1/4 of a mile"	Sacramento River	1860	Von Schmidt 1860
"In some places I should think it would be from one hundred yards to a quarter of a mile"	American to Feather	1860	Buzzell 1860
"Quest. 14th. Since the average width of the good land along the east bank of the Sacramento river to the mouth of the Feather river? Ans. 14th. About half a mile, more or less."	American to Feather	1861	La Rue 1861
"As it was in 1849, the slope ran back about a mile, on an average. Some places it was but half a mile, others over a mile...Right at Washington it was not a half mile to where the tules originally were"	Sacramento River	1881	Hoagland 1881
"Along the border of this river there is a belt of alluvial land varying in width form on-half to a mile or more"	Sacramento River	1884	Hilgard 1884
"Along the Sacramento river, averaging something over half a mile on each side, there were forests of oak, ash, and sycamore"	Sacramento River	1887	McConnell 1887

In comparison to the Sacramento River, the Mokelumne River's riparian forests were relatively narrow. They were not wide enough to be subdivided by the GLO survey like the natural levees of the Sacramento River. We mapped widths of approximately 215 feet (65 m) along the Mokelumne River up to the Cosumnes confluence. This difference is related to the lower height of the natural levees. Unfortunately, this lower height made it challenging to use the USGS topographic maps to map the forest corridor. However, several sources were helpful in estimating width. One point of calibration was found 1.2 miles (1.9 km) above the head of Staten Island on the Mokelumne River, where a GLO survey recorded the "margin of dense tule swamp" 190 feet (58 m) from the river (at a slightly oblique angle). This corresponded with a similarly wide riparian belt mapped by the Debris Commission on the other side of the river (Fig. 5.55, California Debris Commission 1914). Width likely varied, perhaps associated with secondary channels, as is indicated in the Debris Commission map.

EXTENSIONS INTO THE WETLANDS At some points, forest extended much farther into the basins along crevasse splay deposits or other elevated landforms. One of these was Babel Slough, a particularly large crevasse splay that extended into the Yolo Basin (Bryan 1923). Its higher lands were described as wider but lower in elevation in comparison to those of nearby Elk Slough

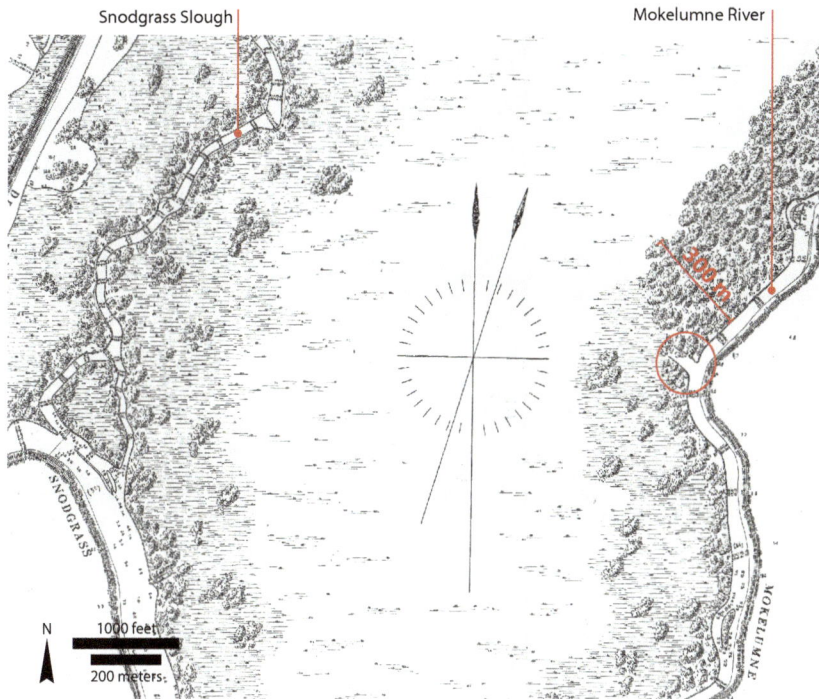

Snodgrass Slough

Mokelumne River

N
1000 feet
200 meters

Figure 5.55. Riparian forest width as shown in 1913. A relatively narrow corridor is shown close to the head of Staten Island, but this widens upstream over a short distance to around 984 feet (300 m). This widening may be associated with secondary channels and their banks (the trunk of one can be seen in the map, circled in red) that branch into what is today the McCormack-Williamson Tract. (California Debris Commission 1914, courtesy of the California State Lands Commission)

(Sprague and Atwell 1870). Its farthest point into the wetlands of the Yolo Basin – over three miles (4.8 km) from the Sacramento River – was a landmark called Willow Point. Here, the extent of tule beyond was narrow and a summer-season road was constructed across the marsh for a mile (USGS 1909-1918). One traveler described this route along Babel Slough to Willow Point as "through a dense grove of large oaks and sycamores" (*Sacramento Daily Union* 1860), and another stated that the "margins of Babel Slough are similar to the banks of the Sacramento River" (Gwynn 1881). The GLO survey of the area includes notes of "oak timber on both sides of Babels [sic] Slough" (Lewis 1858c) and "undergrowth oak, willow and briars very dense" (Lewis 1858b). Though the area is farmed today, a remnant of the forest remains where Highway 84 crosses Babel Slough (Fig. 5.56).

Another place where forest was especially broad occurred at the site of the present-day Delta Meadows State Park. This site is one of few areas in the Delta characterized by surficial eolian deposits as opposed to more recent Holocene alluvial deposits (most are found in eastern Contra Costa County, see page 186; Atwater and Belknap 1980, Atwater 1982). The area of higher land associated with Delta Meadows protrudes into the wetlands from the regular riparian forest edge along the Sacramento River. The area is described by GLO surveyor William Lewis (1859a) as "a strip of high land six chains [396 ft/121 m] in width extends about 20 chains [1,320 ft/402 m]" and it is labeled "timber mound" in an 1869 map (Heynemann 1869). The USGS topographic maps (1909-1918) show significant elevation complexity, complete with a high mound extending 20 feet (6 m) above sea level and several small ponds.

Figure 5.56. Oaks on Babel Slough. The crevasse splay deposits of Babel Slough, which extends into the Yolo Basin, once supported broad riparian forests similar to those along the Sacramento River. Two views are shown of a remnant grove that stands near the tip of Babel Slough today. (A: photo by Alison Whipple March 30, 2011; B: USDA 2005)

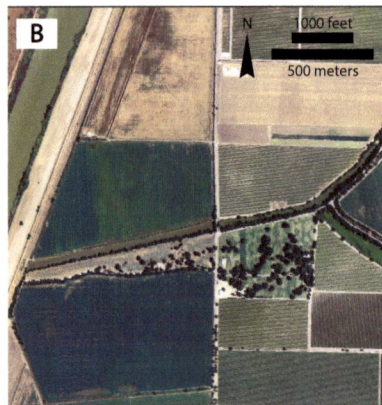

Riparian forest species

Conditions along the rivers of the Central Valley were ideal for establishing a diverse vegetation community and complex forest structure. Trees, shrubs, herbaceous vegetation and vines flourished. Trees of the forest canopy included valley oak (*Quercus lobata*), California sycamore (*Platanus racemosa*), cottonwood (*Populus fremontii*), and Oregon ash (*Fraxinus latifolia*). Interior live oak (*Quercus wislizenii*) and California walnut (*Juglans californica*) were somewhat less common. The sub-canopy and understory consisted primarily of willows (*Salix* spp.), with alder (*Alnus rhombifolia*), button bush (*Cephalanthus occidentalis*), dogwood (*Cornus sericea*), box elder (*Acer negundo*), buckeye (*Aesculus californica*), grape vines (*Vitis californica*), wild rose (*Rosa* spp.), and numerous herbaceous species also present (Jepson 1893, Jepson 1910, Thompson 1961, Thompson 1977, West 1977, Vaghti and Greco 2007). Botanist Willis Jepson (1893) listed most of these species and mentioned relative abundance in his summary:

The major part of the growth is made up of various species of willow [*Salix nigra*, Marsh, *S. lansiandra*, Benth., and *S. longifolia*, Muhl.]. Fine specimens of the Plane Tree [*Platanus racemosa*, Nutt.] are not uncommon. The Cottonwood [*Populus Fremonti*, Wats.] is frequent; while the Button Bush [*Cephalanthus occidentalis*, L.], the Oregon Ash [*Fraxinus Oregana*, Nutt.], the California Walnut [*Juglans Californica*, Wats.], and the Alder [*Alnus rhombifolia*, Nutt.], though not abundant, are to be met with throughout this entire region. The Wild Grape [*Vitis California*, C. & S.] and Blackberry [*Rubus vitifolius*, C. & s.], with various herbaceous and suffrutescent plants. The Box-Elder [*Acer Californicum*, Greene] and Poison Ivy [*Rhus diversiloba*, T. & G.] were noticed near Walnut Grove, as also fine individuals of the Live Oak [*Quercus Wislizeni*, CD.] on the highest river banks. The River Dogwood [*Cornus pubescens*, Nutt.] is fairly frequent.

Due to early conversion of riparian forest to orchards and homesteads, there are few sources providing detailed and comprehensive botanical information. However, narrative descriptions and surveys of the 1800s riparian forest offer valuable information concerning dominant species and the appearance of the assemblages. Early paintings and landscape photography also provides glimpses into what the diversity of the forest looked like (Fig. 5.57).

After following many windings they entered a river, the banks of which were lined with alder, willow, buckeye, and sycamore with wild grape clinging to their branches, while cottonwood, poplar, and oak formed a background.

—WOOD 1941 AS OBSERVED IN 1839

Figure 5.57. Complexity of the riparian forest. Tall trees line the background with a dense undergrowth shielding them in the front and a few snags overhang the channel in this painting, likely of the Sacramento River, by surveyor and engineer Carl Grunsky. (Grunsky ca. 1879, courtesy of The Bancroft Library, UC Berkeley)

One of the first narrative accounts of the Sacramento River forest comes from Spanish explorer Ramon Abella's 1811 expedition into the Delta, where he notes that the trees along the river banks are "of various kinds and very large," specifically mentioning the prevalence of walnuts and grapes (Abella and Cook 1960). The earliest detailed description of the vegetation community along the lower Sacramento River comes from Captain Belcher's 1837 voyage. Upon ascending the river, he found:

> banks well wooded with oak, planes, ash, willow, chestnut [Thompson (1961) notes this is likely buckeye], walnut, poplar, and brushwood. Wild grapes in great abundance overhung the lower trees, clustering to the river, at times completely overpowering the trees on which they climbed.

Farther upstream, the account continues:

> belted with willow, ash, oak, or plane, (*platanus occidentalis*) which latter, of immense size, overhung the stream, without apparently a sufficient hold in the soil to support them, so much had the force of the stream denuded their roots. Within, and at the verge of the banks, oaks of immense size were plentiful.

Frequently, accounts mentioned the trees that overhung the river, like those with exposed roots in the above quote. These characteristics are important to consider regarding the in-channel habitat provided to fish, with channel edge complexity caused by the trees and the input of allochthonous material and woody debris.

Another description of the Sacramento River forest composition is found in the botanical report of the U.S. War Department (1856):

> The banks of the streams are lined with belts, of greater or less width, of timber, which are composed chiefly of the long-acorned oak, (*Q. Hindsii*,), here exhibiting the size and beauty of form not surpassed, if equaled, by the oaks of any other part of the world. Along the water's edge, the sycamore, (*P. Racemose,*) *Fraxinus Oregona*, the cotton-wood, (*P. Monilifera,*) and two species of salix, (*S. Hindsiana* and *S. lasiandra* ?,) are overgrown by grape vines, (*Vitis Californica*,) and form a screen, by which the view of the river is frequently shut out from the traveler upon its banks.

Summarizing from other brief mentions of the forest composition, the typical list of common trees included valley oak, sycamore, live oak, ash, walnut, cottonwood, willow, and alder.

A number of different sources indicate that sycamores and valley oak were the dominant trees of the Sacramento River riparian forest. Many early descriptions specifically list these two species as most prevalent. The sycamore is often specifically discussed in context of the Sacramento River (Jepson 1910). For example, in describing the range of this latter species, one account stated that it "grows in plenty among the Sacramento River" (Bidwell [1842]1937). Another stated that the many farm landings along the Sacramento were "usually in the shade of a wide-spreading sycamore tree" (Muir 1888). Edwin Bryant ([1848]1985) provided a limited list of

The banks were...overhung by large, branching oak and sycamore trees, and were covered by a dense thicket of young oak, wild roses, blackberry and other bushes.

—HOAG 1882

Large sycamore trees then lined the river-front abreast of the infant city of Sacramento; deep sloughs creased its site whose beds were a bramble of grapevines, blackberry bushes and other undergrowth, while big white oak trees bearing heavy crops of long slim acorns dotted the space between these depressions.

—FAIRCHILD 1934
AS OBSERVED IN 1849

dominant trees, recording in his diary that the forest was "chiefly oak and sycamore." Also, though maps rarely explicitly noted species, one map made in 1849 shows the broad forest lining the rivers of the Sacramento Valley and included the words "Sycamore and Oak" near the City of Sacramento (Derby 1849).

Survey data also provides confirmation of dominant trees. The bearing trees used by the GLO surveyors include alder, ash, black walnut, buckeye, cottonwood (also referred to as poplar), live oak, sycamore, willow, oak (likely valley oak), and white oak (valley oak; Fig. 5.58). Selected descriptions from the accompanying field notes of these surveys are listed in Table 5.3. In agreement with other narrative accounts of the forest, they point to a forest dominated by oak and sycamore with a dense understory. The mix of trees used as benchmarks by State Engineer surveys in the 1870s include sycamores, oaks (presumably both live and valley), walnuts, willows, cottonwoods, and alders (Hall 1879). The first soil surveys of the region describe the native vegetation of the natural levees as "originally covered with a heavy forest growth consisting mainly of sycamore, cottonwood, willow, and oak, with a thick undergrowth" (Holmes and Nelson 1915).

The ubiquity of the sycamore along the Sacramento contrasts with that of the cottonwood. While cottonwoods did occur along the Sacramento River, they are mentioned less frequently in the historical record. An 1841 summary of the range and character of a number of species suggests that the cottonwood was not ubiquitous: "on the Sacramento and its branches are more or less of it" (Bidwell [1842]1937). Cottonwoods may have been more common along the upper, less tidally influenced reaches and where disturbance frequencies associated with more active meandering river processes were greater. This is suggested by explorer Wilkes (1845), who found the Feather River banks "lined with sycamore, cottonwood, and oak." Only two cottonwoods (called "poplar" by the surveyors and each

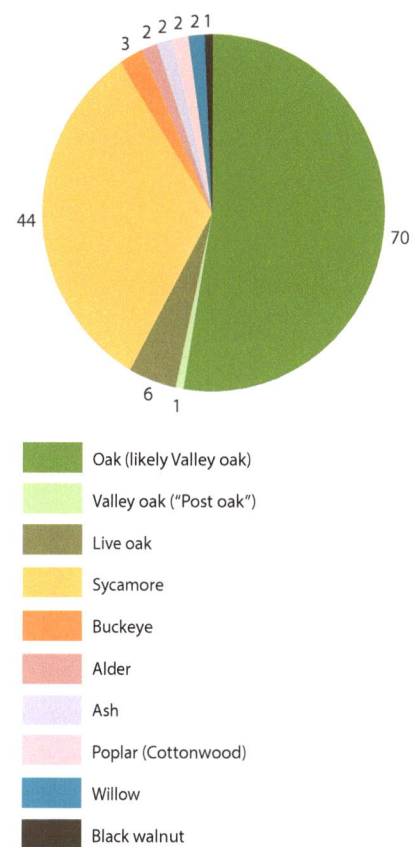

Oak (likely Valley oak)
Valley oak ("Post oak")
Live oak
Sycamore
Buckeye
Alder
Ash
Poplar (Cottonwood)
Willow
Black walnut

Figure 5.58. Bearing trees recorded by the GLO within the riparian forest associated with the Sacramento River and its major distributaries (e.g., Elk Slough). The dominant species are oak and sycamore. A high proportion of sycamores were obtained, in contrast to zero obtained along the San Joaquin River in the south Delta. A total of 133 trees were recorded within non-upland habitat types (e.g., not oak woodland or savanna).

Table 5.3. Selected GLO notes describing the quality of timber within the riparian forest.

Selected quote	Reference
"Mostly first rate land covered with heavy oak timber with a thick undergrowth of briars and vines."	Loring 1851
"Strike oak and sycamore timber…Dense undergrowth of grape vines, briars, and oak."	Lewis 1859b
"Timber chiefly oak, undergrowth same."	Lewis 1858-9
"Timber consists of small oaks, willows, and swamp alders."	Prentice 1870
"Timber live oak, undergrowth oak."	Lewis 1858c
"Timber oak. Dense undergrowth of grape vines, briars, and oaks."	Lewis 1859b
"Timber of oak and sycamore, dense undergrowth of oak and grass."	Lewis 1859c
"Timber on bank of river, oak, sycamore, and buckeye."	Lewis 1858-9
"Timber on river, oak and sycamore."	Lewis 1858a
"Timber, chiefly oak and sycamore."	Lewis 1859a
"Timber, oak and sycamore."	Lewis 1859a

less than a foot in diameter) were included in the GLO bearing tree dataset and were not associated with the Sacramento River: one at the American River confluence (Dyer 1862b) and another at Putah Sinks (see Fig. 5.58; Hays 1852b). By comparison, 44 sycamores and 68 oaks were used as bearing trees along the Sacramento River. Though GLO surveyors were likely preferential to the longer lived oaks and sycamores, the absence cottonwoods is notable. Surveys conducted along the Sacramento by the State Engineering Department did note cottonwoods, but they were far outnumbered by sycamores and oaks (Hall 1879).

The California walnut appears to have been particularly prevalent in the vicinity of Walnut Grove, as the name suggests (Thompson and West 1880). The single black walnut tree in the GLO bearing tree dataset was recorded at the head of Georgiana Slough (Lewis 1859a). Other evidence for this apparently more limited distribution is found in an 1842 description of the species as "confined to a few miles" on the Sacramento (Bidwell [1842]1937). Interestingly, botanist Herbert Mason (n.d.) associated the walnut with the human habitation sites in the Delta. Explorers encountered a number of villages of Delta tribes along the Sacramento River near Georgiana Slough, where Walnut Grove stands (Durán and Chapman 1911).

Beneath the tall tree canopy were willows, blackberry, grape vines and numerous other shrub and herbaceous species (e.g., wild rose, *Rosa californica*; wild pea, *Lathyrus jepsonii*) created the dense understory that many people found virtually impenetrable (Belcher 1843, Kerr 1850, Hoag 1882, McConnell 1887). Common terms used in the historical record to describe this vegetation include "dense thicket," "jungle," "underbrush," and "thick undergrowth" (Loring 1851, Prentice 1870, Bidwell [1884]1904, Abella and Cook 1960, Arguello and Cook 1960). Voyagers on the river picked grapes from the overhanging trees. The vines and other shrubs produced "a most charming effect" (Bryant [1848]1985). The grape vines, in particular, gave the forest the appearance of "a matted jungle of grape and blackberry vines" (Fairchild 1934) and formed a "screen" that blocked the view beyond (Williamson 1856). More recent descriptions of the valley foothill riparian forests of the Central Valley discuss these understory species as creating a complex multi-layered vertical structure (West 1977).

Local-scale complexity and ecotones

Just as vegetation patterns tracked the steadily increasing height and breadth of natural levees upstream, they also reflected latitudal, or cross-levee, changes in height (see Fig. 5.47). Willows and other species tolerant of long periods of inundation lined the rivers, while larger trees stood on the higher ground farther back. The "wide-spreading magnificent oaks" gave way, reads one early description, to "a meager border of willows, poplar, or sycamore, hung with festoons of grape along the water's edge" (Williamson 1856). The pattern is mentioned in numerous other accounts as well (e.g., Taylor 1854, Wells 1909, Sullivan 1934, McGowan 1939, Wood

Abundance of wild rose and everlasting sweet pea alongside the banks growing in great luxuriance.

—KERR 1850

1941). A *diseño* showing land along the Sacramento River presents a depiction of this general pattern (Fig. 5.59; Bidwell 1844). Symbols indicating dense forest line the river, while scattered trees are depicted beyond on what would have been the higher parts of the natural levees. Here, trees were found arranged in groves "like a park" (Durán and Chapman 1911). Captain Belcher's (1843) diary includes a similar description of oaks "disposed in clumps" along the river banks. In such places the dense thickets opened into forests where the understory was herbaceous and more easily traversed.

Grasses and other herbaceous vegetation likely occupied the transition between riparian forest and basin wetlands. This is suggested in an 1851 account stating that "next the River is covered with very heavy timber about half way back to the Tola and this is grass of the best kind" (Browning 1851). Another used the term "prairie land" to describe the area between the "timber land" and the "tules" (Buzzell 1860). In his 1846 journey on foot up to Sutter's Fort, Heinrich Leinhard noted "a grass-covered area between the forest and swamp" (Lienhard and Wilbur 1941). Though it is unclear how extensive this area was, the diseño mentioned earlier suggests a more open transition between riparian forest and the tule swamp: tree symbols end before the line of tule begins and while dense forest lines the river banks, no such symbols are drawn on the back-side of the natural levees (see Fig. 5.59; Bidwell 1844). The transition would have been gradual, occurring as localized patches or unevenly dispersed clumps of vegetation. This is conveyed in testimony for the Sutter land case. Along the river banks, a witness stated, "there was a very large quantity of very elegant timber…and extending back in somewhat detached quantities" (Gillespie 1860). He then identifies a grove of oaks near where the Capitol building stands today.

There also is evidence that willows may have formed dense thickets in places along the transition between forest and tule. One of the more explicit maps is one that shows land ownership in the vicinity of Brytes Bend just above the mouth of the American River (Fig. 5.60). Orchards are shown along the highest elevations (only one small patch of scattered trees remains), while a belt of apparently shrub-dominated vegetation spans the distance between orchards and tule.

Another example of conditions in the wetter, lower-elevation positions at the transition between forest and wetland comes from Willow Point at the tip of Babel Slough. Though sycamores and oaks were well established on the higher parts of this crevasse splay, this place name suggests that willow may have occupied the transition between the higher sycamore and oak forest and the lower wetlands. An article referring to a break in a Sacramento River levee just upstream of Babel Slough, reported that water from the break "will run out into the tule and willow swamps," also suggesting transitional conditions (*Sacramento Daily Union* 1892a). Given the position of the break, the water would have passed right around Babel Slough, so the willow swamps are likely a reference to willows at Willow Point.

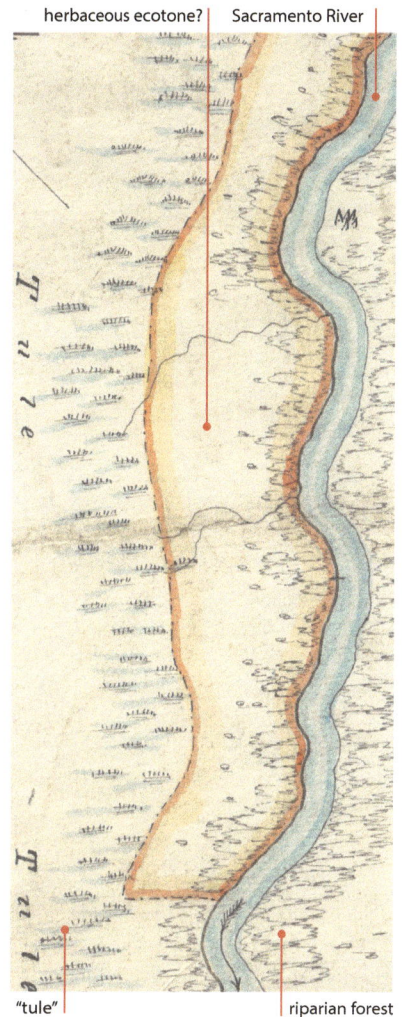

Figure 5.59. A possible herbaceous ecotone between riparian forest and tule is illustrated in this *diseño* of an unconfirmed land grant along the west bank of the Sacramento River, called Nueva Flandria. Dense forest is indicated directly adjacent to the channel, a few scattered symbols continue to the west into a blank area, presumably herbaceous cover or perhaps open oak woodland, before the "tule" is reached. (Bidwell 1844, courtesy of The Bancroft Library, UC Berkeley)

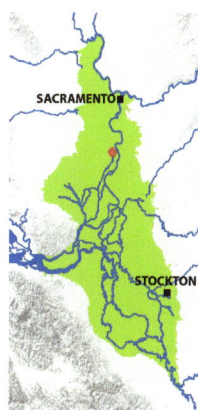

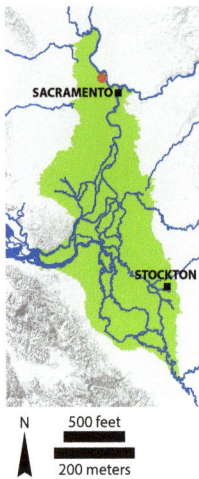

Figure 5.60. Dense scrub cover between oak forest and tule. This map depicts a growth of scrub, likely willow at the back of the natural levee lands which are covered in cultivated fields, and a few scattering trees. (Boyd 1895, courtesy of the California State Lands Commission)

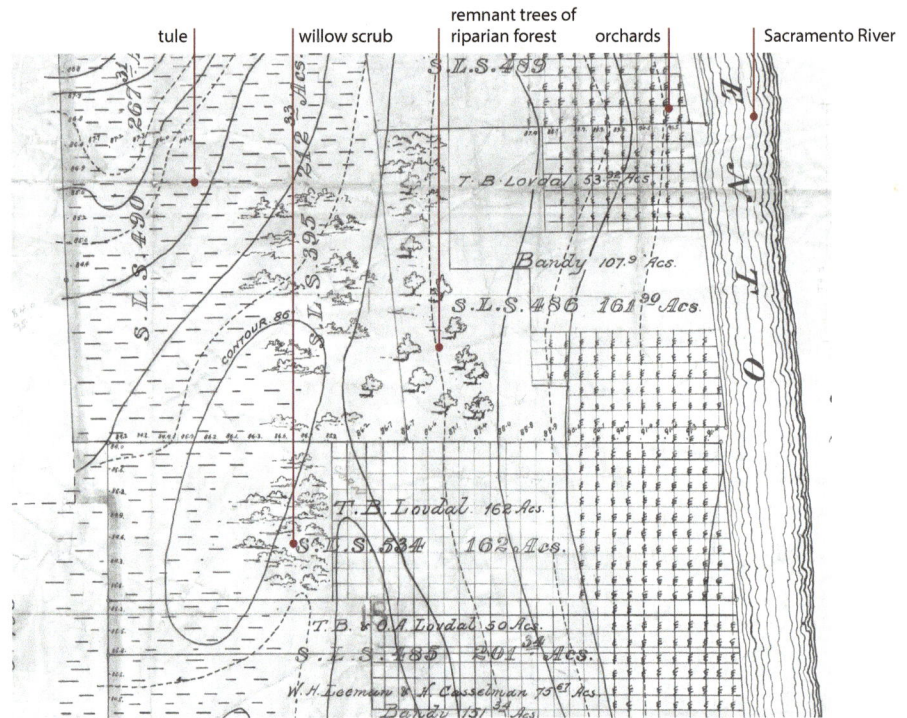

Variability within the riparian forest community occurred at more local scales. Relatively small areas were occupied by a wide range of species, but certain species were more dominant in some locales than in others. Underbrush was encountered in some places, while in others the forest was "clear of growth" (Leinhard and Wilbur 1941). Localized topographic shifts included the small overflow channels along the natural levees. With the complexity of the wetland edge, in some places travelers found "tule-covered marshes that often reached far back in the forest" (Leinhard and Wilbur 1941).

SINKS AT DISTRIBUTARIES

While most large rivers emanating from the Sierra Nevada fed directly into the Sacramento River due to their size and perennial snowmelt-fed flows, smaller rivers and creeks often spread into numerous distributary channels across their alluvial fans before dissipating into the wetlands alongside the river (Moerenhout [1849]1935, Bryan 1923). The area encompassing these distributary networks was known as a "sink;" as early narrative accounts describe the streams sinking or losing themselves in the tule (e.g., *California Star* 1848, *Sacramento Transcript* 1851a, Matthews in Houghton 1862, Flint 1860, *Sacramento Daily Union* 1862a, USGS 1909-1918, Derby and Farquhar 1932). These distributary environments were complex and dynamic places, where floods caused the abandonment of some channels, the formation of new ones, and transported sediment out onto the plain.

Sediment deposits were spread unevenly by the floods and distributary channels, forming localized ridges and depressions. In the summer, flows within the main channels became minimal and ceased in many cases. The sinks supported a dense growth of willows, cottonwoods, oak scrub, and other shrubs, as well as patches of emergent vegetation and seasonal wetlands. Perennial and intermittent ponds were also found within the sinks.

In the north Delta, four sinks were identified in the historical record. Three were found along the western edge of the Yolo Basin, where Cache Creek, Willow Slough, and Putah Creek met the basin's wetlands. The fourth, the Cosumnes Sink, was on the east side of the Sacramento River, within the lower extent of the Cosumnes River just before its confluence with the Mokelumne River. It is possible that other less extensive sinks were found within the north Delta at the base of smaller distributaries, though lack of water may have prevented true swamps from forming. The following discussion includes a few details concerning these larger sinks.

Putah and Cache creek sinks

For most of its extent, the western margin of the Yolo Basin tules adjoined seasonal wetland complexes within what was called "extensive plains" (Verix 1848), an "open plain," or "prairie" (Hays 1852b). The sinks of Putah and Cache creeks, and to a lesser extent perhaps Willow Slough, appear to have been the only extensive areas where the upland edge of the basin wetlands met dense woody undergrowth (Fig. 5.61). The sinks were distinctive features, identified in some of the earliest maps of the area (e.g., U.S. District Court ca. 1840a, Larkin 1848, Bidwell 1851, Eliason 1854, Henning 1871, De Pue & Co. 1879, Eager 1890). Several maps show only the spreading of the distributary channels within the sinks (Fig. 5.62); others depict the sinks using symbols indicating trees or shrubs (see Fig. 5.61). One narrative description likens the Putah and Cache creek sinks to the natural levee lands along the Sacramento River, grouping them together into the category of "made land" or depositional landforms where there is a "great growth of willow and 'underbrush'" (Sprague and Atwell 1870).

An early description of the sinks is found in the diary of fur trapper John Work (Maloney and Work 1943). In 1833, when traveling from Putah Creek

Below it are Grapevine Creek, Carter Creek, Sycamore Slough, Cache Creek and Putah Creek, which in Summer sink in the tule between the river and foot hills on the west, and in time of floods mingle their waters with the overflowing of the main river and debouche through Cache Slough and the tules at the foot of the Montezuma Hills.

- SACRAMENTO DAILY UNION 1862A

Figure 5.61. "Sink of the Rio Putas" is written at the base of the distributary network of Putah Creek where it enters the wetlands of the Yolo Basin. Neither of the two forks of the creek shown on this map are currently the primary route of the creek. In the 1870s, the main channel was diverted south. (Eliason 1854, courtesy of The Bancroft Library, UC Berkeley)

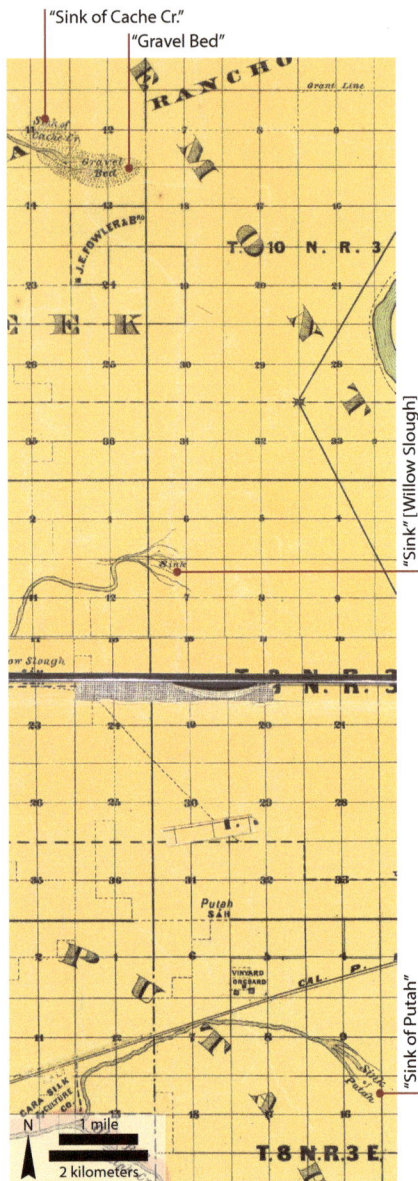

Figure 5.62. Cache Creek, Willow Slough, and Putah Creek sinks are illustrated in this 1871 county map as branching channels and gravel beds (in the case of Cache Creek) as the systems enter the Yolo Basin. (Henning 1871, courtesy of the Library of Congress)

toward the Sacramento River, "where the woods terminate and the country is plain," he became "entangled in such a thicket of willows and other bushes" in his attempts to cross. Testimony concerning Putah Sink identifies "a thick growth of timber principally oak" in the last mile and half of the creek before it spread into the "extensive marsh covered with bulrushes" (Eliason 1854). The descriptions of soils that comprised the areas of the sinks also contribute to the sense of dense woody undergrowth in the area, using such phrases as "overgrown with willow, cottonwood, and alder" for areas not already under cultivation (Mann et al. 1911). Histories of the area also recount the past existence of "thickets" at the terminus of Putah Creek (Larkey 1969, Vaught 2007).

Since these were fertile lands, settlers began occupying the area in the 1840s (Baca 1854; for a full treatment of this region's history, see Vaught 2007). Consequently, GLO surveys of the area from 1852 and 1862 reflect an altered landscape. A few hints are given as to the historical character of the sinks, however. Though boundaries were later disputed, the wetter portions of the sinks were defined as "swamp and overflowed land," suggesting that it was much wetter than areas such as the riparian forests along the Sacramento River (Dyer 1862a, Vaught 2006). GLO surveyor Robert Hays (1852b) noted a "thicket of briars and shrub oak" and described "oak and cottonwood timber" within Putah Sink. Only two bearing trees, one 12 inch (30.5 cm) diameter oak and one 24 inch (61 cm) diameter cottonwood are noted within Putah Sink. Also, a corner of an 1855 swamp and overflow land survey was stated as "standing in the timber" and "25 or 30 feet [7.6-9.1 m] from a large cottonwood" (Martin 1855).

Less is known about Willow Slough, historically known as the Laguna de Santos Calle, which extended north from Putah Creek, terminating between Putah and Cache creek sinks. It may have also received overflow from Cache Creek (Gregory 1913). It likely formed a similar complex mosaic of habitats to Cache and Putah sinks. Trees are shown lining the channel (Eliason 1854) and are referred to in histories of the area (*Pacific Rural Press* 1880). Also, within the testimony for an unconfirmed Mexican land grant in the area is a description of a lake that occupied the downstream extent of the slough:

> The lake called Laguna de Santos Calle…is a lake in the midst of a plain. When the water in it becomes high in the wet season, it overflows and runs into the tulares of the Sacramento. In a dry season there is no outlet to it on communication with the tulares—but there is always water in it. The lake is four or five miles [6.4-8 km] long, but not more than from fifteen to twenty varas [12.6-16.8 m] in width, but it is very deep. Its form is not straight but crooked. (Vaca 1853)

A history of Yolo County states that the perennial waters of Willow Slough are attributable to "a large cold spring," and along the slough "a succession of ponds or springs" (Gregory 1913).

The sinks were inundated during the wet season after large storm events caused overflow of channel banks, and continued receiving water through the late spring and early summer (Bryan 1867). A traveler to Putah Sink noted that the floodwaters "form a lake during the rainy season and as late as July, and it is only when this lake overflows that the water reaches the Sacramento" (Moerenhout [1849]1935). This flood-season lake is likely a description of overflow in the Yolo Basin as a whole, as it is a "lake some 40 miles long, and from 5 to 10 miles wide" described in a newspaper article as being formed by the annual flows Cache and Putah creeks (*Californian* 1848). One article suggested that the flows were usually only great enough to contribute to this broader Yolo Basin inundation for a few days after heavy rainfall (Flint 1860). As a general pattern, "in the wet season it runs into the tulares of the Sacramento, but in the dry season it does not reach as far" (Vaca 1853). In its lower reaches, the channel of Putah Creek contained very little or no water by the end of the summer (Moerenhout [1849]1935, Alexander et al. 1874, Hilgard 1884, Vaught 2006). However, groundwater levels were close to or at the surface year-round, supporting flourishing willows and other riparian-associated trees and shrubs. The sinks were broader and lower in position, and thus wetter, than the riparian forest directly adjacent to the channels extending upstream. They were also set apart from the easterly Yolo Basin, occupied predominantly by tule.

The disturbances caused by the flooding regime undoubtedly contributed to habitat complexity within the sinks. Vegetation communities would have been found at different successional stages within the matrix of branching channels that constantly changed their course and the intermittent and perennially flooded depressions. This non-static environment (promoted by sediment contributions from creeks originating from the fragmented Franciscan formation of the Coastal Range) made the land fertile ground for cultivation, but at the same time challenging to settle and define boundaries of ownership (Vaught 2007).

To map the extent of the sinks, we drew primarily on floodplain soils mapped by early soils surveys (Fig. 5.63) and a few general county maps that depicted the extent of the distributary networks. We mapped about 6,300 acres (2,550 ha) of willow thicket and another approximately 2,100 acres (850 ha) of willow riparian scrub or shrub (a slightly drier habitat type; i.e., no emergent vegetation within the mix, more closely associated with channels) within Putah Creek Sink, and less than 1,000 acres (400 ha) for Cache Creek. The lack of spatially explicit vegetation boundaries in sources means these area estimates are approximate: the extent may have been half as small or much larger in extent. We mapped less than 1,000 acres (400 ha) of willow thicket within the sinks of Willow Slough as well, though we assigned these features with lower confidence levels in interpretation and size given mapping sources and lack of multiple early descriptions and maps of the area. Though the sinks were distinctive features within the landscape, exact boundaries are impossible to define. The sinks should be thought of as a continuum from riparian forest lining the banks of the main channels to willow and other underbrush within the matrix of distributary channels to the tule-dominated wetlands of the basin.

Both of these creeks [Cache and Putah] are sediment-bearing and deliver vast quantities of detritus into the tule basin every year.

—*SACRAMENTO DAILY UNION 1892B*

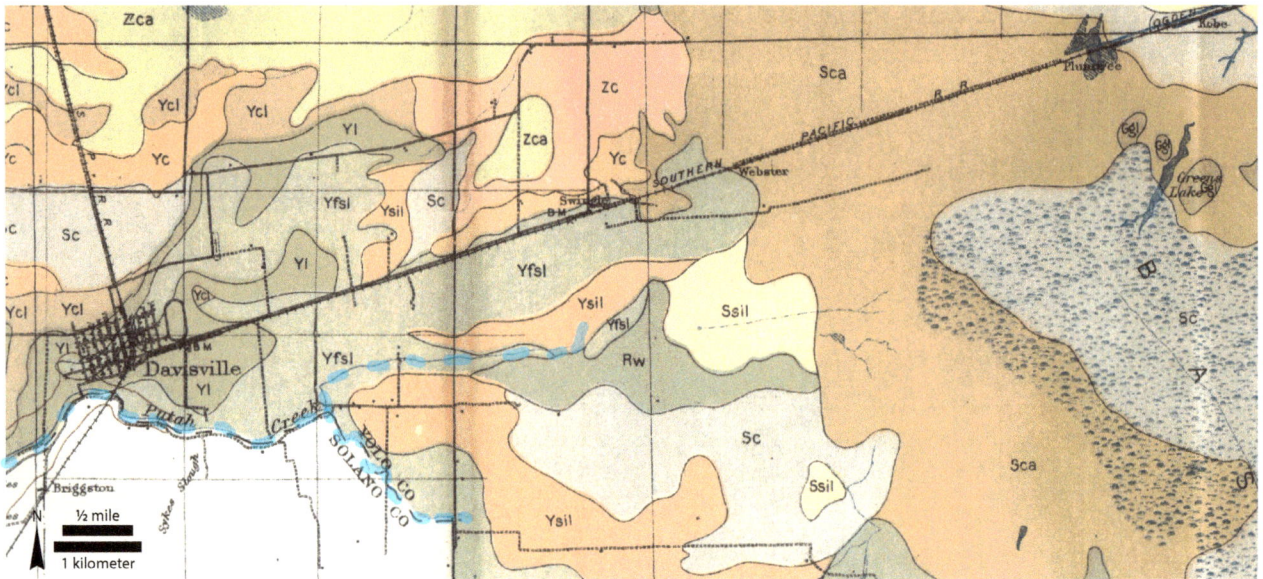

Figure 5.63. The alluvial soils associated with the Putah Creek and its distributaries helped establish the boundary between the more wooded sinks and emergent vegetation of the Yolo Basin (main forks illustrated with dashed blue lines; see Fig. 5.61). These soils (gray-green colors) include the Yolo fine sandy loam (Yfsl), Yolo loam (Yl), Yolo silt loam (Ysil), Sacramento heavy clay (Sca), and Riverwash (Rw) soil types. (Mann et al. 1909)

Mapping the numerous distributary channels also was extremely challenging given the dynamic nature of the area, where channels were constantly shifting positions. The most significant change was the re-routing of Putah Creek in 1871 to the south of the town of Davis into the channel it occupies today (*Sacramento Daily Union* 1892b, Dunn 1915, Vaught 2006). Early histories indicate that this diversion was not an intentional act by settlers, though early ditching efforts in the area likely contributed to the creek seeking a new path. Though we were able to map the primary channel orientation that existed prior to that date, smaller channels are mapped mostly from later sources (e.g., early 1900s USGS topographic maps, 1937 aerial photography), so the pre-1871 orientation of smaller distributary channels is uncertain. Given confirmation from general descriptions and maps of the area prior to this date and a few GLO notes of slough crossings in the 1850s (Hays 1852b, Von Schmidt 1858a), we are confident that the mapping represents the overall landscape patterns of the early 1800s.

Cosumnes Sink

The Cosumnes Sink comprised the lower extent of the Cosumnes River. It shared similar characteristics with the sinks of the Yolo Basin: swamps were laced with myriad distributaries and flooded annually. However, the Cosumnes is a much larger river than Putah and Cache creeks, with different hydrologic and geomorphic variables. For one, rather than spreading and dissipating into freshwater emergent wetlands associated with the Sacramento River, the Cosumnes River's distributary channels coalesced again into a single channel that directly fed into the Mokelumne River.

One of the most detailed pre-reclamation maps showing the Cosumnes Sink illustrates this plexus of channels branching and converging (Fig. 5.64). Testimony from the same land case for which this map was an exhibit contains descriptions of the river here: "It has a distinct channel at low

Cosumnes River

Mokelumne River

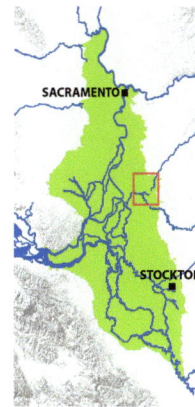

Figure 5.64. The many branching channels of the Cosumnes River at its sinks just a few miles above its confluence are elegantly depicted in the land case map. Unlike distributary networks within the sinks of Cache and Putah creeks, these channels coalesce again before the river flowed into the Mokelumne River. (Von Schmidt 1859, courtesy of The Bancroft Library, UC Berkeley)

water, it spreads out on both sides in the wet season" (Gray 1859). Another witness put it slightly differently, stating:

> The river is lost at the point marked Indian Rancheria so far as relates to its having a distinct channel…The waters spread out during the time of freshets in the wet season and for several months in the year would cover all the part that is colored blue on Exhibit A. (Sherman 1859)

Large and small, perennial and intermittent, channels were present in the Cosumnes during the dry season. Witnesses confirmed that though water could always be found, flow was not always detectable throughout the length of the Sink (Gray 1859, Sherman 1859). Several general early maps show two main branches of the river (e.g., Boyd 1903). Based on GLO surveys, it appears that what was considered the main channel of the Cosumnes is now a series of remnant, functionally disconnected sections of a former waterway about one mile west of the present river and south of Lambert Road.

The distributary networks were a part of the locally variable topography which caused the bottomland to experience varying degrees of inundation frequency, duration, and depth. The larger and more well defined lakes and ponds occupying the small drainages at the edge were important perennial features, and some small ponds and secondary distributary channels maintained standing water through the summer months. Another witness noted, "in the dry season there is a string of lakes connected with each other…These lakes are always full of water. The channel of the stream runs through these lakes" (Gray 1859).

The Cosumnes Sink was an extensive bottomland of "wooded sloughs" (Fremont 1845) and "dense thickets" (Taylor 1854), forming the lower floodplain of the Cosumnes River. This riverine landscape supported swamps of willow, cottonwood, oak, blackberry, wild rose, and wild grape (Cook and Heizer 1951), along with emergent vegetation. A general description of the region by the GLO survey compares the overflow land of the Cosumnes and Mokelumne with that of the Sacramento: in contrast to the tule-dominated marsh along the Sacramento, there was "a wide belt" along the Cosumnes "of a thick and almost impenetrable swamp of tules and willows" (Wallace 1869). Surveyor Edwin Sherman, a witness in the Mokelumne land case, described the overflow land along the lower Mokelumne and Cosumnes rivers as "dense thicket, willows, brush, tules and grass, where the grass now grows was tule when I surveyed it" (Sherman 1859). This quote concerning conversion of tule to grass indicates the early and rapid conversion of the landscape in certain locales to drier conditions.

Evidence also suggests that, in contrast to Cache and Putah sinks, the Cosumnes Sink was occupied by a greater proportion of emergent vegetation; one map labeled "dense tules" within the sinks (see Fig. 5.64). In reference to the boundary line shown on the map north of Dry Creek, a witness testified that "fully one half if not two thirds of the land through which the line passes southerly and westerly is covered with tule" (Sherman 1859). The transition from swamps or willow thickets predominantly in the northern part to a more tule-dominated marsh in the south was gradual, and it is certain that willow and tule were found scattered in patches throughout associated with local topographic and hydrologic gradients. We chose to map that transition between two GLO survey notes: the note to the south reads "to tule," and the one to the north includes a description of "dense thicket of willow and tule" for the section mile (Wallace 1869). Other field notes mention "timbered" sloughs for a number of the small drainages that feed into the Cosumnes Sink from the east. Elevations and several accounts indicate that the southernmost extent may have been subject to tidal influence (Gray 1859, Sherman 1859).

The freshwater emergent wetlands mapped within the Cosumnes Sink would have been interspersed with patches of willow and other woody vegetation following channels and other topographic gradients, while parts of the willow thickets would have also been occupied by emergent vegetation. Throughout were numerous ponds that we were unable to capture in the mapping due to the spatial resolution of sources consulted.

UPLAND ECOTONE

The valley plains that met the wetland basins along the Sacramento River were wet in the winter, often overflowed by numerous small drainages, bloomed in brilliant shows of wildflowers in the spring, and became dry in the summer. The valley was generally devoid of trees, which contrasted with the deep green riparian forests and woodlands bordering the larger streams and rivers (Fig. 5.65). The gently sloping valley gradually adjusted its character over broad expanses. The tules thinned as elevations increased at the edge, giving way to seasonal wetland complexes that experienced spatially and temporally variable inundation patterns. On the west side, along the Yolo Basin edge, vernal pool complexes were common within other seasonal wetlands and grasslands – what early travelers referred to as a treeless plain. On the eastern edge, seasonal wetlands merged into higher-elevation grasslands. Though oaks were found scattered about the plain and in groves in localized areas, expansive oak woodlands and savannas, like those extending south from the Mokelumne River, are not evident in the historical record. Gradients in hydrologic, topographic, and soil characteristics produced spatially and temporally variable inundation patterns across the landscape, supporting mosaics of seasonal wetlands, grasslands, oak woodlands and savannas, as well as occasional perennial ponds or wetland patches (Fig. 5.66).

Seasonal wetland complexes at the basin margins

The broad extent of seasonal wetland complexes along much of the north Delta wetland margin faced great extremes. Within heavy clay soils, vegetation had to withstand periodic inundation as well as extreme seasonal drought. Typical descriptions of the seasonal wetlands drawn from GLO surveyor field notes include "dry plain," "land prairie…dry and baked," "meadow prairie," "low prairie," "open prairie," and "no timber" (Hays 1852a,b, Jones 1855, Lewis 1859d). On the eastern margin, along the Sacramento Basin, this is broken in a few instances by mention of "a few scattering oaks" (Jones 1855). Earlier accounts describe the land south of Putah Creek as "an extensive plain" (Verix 1848), and the land south of Sutter's Fort as "a dry level plain without timber or grass" (Clyman and Camp 1928[1848]).

Seasonal wetland complexes were also quite variable in character at the local scale, following small-scale changes in hydrology, topography and soils. The land was temporarily or seasonally flooded by the intermittent streams that lost definition before reaching the wetlands. As a general

I might state that it is an open champaign country, cut on the east side of the river by numerous beautiful tributaries skirted with timber, and on the west dotted and striped with groves and lakes.

—FARNHAM 1857

On the lower prairie are here and there small lakes or ponds, some of which are supplied by streams and others are stagnant. These are surrounded by a thick underwood interwoven with vines, and being sunk many feet below the surface, render it difficult to obtain the water. There are occasional deep and dry gulches, which are filled by water-courses during the rainy season. Towards the latter part of the dry season (September and October), the lower prairie becomes rent in many places by the continued drought.

—WILKES 1849

Figure 5.65. A sketch of the open plain in the vicinity of Sutter's Fort, looking east toward the Sierra Nevada. Farther south, the plains met the wetlands of the Sacramento Basin. (courtesy of The Bancroft Library, UC Berkeley)

Figure 5.66. A small pond lined with trees among seasonal wetlands. Bare patches indicative of seasonal flooding and possibly alkali can be seen in the foreground. (courtesy of the Center for Sacramento History, Eugene Hepting Collection, 1985/024/5558)

pattern, the land lying closest to the tule margin was also overflowed by extreme flooding of the Sacramento. Some of the more frequently inundated portions were considered to be "swamp and overflowed land" by the GLO surveyors (see Box 2.3). These places that were subject to overflow, but not occupied by tule, were defined by some as the place where "the grass is coarse" (Robinson 1860).

An important component of topographic variability at the local scale on the western edge came from what geomorphologist Kirk Bryan (1923) termed "channel ridges." Bryan distinguishes these from more typical alluvial fans, though both landforms originated through depositional processes as streams entered valley floors. Instead of a fan shape, the channel ridges built up with the bed of the stream until the elevation was too high and the stream broke out of its bed to find a lower course. The Yolo Basin edge, Bryan concluded, was "so largely made up of branching and interlacing channel ridges that they form a distinct type of alluvial slope which may be called a channel-ridged plain." The topographic complexity associated with this type can be seen in the early 1900s USGS maps as well as historical aerial photography, which resulted in the dense, generally parallel network of intermittent streams mapped along the western edge (Fig. 5.67; see Fig. 5.3; USGS 1909-1918, USDA 1937-1939). This contributed substantial local-scale complexity at the wetland margin, what Bryan described as a "crenulated border."

Perhaps more than with the extensive tule-dominated basins, characteristics of the seasonal wetland complexes varied substantially. The boundary between perennial and seasonal wetlands was not a smooth line: patches of tule were found within seasonal wetlands and vice versa. An example of this comes from detailed descriptions in Sutter land case testimony concerning the native vegetation patterns where Sacramento now stands. Scattered patches of tule were found just below the city (Colby 1860). Tule also continued as a narrow strip "as high as Q street" (Keseberg 1860) and "tongues of tule coming up in some cases as far as K street" (McClatchey 1860). Also indicative of the complexity of the edge is testimony concerning "short tule mixed with grass was about a foot high, and rather more grass than tule" at this transition (Keseberg 1860), which is also mentioned by other witnesses (Rhoads 1860). Drier expanses more characteristic of grasslands were covered in annual forbs (Sanford 1860) and perennial needlegrass (Kyburg 1860).

ALKALI IN THE NORTH DELTA In contrast to the alkali seasonal wetland complexes that were common along the wetland margins to the south, there is less evidence that such a pattern continued in the north Delta. The GLO survey makes no mention of alkali north of the Mokelumne River or above the Delta mouth on the west. Also, strongly alkaline soils were not recorded in early twentieth century soil surveys for the eastern edge of the Yolo Basin, though a few types are described as containing "more or less alkali" (Carpenter and Cosby 1934) or "slightly affected with alkali" (Cosby and Carpenter 1932). In some cases, the lesser amounts of alkali is attributed to regular flushing provided by overflow (Holmes and Nelson 1915). Alkaline soils were not absent, however; a few soil types in the larger Sacramento Valley survey that fall within the study area do mention extensive areas affected by alkali (Holmes and Nelson 1915). Also, vernal pools are characterized by alkali, given their isolated and evaporative conditions, but rarely is the entire soil type recognized as strongly alkaline. Also, a few narrative accounts suggest alkali in some locations. For an area north of Putah Creek, the witness describes "an open plain whose general character is that of low wet salt land, and is of but little value" (Eliason 1854). Similar to the relatively narrow strip of alkali land positioned along the freshwater wetland margin to the south, a description of Yolo County notes:

> Where the grain lands join the tules the quality of the soil is frequently very different from that which lies but one section further inland. A narrow belt of lands, often strongly impregnated with alkali, generally unites the two divisions. (Sprague and Atwell 1870)

Early grazing and reclamation may have caused conversion of species at the upland ecotone. Dominant species likely converted to more disturbance tolerant and less palatable species. Also, retreat of the tule boundary associated with grazing could have supported the development of this band of alkali (see Box 5.1).

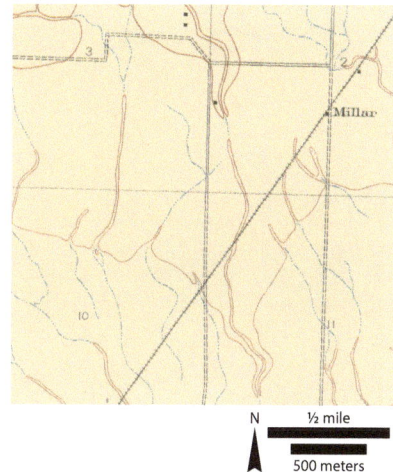

Figure 5.67. Channel ridges along the plain to the west of the Yolo Basin are shown in this 1915 USGS topographic map. (USGS 1909-1918)

Figure 5.68. Vernal pool within the Jepson Prairie Reserve. (photo by Marc Hoshovsky)

The herbaceous plants on the plains are chiefly annual…The wide plain is covered with showy Lupines, Clovers, Calandrinias, Platystemons, Baerias, Gilias, Nemophilas and Allocaryas. The shallow streams and pools are edged with handsome Eunani and curious Bolelias. The tide of plant life reaches its maximum from April 5 to 20. In one, two, or three weeks more the brilliant colors have faded from the landscape and the vernal aspect is succeeded by the dullness and aridity of summer. For months there is nothing to be seen but the grass-whitened plain, only later relieved by tufts of Grindelia and broad areas of the exclusive Hemizonias.

—JEPSON 1893

JEPSON PRAIRIE VERNAL POOLS Vernal pools were found along the edges of the valley troughs, areas of small mounds and depressions typically described as "hog-wallow" land. Extensive research has been conducted to describe and map these ecologically significant habitats (e.g., Keeler-Wolf et al. 1998, Holland 1976, Holland 1978, Holland 1998, Smith and Verrill 1998). The Jepson Prairie Reserve in Solano County is one of the few remaining areas of vernal pools and associated habitats in the Sacramento Valley. Here, one can see the splendor of the spring wildflower shows for which the valley was known (Fig. 5.68). The vernal pools support a number of rare and endangered species endemic to these unique habitats. Within a landscape that was, aside from a few trees, "wholly herbaceous and the herbs are mainly annuals," a whole host of species were found, including popcorn flower (*Plagiobothrys* spp.), *Downingia* spp., button celery (*Eryngium* spp.), *Navarretia*, woolly marbles (*Psilocarphus* spp.) and goldfields (*Lasthenia* spp.; Gregory 1912). As a history of Solano County summarized, these "little vernal pools that have no outlet" slowly dried out and "their contracting margins support in succession a number of peculiar plants" (Gregory 1912).

The landscape late in the season was quite different. As two GLO surveyors reported in September 1852 and 1853, the annual summer drought left dried-out ponds (a few remained with standing water) and soil with very little or no vegetation (Hays 1852a, Denton 1853). In two places, surveyor Denton simply noted "bad land." Describing the vegetation, he states that the land "produces oats, bunchgrass and clover," and in a few places notes a "growth of weeds and small grapes." He also mentions that the land was "interspersed with small hills," which addresses the microtopographic relief characteristic of vernal pool complexes.

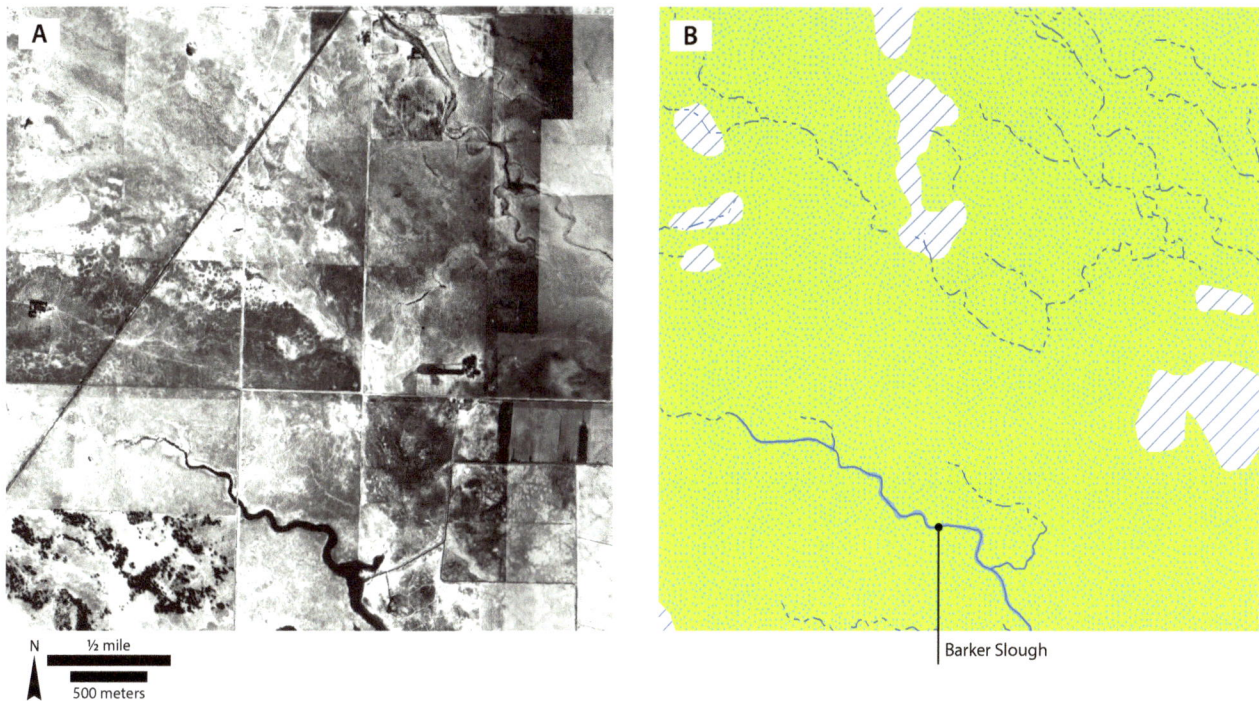

Figure 5.69. Vernal pools near the Jepson Prairie Reserve in Solano County, west of Cache Creek Slough. The 1937 aerial photography shows the signature light and dark pattern of the vernal pool complex. A few of the defined larger seasonal pools (shown in hashed marks in B) can be seen in the aerial photography. Barker Slough, a tidal channel, is seen extending into the area from the southeast. (USDA 1937-1939)

The Jepson Prairie Reserve's area is but a fraction of a larger area that historically extended northward (within the study area, the mapped area – largely based on soil boundaries – is over 25,000 ac/10,120 ha). This area encompassed a wide range of conditions: a matrix of vernal pools, wet meadows, grasslands, and intersecting drainages was found throughout. Within the study area, we mapped 11 vernal pools above the size of five acres (2 ha; classified as intermittent ponds within vernal pool complex), many of which can be seen in the landscape today (Fig. 5.69). Numerous smaller pools can be identified in aerial photography.

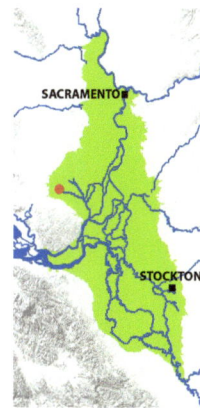

Hawkin's Point: the ridge of New Hope Tract

What was known as Hawkin's Point, extended east from where New Hope now stands and was one of the few places where upland habitats extended far into the tidal wetlands. Hawkin's Point contributed to the complexity at the wetland edge, increasing opportunities for species to seek refuge during floods or to access the marshes (functions primarily served by the riparian forests). This approximately three mile (4.8 km) long and 0.5 mile (0.8 km) wide point of land can be identified in the early 1900s USGS topographic maps, as well as in the recent LiDAR survey, by elevations about five feet (1.5 m) above the surrounding land (Fig. 5.70, Watson 1859b). The intermittent streams that wound along this and another ridge of land extending north from New Hope suggest that these were former routes of the Mokelumne River.

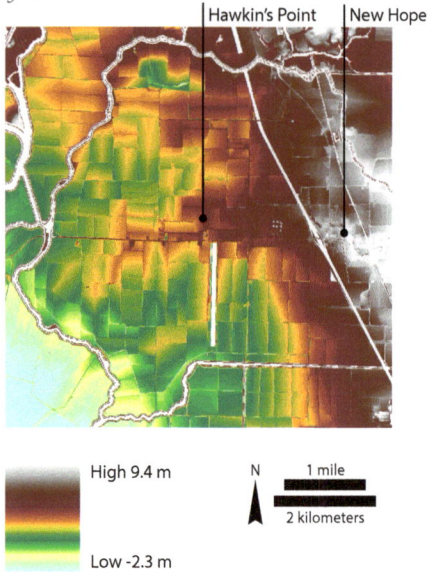

High 9.4 m

Low -2.3 m

N

1 mile
2 kilometers

Figure 5.70. Hawkin's Point in LiDAR imagery is seen as elevated land protruding into the lower lying land to the south and east of the Mokelumne River. (CDWR 2008)

A GLO survey crossed the eastern portion of Hawkin's Point, and where a single oak bearing tree was noted, the surveyor described as "a narrow neck of high land, with a few scattering oak trees, and tule on each side, bears off to the southwest" (Von Schmidt 1858b). Depositions from the Mokelumne land grant case as well as a map produced as an exhibit in the case provide rich descriptions of this point of land. This evidence discusses the fact that this was the only place along the western margin of the land grant that did not overflow (Gray 1859). One witness traveled as much as two miles out along the ridge in a buggy, finding that the higher portions of the ridge were "covered with oak timber" (Sherman 1859). These descriptions support the map from the case, where the words "ridge of land covered with grass and clover" are written along Hawkin's Point, with tree symbols extending for a portion of the distance (see Fig. 5.53). Vegetation cover varied along the elevation gradient, where herbaceous species occupied the intermediate elevations between oaks and tule. The neck of land apparently became quite narrow in the last mile or two of its extent: "not much wider than a trail" (Thayer 1859). The edge was not smooth; at the tip, it "breaks up into detached portions" (Sherman 1859) or "breaks out into little knolls" (Thayer 1859).

SUMMARY

In the south Delta, tidal wetlands transitioned southward to non-tidal wetlands associated with the three main branches of the San Joaquin River. Secondary channels left the river, providing direct flow onto floodplains during high water. Willows along channels, seasonal and perennial lakes and ponds, and seasonal wetland patches made up the matrix of the south Delta.

Floodplain morphology (page 313) • The south Delta floodplain was more topographically complex than the tidal wetlands downstream, owing in part to riverine floodplain features such as oxbow lakes, former channels, and natural levees. The natural levees of the San Joaquin River followed the similar pattern as the Sacramento of decreasing width and height downstream, but were lower and narrower overall (page 316).

Seasonality and flow (page 319) • The San Joaquin River spread across its floodplain during high flows, which tended to be later in the season than on the Sacramento, related to the greater relative influence of snowmelt on the San Joaquin hydrograph (page 320). Floodplain wetlands were often increasingly overflowed through mid-July, though the landscape dried out by late in the season in non-tidal areas (page 321). In times of lowest water, the San Joaquin was fordable near the Interstate 5 crossing today (page 325).

Channels at the fluvial-tidal interface (page 326) • As tidal influence decreased upstream, the width of the three main San Joaquin River branches decreased (page 327). Though the wetland plain was above tidal elevation, the rivers and many of its larger secondary channels were tidal. Secondary channels branching off the river upstream met the southern extents of tidal channels within the wetland (page 333). Overall, secondary channels can be distinguished by whether they ended within wetlands, connected to a tidal channel or returned to the river a short distance downstream (page 336). Upstream, the river was more spatially mobile, with evidence of the river meandering visible in aerial photography and maps (page 342).

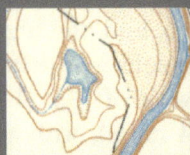

Landscape position and character of lakes and ponds (page 346) • Lakes and ponds in the south Delta were usually positioned within a non-tidal wetland and connected to the river by one or more secondary channels, which sometimes connected to other lakes and ponds (page 348). Though some dried out in the summer, many were perennial (page 350).

Complexity within the wetland plain (page 351) • The floodplain reflected the transition from tidal to fluvial dominating processes, incorporating a variety of features: secondary channels, ponds, seasonal wetland patches of grasses and sedge species, and willow thickets (page 352). Peat deposits thinned at the edge of tidal influence, in upper Union and Roberts islands (page 354). Upstream, the floodplain was comprised of a greater component of willows and even larger trees such as oaks (page 356).

Riparian forest characterisitics (page 357) • As channel banks decreased in height downstream, riparian vegetation transitioned from a forest of tall trees to one dominated by dense willow scrub (page 360). Riparian forest width increased upstream to well over 1,500 feet (457 m) in some locations, usually at the inside of meander bends (page 362). Oaks and willows were the dominant species (page 364). The riparian forest contributed woody debris to the river, causing large obstructions in particular locations (page 366).

Wildflower fields and alkali meadows (page 370) • Sandy soils supporting herbaceous species intersected the alkali seasonal wetlands along the southeastern Delta edge (page 371). Though the plain became quite dry later in the season, a variety of wildflowers could be seen in the spring.

INTRODUCTION

The south Delta is defined by the distributaries and meanders of the San Joaquin River upstream of the central Delta. At the landscape scale, the south Delta historically presented an array of tidal wetlands interwoven with distributary riverine channels and non-tidal floodplains across a broad transitional zone, or ecotone. Early travelers encountered rivers that were fordable only late in the season, often with dense willow and oak riparian forest along their banks. Beyond forested natural levees, the land surface sloped away to meet a matrix of perennial wetlands (dominated by tule, *Schoenoplectus* spp.), patches of sedges and grasses, perennial and intermittent ponds, and overflow channels (Fig. 6.1). This floodplain was challenging to traverse for much of the year, owing to annual inundation. This chapter discusses roughly 120,000 acres (48,560 ha) that once comprised an extensive mosaic of wetlands and adjacent upland habitat types of the south Delta, generally defined as extending from upper Roberts and Union islands to the Stanislaus River (Figures 6.2 and 6.3).

The south Delta was unlike the fluvially-dominated upper San Joaquin River and unlike the tidally-dominated Delta, though elements of both landscapes were found. Moving upstream from the tidal central Delta, peat soils were replaced by clay loams as surface elevations gradually increased above tidal levels in upper Union and Roberts islands. Though the floodplain's distributary channels were formed primarily by fluvial processes, many were also subject to tidal flows at least part of the year. At the landscape scale, the channels presented a recognizable deltaic planform.

The south Delta marked the terminus of the San Joaquin River, a large riverine system that frequently overflowed its banks to fill numerous secondary channels, ponds, and floodplain wetlands. It conveyed floodwaters that spread and inundated land sometimes several feet in depth before much of it entered downstream tidal channels in the central Delta. In contrast to the more rainfall-event driven hydrograph of the Sacramento River, winter floods were less frequent on the San Joaquin, with flooding typically snowmelt-driven. The resulting hydrograph was characterized by fewer peak flood events and exhibited a gradual rise of river stage in the late spring and early summer (Young 1880, TBI 1998). Also different from the northern flood basins, the south Delta floodplains were apparently less isolated from the river by natural levees (presumably related, in part, to the lower flood peaks and sediment supply in comparison to the Sacramento River). This greater hydrologic connectivity was maintained through multiple side channel systems that made floodplain hydrology more responsive to river stages and enabled water to pass through the system with relative speed. Masses of woody debris obstructed the main channels at certain locations, such as Old River near present-day Fabian Tract, affecting flows and habitat complexity. The combination of these factors meant that floodwaters in

We passed during the afternoon several tule marshes, with which the plain of the San Joaquin is dotted. At a distance, the tule of these marshes presents the appearance of immense fields of ripened corn.

—BRYANT 1848

Figure 6.1. South Delta views. (chapter title page) Top, the San Joaquin River at Dos Reis County Park is seen at dusk on March 30, 2011. Bottom, the photograph shows the inundated San Joaquin River floodplain at Durham Ferry Road Crossing in 1938, perhaps from the floods that occurred in February and March of that year. (top: photo by Alison Whipple 2011; bottom: Covello 1938, courtesy of Bank of Stockton Historical Photograph Collection)

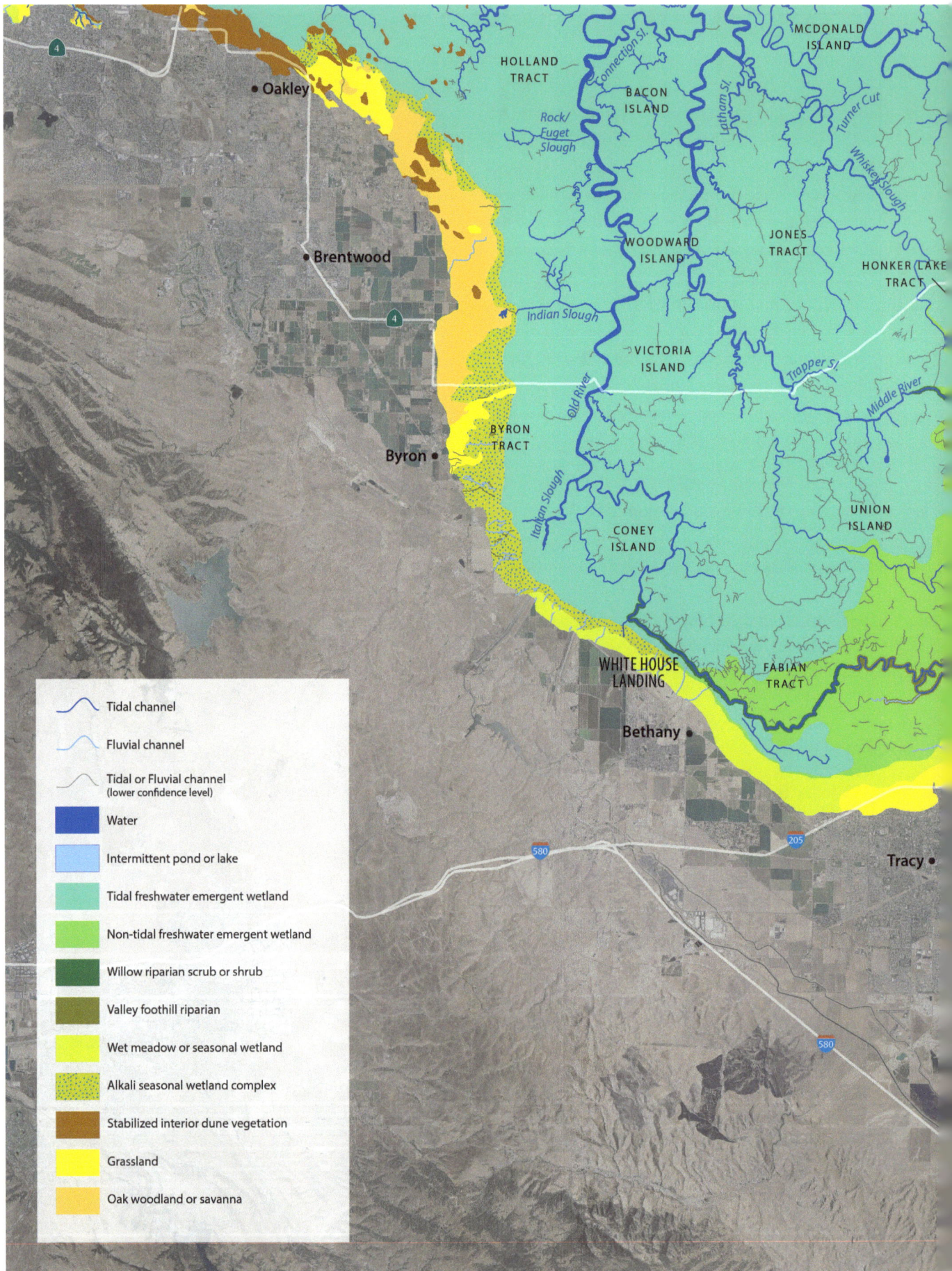

Legend

- Tidal channel
- Fluvial channel
- Tidal or Fluvial channel (lower confidence level)
- Water
- Intermittent pond or lake
- Tidal freshwater emergent wetland
- Non-tidal freshwater emergent wetland
- Willow riparian scrub or shrub
- Valley foothill riparian
- Wet meadow or seasonal wetland
- Alkali seasonal wetland complex
- Stabilized interior dune vegetation
- Grassland
- Oak woodland or savanna

Map labels: Oakley, Brentwood, Byron, Bethany, Tracy, HOLLAND TRACT, BACON ISLAND, McDONALD ISLAND, Connection Sl., Rock/Fuget Slough, Latham Sl., Turner Cut, Whiskey Slough, WOODWARD ISLAND, JONES TRACT, HONKER LAKE TRACT, Indian Slough, VICTORIA ISLAND, Old River, Trapper Sl., Middle River, Italian Slough, CONEY ISLAND, UNION ISLAND, WHITE HOUSE LANDING, FABIAN TRACT, BYRON TRACT

Figure 6.2. Distribution and extent of habitat types within the south Delta in the early 1800s. Over a broad ecotone, the landscape transitioned from a single meandering San Joaquin River channel with associated riparian forest and non-tidal floodplain wetlands to an

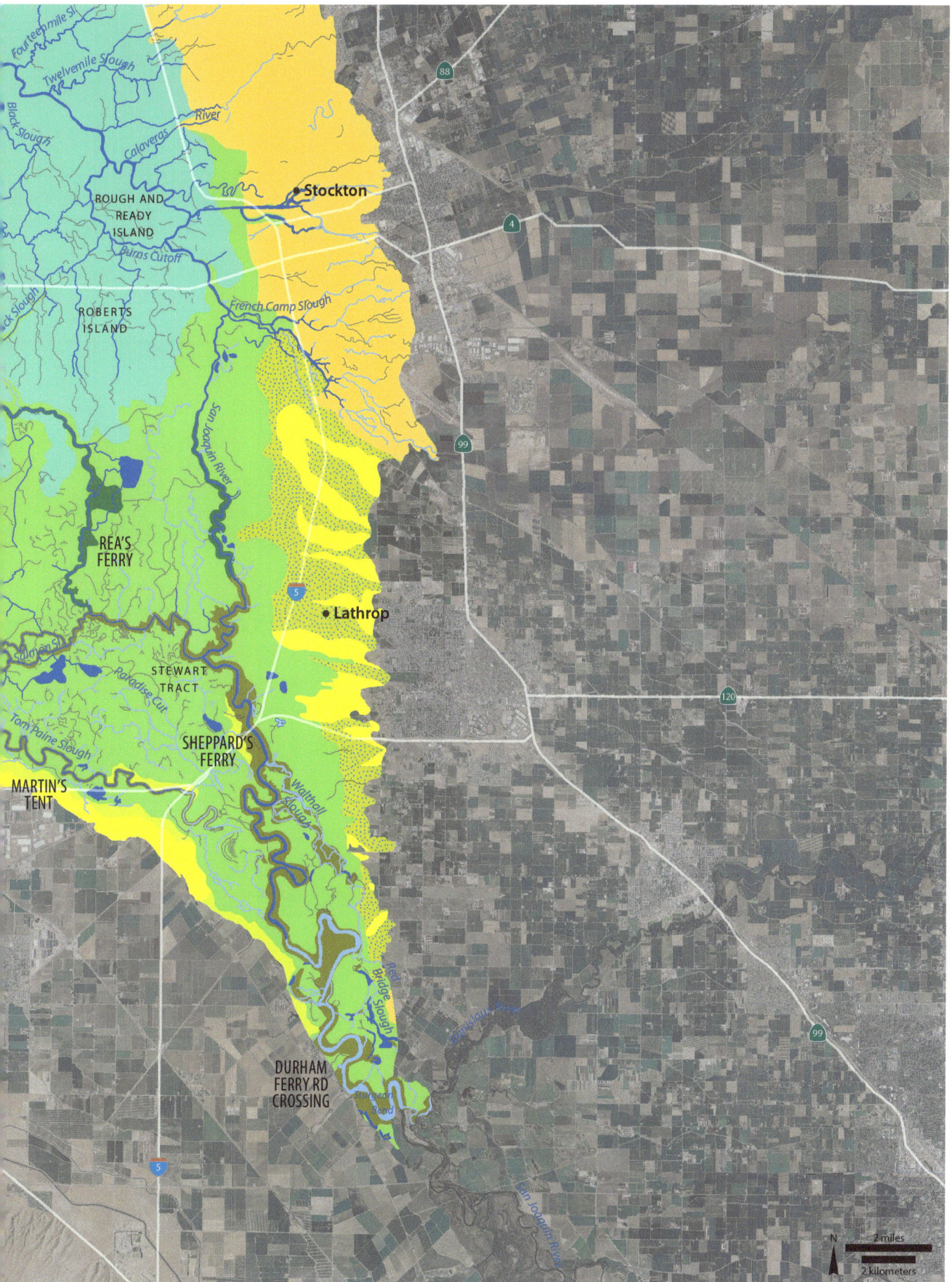

Fourteenmile Sl.

Twelvemile Slough

Black Slough

Calaveras River

River

88

ROUGH AND
READY
ISLAND

Buras Cutoff

● Stockton

4

Black Slough

ROBERT'S
ISLAND

French Camp Slough

99

San Joaquin River

REA'S
FERRY

5

● Lathrop

Salmon Sl.

STEWART
TRACT

Paradise Cut

120

Tom Paine Slough

SHEPPARD'S
FERRY

MARTIN'S
TENT

Walthall Slough

DURHAM
FERRY RD
CROSSING

Red Bridge Slough

Stanislaus River

99

San Joaquin River

5

N   2 miles

2 kilometers

extensive distributary network that met the broad tidal-dominated wetlands of the central Delta. Modern as well as historical place names are included on the map for reference throughout the report. Modern imagery is included for context. (USDA 2009)

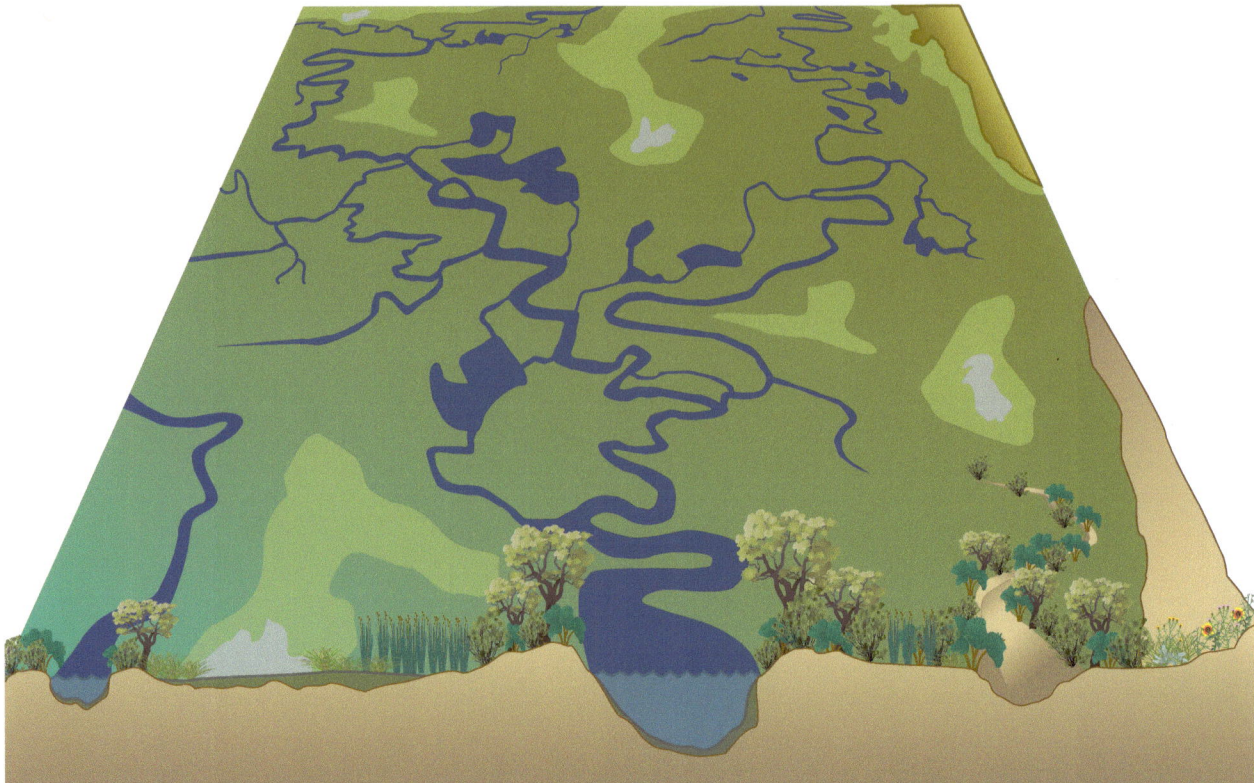

Figure 6.3. Conceptual diagram of the south Delta distributary rivers landscape. In the south Delta, the three distributary branches of the San Joaquin River drove the general pattern of the landscape. From these branches, numerous secondary overflow channels accessed the floodplain, which broadened quickly downstream and merged gradually into tidal wetlands. Patches of different habitat types were interspersed within the emergent wetland, including willow thickets, seasonal wetlands, grasslands, as well as perennial and seasonal lakes and ponds. The relative historical proportions of habitat types based on the map are illustrated in the pie chart.

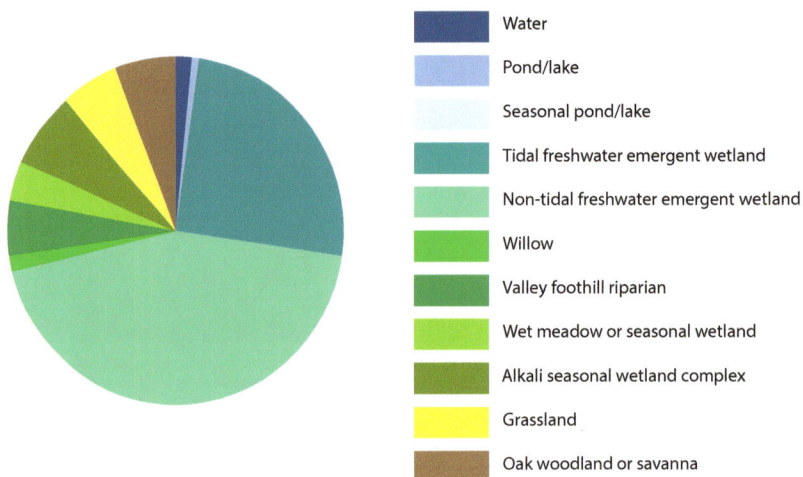

- Water
- Pond/lake
- Seasonal pond/lake
- Tidal freshwater emergent wetland
- Non-tidal freshwater emergent wetland
- Willow
- Valley foothill riparian
- Wet meadow or seasonal wetland
- Alkali seasonal wetland complex
- Grassland
- Oak woodland or savanna

the south Delta were routed and channelized differently from those in the north Delta.

The south Delta floodplain vegetation patterns were likely more variable at a local scale than the basins of the north Delta. Over five miles southeast from the vicinity of Bethany, Spanish explorer Viader described "oak groves, willow thickets, ponds, and lands flooded during the freshets" in 1810 (Viader and Cook 1960). Tule dominated the freshwater emergent wetlands and was most extensive toward the lower elevation and more tidally-influenced portions of the landscape. Willows and oaks became more common along natural levees that increased in height upstream. Particularly in the vicinity of the Stanislaus River, woody vegetation was also found within the floodplain wetland complex, likely associated with the many secondary side channels and oxbow lakes. In comparison to the lower Sacramento River riparian forests, a greater proportion seems to have been composed of willows and other shrubs, as opposed to oaks and sycamores.

Related to this rich habitat diversity, the secondary channels and associated wetlands and riparian forest offered valuable habitat to numerous species (Tockner and Stanford 2002). These species likely included floodplain-associated fish such as Sacramento perch, Thicktail chub, and splittail, as well as out-migrating juvenile Chinook salmon (Sommer et al. 2001, Jeffres et al. 2008, Opperman 2008, Moyle pers. comm.). The San Joaquin River mainstem provided key ecological functions (e.g., fish migration), and the larger surrounding floodplains played a critical role in the life histories of many Delta species. The ecological value of individual habitat types was enhanced by the surrounding landscape. Side channel habitat, for instance, provided benefit through the connectivity to riparian forest and between backwater ponds and the primary tidal river system. Improved understanding of historical conditions contributes to the increasing consensus in the scientific literature of the ecological importance of floodplain habitat.

The following sections provide a sense of the historical hydrologic, geomorphic, and ecological characteristics of the south Delta. We convey historical landscape patterns and offer insight about governing physical factors and related ecological functions. The larger-scale floodplain perspective of the south Delta is covered first, with discussion of the major landforms, geology, and hydrology. Discussions of primary habitat components follow with sections on channels, lakes and ponds, marsh plain habitat complexity, riparian forest, and upland ecotone.

FLOODPLAIN MORPHOLOGY

Floodplains owe their formation to the migration of a meandering river across its valley and frequent flooding with associated sediment deposition (Keller 1977, Leopold 1994). In the south Delta, where the downstream extent of the San Joaquin River floodplain merged with the zone of tidal influence, the floodplain exhibited unique characteristics. The term "upper delta plain" has been used to describe this type of transitional environment where tides have an influence, but riverine processes dominate (Fig. 6.4; Coleman 1976, Brown and Pasternack 2004). The influence of fluvial processes (flooding and sediment dynamics) introduced topographic complexity and hydrologic variability to this part of the Delta, in comparison to the landscape of the central Delta. Common landforms of floodplain environments include natural levees, meander scroll topography, oxbow lakes, and abandoned channels, all found in the south Delta. These landforms affected the spatial variation in overflow, tidal influence, the frequency and depth of inundation, vegetation communities, and consequently the ecological functions provided by the south Delta landscape.

Left the river in good season and departing gradually from its timber – came into large marshes of Bulrushs.

—BIDWELL 1937. TRAVELING WEST IN 1841 FROM THE SAN JOAQUIN TOWARD MARSH ESTATE IN EAST CONTRA COSTA COUNTY

Floodplain: "the flat area adjacent to the river channel, constructed by the present river in the present climate and frequently subject to overflow"

—LEOPOLD 1994

314

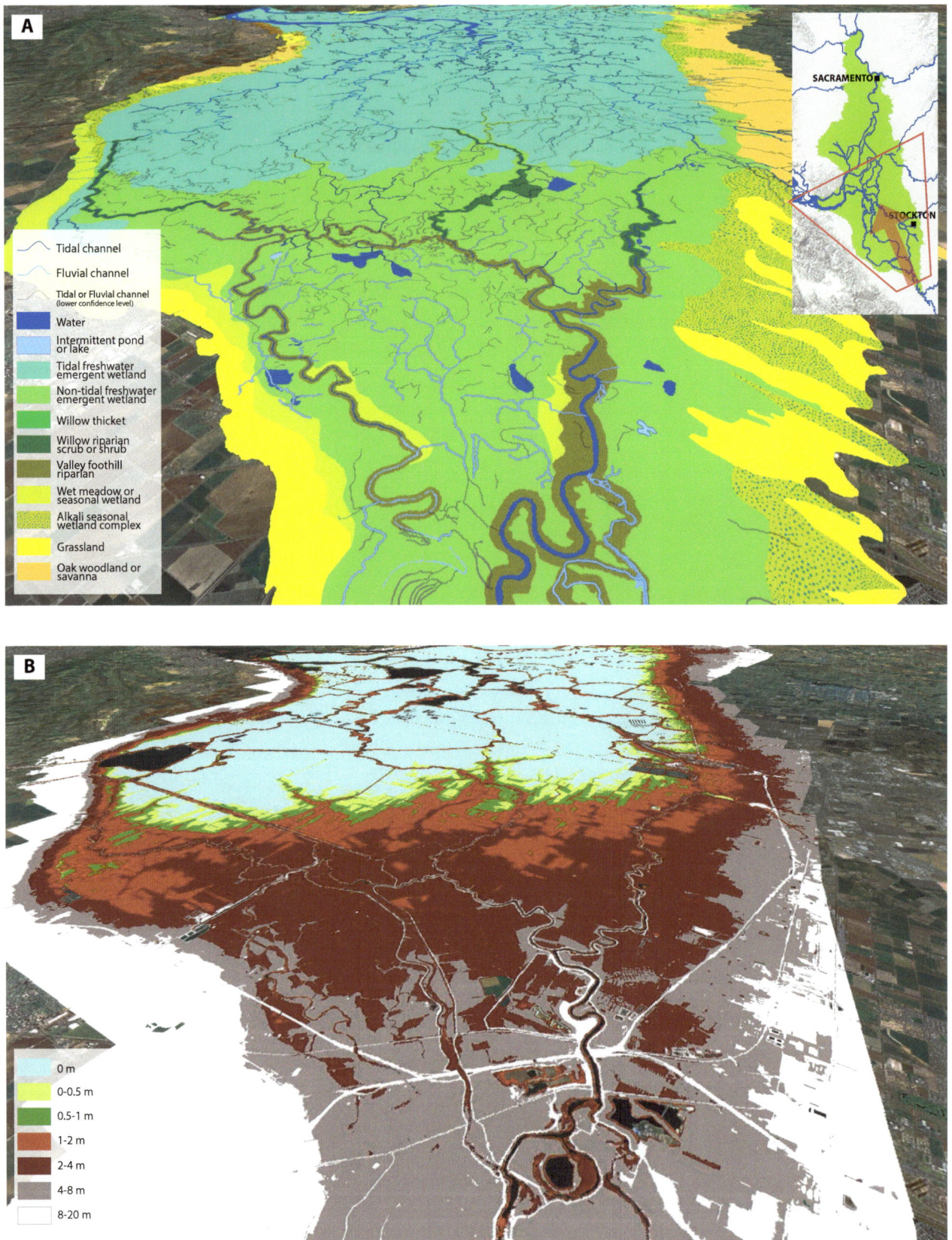

Figure 6.4. Oblique views of the south Delta looking northward illustrate the broad transition from riverine floodplain environment to the tidal central Delta. The distribution of historical habitats in A illustrates the relationship of landscape patterns to the major Delta landforms and topography. Shapes and locations of transitions can be seen to correspond with contemporary topographic forms, as revealed by the LiDAR (B; warm colors depict land above tide level and cool colors below). For example, natural levee topography extends above tide levels well into

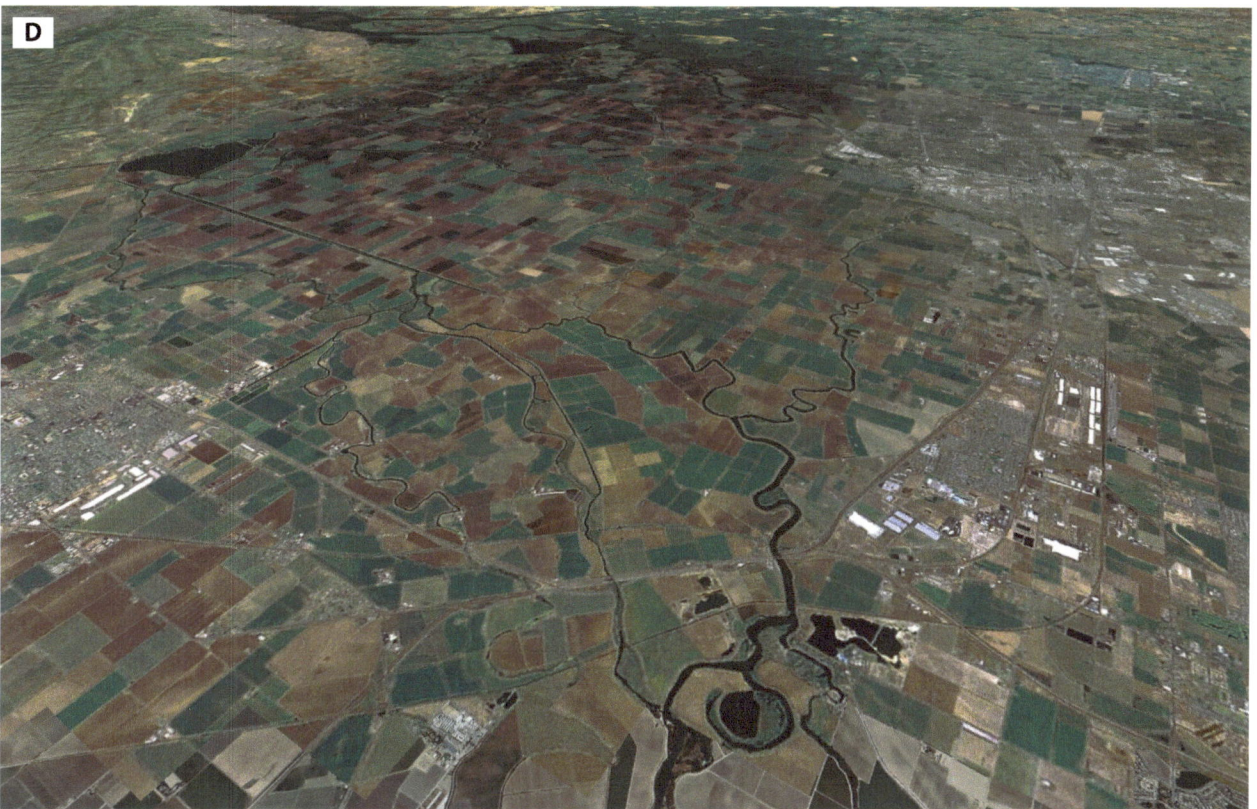

the historical tidal boundary (roughly where colors transition from warm to cool). Remnant signatures of the fluvial-tidal transition can also be seen in the aerial imagery from 1937 (C) and 2009 (D). Towns skirt the floodplain boundary, patterns produced by former channels are visible as tonal shifts between the main river branches in the 1937 imagery, and darker wetter soils in the 2009 imagery are seen in agricultural fields proportionately more in the central Delta (top of image). (B: CDWR 2008; C: USDA 1937-1939; D: USDA 2009)

The subtidal waterways were the lowest elevation landforms of the south Delta floodplain landscape. The San Joaquin River spread into a network of large and small distributaries as it met the tides. At the largest scale, this pattern is observed in the three distributary branches of the San Joaquin, which meet again in the central Delta. Near the town of Lathrop, the river divides into the western Old River branch (formerly Río del Pescadero) and the eastern mainstem (formerly Río de San Miguel). About four miles farther downstream, the Middle River (formerly Río de San Francisco Jabier) branches off Old River, which turns westward.

This distributary pattern occurred at smaller scales as well, adding local-scale topographic variability. Many small channels branched off of the main rivers, some of which reconnected again further downstream (U.S. War Department 1900). Where land surfaces were above tidal elevations, non-tidal secondary channels, ponds, and lakes held higher positions than the nearby river channels. Most of these features were connected to the San Joaquin River only during high river stages and dried out before winter rains. For such channels, bed elevations would have been several feet above the low water elevations in the river (U.S. War Department 1900). Some of the more substantial channels, such as Walthall Slough, likely remained at least partially tidally connected to the San Joaquin.

Just above these water features, the marsh plain gradually increased in elevation from the central Delta tidal wetland plain surface elevation to over five feet (1.5 m) above the reach of tides near the mouth of the Stanislaus River. Though its extent is only evident in the most severe floods today, the San Joaquin River floodplain within the south Delta broadened as it transitioned into a tidally dominated wetland. This expansion began where the three main distributary channels of the San Joaquin part just downstream of the present-day I-5 crossing (Fig. 6.5). While the distance between the east and west branches of the San Joaquin is over 13 miles (21 km) between French Camp Slough and Clifton Court Forebay, it is about half that five miles upstream. Farther upstream, where the San Joaquin maintained a single primary channel, the floodplain corridor was a relatively narrow few miles wide. This contrasts against the flood basins that occupied the Sacramento Valley upstream of the Delta, which were over five miles wide in places on both sides of the river.

Natural levees, built by inorganic sediments deposited during high river flows, offered the greatest topographic variety in the floodplain landscape. They were more substantial along the larger waterways, gradually increasing in height upstream away from the central Delta, although lower places did occur (*Daily Alta California* 1852, Kleugul 1878). Natural levee heights generally reached elevations just below the maximum height of floods. Flood heights reached around 10 feet (3 m) in the most southern part of the Delta, but were only several feet high in the central Delta (De Mofras and Wilbur 1937). In comparing the natural levees at similar positions along the fluvial-tidal gradient, the San Joaquin's levees were smaller than those of the Sacramento (Alexander and Mendell 1874, Rose et al. 1895).

On the upper portion of Old River the banks are high and similar to those of the upper Middle River described on page one. They gradually decrease in height for a distance of 20 miles down the River, where they continue at the general level of tule lands.

—KLUEGUL 1878

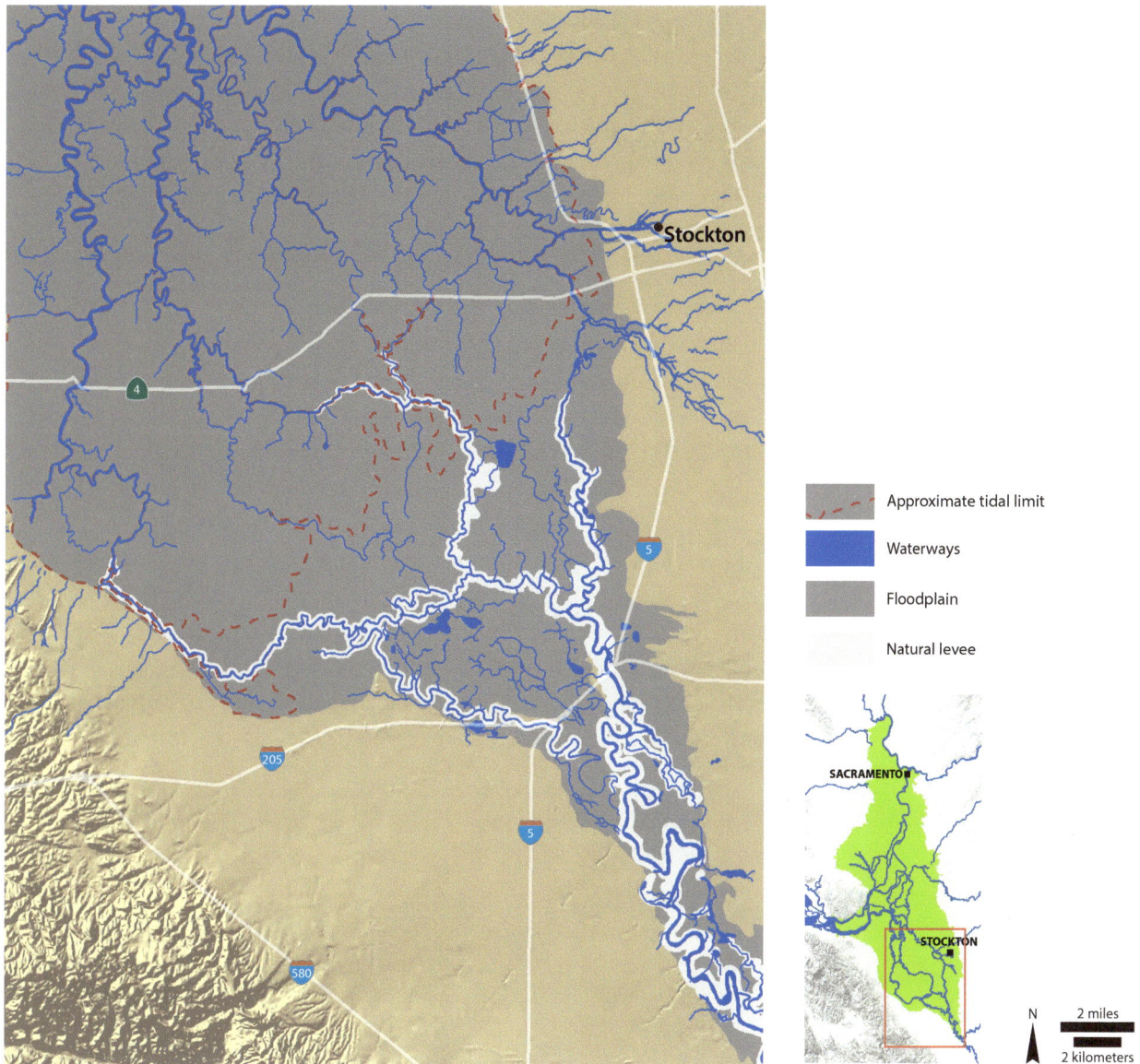

Figure 6.5. Floodplain extent in the south Delta broadened northward toward tidally influenced wetlands. The San Joaquin split into its three main distributary channels where the floodplain widened dramatically. At the I-5 crossing, the floodplain is only 4 miles wide, while at Stockton, it is 18 miles wide. The floodplain (in gray) is constricted by alluvial fans of the valley and dissected by the natural levees (in light gray) of the San Joaquin River.

The natural levees became notable features in the vicinity of French Camp on the San Joaquin River, Duck Slough on Middle River, and Bethany on Old River (*Daily Alta California* 1870, Kleugul 1878, Naglee 1879). In the vicinity of upper Roberts and Union islands, the natural levees were described as being from two to five feet (0.6-1.5 m) above the marsh plain or several feet above the extent of high tides (Gibbes 1850b, Alexander 1870). The ridge associated with Duck Slough was described as generally six feet (1.8 m) above the marsh surface (*Pacific Rural Press* 1883). Along Old River in upper Union Island (Fabian Tract), the natural levee was described as "one to seven feet higher than the land inside of it" (Naglee 1879). Farther upstream, near where I-5 crosses the river today, the banks were noted as being 13 feet (4 m) high (Norris 1851b). In the southernmost reaches of the study area, the natural levees extended up to 18 feet (5.5 m) above low water in Suisun Bay (Rose et al. 1895). The levees were characterized by relatively steep banks at the water's edge and sloped more gradually on the backside toward the tules, with most of the elevation change occurring within the first several hundred feet (Kleugul 1878, Naglee 1879).

Landscape character in a first-hand account

Habitat types in the south Delta are best understood within the context of mosaics and overall landscape patterns. A passage from the Gold Rush-era account of Jacques Moerenhout ([1849]1935) provides a view of individual features or habitats within the south Delta landscape, consisting of "meadows and swamps…as far as the eye could see." Moerenhout's group passed through the south Delta on July 13, 1848, likely in the vicinity of Sheppard's Ferry, one of the most commonly used early fords on the San Joaquin. The narrative of this San Joaquin crossing speaks to the ubiquity of the side channels and backwater lakes as part of the rich complexity of the south Delta floodplain environment. Local scale topographic complexity translated to varying inundation levels and vegetation patterns characteristic of this area:

> To approach it [the San Joaquin crossing] there were more ponds, swamps and sloughs, very difficult and very dangerous to cross, but it had to be done for there was no other way…
>
> The first of these places had about three feet of water, but the bottom was solid and we crossed it without difficulty. The second was a slough more than fifty meters long where one went at random (au hazard)… But it also we crossed without accident. The third was a little lake. There we were lucky enough to find a balsa of tules or an immense bundle of reeds or bullrushes tied together, on which we took over our saddles, our baggage and ourselves. The horses were forced into the water and swam across. After this lagoon we still had [to pass over] several more very difficult sloughs in which animals which had perished there were to be seen all about. We crossed them all safely…
>
> Here [on the other side of the river] also there were sloughs to cross and it was on this side that two Americans had lost their lives…We passed through several bad places without much difficulty. Towards two o'clock we reached the lagoon where an American had perished a few days before…The night was clear, we went carefully, following as nearly as possible the crossing marked by the broken reeds, [and] in less than ten minutes we were on the other side and out of all danger. The place where we then were being quite high and dry, the horses were tethered near the lagoon where there was some grass, and each of us, worn out by fatigue, made himself as comfortable as he could on the sand to wait for daylight and to continue on our way. That where we were now was but a tongue of land between two great pools or lagoons, and all the portion which we had crossed, toward the San Joaquin, seemed the same as that on the other side of the river, although less wooded--that is to say, [it consisted of] meadows and swamps which extended as far as the eye could see. (Moerenhout [1849]1935)

We came to the Tulares now overflowed with water skirted these some miles, stopped to bathe…spent some time in passing a slough… Found the plain so overflowed as to prevent reaching the river today. Weather very hot and mosquitoes miserable…passed 2 or 3 sloughs… Encamped at river – beautiful place, cool – few mosquitoes. This is called the Piscadero crossing place [likely present-day I-5 crossing] – the usual crossing place in dry weather. Fine day – river rising – felt it necessary to cross as soon as possible.

—LYMAN 1848, EXCERPT FROM DIARY FOR JUNE 9-11

The assemblage of sloughs, lakes, lagoons, patches of reeds, and grass in this account suggests landscape elements melded to present a complex and seasonally dynamic place (Fig. 6.6).

SEASONALITY AND FLOW

Like most valley reaches of California rivers, the San Joaquin River frequently overflowed its banks (Gibbes 1850b, Gilbert 1879, U.S. War Department 1895, U.S. War Department 1900). Much of the lower elevation lands consequently experienced annual flooding, often for many months at a time (Fig. 6.7). As river stages rose, the lowest ponds and sloughs filled and overflowed to flood the surrounding wetlands up to several feet deep (Swan [1848]1960). The south Delta in flood was described by some as a "vast assemblage of lakes" (Farnham 1857) or "low flaggy ground which was covered with water" (Sullivan 1934). Only the higher landforms – natural levees and low mounds – remained dry in most circumstances (Swan [1848]1960).

Flood flows spread through multiple channels and across extensive wetlands, such that the actual rise in river stage was much less than if the flow had been contained solely within the channels. This relationship was noted by many, including the geologist J.C. Gibbes, who recommended against attempts to confine all the floodwaters within the river channels (Gibbes 1850b). A newspaper article in 1852 estimated that, if the rivers

Figure 6.6. The complex landscape of waterways, ponds, grasslands, riparian forest, and seasonal and perennial wetlands composes the scene of this 1873 print by Thomas Moran. (image from Bryant 1874)

Figure 6.7. Lower lands flooded along the San Joaquin River at Durham Ferry Road crossing in 1938, likely during the February and March floods. Only tufts of taller brush and trees within the floodplain can be seen, particularly along channels or artificial levees. (Covello 1938, courtesy of Bank of Stockton Historical Photograph Collection)

In June, 1847, the Joaquin was nowhere fordable, being several hundred yards broad…and scattered in sloughs over all its lower bottoms.

—FRÉMONT ET AL. 1849

were leveed completely, "it would raise the water from one to two or three feet higher than it now rises" (*Daily Alta California* 1852).

Owing to snowmelt-driven high flows, the annual rise and overflow into the surrounding floodplain was less extreme, on average, than the flashier rainfall event-driven floods on the Sacramento. It was observed in 1878 that flood heights above low water reached 10 feet (3 m) on the San Joaquin, but over 20 feet (6 m) on the Sacramento (*Daily Alta California* 1878). The generally larger winter (rainfall-driven) floods were less frequent than on the Sacramento. A state engineering document from 1880 claimed that while destructive floods occurred nearly every year on the Sacramento, those on the San Joaquin occurred only once every four years (Young 1880).

As a largely snowmelt fed river, the San Joaquin River water levels rose gradually and peaked well after winter storms, from March through June (Fig. 6.8; Thompson 1957). The function of the historically immense Tulare Lake in the southern San Joaquin Valley had an additional effect of retarding flows by being filled by southern Sierra Nevada rivers before eventually spilling to the San Joaquin River. It is perhaps counterintuitive, considering California's Mediterranean climate of wet winters and dry summers, that water within the south Delta floodplains was more abundant in the late spring and early summer than in the winter. With water often a limiting factor for ecosystem productivity and habitat availability, it is ecologically significant that the river, riparian forest, and floodplain wetlands experienced greater water availability after rainfall had ceased and as temperatures climbed in the summer months.

December and early January were included as part of "the season of low waters" on the San Joaquin River (Frémont et al. 1849). Until mid-July the floodplain was heavily overflowed by high river stages caused by snowmelt in the Sierra Nevada. During the times of high river stages, one could travel widely in boats or the "tule balsas" used by the indigenous tribes of the Delta (Moerenhout [1849]1935). As explorer Viader put it, observing the vicinity of Fabian and Stewart tracts: "All this place and its surroundings are inundated during the high water of the rivers, which is in the summer" (Viader and Cook 1960). Once reclamation had begun, late spring and early summer was the time farmers faced the greatest challenges from levee failures and flooding.

During the hottest and driest parts of the year in the valley, the landscape was at its wettest. According to early travelers in the area, the most difficult time to cross the south Delta was during the few months following the first of July. During this period, water levels had usually begun falling and flows were returning to the river channel (*Daily Alta California* 1852). This meant, as a gold-miner explained, that the San Joaquin "leaves lagoons and swamps on all sides and very dangerous *atascaderos* [muddy areas] or sloughs which in many places absolutely prevent approaching it" (Moerenhout [1849]1935). Another traveler described crossing deep sloughs and navigating around ponds, and even encountering difficulty coercing cattle "across a little water not more than knee deep" in mid-July 1837 (Edwards [1837]1890). It was made more unpleasant to travelers by mosquitoes that flourished (see Box 4.4; Taylor 1854) in the "stagnant pools of putrid water, which send out most pestilential exhalations" (Farnham 1857).

While the daily tides and maritime influences muted seasonal dynamics within the central Delta, the south Delta was characterized by more dramatic seasonal variation. Much of the south Delta dried out for several months in the early fall. As freshwater inflow decreased late in the season and water levels dropped, much of the land surface (and its many associated habitat types) above high tide levels became functionally disconnected from the river. We estimate that the majority of the approximately 47,000 acres (19,020 ha) of non-tidal freshwater emergent wetlands mapped within the south Delta were characterized by such dynamics.

Large parts of the "tule land" within the south Delta (primarily those areas mapped as non-tidal freshwater emergent wetland) dried out late in the season, including many of the smaller waterways that served as overflow channels at high river stages (Bryant [1848]1985, Lyman 1848). This drying would have made fires posssible (Box 6.1). A tale of an attempted stagecoach robbery relates how the road was "dry and the dust thick" in the "tule flats" of present-day Stewart Tract (Williams 1973). Some areas

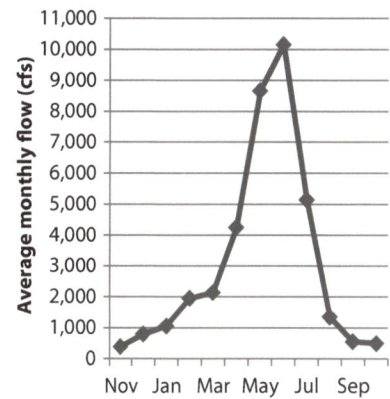

Figure 6.8. San Joaquin River average monthly flow at Friant (in the Sierra Nevada foothills near Fresno) between 1878 and 1884. Maximum flows were recorded in May and June. It should be noted that there were no significant winter floods during this short period of record and it does not represent a long-term average monthly hydrograph. (Newell 1896)

But from the first of July to the fifteenth of August the crossing was considered impracticable even for horsemen on account of the swamps and quagmires.

—MOERENHOUT [1849]1935

BOX 6.1. EVIDENCE OF FIRE IN THE TULE

The phenomenon of repeated, seasonal burning of Delta wetlands has profound implications for expected habitat mosaics and species assemblages, as well as organic matter accumulation, available moisture, and evapotranspiration levels. Numerous historical accounts indicate that fires in the tule were quite common during the period of rapid settlement during and after the Gold Rush. However, it is less clear whether these fires represent an increase over early 1800s conditions. Some researchers have speculated that fires were uncommon in the Delta prior to European contact (e.g., Fox 1987a). However, due to the paucity of written records prior to the Gold Rush and the absence of comprehensive research in this regard, it may be premature to discount the role of fire in shaping landscape characteristics in at least parts of the early 1800s Delta.

Indigenous management practices may have involved the burning of Delta wetlands in ways similar to the well documented use of fire to manage California grasslands and scrub (Milliken 1995a, Anderson 2005, Minnich 2008, Lightfoot and Parrish 2009), and to clear wetland vegetation in other parts of the U.S. (Lewis 1982). Some pre-reclamation accounts suggest linkages between fires in the tule and management by the tribes. For example, a late 1830s account states that "during the dry season the natives burn this down" in reference to "flag grass, roses, arbutus, and other small shrubs, pasturage" in the Delta (Belcher 1843). Unfortunately, it is unclear if tule was targeted specifically. Reports in the late 1800s, referring to past conditions, discuss these practices as well and provide suggestive, though often unsubstantiated, evidence. One states, for instance, that "for centuries these tules have been burnt off, more or less regularly during the dry season, by the Indians in search of game" (Whitney 1873). Other potential connections are found in journals of residents at Sutter's Fort in Sacramento in 1847, which suggest widespread burning possibly attributable to tribes (Anderson pers. comm.). A rare primary account of fire in tule prior to European settlement is found in

Pedro Font's diary describing his route east of Byron in April 1776: "Going with some difficulty in the midst of the tulares, which for a good stretch were dry, soft, mellow ground, covered with dry slime and with a dust which the wind raised from the ashes of the burned tule" (Font and Bolton 1930). While clearly suggestive, it should be noted that this observation was made at the margin of tule, so whether the fire was intentional or not, its extent, and its representation as a commonly-used practice remains uncertain.

In a recent sediment core study on the McCormack-Williamson Tract on the Mokelumne River, researchers concluded that fire was not a significant disturbance factor there because charcoal was not found in the cores from tidally-influenced areas (Pasternack and Brown ca. 2006). It is unknown whether tides or flood events could have kept charcoal layers from developing. To make conclusions about the greater Delta region, additional research is warranted.

Compilation of newspaper clippings, diaries and other accounts suggests that most fires in the early settlement period occurred annually and primarily in the winter months, prior to widespread flooding. While it is possible that some of the fires reported during this time period were products of indigenous management practices, fires within tule in the mid- and late-1800s were often attributed to either exposure of game for hunting, purposeful ignition as entertainment for passersby, accidental ignition from firing guns, abandoned fires, sparks from steamers, etc. In addition, fire was used as a method of reclamation after leveeing had allowed wetlands to drain. Burning tule for reclamation is a documented practice in the historical record, and is discussed in more detail in one of John Thompson's (in press) more recent works. In it, he discusses the profound effect this practice had on dramatically lowering the levels of the land almost immediately after reclamation (by many feet in places). Field notes on early reclamation documented burning in this manner for Andrus Island, Bouldin Island, Bradford Tract, Grand Island, Mandeville Island, Roberts Island, Staten Island, Twitchell Island, and Webb Tract (Tucker 1879a, c, e, f). Most reports of fires, however, do not specify the cause of ignition. Representative quotes discussing fire in tule are listed in Table 6.1 (following page) and images of tule fires are shown in Figure 6.9.

Further research is needed to address questions of frequency and prevalence of burning by indigenous tribes in the Delta and to infer the effect these practices may have had on the habitat patterns and wetland functions of the Delta circa 1800. Oral histories, soil cores, and careful extrapolation based on practices in other wetlands in places such as Tulare Lake, Klamath Basin, and the Puget Sound could potentially shed light on these questions.

Box 6.1 continued on *page 324*

A

B

By Pass - Burning tule, Kercheval Nov. 1918

Figure 6.9. **Plumes of smoke** can be seen rising from the Delta tules in an1860s-era engraving (A) and in a 1918 photograph of reclamation in the Yolo Basin (B). (A: Hutchings 1862; B: Kercheval 1918, Holland Land Co., D-118, courtesy of Special Collections, University of California Library, Davis)

BOX 6.1. EVIDENCE OF FIRE IN TULE (CONTINUED)

Table 6.1. Selected pre-1900 quotes related to fire in tule present a picture of frequent burning during the Delta in the early settlement period. While many fires are attributed to the settlers' activities, some may relate to indigenous management activity.

Year	Month	Quote	Cause	Citation
1776	Apr	"And so we traveled more than three leagues, which in general may be estimated as to the southeast, going with some difficulty in the midst of the tulares, which for a good stretch were dry, soft, mellow ground, covered with dry slime an with a dust which the wind raised from the ashes of the burned tule"	unknown	Font and Bolton 1930
1847	Nov	"The Tular on the left bank of the Sacramento in fire"	unknown	Sutter and Brancroft 1876
1847		"We continued our progress up the river, occasionally stopping and amusing ourselves by firing the woods on either side, and watching the broad flames as they spread and crackled through the underbrush."	for entertainment	Buffum 1850
1848		"Occasionally an opening would be found which had been burned off by Indians at get at the elk, which frequented them in large numbers."	hunting by Indians	Anonymous, 5 in Fox 1987b
1849	autumn or spring	"In the autumn before the rains, or in Spring before growing up again, they are frequently set in a blaze from the camp fires of the Indians or others, causing most extensive and long-continued conflagrations. "	accidental	Johnson 1849 in Fox 1987b
1849	Aug	"the very beds of the tule marshes were beginning to dry up. The air was thicker than ever with the smoke of burning tule"	unknown	Taylor 1854
1850	Jun	"Tule Plains on Fire: It is said by passengers who arrived from San Francisco, yesterday morning, that the Tule plains, on the San Joaquin, were on fire, Saturday evening, and that the flames could be seen from the Sacramento, lighting up the whole heavens. The appearance is described as brilliant in the extreme."	unknown	*Sacramento Transcript* 1850b
1850	Dec	"Tule plains on Fire. The sky in the SSE was beautifully illuminated last evening by the burning of large tule's"	unknown	*Sacramento Transcript* 1850c
1851	Oct	"The tule marshes are again on fire away off in the middle of Yolo county."	unknown	*Sacramento Daily Union* 1851b
1854	Feb	"The dry tules which cover the marshes are thus burned over every season. Any accident which starts the fire – the carelessness of a party camping out, or even the sparks from a passing steamer, begins a conflagration which spreads over a wide extent of country."	accidental	Kip 1892
1861		"On the plain below camp, fire was in the tules and in the stubble grounds at several places every night, and int he night air the sight was most grand – great sheets of flame, extending over acres, now a broad lurid sheet, then a line of fire sweeping across stubble fields. "	unknown	Brewer 1974
1862		"An apparently interminable sea of tules...when these were on fire, as they not unfrequently are, during the fall and early winter months"	unknown	Hutchings 1862
1868	fall	"Late in the season, however...large sections of these [tule] lands becoming dry on the surface...the latter often take fire, and burning with terrific fierceness for days in succession, many thousand acres are burned over and stripped of both the dead and living tules. In all the counties containing large tracts of tule lands, these fires are common, generally occurring in the fall and winter. "	unknown	Cronise 1868
1871	Jan	"Tule fires, extending over a space of two or three miles, are now burning in Yolo County"	unknown	*Daily Alta California* 1871
1873		"The plan of substituting fire for the plow and sheep for the harrow is a novel one, and one which seems to have originated in these islands. It has been customary for many years, in other parts of the State, to burn off the tules, and the fire consumes portions of the root"	for reclamation	*Sacramento Daily Union* 1873
1875		"Andrus Island...In 1875 nearly all the land was burned"	for reclamation	Tucker 1879f
1879		"There has been some trouble from fires in the hunting season; but a close watch is kept to prevent the spread of fires, whether in the levee or tules"	hunting, accidental or on purpose	Tucker 1879a

at the wetland margin may have dried at other times of the year as well. In one of the earliest written accounts of the Delta, Spanish explorer Font encountered in April 1776 "dry, soft, mellow ground, covered with dry slime and with a dust which the wind raised from the ashes of the burned tule," after traveling "with some difficulty in the midst of the tulares" east of Byron in Contra Costa County (Font and Bolton 1930). Both the dry ground and the burned tule in this account suggest seasonal drying, though the spatial extent is uncertain.

The past 160 years of flow alterations and channel modifications have profoundly affected the timing and magnitude of flow in the San Joaquin River today. The hydrologic connectivity between the river and its floodplain once provided by the late spring and summer floods supported high productivity, ecological function, and ecosystem services (Sparks 1995, Benner and Sedell 1997, Jassby and Cloern 2000, Tockner and Stanford 2002, Opperman 2008). Understanding the role of floodplains and related hydrology can help address current and future challenges related to climate change.

Low flow conditions on the San Joaquin River

During the late summer and fall, the San Joaquin River receded to its lowest flows. Dry season baseflow steadily increased northward for the length of the San Joaquin Valley, as the mainstem accumulated inputs from the tributary Sierra Nevada rivers such as the Merced, Tuolumne, and Stanislaus. Downstream of the Stanislaus, tides directly and indirectly (through groundwater levels) helped maintain the river's water levels (CDPW 1931).

Most early assessments of the river's flow conditions relate to the effect on travel in the San Joaquin Valley. According to available reports, the San Joaquin was a navigable channel for about 40 miles upstream of Stockton, and for parts of the year even farther (McCollum [1850]1960, Heuer 1892, Clark ca. 1905). While the ability to travel by boat implies significant year-round flow, the term "navigable" can encompass a relatively wide range of conditions. An 1854 report offers one explanation, specifying that the river "would be navigable if the snags were taken out, from its mouth to the mouth of the Merced river, for the largest class of steamboats, eight months in the year" (Marlette 1854), while another account specifies that "from April to the end of August ships of a hundred tons could go up it [the San Joaquin] for thirty leagues [~90 mi] into the interior" (Moerenhout [1849]1935). Upstream of this point (between Hills Ferry and Firebaugh), one report stated that some reaches could become pools at lowest water (U.S. War Department 1895). Decreased flows as a result of water withdrawals were noted relatively early. For example, a 1916 flood control report states that the river was then navigable only 15 miles above Stockton, a change attributed to water supply demands (U.S. Congress 1916).

Though the river was deemed a navigable channel, water levels became quite low late in the summer and early fall near the Stanislaus confluence. Sometimes water was only a few feet deep, and fords existed but a few miles south of the head of Old River (U.S. War Department 1895). The farthest downstream ford (commonly referred to as "the ford of the San Joaquin") was apparently at Sheppard's Ferry, in the vicinity of the current I-5 crossing. Accounts indicate that this well known crossing became fordable by the end of September (Moerenhout [1849]1935). One traveler detailed the conditions at the ford during this period of low flow: "The stream at the ford is probably one hundred yards [91 m] in breadth, and our animals crossed it without much difficulty, the water reaching about midway up their bodies" (Vizetelly 1849). At the same time of year (October), just two and a half miles downstream at the head of Old River, the river was reportedly too deep to cross, likely related to tidal influence (Abella and Cook 1960).

It is unclear exactly where tidal influence became a significant factor in terms of water levels, but the head of tide may correspond with this lower ford at Sheppard's Ferry. This interpretation is supported by a Spanish explorer, who noted that tide was slight at the mouth of Old River in October 1811 (Abella and Cook 1960). On Old River, near Bethany, the same expedition found that passage within the channel was only possible at high tide. The boats had run aground at low tide when the channel appeared "to carry about as much water as the river at the ranch at Monterey [translator notes this is the Carmel River]." Incidentally, this place of "shoal water," or shallow water, was apparently a relatively persistent feature as it was reported in 1892 as the first place of low water on Old River (traveling upstream), where the depth of water on the bar was only 28 inches (0.7 m), with tides adding 14 to 20 inches (0.36-0.5 m; U.S. War Department 1892).

CHANNELS AT THE FLUVIAL-TIDAL INTERFACE

A visual inspection of the historical channel networks of the south Delta reveals a river distributary system of numerous small tidal and non-tidal channels crisscrossing a land surface that gradually decreases in elevation northward toward mean high tide levels. The morphology of these channels was largely driven by the landscape position between dominating tidal and riverine processes. As a result, the characteristics and related ecological functions of these channels at the tidal-fluvial interface differed from those in the more tidally-dominated central Delta. The numerous secondary channels served to conduct flow from annual snowmelt events onto the floodplain and into the central Delta. Channel characteristics varied substantially over space and time, in part related to the seasonally dynamic landscape. Through the leveeing of the main rivers, damming and filling of secondary channels, and reductions in flood flows, the river and its floodplain – as well as the expression of the north-south tidal to fluvial gradient – are mostly disconnected today. These changes mask the existing channel morphology created by historical physical processes. The following discussion focuses first on the morphology of the San Joaquin River mainstem, particularly evidence of width and depth. The discussion then covers selected characteristics of the secondary overflow channels.

San Joaquin River channel geometry

In comparing the three main branches of the river, historical evidence indicates that like today, the east branch of the San Joaquin River had the greatest flow capacity: Old and Middle rivers were generally narrower and shallower. The name "Old River" suggests that this western branch may have once been the main river channel, though this does not appear to have been the case in the recent past. A 1796 Spanish expedition that crossed over the three branches provides on of the earliest known written records of the basic differences in flow capacity between branches. Explorer Hermenegildo Sal compared the three, noting that Old River had "good water, depth and current," Middle River was "wider than the preceding and with more water, for the latter reaches to the bottom of the saddle pad," and that the San Joaquin main branch was "larger than the two others, and deeper, for the water reaches to the back bow of the saddle" (Sal and Cook 1960). Unfortunately, it is unknown where they crossed exactly or at what point in the tidal cycle the observations were made.

Additional evidence concerning the differences between the San Joaquin channels comes from a report made as major reclamation works were underway. In it, flow capacity of the three branches was compared using differences in width. At the head of Old River, the mainstem width decreased from 300 feet (91 m) just upstream to 180 feet (55 m) just downstream, while the Middle River "has but 47 feet [14 m] two miles [3.2 km] below Rea's Ferry (~2 mi downstream from the head of Middle River), and Old River has but 81 feet [25 m] below the mouth of Tom Payne [sic] Slough" (Naglee 1879). General Land Office (GLO) surveys also note width in a few places, in four out of five cases indicating that the channel may have been wider than mapped (15-65%; Norris 1851b, Stratton 1861, Benson 1877). At one crossing, on Old River just upstream of Coney Island where banks were relatively low, a GLO survey recorded a 100 foot wide channel (30 m; Fig. 6.10). These early observations generally support the mapping synthesis: based on an average of width measurements taken every mile along the San Joaquin branches within the south Delta: the San Joaquin River was on the order of 180 feet (55 m) wide, Middle River 90 feet (27 m), and Old River 100 feet (30 m).

As a general pattern, channel width decreased with decreasing tidal influence upstream, with localized widening or narrowing occurring at meander bends and where distributary channels entered and diverged. The branches of the San Joaquin were well over three times wider at their downstream end than at their upstream point of divergence. The trend of decreasing width and depth upstream is apparent in detailed early 1900s soundings from Debris Commission maps and profiles (Fig. 6.11). It is also supported by a much earlier 1850 navigational survey of the San Joaquin. In describing the Middle River, surveyor Gibbes reported that in the more

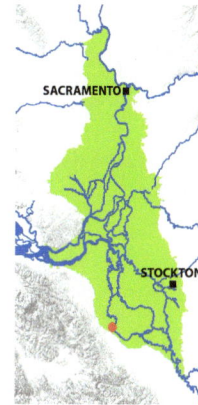

Figure 6.10. A 100 foot wide channel is recorded by an 1861 General Land Office survey. The GLO reported width is slightly narrower, but within expected error, of our mapped channel, which is about 130 feet wide. This and other early sources supports that our mapping is representative of historical conditions. Due to spatial accuracy errors of these particular survey lines, the point is over 400 feet from its actual location on the east bank of the mapped river. (Stratton 1861)

Water

Tidal freshwater emergent wetland

Willow riparian scrub or shrub

Alkali seasonal wetland complex

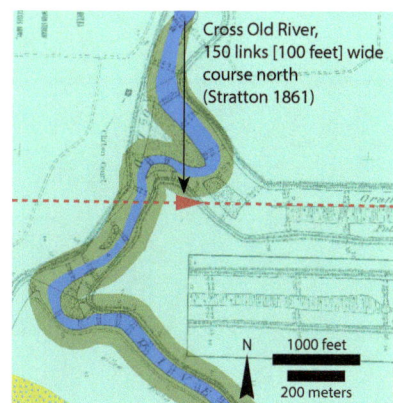

Cross Old River, 150 links [100 feet] wide course north (Stratton 1861)

N   1000 feet

200 meters

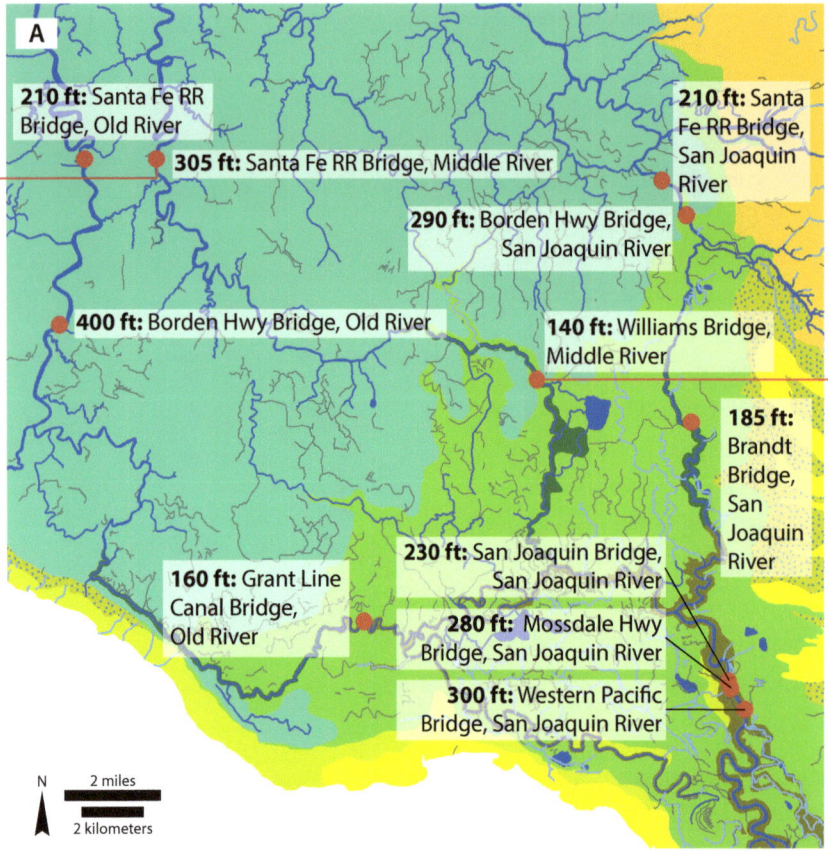

A

210 ft: Santa Fe RR Bridge, Old River

305 ft: Santa Fe RR Bridge, Middle River

210 ft: Santa Fe RR Bridge, San Joaquin River

290 ft: Borden Hwy Bridge, San Joaquin River

400 ft: Borden Hwy Bridge, Old River

140 ft: Williams Bridge, Middle River

185 ft: Brandt Bridge, San Joaquin River

230 ft: San Joaquin Bridge, San Joaquin River

160 ft: Grant Line Canal Bridge, Old River

280 ft: Mossdale Hwy Bridge, San Joaquin River

300 ft: Western Pacific Bridge, San Joaquin River

N
2 miles
2 kilometers

Tidal channel

Fluvial channel

Tidal or Fluvial channel
(lower confidence level)

Water

Intermittent pond or lake

Tidal freshwater emergent wetland

Non-tidal freshwater emergent wetland

Willow riparian scrub or shrub

Valley foothill riparian

Wet meadow or seasonal wetland

Alkali seasonal wetland complex

Grassland

Oak woodland or savanna

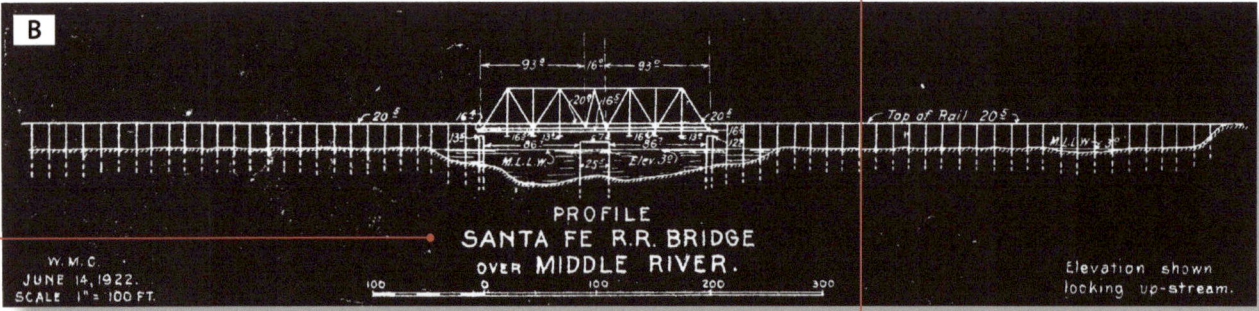

B

PROFILE
SANTA FE R.R. BRIDGE
OVER MIDDLE RIVER.

W.M.C.
JUNE 14, 1922.
SCALE 1" = 100 FT.

Elevation shown
looking up-stream.

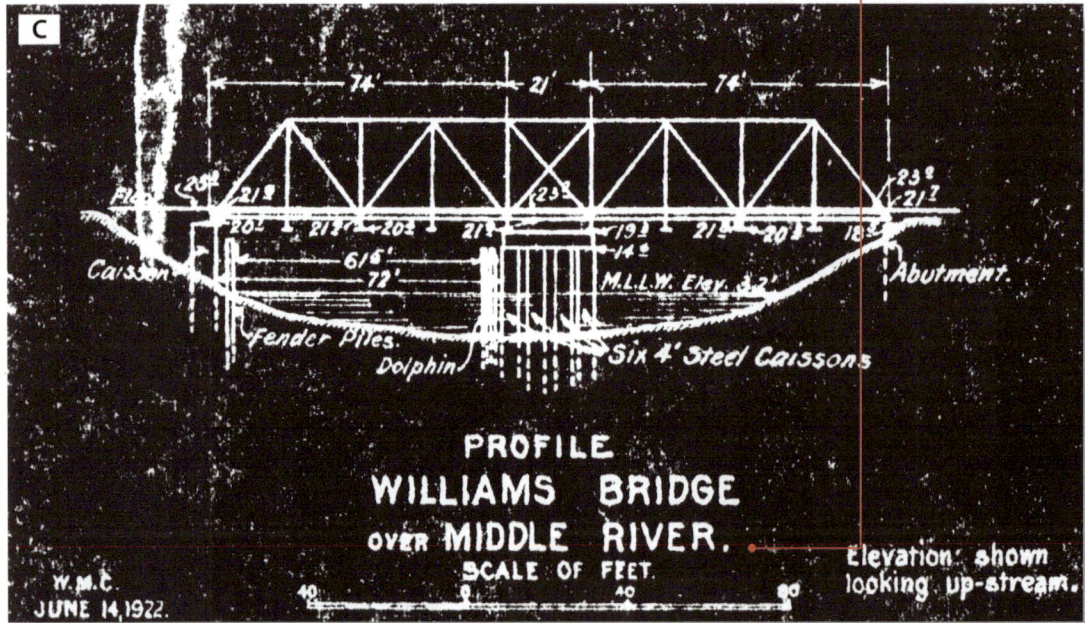

C

PROFILE
WILLIAMS BRIDGE
OVER MIDDLE RIVER.
SCALE OF FEET

W.M.C.
JUNE 14, 1922.

Elevation shown
looking up-stream.

Figure 6.11. Channel widths of the main branches of the San Joaquin decrease upstream as relative tidal influence decreases (left). The measurements shown in the map (A) and the two example profiles from Middle River (B and C) were taken by the Debris Commission in the 1920s at bridge crossings. The Santa Fe Railroad crossing on Middle River is just over 300 feet wide, while upstream at Williams Bridge crossing, the channel is only 140 feet wide at MLLW. At the downstream profile, raised natural levees are not visible, while at the upstream profile, channel banks are much higher than MLLW (height may be augmented by artificial levees). At this upstream crossing we mapped willow riparian scrub along the natural levees of the river. All width measurements are at MLLW. A potential complication lies in the possibility that bridges were located at natural narrow points in the channel. (U.S. Army et al. 1913, courtesy of the California State Lands Commission)

tidally-dominated lower half it is a "good sized river for navigation," but upstream the channel was found to narrow to 50-60 feet (15-18 m) with decreased depth to about 12 feet (3.6 m soundings were taken at high river stage in June; Fig. 6.12; Gibbes 1850b). In a later survey on Old River, this same pattern was noted, where the channel became difficult to navigate upstream of Union Island Landing due to tighter river bends and variable width between 100 and 300 feet (30-91 m; U.S. War Department 1892). This pattern is reflective of the decreasing tidal volume and increasing influence of riverine processes and is consistent with tidal marsh channels in San Francisco Bay (Atwater et al. 1979). The relatively low channel capacity of the main river channels in the less tidally-dominated reaches implies that substantial volumes of the river's flood flows communicated with the tidal Delta through the south Delta's side channels and floodplains rather than the mainstem river.

Unlike the central Delta where historical channel widening (due to reclamation) was evident along a number of the major waterways, we found no marked change in channel width along the south Delta mainstem branches bordered by natural levees. Similar to the Sacramento River in the north Delta, natural levees were the ideal place to build artificial levees. For the most part, this locked the historical channel width and location in place. However, localized impacts – including both widening and narrowing – can be identified.

Though not evaluated, the channel bed profile (i.e., the shape of the channel rather than its width) may have sustained comparatively more changes through time, because most activities in the channels tended to have the effect of homogenizing the channel, reducing the complexity of the cross-section profile, and removing longitudinal changes in bed elevation (e.g., shoals). Dredging (both for purposes of navigation as well as levee building), scouring due to containment of flood flows within artificially leveed banks, and snag removal were likely the predominant mechanisms for these changes. While hydraulic mining debris should be considered as a factor that could have raised river bed levels, these sediments were deemed, for the San Joaquin and its tributaries, "so small as to produce little or no effect on the navigability of the rivers" and would have reached their peak by the early 1900s (Gilbert 1917).

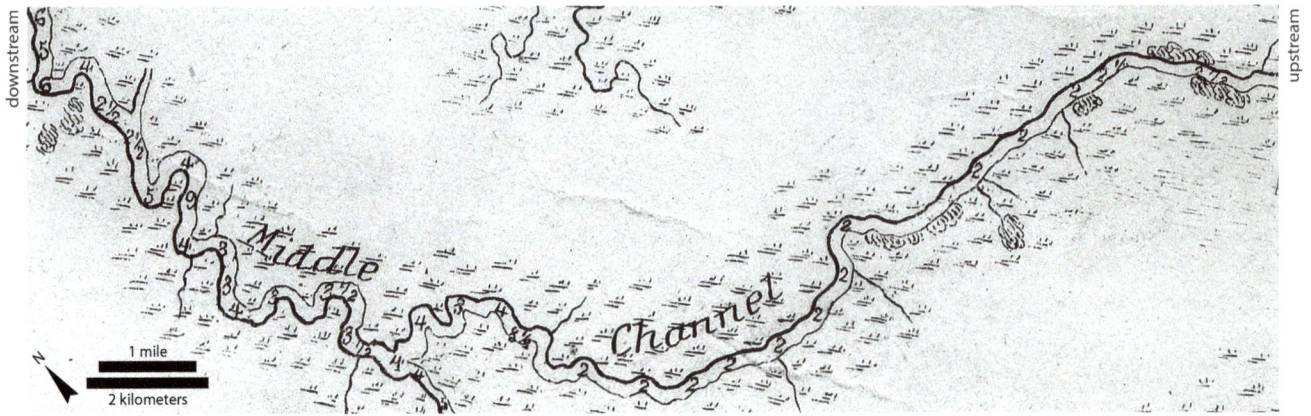

Figure 6.12. Channel depths on the Middle River decrease upstream. This 1850 map shows soundings in fathoms decreasing upstream (towards the right) from 3 to 9 fathoms (18-54 feet / 5.5-16.5 m) close to its mouth on the San Joaquin and about 2 fathoms (12 feet / 3.7 m) close to its head on Old River. The surveying was performed during periods of high flow in June. (Gibbes 1850a, courtesy of the Map Collection of the Library of UC Davis)

Since channel geometry was affected early by levee building, dredging, hydraulic mining, and snag removal, relatively little detailed and spatially accurate information concerning channel bathymetry prior to these changes exists (see Box 6.3). However, maps and surveys from the early 1900s, particularly those of the California Debris Commission, do provide extensive quantitative measurements of the channel geometry of the major rivers (e.g., Sacramento, San Joaquin, Mokelumne) which can be used to interpret natural channel geometry and change over time, at least over the past century (e.g., Wadsworth 1908a, Wadsworth 1908b, California Debris Commission 1914). We found these to be generally supported by the limited number of pre-1900 narrative accounts and surveys. Further research and modeling efforts may benefit from a grid-based bathymetric reconstruction.

Low order channels of the floodplain

Seasonal flooding of meandering rivers into their floodplains naturally results in a landscape intersected by numerous active and abandoned secondary channels that distribute flood flows onto and off of the floodplain surface. These secondary or side channels of the south Delta (classified as low order channels in the historical habitat type map) were positioned at the upstream end of tides and the downstream end of a major river. They expressed that intersection through widely varying morphologies. Many of the secondary channel functions, including the dispersal of flood flows across the broad Delta, were shaped by interacting with the tidal central Delta as well as the inflows from upstream. Historical maps depicting channels and topography and texts that discuss the nature of flow aid our understanding of the wide range of channel characteristics that existed in the early 1800s at this fluvial-tidal interface.

CHANNEL MAPPING Given the complexity of the system and the diverse available data, mapping historical channels involved some uncertainty (see page 51). We mapped 230 miles (370 km) of low order channel that we classified with high confidence of early 1800s presence, with an additional 310 miles (500 km) of probable and 80 miles (130 km) of possible early 1800s presence within 150,000 acres (60, 700 ha) of emergent wetland.

For the purposes of this summary, the area lies south of the northern boundaries of present-day Roberts Island and Jones, Woodward and Byron tracts. Based on the mapping performed, we estimate that this represents overall channel density between 8 and 20 feet per acre (0.6-1.5 km/km^2), not including mainstem channels. We mapped approximately 37% as exclusively fluvial channels, which naturally fall within the most southern parts of the study area.

The number of channel segments mapped per unit area is a visible landscape-scale difference when compared to the central Delta. However, when comparing metrics such as channel density between the central and south Delta, landscape context and differing formative processes should be considered. The mapping process produced channel configurations similar to those mapped by Atwater (1982), who used 1970s aerial photography and historical USGS topographic maps. Total mapped miles were comparable as well. For example, on Union Island, we mapped 156 miles (251 km) of channel of either high or medium certainty with an additional 36 (58 km) of low certainty in comparison to 158 miles (254 km) in Atwater's mapping.

Atwater identified those channels that were dominated by fluvial processes with a classification of "chiefly or wholly subject to non-tidal flow." Since we classified a channel as tidal or non-tidal depending on whether we understood water levels to be affected by tides as opposed to whether fluvial processes were dominant, many of the secondary channels that we mapped at the tidal wetland boundary (i.e., where we believe channel beds fell below tide level) were classified as possibly tidal (medium or low certainty level). This means that quite a few secondary channels classified as tidal in the mapping are "chiefly or wholly subject to non-tidal flow" in Atwater's mapping. Consequently, tidal classifications between the two mapping efforts should not be compared.

Another important consideration that has implications for mapping interpretation is that not all visible channel remnants in aerial photos (i.e., the inorganic sediment banks set against darker peats that show up in aerial photos) were active secondary channels in the early 1800s. We found that a substantial portion of the channels mapped from historical aerial photography in this region may be pre-1800 channels that were exposed in the process of reclamation (Fig. 6.13). That is, as the relatively thin peats of the south Delta have oxidized, pre-1800-era channel deposits beneath are exposed at the surface (Lajoie 2010, Atwater pers. comm.). Atwater (1982) discusses this issue concerning his mapping of non-tidal channels in the south Delta, explaining that "many…traces were covered with tidal-wetland deposits within the past 5,000 years and then exhumed in historic time." Applying this uncertainty generously, we estimate as much as 50% of the total mapped low order channel length in the south Delta may be exhumed in this manner. This is particularly an issue in the south Delta, where peats are shallow and the river well connected to its

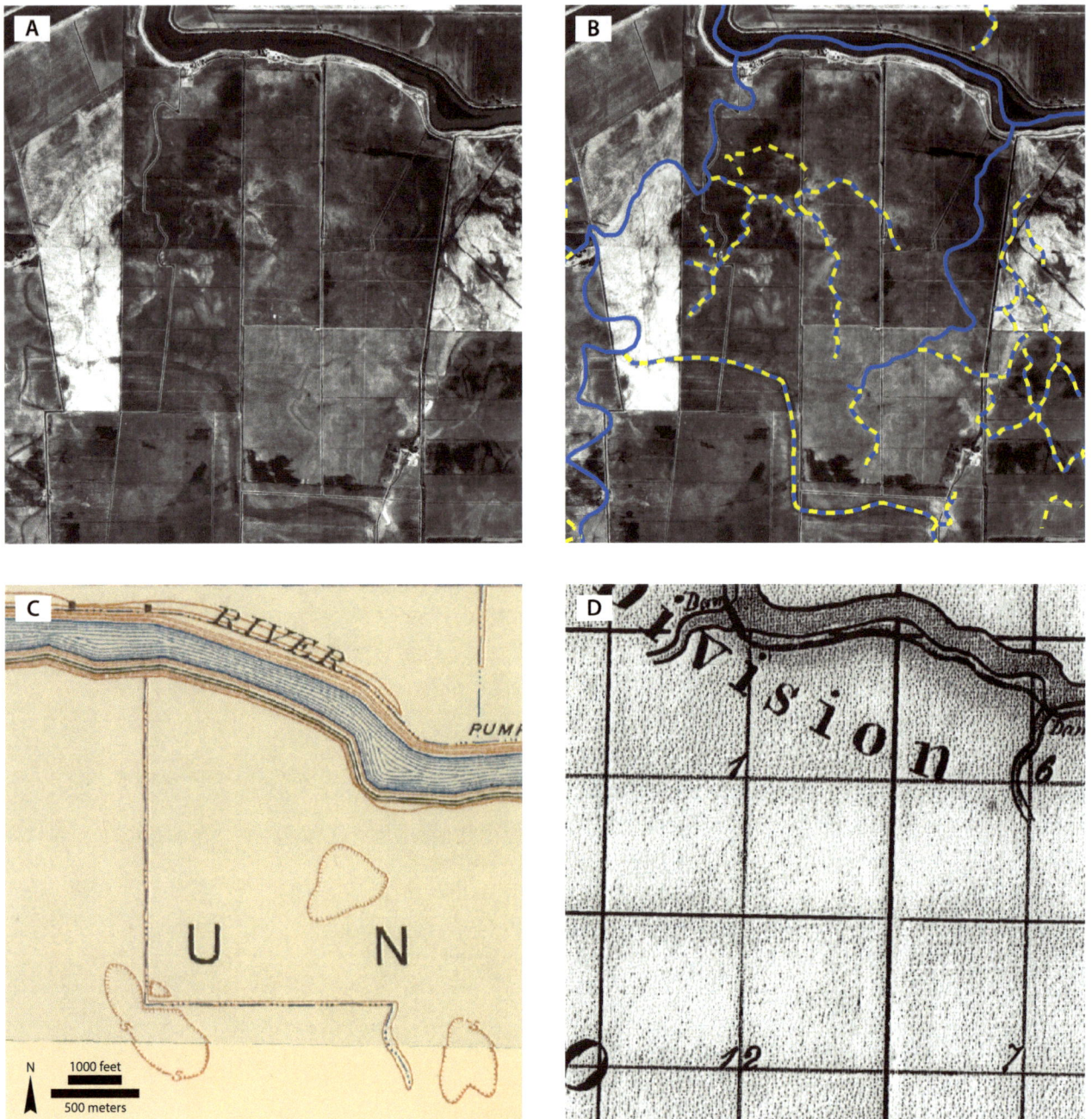

Figure 6.13. Numerous remnant channel signatures in the south Delta can be seen in historical aerial photography (A). Some of these may be older (pre-1800s) channels that were exhumed in the process of reclamation, which gradually removed the peat layer through unintentional oxidation. Those channels confirmed by reclamation era sources (and therefore not exhumed) are shown in B as blue lines and possible ancient exhumed channels are shown as dashed yellow and blue lines. An early 1900s USGS topographic map (C) shows low positions and remnant channels associated with early 1800s channels in an already modified landscape, and a reclamation map (D) confirms the presence of two tidal channel networks that were functional prior to reclamation. They are already dammed by the time of this 1876 map. (A, B: USDA 1937-1939; C: USGS 1909-1918; D: Wallace 1876, courtesy of the California State Lands Commission)

floodplain. While this uncertainty suggests that the total mapped channel density may over-represent what was present in the early 1800s, the morphology differences between the south Delta low order channels and the blind tidal channels of the central Delta still hold true and can be evaluated using the historical mapping, particularly those channels of high certainty. More research to determine the mechanism by which the channels have seemingly appeared and disappeared through time and to identify possibly exhumed channels, such as studying and dating the channel deposits, could help address uncertainties associated with interpreting these channel signatures.

INTERPLAY BETWEEN TIDAL AND FLUVIAL PROCESSES One characteristic with far-reaching implications for landscape function was that the San Joaquin River and its larger distributaries were tidal while many of its secondary channels and much of the surrounding floodplain were not. However, many secondary channels of the San Joaquin River south of the confluence with the Stanislaus were at least in part influenced by the tidal Delta. This influence was either through direct changes in water levels in the channel due to tides or through indirect effects from being part of a larger wetland that was tidal at the downstream end (e.g., through moderated water levels). Some were tidal at higher river stages or were tidal for only portions of their reaches. The gradual decrease in height of natural levees along the major distributary channels reveals this interaction between tidal and fluvial processes: large subtidal secondary channels are rare where substantial levees are present. This relationship can be seen in an 1876 reclamation map of Union Island, where channels branching off the mainstem that are large enough to be dammed only become common where natural levees become almost level with the surface of the marsh plain (Fig. 6.14). For the most part, south Delta channels were unlike the tidally-dominated channels of the central Delta: morphology and hydrology of many low order channels in the south Delta were driven primarily, though not exclusively, by fluvial processes.

The significant influence of fluvial processes is revealed in the mapping, where planforms are distinct from those in the central Delta. In comparing the channel planform of upper Roberts and Union Islands to their lower, tidal portions, a distributary pattern suggesting dominant flow northward meeting southward branching tidal channels is visible at the landscape scale (Fig. 6.15). For example, using historical survey and cartographic evidence (e.g., Gibbes 1850a, Gilbert 1879, USGS 1909-1918, Unknown 1917a, USDA 1937-1939), six channels were mapped branching northward along the over four mile reach along Old River between the San Joaquin and Middle River. Their common direction and straighter planform in comparison to tidal channels of the central Delta is indicative of a northerly flood flowpath from the higher elevation upper Roberts Island (generally between five and ten feet/1.5-3 m above sea level) to the tidal plain lying about five miles (8 km) to the north.

Exhumed channels in the south Delta: If we assume an annual historical accumulation rate of peat at 1-2 mm per year and we measure subsidence of 2 meters in a particular location, then we should be seeing a land surface (as well as the mineral deposits from channels in between) from approximately 1,000-2,000 years ago.

—ATWATER PERS. COMM.

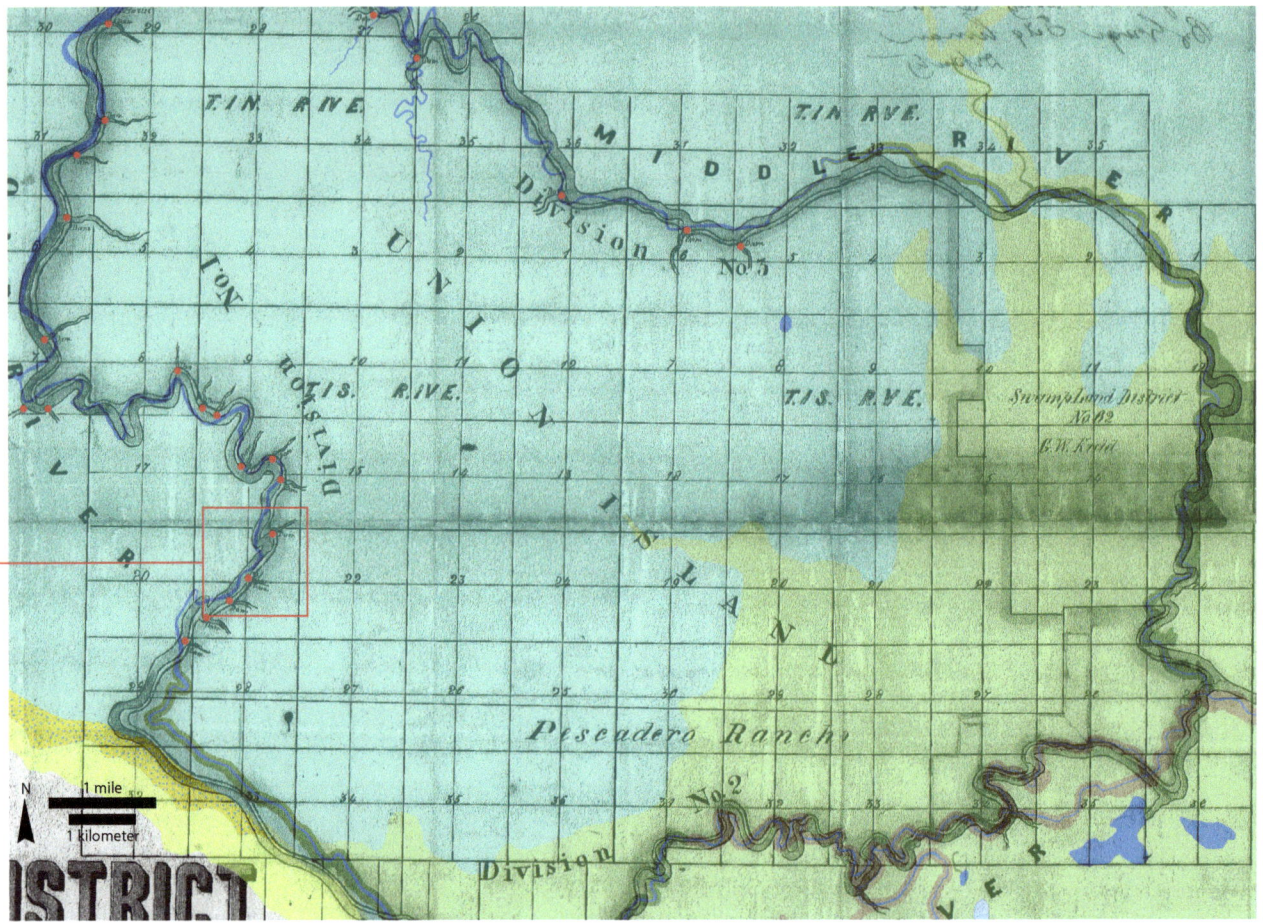

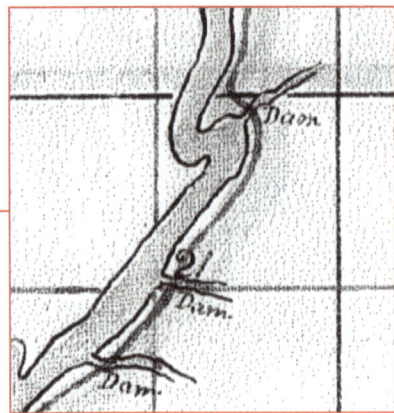

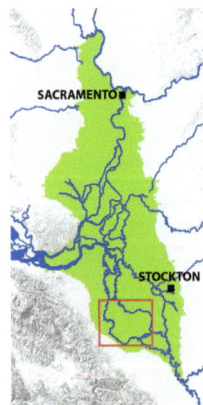

Figure 6.14. Channels large enough to dam become sparse where rivers are bordered by natural levees. These channels (red dots) are found only along lower (or the north part of) Union Island, as shown in this 1876 reclamation map. Natural levees became more substantial and limited the formation of these networks. Also, tidal energy was lower and therefore did not establish a dense network of channels. This transition is represented in the habitat type map with the appearance of willow riparian scrub along the mainstem channels and non-tidal wetland on the floodplain. These were significant tidal channels, as suggested by the dams that are indicated at the mouth of each channel, as shown below. (Wallace 1876, courtesy of the California State Lands Commission)

Water

Tidal freshwater emergent wetland

Non-tidal freshwater emergent wetland

Wet meadow or seasonal wetland

Alkali seasonal wetland complex

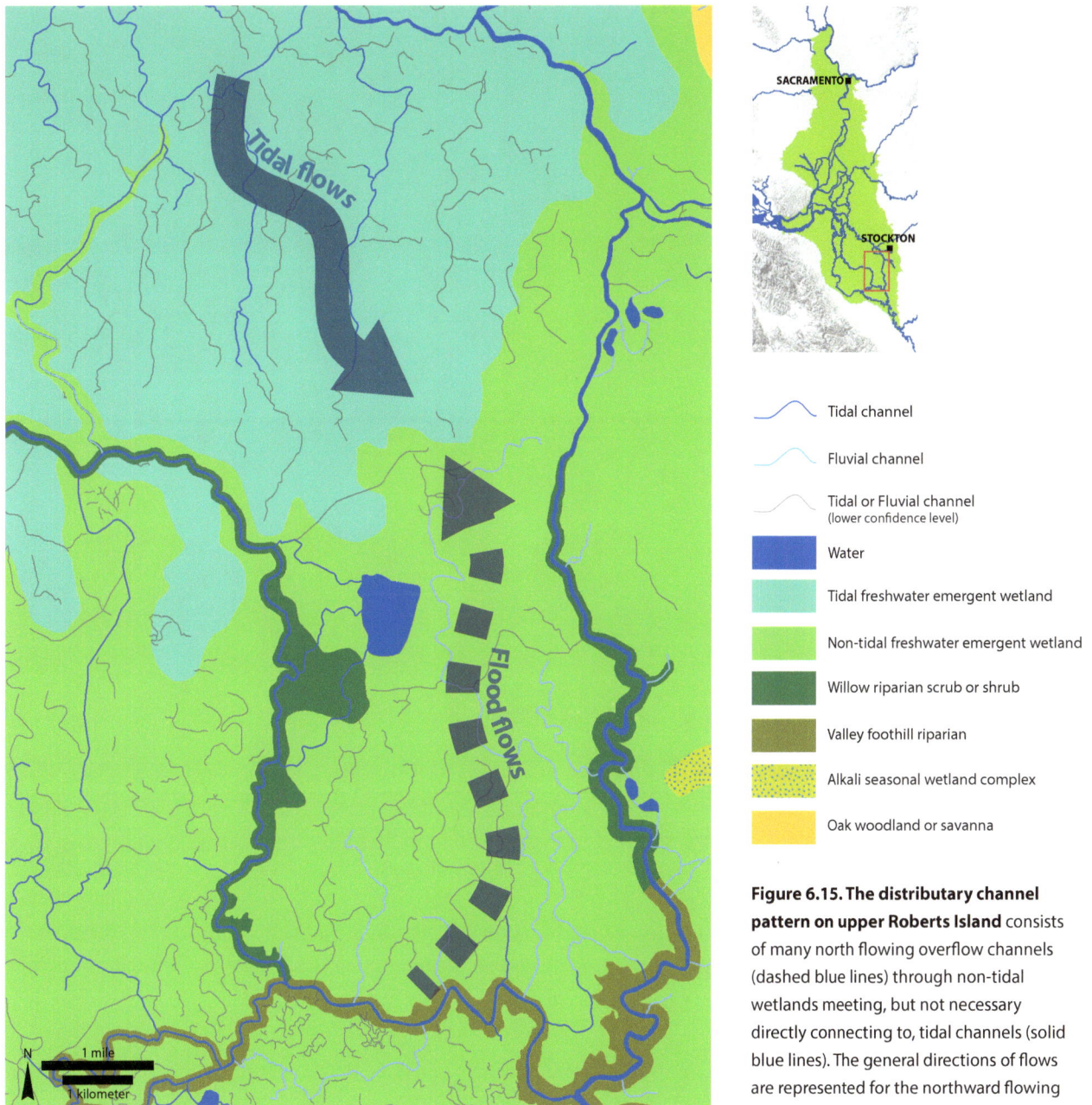

Tidal channel

Fluvial channel

Tidal or Fluvial channel
(lower confidence level)

Water

Tidal freshwater emergent wetland

Non-tidal freshwater emergent wetland

Willow riparian scrub or shrub

Valley foothill riparian

Alkali seasonal wetland complex

Oak woodland or savanna

Figure 6.15. The distributary channel pattern on upper Roberts Island consists of many north flowing overflow channels (dashed blue lines) through non-tidal wetlands meeting, but not necessary directly connecting to, tidal channels (solid blue lines). The general directions of flows are represented for the northward flowing overflow channels in a large dashed blue arrow and the tidal channels in a large solid blue arrow.

Another landscape element illustrating the influence of fluvial processes is the presence of inorganic sediment banks along many of these channels, including some of which may have been historically affected by tidal flows at certain times of the year. These banks are visible as lighter tonal signatures in the aerial photography and shown as linear topographic features in the early USGS topographic maps (see Fig. 6.13; USGS 1909-1918). Despite the fact that bank topography may be more pronounced in the topographic maps than it was in the early 1800s due to peat loss, this information is still relevant to inference of formative processes. Also, the trend of decreasing height northward into tidal range reflects the dominant flood flows in the direction away from the mainstem rivers.

Legend:
- Tidal channel
- Fluvial channel
- Tidal or Fluvial channel (lower confidence level)
- Water
- Intermittent pond or lake
- Tidal freshwater emergent wetland
- Non-tidal freshwater emergent wetland
- Willow riparian scrub or shrub
- Valley foothill riparian
- Wet meadow or seasonal wetland
- Alkali seasonal wetland complex
- Grassland
- Oak woodland or savanna

N
1 mile
1 kilometer

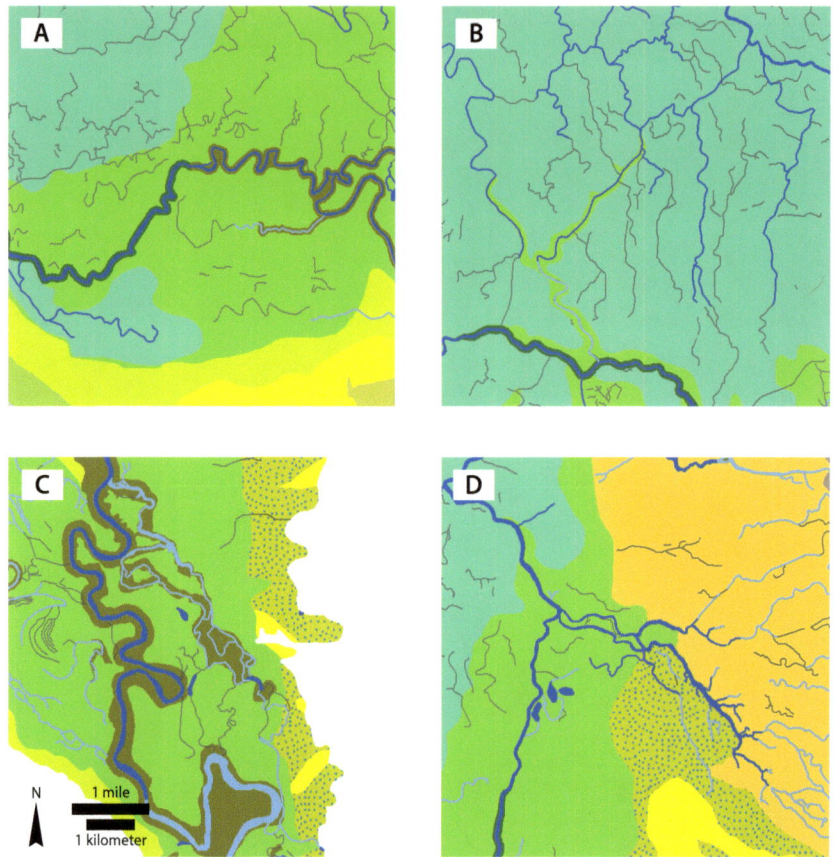

Figure 6.16. Examples of three different types of secondary channels are shown in our mapping. In A, the most common type of channel is highlighted, one that terminates within the perennial (in this case, non-tidal) wetland. The channel highlighted in B, Duck Slough in Roberts Island, is an example of only a few larger secondary channels that begin as a fluvial-dominated channel upstream and directly connect to a tidal-dominated channel downstream. The third example (C), Walthall Slough, is a type of channel common to fluvial floodplain systems, where a fluvial-dominated channel exits and then returns back to the same mainstem channel downstream. French Camp Slough (D) is an example of a relatively rare type of channel that directly connects to upland drainages. These graphics also depict in dashed lines those channels mapped from historical aerials that are associated with lower certainty due to absence of mid-1800s era confirmation. For many of these channels, it is often difficult to determine which primary type of channel they belonged to historically. However, as is supported by historical accounts, one can assume that most secondary channels dissipated into the wetland plain rather than directly connecting back up to a main channel.

DISTINGUISHING CHARACTERISTICS The secondary channels of the south Delta floodplain can be grouped according to whether they spread and terminated within wetlands, directly connected to a tidal channel of the central Delta, or returned to the river a few miles downstream (Fig. 6.16). A fourth and relatively rare type not discussed here are secondary channels that connect directly to an upland drainage, such as French Camp Slough (see page 164). Away from the mainstem river, it is likely that only directly tidally connected channels maintained flows year-round. However, those channels that were seasonally isolated from tides or subject solely to flood flows remained as wetter features in the landscape, becoming standing pools of water or depressions late in the season.

Most secondary channels terminated within wetlands. With the higher banks of the San Joaquin River distributaries, river water levels at a rising stage were usually higher than the land surface on the other side of the bank. Dips and breaks in the natural levee allowed floodwaters to escape into the lower lying wetland, forming the secondary, or overflow, channels (*Daily Alta California* 1852). Many of these breaks, however, were not so deep as to intersect low water elevations (U.S. War Department 1900). The small channels thus transported water laterally across the river's natural levees only during high flows.

These channels either terminated immediately upon reaching the wetland beyond or became lower swales within the wetland complex that dried out later in the season. Surveyor Gibbes' report on his 1850 survey of the San

Joaquin includes a discussion of this pattern, one that would not have been found in the central Delta. He states that the land on upper Roberts Island is "two to five feet lower than the banks of the river and when the water is high most of the small slues [sic] afford fine water power" (Gibbes 1850b). He continues with: "these discharge into small lakes or spread out into the tule, and are drained off by the slues." Here, his reference to sloughs probably means the larger tidal sloughs such as Duck Slough that carried the water into the tidal network of the central Delta. This description corroborates the mapping that was based on cartographic and photographic evidence, where many non-tidal secondary channels lose defined beds and banks at some point within the floodplain.

Court transcripts from a case related to upper Roberts Island pertaining to flooding caused by a dammed slough provide characterization of the secondary channels. The transcript documenting one of these features describes: "by said natural way and depression, all of the said waters which so accumulated on said lands were naturally conducted and carried away from the lands of plaintiff except a small amount estimated at 34 acres" (Fig. 6.17). The sheer number of these "natural ways" gives the sense that significant volumes of water at high river stages made their way across the floodplain within slower moving swales, filling depressions along the way. Only during the highest flows would all of the land overflow. The localized nature of this flooding is suggested by topographic variability evident in early USGS topographic maps (USGS 1909-1918). Some of this water reached tidal wetlands to be conveyed eventually to San Francisco Bay, while a portion remained on the surface to evaporate over the course of the dry season.

Other evidence found in reclamation documents affirms that numerous channels bisected the natural levees of the major waterways, flowing only

With reference to the sloughs, it is known that where some of them leave the river the bottom of the sloughs are 5 to 6 feet above low-water surface of the river, and hence water cannot flow from the river into such sloughs until the river is about 5 feet above its low-water stage.

—U.S. WAR DEPARTMENT 1900

I have seen the water in some of them a foot lower than the river, and rushing like a mill stream; these discharge into small lakes or spread out in the tule, and are drained off by the slues...

—GIBBES 1850B

Figure 6.17. A seasonal floodway is shown in this photograph from May 26, 1907. The court case that resulted from the alleged flooding caused by the dam seen in the photo included testimony stating that the secondary channels, or "natural ways," allowed for most of the water to move off the floodplain during high flow events. (Unknown 1917c)

338

during high stages. For example, on Tom Paine Slough efforts were made to dam "small sloughs or 'cuts'" (Tucker 1879d). It was also specified that "the banks of Tom Paine Slough are very high and ordinary floods do not get over them except in some few low places." Presumably these low places were the start of the small sloughs referred to. For this reason, although Tom Paine Slough was tidal, its secondary channels carried flood flows exclusively. They appear to have been well established features, however, as indicated by topography shown on several detailed maps (Fig. 6.18; Herrmann 1921). We were able to map nine such secondary channels along a 10 mile (16 km) stretch of the right bank of Tom Paine Slough with high confidence. A few others were mapped with lower confidence as they were located exclusively using historical aerial signatures. Additional textual discussion of "small sloughs" was also found for the reach of the San Joaquin that extends along present day Stewart Tract, where three to four channels were reported upstream of Paradise Cut and "a few" between it and the head of Old River (Naglee 1879). Through the mapping process, we found that this pattern occurred along other similarly positioned channels (i.e., along tidal mainstem channels with natural levees and a wetland plain above mean high tide levels).

It appears that some low order channels bisecting natural levees and entering non-tidal wetlands may have experienced tidal flows at low water stages. Connections like these would have been important avenues of exchange between the non-tidal wetland plain habitats and the tidal rivers. A few early accounts suggest that some secondary channels were deep enough to maintain tidal flows at low river stages:

> Besides these low places, there are occasional narrow, deep breaks of twenty to thirty feet wide, and from five to ten feet deep; these all lose themselves in reaching the low lands, which are from ten to two hundred yards from the river, with occasional exceptions of greater extent. There are also large outlets or branches of the main river, with continuous deep channels, many of which continue to flow at the lowest stages of the river,

Figure 6.18. Secondary channel banks built by floods of Tom Paine Slough are shown in this detailed topographic map (1-ft contour interval). They flowed only during high water, but offered substantial topographic complexity within the landscape. (Herrmann 1921, 1966.x-335.001, San Joaquin County Historical Society, Lodi)

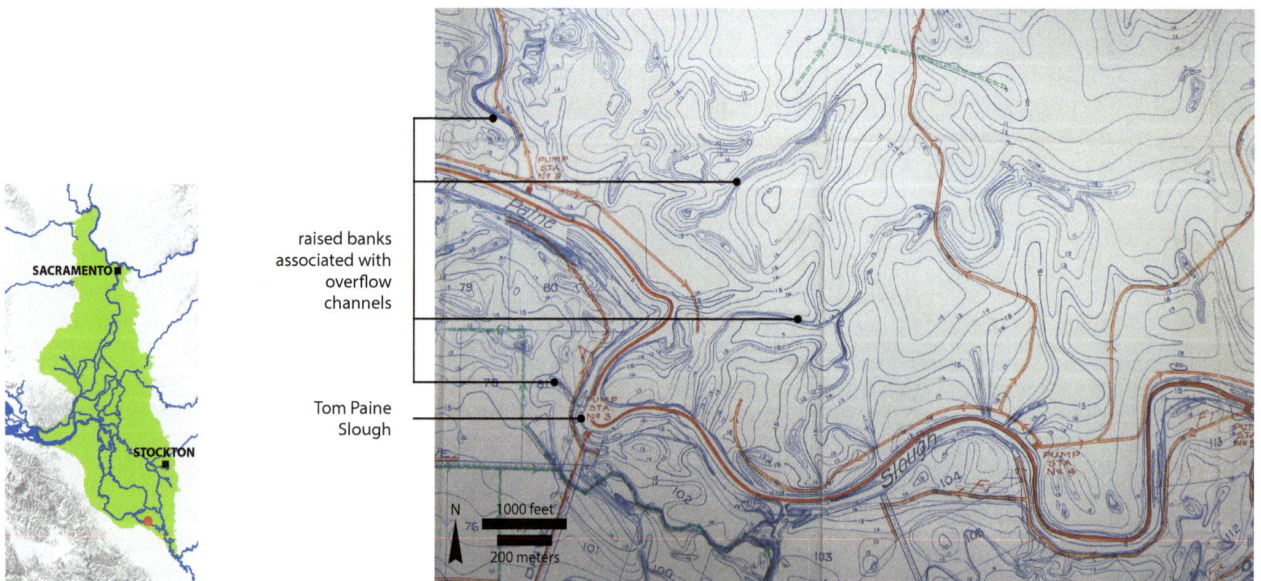

raised banks associated with overflow channels

Tom Paine Slough

while others during the higher stages only; most of these outlets unite again with the main channel, and thus form extensive islands. (*Daily Alta California* 1852)

Unfortunately, this quote refers to San Joaquin County in general, so it is unclear how prevalent these deeper branches were in the south Delta. Though the number and extent of such channels was challenging to determine given limited spatially-explicit evidence, we were able to identify several using double line channels shown in the 1850 Gibbes map. These were interpreted as navigable channels that likely maintained tidal flow at lower river stages (unfortunately, the fact that surveying was conducted during high river stages casts some doubt; Fig. 6.19). We found that other early maps supported that these were the larger, more well established channels and may have been affected by tidal flows (Wallace 1870, Hall ca. 1880a, Compton ca. 1894). While no one map or account provided clear evidence, the synthesis of multiple independent maps and accounts offered reasonable support for mapping these features as possibly subject to tidal flows.

The second type of secondary channels connected directly to tidal channels of the central Delta. Notable examples include Duck Slough, possibly Whiskey Slough, and an unnamed channel branching into the interior upper Union Island. In the case of Duck Slough, a number of maps establish the early presence of the waterway, which was clearly tidal at its downstream end at Rough and Ready Island (Hall ca. 1880b, San Joaquin County Surveyor 1882, Tucker and Smith 1883). Its connection to Middle River is confirmed by evidence of sediment banks extending to meet the tidal channel and the appearance of a natural sinuosity in the artificial levees visible today. It was also mapped independently by geologists Brian Atwater (1982) and Ken Lajoie (2010). Lajoie (2010) asserted that "the size and complexity of the Duck Slough and its numerous distributary and tributary channels…indicate this is the primary drainage system between Middle River and the San Joaquin River." The channel does not appear to

Figure 6.19. **Well established secondary channels,** depicted as double-line channels, are shown leaving the main branches of the San Joaquin River. In contrast to the single line channels, these likely maintained tidal flows through relatively deep cuts in the natural levees along the rivers. At the transition between tidal and fluvial landscape, much of the land surface in this part of upper Union and Roberts islands and present-day Stewart Tract was above the reach of tides, though some secondary channels like those shown here likely intersected below tidal elevations like the river channels. (Gibbes 1850a, courtesy of the Map Collection of the Library of UC Davis)

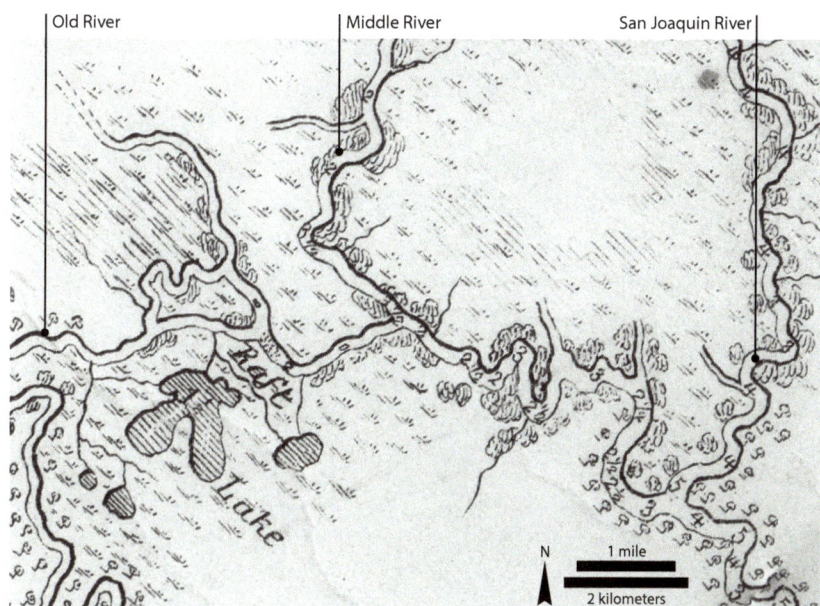

have been tidal for its full extent: it was characterized by high banks and held a position within a non-tidal wetland at its upstream end on Middle River. This is supported by the written record: reclamation documents refer to the "head of Duck Slough" at Honker Mound, which was located about four miles (6.4 km) downstream on Duck Slough from its divergence from Middle River (Tucker 1879b). This is likely a reference to the point where tidal action ends. It is also worth noting that Gibbes (1850a) did not map a channel connecting all the way through (he seems to have mapped both ends instead).

It is possible that other such connections in the south Delta were present as well. It was challenging to determine how the channel network linked together, as can be seen in the many fragments of channels mapped in the south Delta. We do not believe, however, that these were substantial sub-tidal channels. Otherwise, they would have been more well established features in the historical record and it is unlikely that Union and Roberts islands would have been referred to as single islands.

Lastly, a pattern particularly common in the southernmost extent of the Delta (i.e., in the vicinity of the San Joaquin Bridge) was for secondary channels to branch off of the main river, only to return to the same channel just a few miles farther downstream. This pattern is common to many floodplain landscapes. The most notable and spatially extensive is that of Walthall Slough, which lies along the east bank of the San Joaquin River downstream of the Stanislaus confluence (Fig. 6.20). Today, it is confined to a single channel and has no upstream connection. Historically, however, a maze of floodplain channels received overbank flows and coalesced into Walthall Slough, where "a large portion" of the overbank and side channel flows then re-entered the river just above the San Joaquin Bridge (Kluegul 1878, USGS 1909-1918, U.S. Army et al. 1914-1915). This convergence of flow at the mouth of the slough is evident in historical maps (e.g., Fig. 6.20b), but most lack the spatial resolution to show the smaller upstream connections (Gibbes 1850a, Gibbes 1869, Wallace 1870, Secretary of State 1866-1877, Unknown 1915). They are visible, however, in more detailed mapping of the early 1900s (USGS 1909-1918, U.S. Army et al. 1914-1915). The sharp angle in the downstream flow direction at which Walthall Slough enters the river also indicates that the flow in Walthall Slough was primarily directed into rather than away from the downstream mainstem flows at that point. An earlier 1861 Swampland District map along the east bank of the San Joaquin River shows this pattern occuring at several other points

Along the edge of the lowland just below this terrace a string of lakes connected by sloughs extend throughout the greater part of the area.

—SWEET ET AL. 1908

Figure 6.20. Overflow channels along the San Joaquin are shown in historical maps (at right, of different scales) that depict the channel network that comprised Walthall Slough. In A, an 1887 general map of the valley shows Walthall Slough exiting and then re-entering the San Joaquin River downstream. Other general maps, like that in B, just show the larger downstream part of the channel, where overflows coalesced into a single channel. Greater channel detail can be found in maps of larger scales such as the historical USGS topographic maps (C). The historical habitat mapping, with the Walthall Slough network depicted, is shown in D, and can be compared to remnants confined to only a few channels today (E). (A: Hall 1887, courtesy of the Map Collection of the Library of UC Davis; B: Unknown 1915, courtesy of the Earth Sciences & Map Library, UC Berkeley; C: USGS 1909-1918; E: USDA 2009)

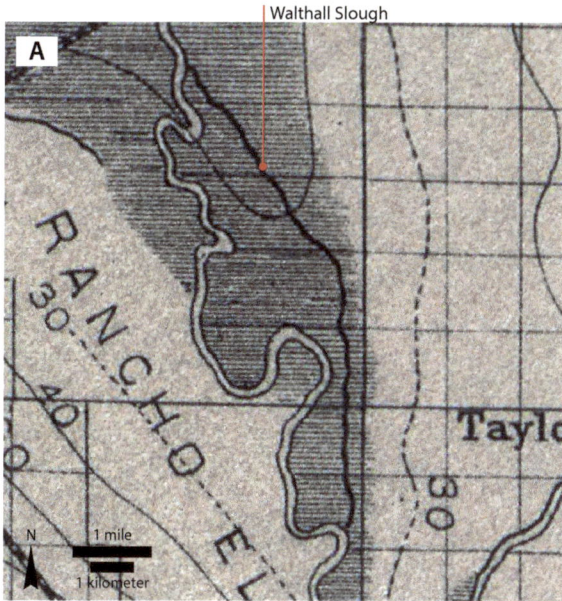

Walthall Slough

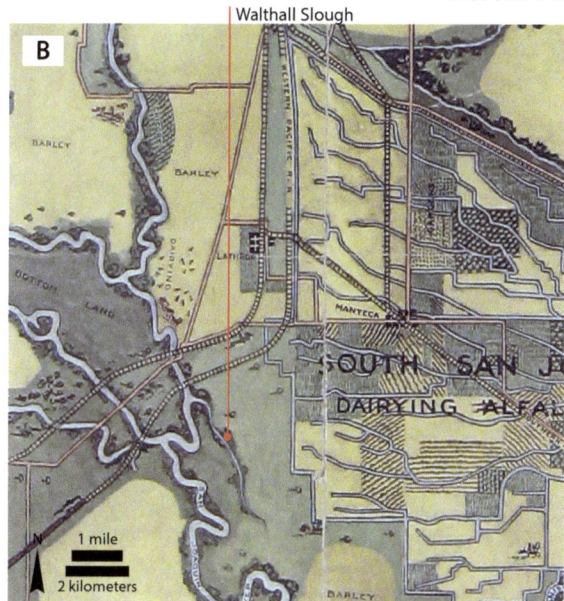

Walthall Slough

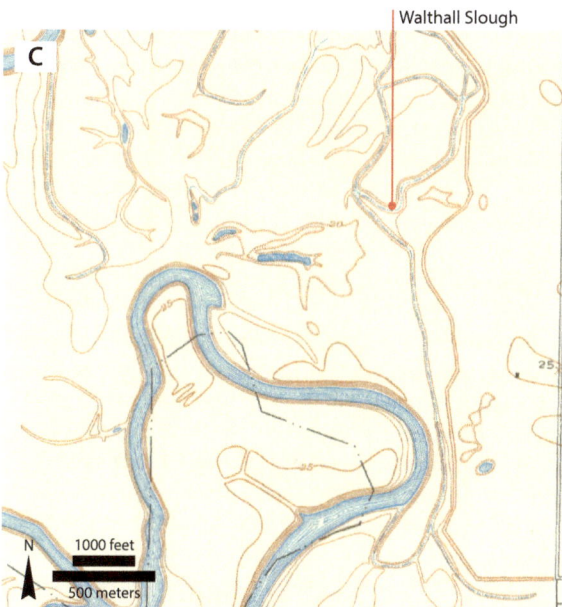

Walthall Slough

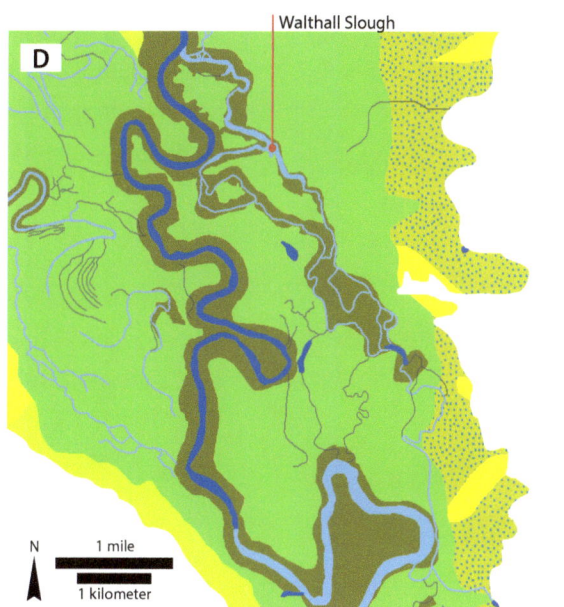

Walthall Slough

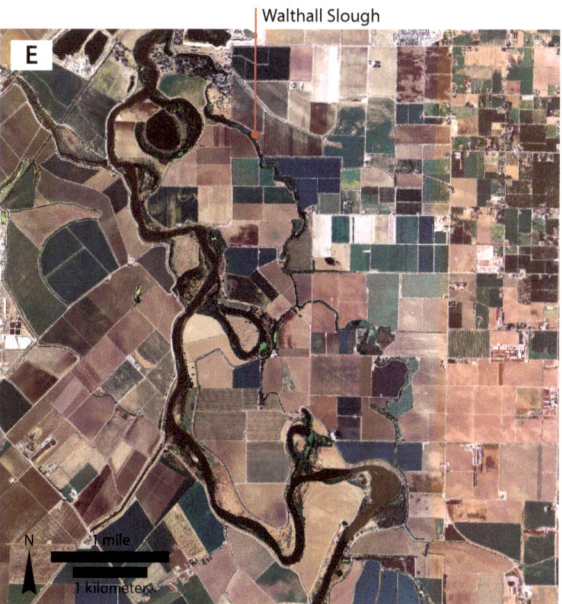

Walthall Slough

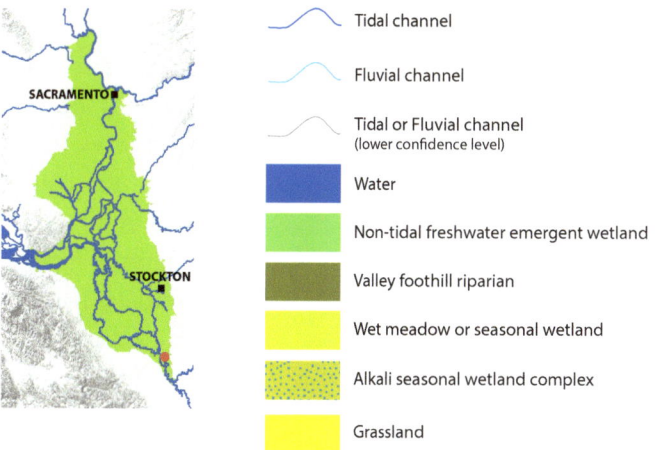

Tidal channel

Fluvial channel

Tidal or Fluvial channel
(lower confidence level)

Water

Non-tidal freshwater emergent wetland

Valley foothill riparian

Wet meadow or seasonal wetland

Alkali seasonal wetland complex

Grassland

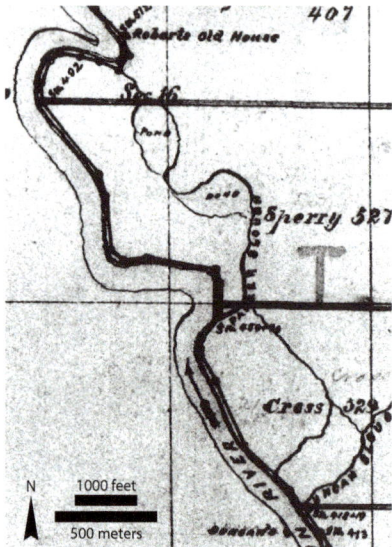

Figure 6.21. The pattern of overflow channels exiting and re-entering the San Joaquin is shown in this 1861 reclamation map. This is located a few miles downstream of the head of Old River. (Beaumont 1861a, courtesy of the California State Lands Commission)

downstream, though these are much shorter networks than Walthall Slough (Fig. 6.21; Beaumont 1861a).

Overall, the overbank channel networks that bisected natural levees and passed into the wetland likely carried a substantial fraction of San Joaquin flows during high stages. Floodplain connectivity and much of the habitat complexity present in the landscape depended upon the presence and functions of these features.

Channels through time

Meandering river landscapes change through time, as riverine forms adjust to changing flow and sediment regimes (Leopold 1994). As rivers meander across their floodplains on the scale of hundreds of years (with perceptible shifts occurring on a decadal scale), they create characteristic landforms of alluvial deposits, crevasse splays, secondary channels, meander scroll topography, and oxbow lakes (Florsheim and Mount 2002, Burow et al. 2004, Singer et al. 2008). Where tidal influence diminished on the San Joaquin River, in the vicinity of Sheppard's Ferry and upstream, the San Joaquin could be characterized as a meandering river with a single mainstem channel that remained fairly stable from year to year (Lyons in Houghton 1862, Mount 1995). Similar meandering river signatures are observed on the Sacramento above the Feather River confluence. On each river, this visible change in channel morphology occurs where tidal influence becomes minimal. As a representation of a shift in dominant processes, these positions also signify a transition from a riverine landscape with a greater disturbance frequency to one where components were fixed over space and time by tidal processes. The landforms and resulting habitat mosaics, both in the north and south Delta, are due in part to the interaction of processes at this transition.

Evidence of the lateral migration of the San Joaquin is most visible south of the head of Old River in early USGS topographic mapping and aerial photography. For example, near the head of Walthall Slough, a portion of a distinctive bend in the river has been cut off and the former bend appears in the process of becoming an oxbow lake (Fig. 6.22). Point bars and scroll topography are mapped in the early 1900s Debris Commission maps, plainly visible in the 1937 aerials, and sometimes visible in modern imagery as well. In an example just downstream of Sturgeon Bend, the channel has shifted from its pre-1900s position (labeled "Old Channel" in the Debris Commission maps), and scroll topography is visible (Fig. 6.23). Today, that channel bend has migrated northward. Also, oxbow lakes formed by previous meander cutoffs are positioned to the west and south of this bend. This is an example where the mapping synthesis was based on early 1900s sources, so the true position of the early 1800s channel may actually correspond to where the oxbow lakes are shown in the early USGS topographic maps (Von Schmidt 1855). Despite this uncertainty in exact location and timing, the mapping conveys the overall meandering river character of the early 1800s landscape.

A

B

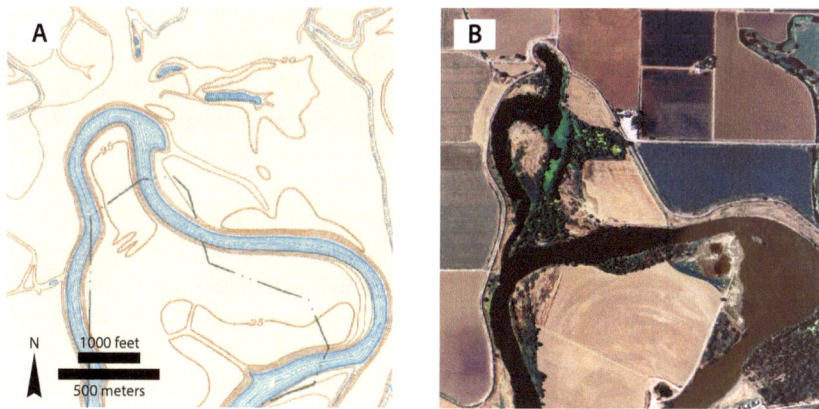

Figure 6.22. A bend on the San Joaquin River changes through time. In (A), the bend seen in the 1915 USGS topographic map has been cut off by the time the aerial photography was taken in 2005 (B). The former meander bend is now becoming an oxbow lake. (A: USGS 1909-1918; B: USDA 2005)

Fig. 6.22
Fig. 6.23

Another striking example of meander scroll topography is found at the head of Tom Paine Slough, which suggests that at one time it may have been a more significant distributary of the San Joaquin (Fig. 6.24). These features, mapped in the early USGS topographic maps as intermittent streams, show the lateral progression outwards of a meander bend through time. By the early 1800s, it was only connected seasonally to the San Joaquin River upstream (Kluegal 1878). Many early regional maps do not show an upstream connection (e.g., Hall 1887) at all, while the more detailed maps show connections via relatively small channels (e.g., Gibbes 1869, USGS 1909-1918). Downstream, however, it was a more substantial channel that was tidally influenced: Gibbes reported from his 1850 survey that Tom Paine Slough was navigable up to Martin's tent, where Paradise Road crosses the slough today (Gibbes 1850b).

The more frequent shifting of channel alignment through time makes it challenging to interpret the features that represent the early 1800s channel configuration from circa 1900 sources. Though our goal was to map those channels that were likely functional in the early 1800s, we may have

Figure 6.23. Evidence of the dynamic nature of the San Joaquin River is found in the delineation of the "Old Channel" in the 1913 U.S. Army Corps mapping (A). Scroll topography (circled in red) visible in the 1937 aerial photography (B) reveals other past locations of the river channel as the central bend had been migrating north. (A: U.S. Army et al. 1913, courtesy of the California State Lands Commission; B: USDA 1937-1939)

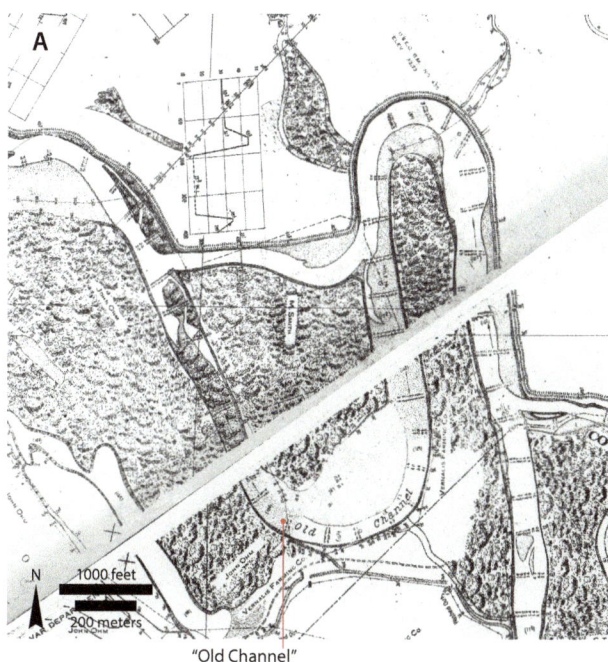

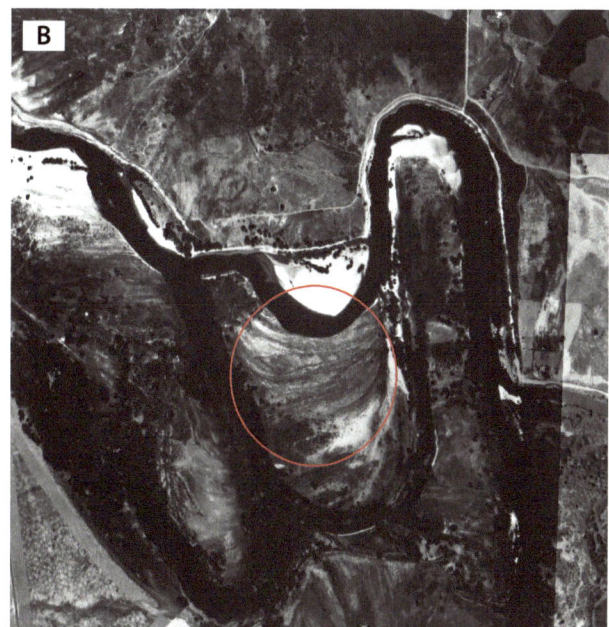

A

"Old Channel"

B

344

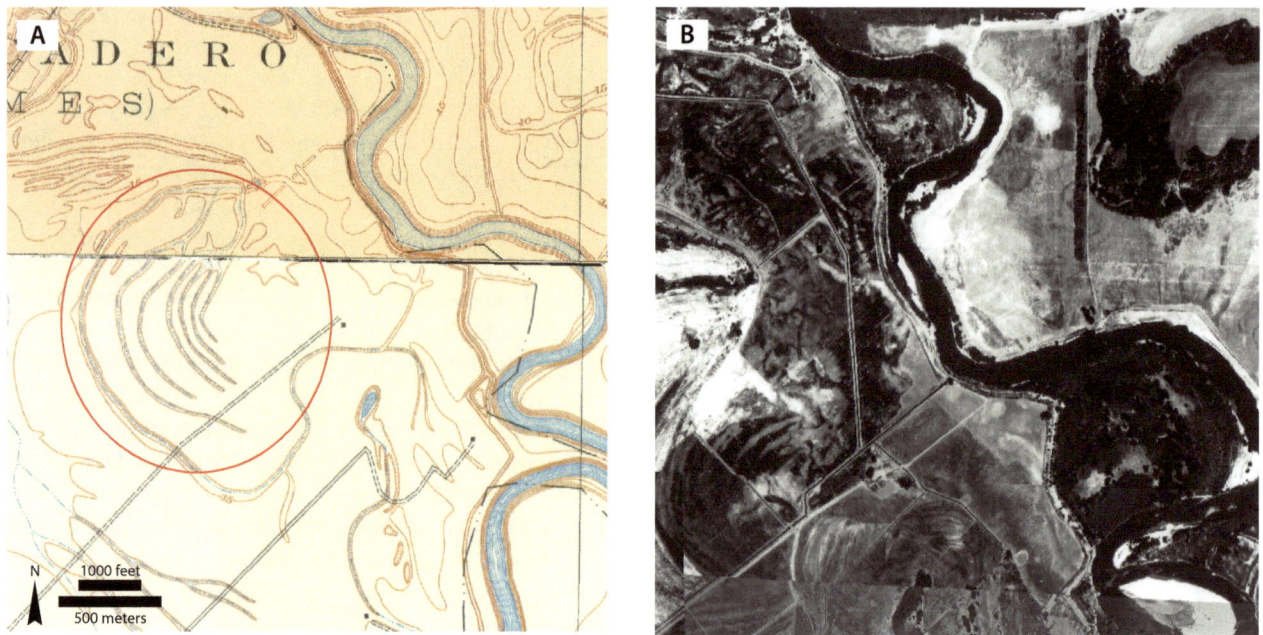

Figure 6.24. Meander scroll topography (circled in red) reveals the progression of a former meander bend on Tom Paine Slough as it moved outwards through time. (A: USGS 1909-1918; B: USDA 1937-1939)

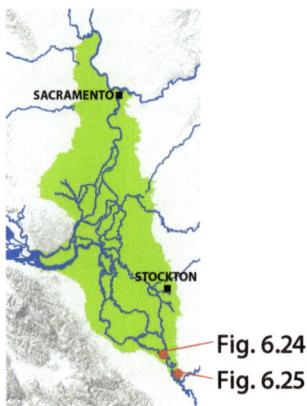

Figure 6.25. Different secondary channel orientations are visible in two mapping efforts from the early 1900s, possibly indicating frequent changes in morphology. Determining likely early 1800s channels is more challenging in the dynamic southern extents of the Delta region, where fluvial influence dominates. These maps are both generally accurate enough that we would not expect the differences to be a result of spatial errors. (A: USGS 1909-1918; B: U.S. Army et al. 1914-1915)

overmapped the channel network of the south Delta given our heavy reliance on early 1900s sources (mostly post-reclamation, post-onset of peat oxidation). Even independent sources mapped during the same decade did not always agree on channel orientations (Fig. 6.25). Particularly for an area like the south Delta, where channels can come and go naturally over the course of a decade, confirmation either through early maps or textual evidence is especially important for establishing early 1800s presence. An example of this is Paradise Cut, which apparently did not carry substantial flood flows prior to 1859 (Naglee 1879), despite appearing as a historical channel in the early USGS topographic maps and aerial photography (Box 6.2). While the alignment or early 1800s presence of a typical south Delta channel may be associated with greater uncertainty than elsewhere in the Delta, textual accounts and early maps suggest that the overall pattern mapped is representative of early 1800s conditions and processes.

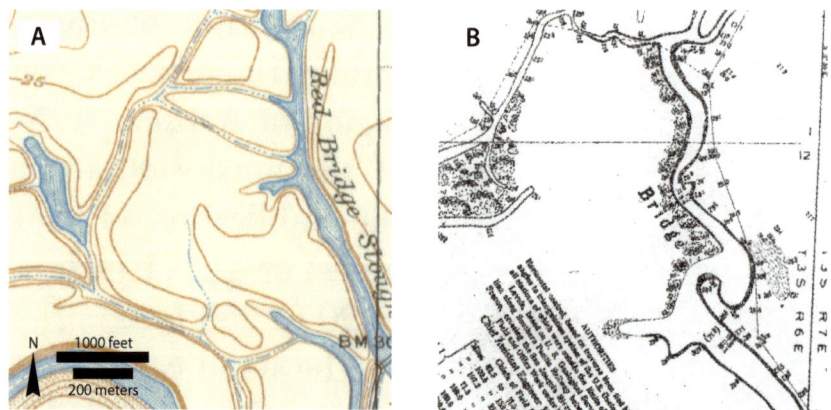

BOX 6.2. EVIDENCE OF EARLY CHANGE AT PARADISE CUT

Today, Paradise Cut forms the western boundary of Stewart Tract, located just south of Union and Roberts islands, and is dammed at its head on the San Joaquin mainstem. Paradise Cut as shown by early 1900s sources (e.g., USGS 1909-1918, U.S. Army et al. 1913) was apparently more similar to conditions today than to early 1800s conditions. Before the early 1860s, the secondary channels occupying this area were less substantial and did not carry perennial water like they do today. When a GLO surveyor crossed immediately northwest of the railroad in 1851, he reported several dry channel beds, one of which was 65 feet across (Norris 1851b). The historical channels in this area also lacked the capacity to convey "at least one third of the San Joaquin River," as Paradise Cut did by 1879 (Tucker 1879d). Paradise Cut is also markedly absent from early maps (Fig. 6.26; Gibbes 1850a, Gibbes 1869, Secretary of State 1866-1877, Wallace 1870, Gilbert 1879, Hall ca. 1880a). Also, GLO surveyor Jeremiah Whitcher noted crossing Tom Paine Slough, but did not remark upon any channel between this and the San Joaquin River to the northeast as he traced the El Pescadero grant line between the Southern and Western Pacific railroad lines (Whitcher 1857a).

The initial break in the bank of the San Joaquin that formed Paradise Cut was reported to have occurred twenty years prior (Naglee 1879). Whether this cut occurred naturally or due to early channel modifications and other reclamation efforts is unknown, though crevasse splays formed through breaks in natural levees are not uncommon to meandering rivers and deltas (Allen 1965, Coleman 1969, Smith and Perez-Arlucea 1994, Singer et al. 2008). The subsequent flooding delivered new sediment onto part of Stewart Tract, which was noted in a reclamation document: "one can see by riding through the bed of the stream that hundreds of acres have been covered with sand and rendered valueless for agricultural or grazing purposes" (Fig. 6.27; Tucker 1879d).

continued on **page 346**

Figure 6.26. Paradise Cut is not shown in early maps as the major channel it is today. (Gibbes 1850a, courtesy of the Map Collection of the Library of UC Davis)

Figure 6.27. Splay deposits spreading northeast from Paradise Cut can be seen in historical aerial photography. The numerous flood events through Paradise Cut after its initial break circa 1859 likely generated these splay deposits. (USDA 1937-1939)

BOX 6.2. EVIDENCE OF EARLY CHANGE AT PARADISE CUT (CONTINUED)

Over the subsequent decades, numerous attempts were made to dam Paradise Cut to "keep the water in its natural channel" and prevent the flooding that often occurred on upper Union Island (Tucker 1879c). It was dammed in 1876, but broke in 1878 (Tucker 1879c). In 1878, it was "the largest opening in the West bank of the River" (Kluegul 1878). Rebuilding the dam was justified in the following manner:

> In consequence of the divergence of the waters of the San Joaquin River through the Paradise Cut, the capacity of that River has been very much lessened, and the navigation of it has been seriously injured; and for the same reason the navigation and capacity of Old River has been entirely destroyed. And unless the Paradise Dam be repaired, and the water be confined where it formerly flowed in the old channels of the San Joaquin Old and Middle Rivers, nothing can be satisfactorily accomplished. (Naglee 1879)

A government-built dam broke again in 1890 (*Los Angeles Herald* 1890). A summary of seasonal high flows on the San Joaquin in June 1895 reports that 18,260 cfs (517 cms) flowed in the San Joaquin below Paradise Cut, 10,000 cfs (283 cms) flowed through the cut, and another 6,818 cfs (193 cms) flowed in other minor channels (USDI 1896).

An unintended consequence of the dam and a lesson in the importance of hydrologic connectivity is recorded by a 1905 newspaper article concerning the hindrance to salmon migration caused by the dam. In early March, "thousands of large salmon" were found dying in the vicinity of Paradise Cut, which was attributed to their inability to leap over the dam (*Pacific Rural Press* 1905). Without this obstacle, salmon migrating across the floodplain and secondary channels of Stewart Tract would have found a path upstream. This stranding occurred because the floodplain was disconnected from the river (through dams, levees, etc.). Since the flow escaping through or around the dam was evidently enough to trigger salmon migration, it seems likely that the numerous secondary channels of the historical Delta provided fish passage upstream. Historically, migrating salmon would have been able to pass through to the mainstem channel at the upstream end of the floodplain due to secondary channel connectivity at high river stages.

LAKE AND POND LANDSCAPE POSITION AND CHARACTER

Lakes and ponds are common features of floodplain environments, including oxbow lakes, remnants of former channels, beaver ponds, backwater areas formed by woody debris obstructing flow, or other off-channel depressions. The south Delta's spatial and temporal habitat complexity can be attributed in part to the myriad lakes and ponds that occupied the floodplain (see Fig. 6.2). The ponded water increased the retentive capacity of the system, providing needed habitat for aquatic and riparian species (Beechie et al. 2001). Some features maintained connections to flowing water year-round while others dried seasonally, offering a wide array of species support functions in different places and times of the year. Numerous waterfowl frequented the perennial and seasonal lakes of the floodplain, making these popular hunting grounds (Fig. 6.28; *Pacific Rural Press* 1883). The "freshwater lagoons" of the San Joaquin Valley were rich in fish such as Sacramento perch, Thicktail chub, and salmon, which naturally attracted predators such as otter and bear (De Mofras and Wilbur 1937).

A

B

Figure 6.28. Hunting on ponds amongst tule in the vicinity of Stockton. The caption for (A) reads, "Duck hunting near 'Head Reach' in the Delta west of Stockton ca. 1894 or 1895." These views give the sense of these ponds' character and surrounding emergent vegetation. Historical depictions of ponds suggest that lakes and ponds were most common at the margins of tidal extent and within non-tidal wetlands. Smoke can be seen in the background of B, likely related to reclamation practices. (Unknown ca. 1894a and b, courtesy of The Haggin Museum, Stockton)

The lakes and ponds of the south Delta were generally positioned at the margins of tidal influence, where land surfaces were generally above tidal range with the adjacent deep waterways experiencing tidal influence. Lakes and ponds were more prevalent in the south Delta than in the more tidally-dominated central Delta and were, on average, smaller than those in the north Delta. Virtually all mapped lakes and ponds fell outside of the tidal wetland boundary, though some may have been connected via tidal channels for most if not all of the year. They held low-elevation positions within the perennial wetland complex, surrounded by large expanses of tule, and were sometimes referred to as "tule ponds" (Edwards [1837]1890). Their connectivity to major channels, their size, and primary formative processes differed from north Delta lakes and ponds.

Lakes and ponds were usually found a short distance from the major distributary channels of the San Joaquin River and were connected via secondary channels (Gibbes 1850b). San Joaquin overflow passing through secondary channels filled these depressions, which then held standing water long after flows had ceased. Court case transcripts concerning upper Roberts Island, where it was stated that water was carried off the land "except a small amount estimated at 34 acres," which would then slowly dry out (see Fig. 6.17; Unknown 1917b).

Many of the lakes and ponds merged almost imperceptibly with the secondary channel networks that laced the floodplain. Often, they were described as a part of a larger-scale pattern of "a string of lakes connected by sloughs" that ran parallel to the main San Joaquin River (Fig. 6.29; Sweet et al. 1908, USGS 1909-1918, U.S. Army et al. 1914-1915). Distinctions between lakes, ponds, and channels were relatively fluid: what was a slough to one person may have been called a lagoon or pond by another, and what was called a slough in the early spring may have been called a series of ponds by late summer. As such, it may not be entirely appropriate to separate the discussion of secondary channels from lakes and ponds. The landscape position of ponds, lakes, and secondary channels suggests significant hydrologic connectivity between these floodplain features and the river, particularly at high river stages.

We mapped a total of 35 lakes and ponds each over 5 acres (2 ha) in the south Delta. Together, this amounts to an estimated 890 acres (360 ha) of lakes and ponds. Only eight were greater than 20 acres (8 ha). Based on connections to channels that were likely tidal, we estimate a little less than half of this acreage may have been influenced by tides. Only 34 acres (14 ha, of the total 890 acres) of seasonal ponds were identified. This is a conservative estimate given the scarcity of sources documenting seasonality. Consequently, we believe that many of the features we classified as perennial were actually seasonal. It should also be emphasized that though these were distinct features, their size changed depending on the time of year. When the south Delta was overflowed in the late spring and early summer, it is likely that these features merged with the surrounding inundated floodplains.

Reaching the first spot where there was water we found that instead of the river it was only a large pond, that the river flowed a half league to the east, that it was impossible to approach it at this point.

—MOERENHOUT [1849] 1935

It was impossible to go [to] the river with a horse for several miles above and below my camp in consequence of the low flaggy ground which was covered with water. Some of the Ponds have Beaver along their flaggy banks.

—SULLIVAN 1934

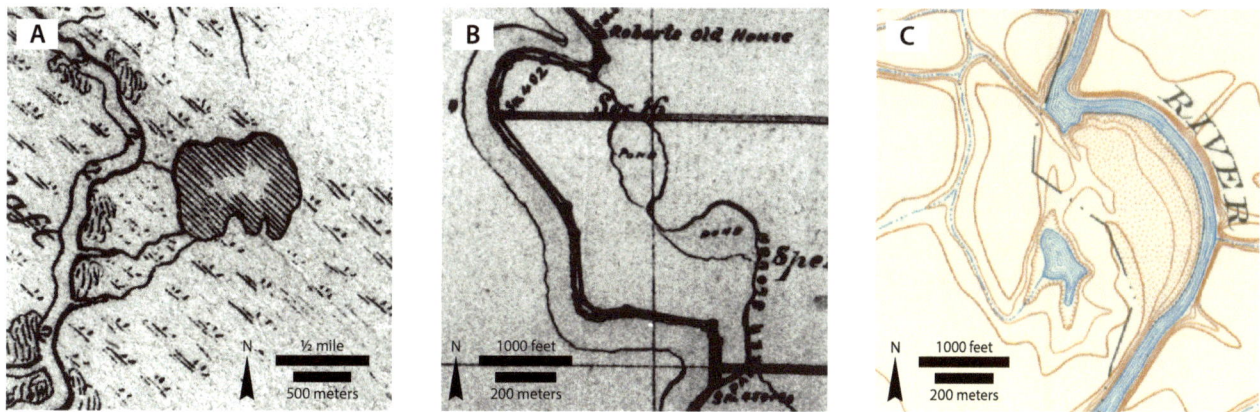

Figure 6.29. Lakes and ponds were often connected to the river via several rather short secondary overflow channels through the perennial wetlands. In A, a lake (175 ac/71 ha) in upper Roberts Island is mapped as connected to the mainstem via three secondary channels. The map in (B) illustrates the pattern of a "string of lakes connected by sloughs" described for this area in the 1908 soil survey (Sweet et al. 1908). A USGS topographic map (C) shows a circuitous channel connection to a smaller pond (<5 ac/2 ha) south of Walthall Slough. A chain of ponds (D) was also found on the west side of Tom Paine Slough. (A: Gibbes 1850a, courtesy of the Map Collection of the Library of UC Davis; B: Beaumont 1861a, courtesy of the California State Lands Commission; C: USGS 1909-1918)

Fluvial channel

Tidal or Fluvial channel (lower confidence level)

Water

Non-tidal freshwater emergent wetland

Valley foothill riparian

Wet meadow or seasonal wetland

Grassland

One possible example of the challenges associated with determining lake size can be found in Gibbes's 1850 map and his accompanying description that was reported in the *Stockton Times*. He maps a lake on the order of 150 acres (61 ha) in size, but states that it was "about one and a half miles long and three to four broad," which is about 20 times larger than the mapped lake and almost a third of the size of present day Stewart Tract, where the lake was located. This discrepancy (other than human error) could be explained by the fact that Gibbes was surveying in June "at nearly the highest state of the water" and thus may have included part of the inundated floodplain and other smaller lakes and ponds in his description, but only mapped the more distinct, smaller features in his map.

We mapped 65% of the lake and pond features (and 85% of the total area) as definitely present in the early 1800s, most of which were supported by pre-reclamation sources. A few particularly early maps, the spatially limited but detailed GLO surveys, and county surveys from the late 1850s and early 1860s provided valuable early mapping of lakes and ponds (Norris 1851, Drew 1856-1857). Most other lakes and ponds were mapped using later sources, and so were classified as probably present in the early 1800s. Since early cartographic sources generally did not document small ponds, later sources such as the early 1900s USGS topographic maps were often used. As a result, these features have a lower associated interpretation certainty. Also, the more dynamic floodplain landscape contributed to uncertainties related to using later sources to map historical features. Supported by general descriptions of the area, we believe the habitat type mapping is representative of early 1800s conditions in terms of general distribution and extent of lakes and ponds. An additional 67 ponds less than 5 acres (2 ha) each and totaling over 80 acres (32 ha) were identified in the USGS topographic maps and other later sources, but not included due to our

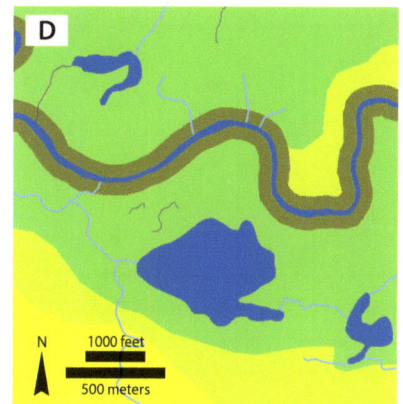

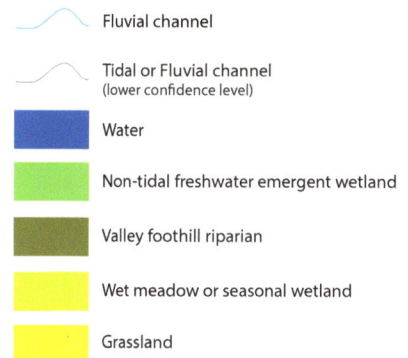

minimum mapping unit (5 ac/2 ha) and associated uncertainties. On average, south Delta lakes and ponds were substantially smaller (~20 ac/8 ha) than those found in the north Delta (~95 ac/38 ha).

Information concerning the depth of lakes and ponds is limited. They clearly posed significant obstacles for travel, though some descriptions of "knee-deep" water suggest relatively shallow features. This is supported by evidence that some depressions dried out by the end of the summer. The 1850 Gibbes map, which offers the earliest known soundings for the San Joaquin, includes soundings of six to nine feet (1.8-2.7 m) of water in a lake in the north end of present-day Stewart Tract (Gibbes 1850a). These soundings were taken during the high water season, so that depth was likely reduced substantially by the end of the season.

Early textual descriptions of the south Delta often discuss ponds as distinct features within the matrix of other floodplain habitat types. Even at high river stages, textual accounts distinguished ponds from the surrounding overflowed floodplain (Moerenhout [1849]1935). At a time when water levels would have been at their lowest, in late October 1810 (a lower than average rainfall year), Spanish explorer Viader recounted passing through a landscape of "oak groves, willow thickets, ponds, and lands flooded during the freshets" in the vicinity of present-day Stewart Tract (Viader and Cook 1960). A year later (a wetter year), also in October, "ponds and tule swamps" were found near present-day Highway 4 (Abella and Cook 1960). In the same season, but in 1851, GLO surveyor Norris encountered a "dry bed of pond" near the east bank of Tom Paine Slough close to its mouth. Further east he met the "south side of pond with water," and then came to another pond with water as he neared the San Joaquin River (Fig. 6.30; Norris 1851). The dry pond mentioned near Tom Paine Slough appears to be coincident with a pond mapped in 1850, which helps establish its

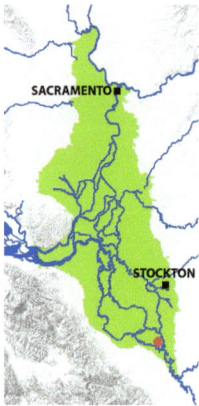

Figure 6.30. One of several ponds with water that GLO surveyor Ralph Norris encountered on his path across Stewart Tract in October of 1851 is shown in a map (A) made from this survey. Only those GLO field notes discussing the pond are included. Other field notes in this vicinity describe patches of tule, willow, and riparian forest along the San Joaquin River. The modern aerial photography (B) is included for comparison. Only a slight signature indicating the position of the pond can be seen. (Fisher 1854, USDA 2005)

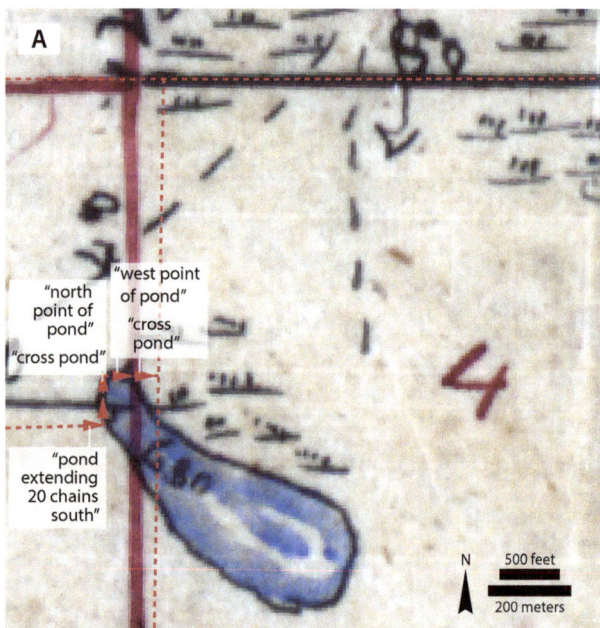

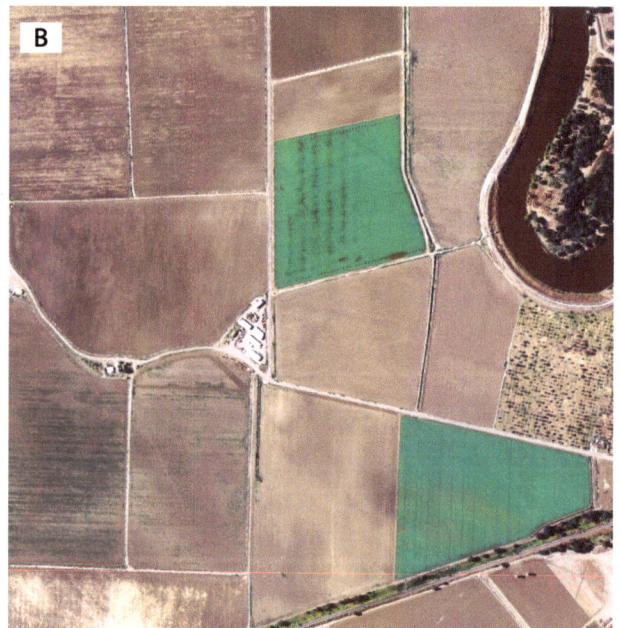

"north point of pond"
"cross pond"
"west point of pond"
"cross pond"
"pond extending 20 chains south"

N
500 feet
200 meters

historical presence and enabled us to classify this feature as intermittent (Gibbes 1850a).

While some water bodies were oxbow lakes or otherwise associated with deeper depressions along the floodplain's secondary channels, several of the largest lakes mapped were found next to mainstem channels occupied by masses of woody debris, or rafts. One was found along Middle River in upper Roberts Island (possibly Willow Lake) and several others were found at the upper end of present-day Stewart Tract (Gibbes 1850a and b). This coincidence points to a possible mechanism of formation: water during high flow was directed out of the channel to form these backwater areas that were connected via apparently relatively well established channels (surveyor Gibbes took a boat through several of them to reach the large lake at the mouth of present-day Paradise Cut). More discussion of woody debris is found on pages 366-369.

We were unable to locate and map several lakes mentioned in the historical record. One account from Spanish explorer Viader's 1810 diary explains that after traveling south along the west boundary of the tules and near the "village of the Cholvones, or Pescadero," (likely in the vicinity of present-day Bethany), they "arrived at a lake in the middle of an oak grove where we could neither get to the river nor turn back" (Viader and Cook 1960). Another historical feature without sufficient evidence to map is Honker Lake, located within Honker Lake Tract. This area forms the triangle of land between Whiskey and Duck sloughs in the middle of Roberts Island. However, we found little direct evidence of this lake. An 1883 newspaper account refers to a former lake that occupied "the greater portion" of the tract (*Pacific Rural Press* 1883). One possible explanation is that the ridges of Whiskey and Duck sloughs and their distributaries caused annual overflows to be retained for greater periods of time, such that the area was referred to as a lake, but not persistent enough to be mapped by early sources. However, pollen analysis from a core of this tract revealed yellow pond lily (*Nuphar* spp.) and water fern (*Azolla filicoides* and *A. Mexicana*), which suggests more permanent limnetic conditions in the vicinity (West 1977).

COMPLEXITY WITHIN THE WETLAND PLAIN

In the south Delta, the floodplain surface of the San Joaquin River hosted complex habitat mosaics controlled by localized differences in topography, soils, and hydrology (Fig. 6.31). Here, tidal processes that maintained water levels, affected channel planform and flows, and promoted ecosystem exchange met riverine processes that brought inorganic sediments, built and shifted secondary channels, and shaped topographic depressions. Organic matter accumulation signified a highly productive system (Sedell and Froggatt 1984). These interactions at the edge of the Delta affected overflow patterns, water velocities, inundation depths, and hydroperiod. During periods of overflow, the topographic variability provided "patches over which a person can with difficulty wade out" (Whiting 1854). This created opportunities for a diverse range of habitat types arranged along localized physical gradients, which in turn provided a high degree of habitat connectivity.

Figure 6.31. Habitat complexity of the south Delta is shown supporting a diverse range of species in this piece by artist and naturalist Laura Cunningham.

The lower portion of the San Joaquin river is bordered by numerous sloughs, and winds about through low marshy ground, covered with rushes and willows. Such portions of these marshes as are only temporarily overflowed, during the winter months, support a growth of coarse grass and other plants.

—BLAKE 1858

Habitat mosaics consisted of large expanses of tules and reeds broken up by secondary channels, ponds, and lakes occupying low-elevation positions, wet meadows of grasses and sedge species in the more well drained areas, and willows particularly associated with secondary channels. Along natural levees, riparian forest contributed additional habitat complexity. The floodplain landscape was captured in a Spanish explorer's description of "oak groves, willow thickets, ponds, and lands flooded during freshets" (Viader and Cook 1960) and by a gold miner as "meadows and swamps which extended as far as the eye could see" (Moerenhout [1849]1935).

Maps, surveys, and textual descriptions indicate that emergent vegetation (primarily tule species) persisted throughout the area and likely dominated the floodplain (Fig. 6.32). Reclamation of upper Roberts Island involved "destroying the dense growth of tules" (*Pacific Rural Press* 1878). In upper Union Island, traveling north from the vicinity of Salmon Slough at Old River, GLO surveyor Norris passed through several patches of tule (also mentioning "switch cane" in one location), before coming "to thick tule" less than half a mile before reaching the present location of the Grant Line Canal, where he was unable to continue. This point is currently positioned about a foot or two above tide elevations. Furthermore, early general maps tend to use the words "tule" or "Tulare" in these areas and use symbols commonly used to represent wetlands.

However, vegetation other than tule comprised a substantial portion of the mapped non-tidal emergent wetland in the south Delta. Willow thickets appear to have been common; several accounts and surveys mention brush, willows, underbrush, and briars associated with secondary channels as well as the major rivers (Lyman 1848, Norris 1851, Alexander 1877). In southern Stewart Tract, a portion of a GLO survey line was described as "covered with willow undergrowth" (Hays 1853). The earliest and most detailed information concerning local-scale habitat complexity comes from a single

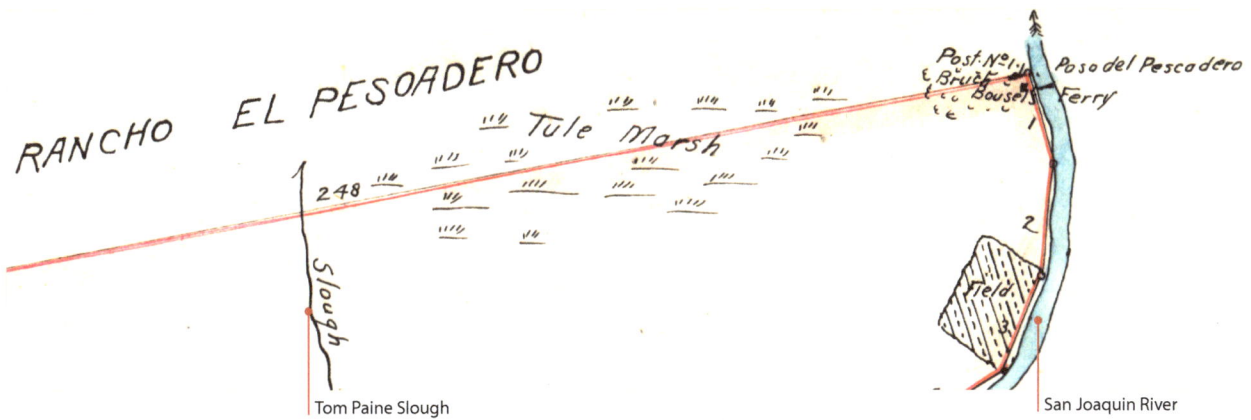

Tom Paine Slough   San Joaquin River

GLO survey conducted by Ralph Norris in 1851 that crosses present-day Stewart Tract. In nine miles of that survey, Norris noted only slightly more than half of that distance as tule, while channels, ponds, bare ground, willows and grass made up the rest. One mile of that line is shown in Figure 6.33, illustrating local-level complexity within the south Delta landscape.

Large trees such as oaks and sycamores were absent in the lower regions of upper Union and Roberts islands and the greater Stewart Tract area. The only GLO survey bearing trees (tree marked to establish survey corners) in this vicinity were a few oaks within the riparian forest lining the natural levees. A corner of the El Pescadero land grant in upper Union Island used a "swamp oak" located about a mile west on Middle River as its bearing tree, indicating that no well established trees could be found nearby. Also, no trees were found between Tom Paine Slough and the San Joaquin River in Norris's 1851 survey across Stewart Tract.

While our mapping captures many of the larger features that are spatially explicit in early sources (e.g., lakes) as separate habitat features, the complexity described here should be taken as representative of the

Figure 6.32. "Tule marsh" covers much of the non-tidal floodplain between Tom Paine slough (marked "slough") and the San Joaquin River. Early maps such as this suggest that tule dominated the wetlands of the south Delta. Though perennial freshwater wetland may have predominated, the floodplain landscape was mixed with patches of willow and other underbrush, seasonal wetlands, and ponds. This map and associated field notes were used to delineate a transition zone of seasonal wetland between the "tule marsh" and the "brush." (Whitcher 1857b, courtesy of the Bureau of Land Management)

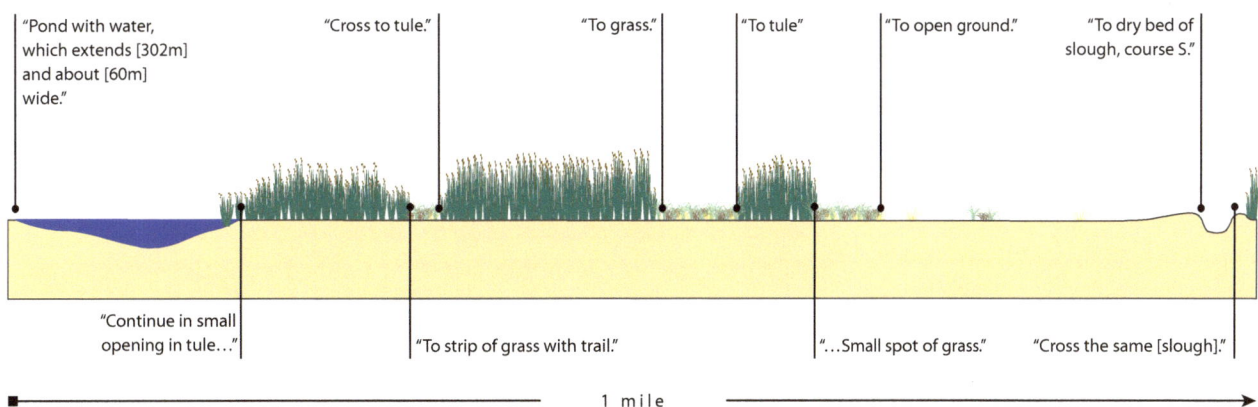

Figure 6.33. A reconstructed survey line of the GLO reveals local-level complexity within present-day Stewart Tract near Paradise Cut and just northwest of I-5. The pond, patches of grass, bare areas, and dry channel beds intermixed with tule along the survey line give an early close-up view of what the floodplain landscape looked like in late October of a dry year prior to significant Euro-American modification. (Norris 1851)

character of the non-tidal freshwater emergent wetland habitat type mapped in the south Delta.

Soils at the interface

Spatial variability in soil type is an important physical factor governing vegetation communities. Transitioning upstream from the tidal Delta, peat soils became thinner and alluvial soils (i.e., sandier, inorganic) became more dominant (Thompson 2006). Peats characterized by high organic content and high water holding capacity interwove with alluvial soils, which were more well drained, but tended to "puddle…in the heavy, low spots" (Nelson et al. 1918). This produced high spatial variability that affected vegetation patterns. The 1918 soil survey distinguishes these peat soils at the Delta margins from the central Delta peats due to their "containing large quantities of sediments, in places more than half the soil mass. In some places these sediments have been deposited over the surface, while in others the typical Muck and Peat is but a shallow layer over the mineral or alluvial material" (Nelson et al. 1918).

At this tidal edge, the land had only recently (in geologic time) come under tidal influence, so peat soils had been building for only a limited time before reclamation. The peat boundaries in early soil survey present a minimum extent of tidal wetlands given subsidence by that point in time and that the surveys did not map the thinnest peats (Fig. 6.34; Lapham and Mackie 1905, Nelson 1915). The early peat losses at this boundary are estimated to have been up to 2.7 in/yr (6.8 cm/yr) in some locations, based on land surface elevations in the early 1900s USGS mapping, 40 years of subsidence (see Fig. 1.16), and an assumed 1850s land surface of 3.5 feet (1.1 m) above sea level.

Much of these thin layers of peat and localized patches at the tidal margin likely disappeared within the first several decades of reclamation. This is the most plausible explanation for why an 1879 reclamation report for present-day Fabian Tract stated that "there is no peat land and very little land with tules growing on it" (Tucker 1879d). This runs contrary to earlier GLO surveys and other evidence that indicates a significant presence of tule and other emergent vegetation. By that time, cattle had grazed on the land "for a number of years," which could have greatly affected vegetation patterns (Tucker 1879d). The land was naturally more amenable to stock-raising than the tidelands of the central Delta, and it is possible that the early 1800s grass patches expanded in size in the early decades of heavy grazing to where the area could be described as having "grass growing in great abundance" (Tucker 1879d).

At the tidal interface, the more peaty soils adjoined the alluvial loamy soils, which were "underlain by the partially decomposed peat" (Lapham and Mackie 1906) and "where a number of winding sloughs and erosion by flood waters give it a more or less uneven and pitted appearance" (Nelson et al. 1918). Only limited areas with a peat layer would have been found

The Sacramento clay loam owes its origin to the admixture of the fine river silts, derived from a variety of rocks and distributed by the San Joaquin River and its tributaries and branches, with the fine alluvial and decomposed organic matter of the tidal fresh-water marshes or peat lands. The material from these two sources is either intimately mixed or deposited in alternating strata.

—LAPHAM AND MACKIE 1906

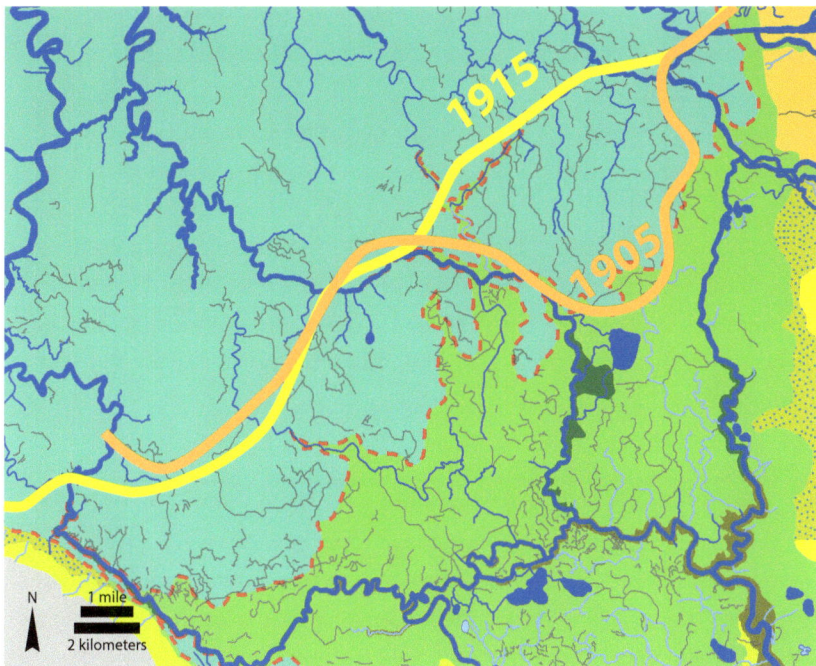

outside the limit of tidal influence. Upper Union and Roberts islands as well as the eastern edge upstream of French Camp Slough are described as containing only "sediment land" (Tucker 1879b). Some sediment land was also apparently found just south of present-day Clifton Court Forebay (Tucker 1879c). The local variations in soil patterns associated with secondary channels can be seen in aerial photography as well as in the topographic variability caused by the sediments on the banks (see Fig. 6.13). In some locations, the alluvial soils were quite sandy. Along the east bank of the San Joaquin upstream of French Camp Slough, peat soils were absent and "soft light loam with some sand" was found within the floodplain, while "heavier soil, principally adobe" lined the upland edge (Tucker 1879b).

Comparison to the north Delta flood basins

Small ponds, tule, willow thickets associated with sloughs, and wet meadows formed a landscape of apparently greater local-scale complexity than other parts of the Delta. Whereas north Delta basins had the appearance of expanses of dense, continuous, tule broken up by occasional ponds, lakes and sloughs, the south Delta floodplain was occupied by smaller mosaics of many different vegetation communities of variable patch sizes. Tule stands with a range of density persisted within the floodplain habitat matrix, in places appearing as if the tule patches were scattered about the plain (Bryant [1848]1985, Dawdy 1989).

A possible explanation for these differences lies in the contrasting scale and position of the landforms between the two landscapes. Large flood basins like those of the Sacramento River with relatively confined boundaries and defined drainage points for water were absent along the San Joaquin (Thompson 1957). Floodplain surfaces in the south Delta were connected

Figure 6.34. The edge of peat in two early soil surveys, represented by orange and yellow lines, lies mostly downstream of the likely limit of historical tidal influence on the marsh plain (shown with a dashed red line). Since peat soils accumulate under tidal influence, these boundaries represent a minimum extent of tides. Soils by this time had already been affected by subsidence. (lines from Lapham and Mackie 1905, Nelson 1915)

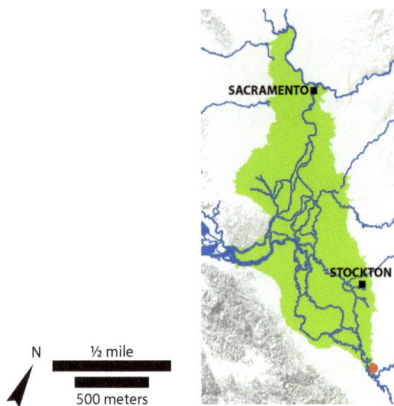

N ½ mile
500 meters

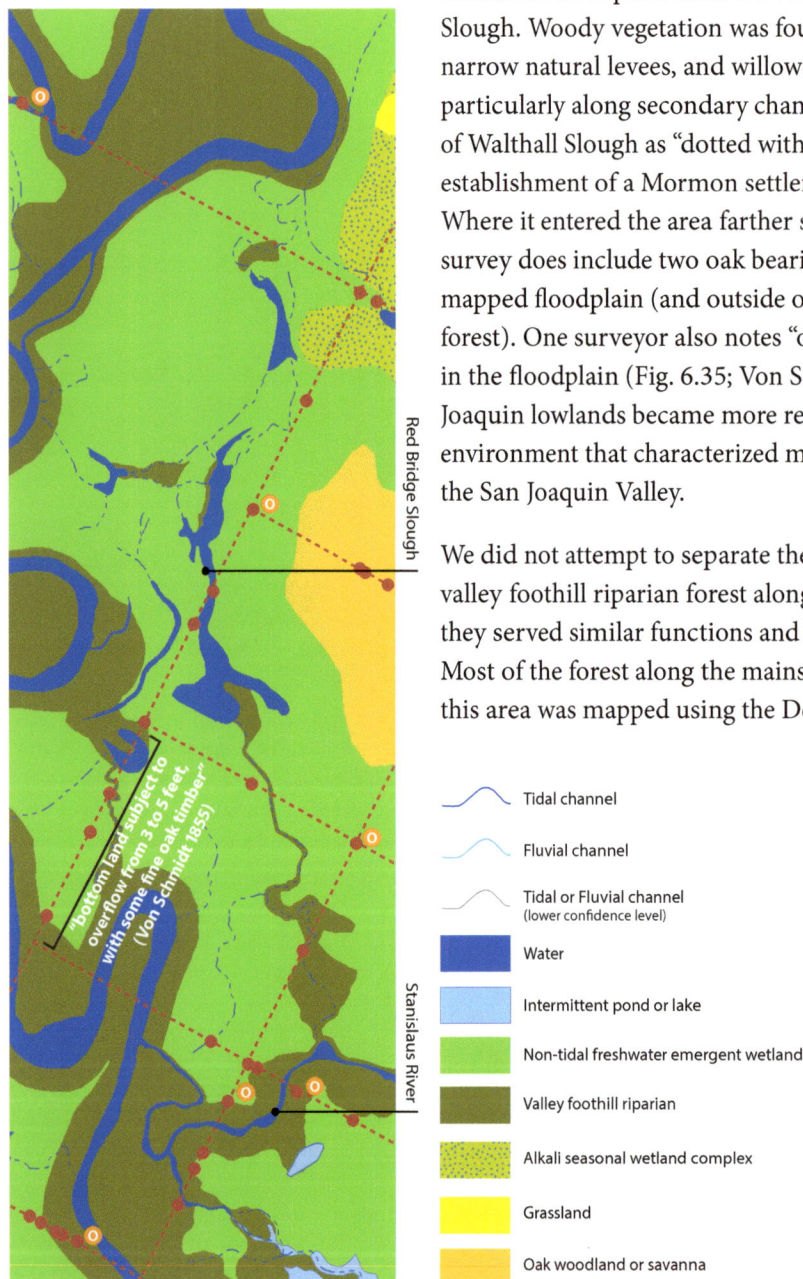

Red Bridge Slough

"bottom land subject to overflow from 3 to 5 feet, with some fine oak timber" (Von Schmidt 1855)

Stanislaus River

Tidal channel

Fluvial channel

Tidal or Fluvial channel (lower confidence level)

Water

Intermittent pond or lake

Non-tidal freshwater emergent wetland

Valley foothill riparian

Alkali seasonal wetland complex

Grassland

Oak woodland or savanna

to the river at numerous locations and water was stored within small depressions. The variability in vegetation communities thus reflects this greater local-scale complexity in landforms. The hydrologic and climatic differences between the north and south Delta also contributed to the differences in relative complexity. The south Delta faced greater extremes, land was drier in terms of climate and freshwater inflows. Although the non-tidal wetlands of north Delta flood basins were markedly different in hydrologic regime and in the mix and landscape pattern of vegetation communities, we classified both as non-tidal freshwater emergent wetland. The non-tidal freshwater emergent wetlands of the north and south Delta should be thought of as different subtypes.

A SHIFT TO WOODY VEGETATION ON THE FLOODPLAIN The relative mix of vegetation within the floodplain shifted toward woody vegetation upstream of Walthall Slough. Woody vegetation was found extending beyond the relatively narrow natural levees, and willows and oaks became more common, particularly along secondary channels. One historian described the vicinity of Walthall Slough as "dotted with ancient live oak trees" in relating the establishment of a Mormon settlement in the area in 1846 (Williams 1973). Where it entered the area farther south near Red Bridge Slough, the GLO survey does include two oak bearing trees out of eight points within our mapped floodplain (and outside of the main San Joaquin River riparian forest). One surveyor also notes "oak timber" along one mile of the survey in the floodplain (Fig. 6.35; Von Schmidt 1855). At this point, the San Joaquin lowlands became more reflective of the riverine floodplain environment that characterized much of the length of the river upstream in the San Joaquin Valley.

We did not attempt to separate these bottomland forested areas from the valley foothill riparian forest along natural levees in our mapping, since they served similar functions and boundaries were challenging to define. Most of the forest along the mainstem and along secondary channels in this area was mapped using the Debris Commission maps from the early

Figure 6.35. Trees within the San Joaquin River floodplain apparently became more common in the most southern part of the Delta region, near the mouth of the Stanislaus River. Compared to downstream, trees were less confined to the higher natural levees. This is apparent in sources such as the GLO survey, which recorded several oak bearing trees (orange symbol) and describe the "bottom land" as having "some fine oak timber" (observation made for the bracketed line, Von Schmidt 1855). The wetland type mapped in this area should thus be interpreted to include a greater proportion of scrub and trees in comparison to wetlands downstream.

1900s and supported in several locations by GLO surveys of the 1850s. By roughly approximating mapped riparian forest that fell within the lower floodplain along the east side of the San Joaquin between Walthall Slough and the Stanislaus, we estimate that more than 675 acres (273 ha, roughly 10% of the area) of riparian forest was not directly associated with the main natural levees of the San Joaquin River. In contrast, we mapped less than 50 acres (20 ha) of forested habitat unassociated with mainstem channels downstream of Walthall Slough.

However, our mapping likely captures only a portion of this increase in the proportion of forested floodplain due to early change and the few cartographic sources that directly mapped these floodplain vegetation patterns (Box 6.3). The mapped freshwater emergent wetland upstream of Walthall Slough should therefore be understood to include a greater proportion of trees and brush than this habitat type further downstream. In some places, woody vegetation may have been dominant over large areas, which diminished our overall certainty of the freshwater emergent wetland classification in this area (see page 363).

RIPARIAN FOREST CHARACTERISTICS

The higher natural levee bank lands and point bar alluvial deposits of river meanders were formerly occupied by biologically productive dense riparian vegetation, in some places as densely impenetrable scrub and elsewhere as thick timber or majestic oak groves. Behind these corridors lay the tule-dominated wetlands (Fig. 6.36; Sands 1977). Most early accounts of the lower San Joaquin River in the vicinity of Stewart Tract and the first several miles along upper Union and Roberts islands describe the banks of the

---

BOX 6.3. EARLY CHANGES TO THE LANDSCAPE

Although large-scale early reclamation efforts were underway by the 1870s, considerable modifications had occurred by over a decade earlier (see Fig. 1.14). In the north and south Delta, many of these alterations consisted of dams on smaller secondary channels that intersected the river's natural levees and small hand-built levees on top of natural levees. The State Engineering Department field notes of John Tucker provide detailed information regarding these early reclamation attempts. For example, work done in Reclamation District 17 (lying east of the San Joaquin River and south of French Camp Slough) began in February 1863 when McCloud's and Wood Duck sloughs were dammed. In 1868, owners "began the construction of a new levee, the old one was located so near the river that it was not considered advisable to repair it; and it was abandoned" (Tucker 1879b). As another example, on Roberts Island, initial hand-built levees were constructed as early as 1856. Overall, these changes impacted the region's hydrology, including the reduction of hydrologic connectivity. Other early impacts, spurred by the Gold Rush, include tree cutting and brush clearing along the river bank land. Such changes are reflected in a late 1800s San Joaquin County history that recalls an area once "thickly covered with timber" in the vicinity of the present-day I-5 crossing (Lewis Publishing Co. 1890).

June 25 8⁰ am. 1/2 S 64°

San Joaquin River + Bridge 1½ miles north of Lanes Station

Figure 6.36. This 1901 view from the San Joaquin Bridge shows the narrow riparian corridor and the flat plain of wetlands behind. This photograph was taken on June 25 at high river stages. (Mathews 1901, courtesy of the California History Room, California State Library, Sacramento)

The Courier is wrong in saying that the "San Joaquin River for its whole length is through an unbroken prairie, and its banks present nothing but a mass of tules." Such may be the case from Suisun Bay to Doak's Ferry [I-5 crossing] but there the San Joaquin is a broad and magnificent stream, whose banks are well and thickly timbered.

—SACRAMENTO DAILY UNION 1851A

rivers as covered with oak trees (Abella and Cook 1960, Gibbes 1850b, *Sacramento Daily Union* 1851a). Typical accounts described higher elevation land "which had a number of oak trees but was entirely surrounded by tule swamps" (Abella and Cook 1960). Thick underbrush also was commonly described along the banks (Tucker 1879b). Farther upstream, the point where I-5 crosses the river today was surrounded by land "thickly covered with timber" (Lewis Publishing Co. 1890). Few trees remain there today. Textual accounts of this nature are corroborated by early maps, GLO surveys, and aerial photography as well as the natural levee landforms observable in topographic maps, soil survey maps, and LiDAR (Fig. 6.37).

We mapped a total of 8,000 acres (3,240 ha) of riparian forest in the south Delta, with 6,200 acres (2,510 ha, 78%) as valley foothill riparian and 1,800 acres (730 ha, 12%) as willow riparian scrub. The extent of similar habitat types within those areas today is only 2,800 acres (1,130 ha) of valley foothill riparian and coastal scrub habitat types, much of it located on artificial levees. This decline indicates at least a 65% loss of riparian forest cover (WHR; Hickson and Keeler-Wolf 2007). We mapped willow riparian scrub along the transition between the emergent vegetation on the channel banks in the central Delta and the oak-dominated forest upstream. Oaks became more common at approximately the latitude of Bethany today, though the riparian forest remained limited by natural levee width (but far

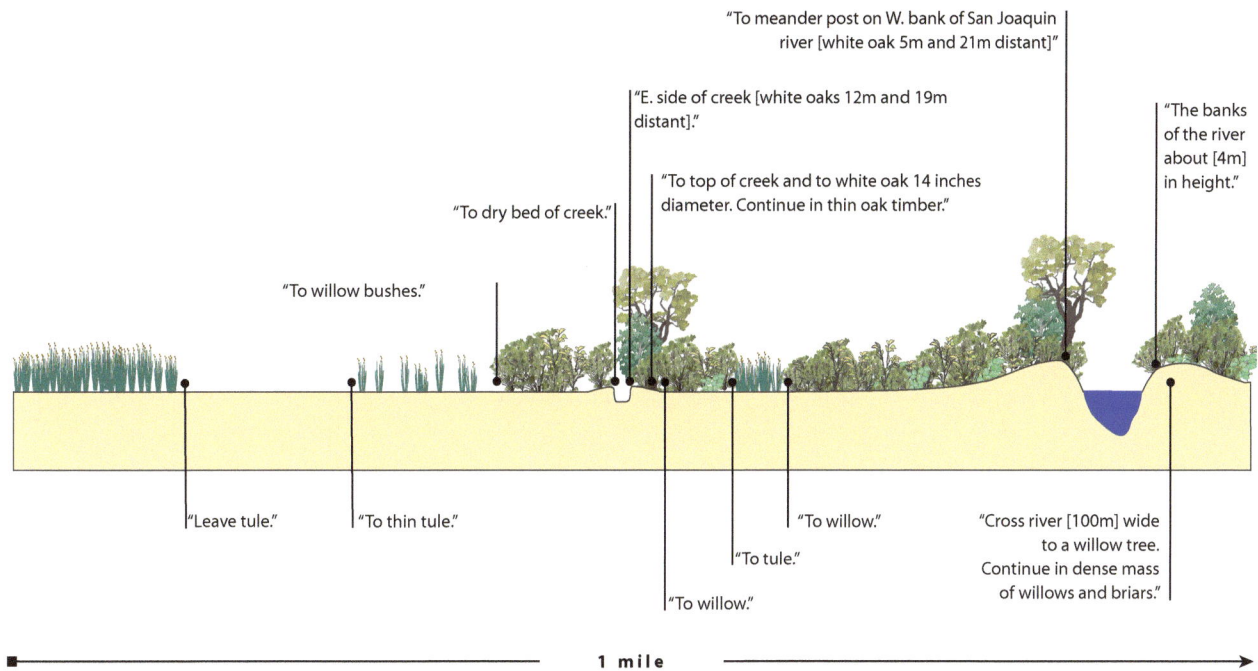

Figure 6.37 annotations:

"To meander post on W. bank of San Joaquin river [white oak 5m and 21m distant]."

"E. side of creek [white oaks 12m and 19m distant]."

"To top of creek and to white oak 14 inches diameter. Continue in thin oak timber."

"The banks of the river about [4m] in height."

"To dry bed of creek."

"To willow bushes."

"Leave tule."

"To thin tule."

"To willow."

"To tule."

"To willow."

"Cross river [100m] wide to a willow tree. Continue in dense mass of willows and briars."

1 mile

wider than today). As discussed in the previous section, upstream of the head of Old River, riparian forest was less restricted to the natural levees and more part of the floodplain habitat matrix of emergent wetlands, secondary and abandoned channels, and oxbow lakes. Along a GLO survey in the "bottom land subject to overflow," surveyor Von Schmidt noted that it was occupied by "some fine oak timber" (Von Schmidt 1855). The greater width and additional patches associated with secondary channels, such as Walthall Slough, in combination with the narrower floodplain here meant that riparian forest habitats made up a significant portion of the floodplains (see Fig. 6.35).

Our mapping likely represents a minimum estimate of forested area given the paucity of spatially explicit pre-1900 sources from which to map. For example, riparian forest associated with secondary channels is likely missing in many locations. Also, the width of the forest was difficult to determine in some locations, prompting the use of buffer widths established through interpretation of natural levee width from topographic maps and LiDAR (see page 68). Therefore, we believe that riparian forest may be under-represented, particularly the willow riparian scrub that likely persisted as patches within the emergent wetland matrix. Our mapping produced different spatial distribution and total area estimates from previous mapping efforts. Riparian forest mapping from 1977 used the early soil surveys to designate large areas as historical riparian forest, which amounted to approximately 21,000 acres (8,500 ha; Roberts et al. 1977). A later mapping effort by The Bay Institute assigned much of this area (approximately 23,000 ac/9,310 ha) as "wetlands mapped within riparian zone" where no riparian forest was mapped downstream of Ripon Road

Figure 6.37. Riparian forest, comprised of dense willows as well as larger oaks, formed a corridor along the San Joaquin River that was, at this 1851 GLO survey line, over 350 meters wide. In contrast to the forest represented here, no trees – only a narrow 40-meter wide span of "willows and briars" – were mentioned at another GLO line that crosses the San Joaquin much farther downstream along upper Roberts Island. This information supported the shift in mapping from valley foothill riparian forest to willow riparian scrub or shrub. (Norris 1851)

360

(TBI 1998). Our mapping represents a refinement of these mapping efforts by bringing together multiple sources that show where tule and other emergent vegetation occupied the lower-elevation portions of soil units otherwise used as typical riparian forest indicators. Our evidence suggests that not all these soil types should be presumed to have supported riparian forest historically.

Transitions along physical gradients

Riparian forest extent, width, and vegetation communities reflected the physical gradient between dominant tidal and fluvial processes. In the southern portion of the Delta, relatively high natural levees meant more inorganic sediments and lower water table levels, conditions favorable to the larger riparian trees. The wide and complex forest that characterized much of the San Joaquin River upstream of tidal influence transitioned gradually downstream into a narrower, willow scrub dominated community before becoming a part of the surrounding emergent wetland matrix where the natural levees neared the general level of tidal extent. Riparian forest characteristics followed a similar pattern with decreasing size of the channel. At positions where the mainstem river channels were associated with a complex forest with tall trees such as oaks, the smaller secondary channels tended to be more dominated by scrub or emergent wetland species (see Fig. 2.16). Vegetation patterns varied laterally across the natural levee as well, as the lower elevation zones were occupied by willows and the highest elevations of the natural levees were occupied by trees such as oaks (Sweet et al. 1908).

Soil surveys and LiDAR imagery illustrate the changes in physical characteristics of the channel banks moving upstream (Fig. 6.38). Peat soils are mapped to the edge of channels in the central Delta, while along upper Union and Roberts islands, Hanford loams and sandy loams begin to appear along the relatively narrow strip of natural levee. These soil types gradually become wider upstream until they comprise the floodplain bottom completely (Nelson et al. 1918). On Middle River, a state engineer identified the transition moving downstream from fine sediment natural levees that "are much higher than the adjacent land" to more peaty banks as occurring three miles below the Union and Roberts islands cross-levees (Kluegul 1878). This transition was related to elevations as well: on Old River, a surveyor wrote that the banks "gradually decrease in height for a distance of 20 miles," at which point they became level with the tidal marsh plain (Kluegul 1878). An early county history characterized this transition with the note that "the pleasant green timber has gone and tule is everywhere" (Smith & Elliot [1879]1979). Because of the more well drained inorganic soils and the concomitant increase in land elevations (with related lower water tables), early settlers found these "bank lands" relatively easy places to settle and grow crops (Sands 1977). Consequently, the early decades of settlement saw "a great many fine orchards and vineyard on the bank land" with little attention initially paid to the lower elevation floodplains lying in back, save as pastures for stock (Tucker 1879b).

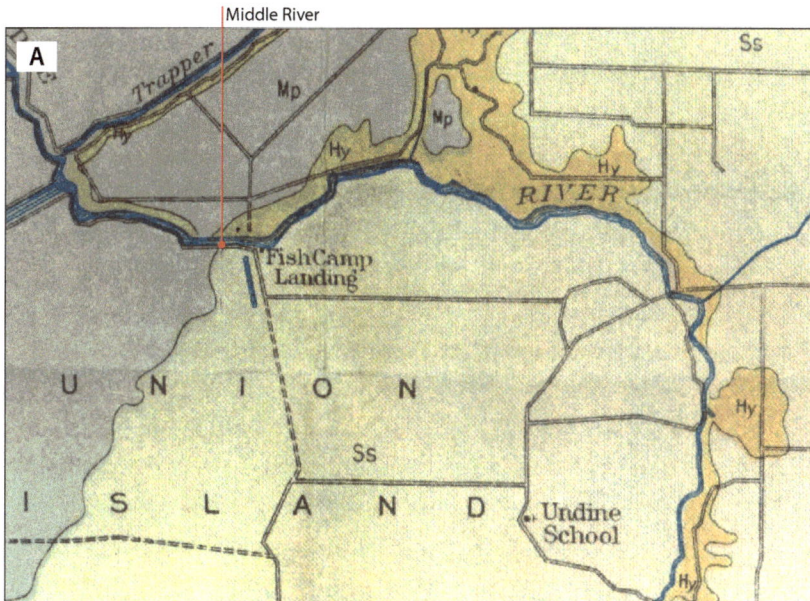

Figure 6.38. Natural levee deposits diminish downstream along this segment of Middle River. Natural levee deposits are also seen along several of the secondary channels branching off of the river. The 1915 soil survey for the San Joaquin Valley (A) shows the Hanford sandy loams (Hy) associated with the natural levees, surrounded by the Sacramento clay loams (Ss) and the Muck and Peat (Mp) of the wetland interior. The natural levee's gradual sloping away from the river is evident today in the LiDAR imagery (B). It is often difficult to detect the subtle sloping of the natural levees when looking at modern imagery (C). (A: Nelson 1915; B: CDWR 2008; C: USDA 2005)

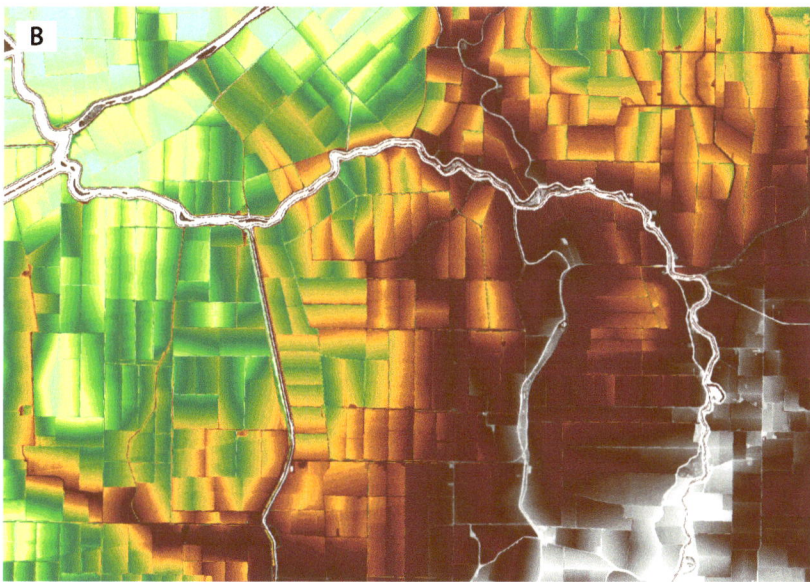

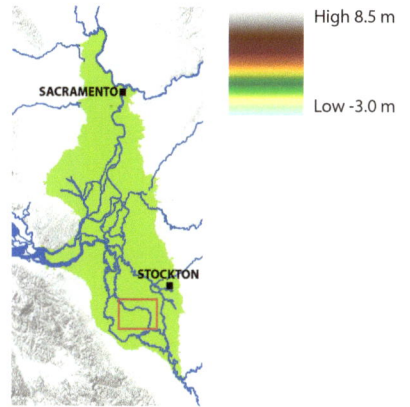

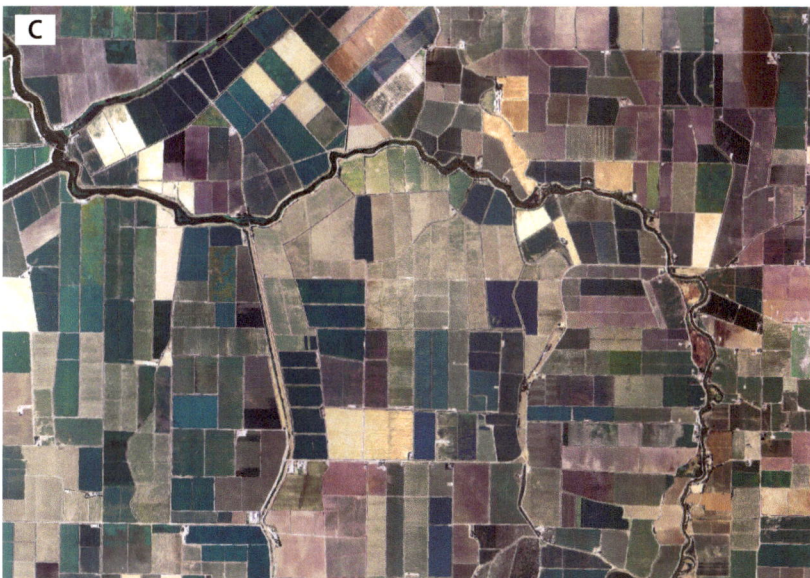

Several accounts discuss vegetation characteristics as banks increased in height upstream. Along Old and Middle rivers, one observer noted: the banks "become higher and firmer," and they became "covered to a considerable extent with willow and other bushes" (*Daily Alta California* 1870). The banks at these points reached several feet above high tide levels. An explorer in 1811 noted that, while the banks were becoming higher on Old River near Byron Tract, it was "still bare of trees" (Abella and Cook 1960). Willow scrub dominated reaches along upper Union and Roberts islands are also suggested by reclamation documents. Those building the first levees faced the challenge of clearing the land of thick underbrush (Tucker 1879b). It was recorded that before enclosing the El Pescadero Grant of Union Island (present-day Fabian Tract), it was first "necessary to have a gang of chinamen clear and burn the brush along the banks as it was so dense as to render it almost impenetrable" (Tucker 1879d). Surveyors also complained about the dense undergrowth as it prevented accurate leveling (Handy n.d., Tucker 1879b).

Oaks became present along the banks – shown by the transition from willow riparian scrub to valley foothill riparian forest in our mapping – near present-day Bethany on Old River and south of Howard Road on the Middle and San Joaquin rivers. The approximate location of this transition is shown in Gibbes' 1850 map of the San Joaquin (Fig. 6.39). These transition points on the map generally coincide with early 1800s explorer accounts: upstream of French Camp Slough on the San Joaquin River, explorer Abella wrote that the river bank "still has some oak trees, but from here downward the tule swamps begin again" and in the vicinity of Bethany on Old River referred to "the place of the oak trees" and found that oaks and other trees continued along the banks upstream (Abella and Cook 1960). Another account from the Bethany area on Old River reported that "all this country is good and has firewood," but pointed out that it was annually overflowed (Viader and Cook 1960). Tom Paine Slough was also bordered by riparian forest, referred to as "a slight strip of timber along the creek" (Norris 1851) and "scattering oaks on slough" (Hays 1853), and bordered by tree symbols in Gibbes map of the area (1850a). The transition from willow to oak dominated forest is also suggested by the difference between two GLO survey crossings on the San Joaquin mainstem, one near present-day Ott Road that only mentioned willow (Benson 1877), and the second farther upstream near present-day I-5 crossing that used oak bearing trees and remarked on the timber in addition to willow brush (Norris 1851; see Fig. 6.37).

Width variability

Following the trend of increasing height and breadth of natural levees, riparian forest width generally increased upstream, to where the typical width was on the order of 500 feet (152 m) with some places over 1,500 feet (457 m). Since localized direct detail indicating riparian width usually is unavailable, we inferred width from topographic maps, soil survey maps, and LiDAR, calibrated by texts. Where we used topographic inference to map width, we employed a continuous buffer for different width classes (see page 68). The buffer width represents our understanding of natural levee width, developed from narrative descriptions as well as maps (e.g., the Debris Commission maps) and historical aerial photography. Though the Debris Commission maps were made well after reclamation, we used the forest shown remaining as a minimum historical extent, aside from areas of known dramatic changes such as Paradise Cut (see Box 6.2).

In one of the early accounts used to calibrate our mapping, a traveler crossing the San Joaquin in 1848 commented that everything was overflowed save for the river banks extending about 300 feet (91 m) out. This less frequently inundated strip would have likely been occupied by riparian forest (Swan [1848]1960). Near Bethany on Old River, a surveyor described the banks as sometimes 150 to 300 feet wide (46-91 m, Naglee 1879) and on the mainstem of the San Joaquin, another surveyor reported that most of the drop in elevation of the banks occurred within "the first few hundred feet from the River" (Kluegul 1878). We were also able to use the GLO survey data to calibrate width in several instances where it crossed the San Joaquin River (e.g., see Fig. 6.37). As discussed earlier, this method raises uncertainty to riparian width in localized areas. Overall, we accept that we captured the landscape pattern, though undermapping and localized inaccuracies likely exist. (These uncertainties are addressed in the mapping, where 63% of the riparian forest is attributed with "medium" in size certainty).

The GLO survey also allowed us to map additional spatial detail in a few locations. For example, in present-day Stewart Tract, a GLO surveyor noted leaving tule marsh and then 984 feet (400 m) later entering brush along the San Joaquin River, indicating an ecotone of seasonal wetland at the transition between tule-dominated freshwater emergent wetland and riparian forest (see Fig. 6.32, Whitcher 1857b). It is likely that such transitions were found at varying widths between most riparian forest and perennial wetlands, though our mapping only includes those associated with specific historical evidence.

Aside from several wider patches associated with secondary channels, riparian forest became noticeably broader upstream of the head of Old River, likely related to the larger natural levees and a floodplain surface above tide elevations. At this transition point, the San Joaquin becomes more sinuous, and scroll topography, abandoned channels, and oxbow lakes are clearly visible in topographic maps and aerial photography. Important features of this riverine morphology are point bar deposits that build on the inside of meander bends, upon which the dense riparian forest established. Consequently, riparian forest was wide on the inside of meander bends (sometimes over 1,300 feet [400 m]) and narrower on the outside or bank cutting edge of bends.

The floodplains, or "bottomland," of the Stanislaus River were apparently almost entirely forested, as is suggested by a GLO survey that noted

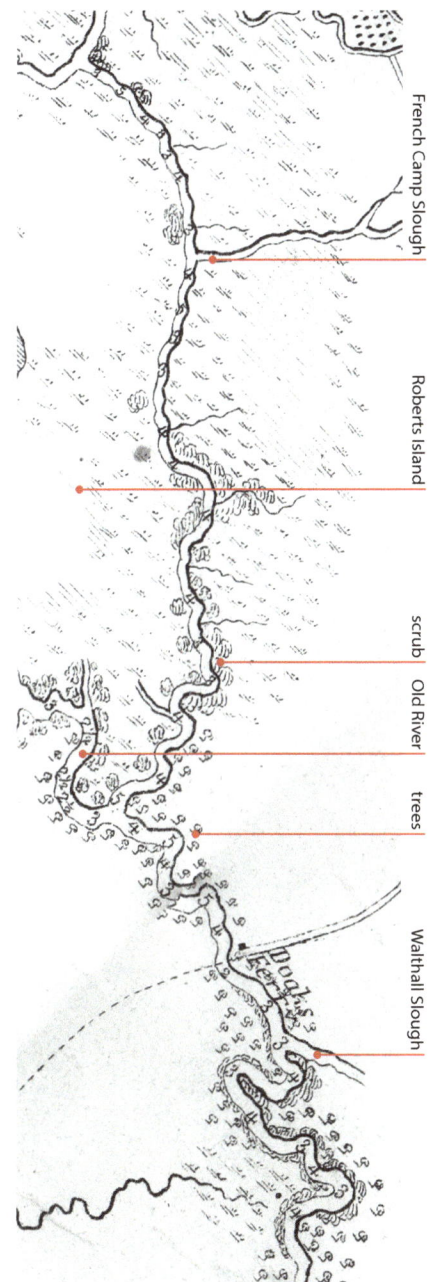

Figure 6.39. This map shows riparian forest shifting from tree to scrub dominated near the head of Old River. Riparian scrub dwindles several miles upstream from French Camp, at which point the river banks are dominated by emergent vegetation. (Gibbes 1850a, courtesy of the Map Collection of the Library of UC Davis)

We came again among innumerable flowers; and a few miles further, fields of the beautiful blue-flowering lupine, which seems to love the neighborhood of water, indicated that we were approaching a stream. Here we found this beautiful shrub in thickets, some of them being 12 feet in height. Occasionally three or four plants were clustered together, forming a grand bouquet, about 90 feet in circumference, and 10 feet high; the whole summit covered with spikes of flowers.

—FREMONT 1845

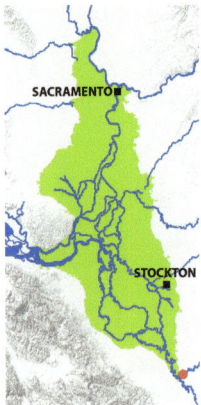

Figure 6.40. Dense forested bottomland of the Stanislaus River close to its mouth. The riparian forest in this 2005 aerial imagery occupies the meander belt width of the river. An even wider bottomland forest is suggested by the GLO survey notes overlain on top (e.g., notes of where the main body of timber begins is outside the current extent). Oaks were also found as bearing trees (orange symbol) in the GLO survey. In contrast, the floodplain of the San Joaquin River in its lower reaches was not as continuously forested as that of the Stanislaus. There, dense forest was primarily concentrated along the natural levees and associated with secondary channels. (USDA 2005)

entering timber at the point where the elevation drops from the plain above to the Stanislaus floodplain (Frémont 1845, Von Schmidt 1854-1855). This forest continued for about a half a mile across the river's floodplain. Some of that meander belt width remains today and is still covered in riparian forest (Fig. 6.40). This pattern is different from that of the historical San Joaquin River floodplain, which was not as continuously covered in forest downstream of the Stanislaus confluence.

Riparian vegetation

The riparian scrub occupying the banks along upper Union and Roberts islands was dominated by willow species, likely including arroyo willow (*Salix lasiolepis*) and yellow willow (*S. lutea*; *Daily Alta California* 1870, Alexander 1877, Jepson 1910, Jepson 1913, Sands 1977). As the natural levees became more substantial upstream, this dominance was replaced by valley and live oaks, with willows comprising the dense understory layer (Fig. 6.41; Hilgard 1884, Sweet et al. 1908, Grinnell 1911, Bidwell [1842]1937). In some locations, however, descriptions of groves of oak trees suggest that the understory may have been a more open herbaceous cover (Lewis Publishing Co. 1890, Williams 1973).

Of 55 GLO bearing, witness, and line trees in the south Delta floodplain (e.g., within the riparian forest), 53 were oaks and only three were willows. The willows were all located downstream of the head of Old River and were quite small (4, 5, and 6 inches/10.2, 12.6, and 15.2 cm in diameter). It is unlikely that such small trees would have been used as bearing trees had other larger and more well established trees been present (White 1983). Most of the oaks were between two and three feet in diameter (61 and 91 cm; Fig. 6.42). A number of oaks and willows, usually over two feet in diameter, were recorded by William Hammond

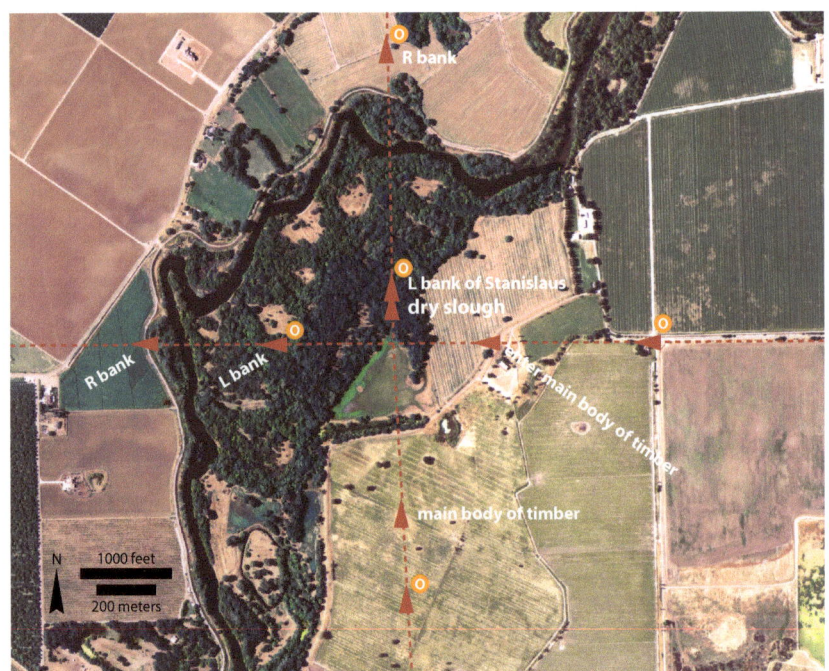

Hall's state engineering surveys along the San Joaquin and Old rivers (Hall n.d.).

Although oaks were by far the most commented-upon species of riparian tree along the San Joaquin, other riparian tree species were also likely present, such as white alder (*Alnus rhombifolia*), box elder (*Acer negundo*), California sycamore (*Platanus racemosa*), California dogwood (*Cornus sericea*), and Fremont cottonwood (*Populus fremontii*; Bryant [1848]1985, Hilgard 1884, Jepson 1910, Riley in Derby and Farquhar 1932). Within the understory layer, lupine (*Lupinus formosus* var. *robustus*), buttonbush (*Cephalanthus occidentalis*), blackberry (*Rubus* spp.), wild rose (*Rosa* spp.), Delta button-celery (*Eryngium racemosum*), hedge nettle (*Stachys ajugoides*), California wild grape (*Vitis californica*), and other species intermixed with willows (Dumas [1852]1933, Grinnell 1911, Abella and Cook 1960, Fox 1987a, Consortium of California Herbaria 2009, CNDDB 2010). Species recorded by Jepson on Middle River in 1913 are listed in Table 6.2.

While sycamores were commonly used as bearing trees by GLO and State Engineer surveyors and remarked upon by numerous travelers along the Sacramento, relatively few sources note sycamores along the San Joaquin (see Fig. 6.42; Bryant [1848]1985, Hilgard 1884, Jepson 1910, Riley in Derby and Farquhar 1932). Cottonwoods were mentioned by only one of the historical sources examined for this study, which was a general description of the large streams of the Central Valley (Hilgard 1884).

Figure 6.41. A mix of riparian trees and scrub can be seen along an unidentified waterway. Willows and oaks were likely the dominant species of the south Delta riparian forest. The relative mix of trees and scrub varied at the landscape scale depending on height of natural levees. Trees dominated the upstream higher levees and scrub occupied the more frequently inundated banks downstream. The photo is part of a collection of San Joaquin reclamation photos. (ca. 1900, courtesy of The Bancroft Library, UC Berkeley)

The timber on the banks of the San Joaquin, Tuolumne, and Stanislaus rivers is composed almost entirely of the holly-leafed oak, a species of white oak, willow, and sycamore. The timber is low, dwarfish, apparently hard to work, and unfit for building purposes.

—RILEY 1849 IN DERBY AND FARQUHAR 1932

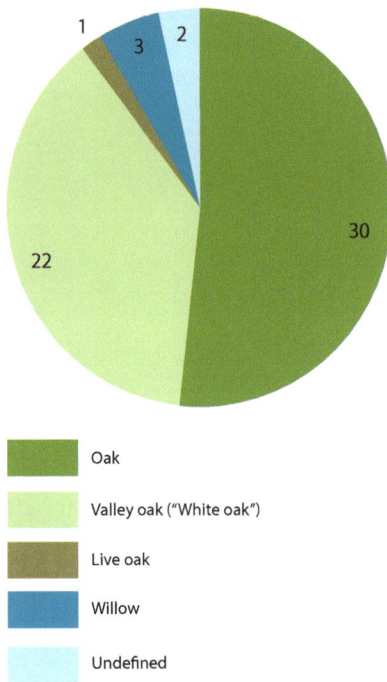

Figure 6.42. Bearing trees recorded by the GLO. The dominant species are oaks, with only three willow bearing trees obtained. These willows were recorded downstream of the head of Old River, suggesting that surveyors used willows when the longer-lived oaks were in short supply. Interestingly, no sycamores were recorded, which contrasts with sycamores comprising 33% of bearing trees used in surveys on lower reaches of the Sacramento River. Only 58 bearing trees were recorded by the GLO survey in the south Delta.

On the return trip he should not take the entrance to the left, which is the one we have just come from, because the river is full of logs. The other one, even if it contains no logs, runs in the middle of the tule swamps, and in that region nothing can be accomplished unless it be salmon fishing and beaver [trapping].

—ABELLA AND COOK 1960 IN OCTOBER 1811

Table 6.2. Species listed in botanist Willis Jepson's field notes for Middle River in 1913. (Jepson 1913)

Species	Common name
Phragmites australis	common reed
Cephalanthus occidentalis	button bush
Euthamia occidentalis	goldenrod
Polygonum amphibium	smartweed
Cornus californica	California dogwood
Stachys albens	white hedge nettle
Salix gooddingii	black willow
no *Populus fremontii*	no Fremont cottonwood

Woody debris

Historical evidence of woody debris derived from the riparian forest and occupying the primary river channels within the upper limits of tides points to this as a potentially ecologically significant structural element of south Delta habitats. Such debris can obstruct flow and cause new channels to develop alongside the old. This may have contributed to the extensive latticework of active and abandoned channels within the south Delta landscape. Much of the material may have originated from the riparian forest nearby as it is unlikely the river would have carried large amounts of debris from far upstream. Woody debris appears to have accumulated in Old River near the head of Middle River (Gibbes 1850a, Tucker 1879d, Abella and Cook 1960). Rafts of debris and other obstructions were reported on Middle River as well (Gibbes 1850a, De Mofras and Wilbur 1937). Reports of individual snags hindering steamboat travel on the San Joaquin mainstem also exist, particularly upstream of the head of Old River (Marlette 1854, Higley 1859, Payson 1885, Williams 1973).

Historical information concerning woody debris in the south Delta comes from four distinct points in time during the 1800s. In the earliest of these, from 1811, a Spanish explorer on the Old River in the vicinity of Salmon Slough reported "the stream bed is full of logs" (Abella and Cook 1960). The channel was so full of woody debris that the explorers sent scouts ahead to see if it was worthwhile to continue. Forty years later in the high water season, surveyor Gibbes found a large raft of timber in the same location:

> I came to a raft of large timber, and after some hard work in cutting and sawing logs, we succeeded in dragging our boat through. At the foot of the raft the river divides, taking the left, which is the largest, although much smaller than the main channel, and filled with floating drift wood that made it difficult to proceed. I came to where it again divides, the right being stopped with drift wood. (Gibbes 1850b)

The map from his survey shows the position of the raft (Fig. 6.43).

The position of this raft also coincides with the location where reclamation efforts in the late 1870s diverted the main flow of Old River into Salmon Slough because of the woody debris occupying the main channel (see Fig.

6.43; Kluegul 1878, Naglee 1879). An 1877 document stated that an "opening of a new channel for Old River around the Raft near Salmon Slough" was created (Naglee 1879). Tucker (1879d) detailed these activities:

> There were a great many old logs and an immense amount of driftwood and rubbish in Old River, and we removed most of it.
>
> Below Salmon Slough the river was very narrow and so badly choked up with driftwood that it was deemed advisable to build a dam on it at the head of Salmon Slough and turn the water through a new channel.
>
> A canal, 1600 feet long, was cut from the head of Salmon Slough, across a low piece of land to a part of the slough that was comparatively wide and deep. (Tucker 1879d)

Rafts located on Middle River coincided with the position of several side channels leading to a backwater lake (possibly Willow Lake; see Fig. 6.43; Gibbes 1850a). Gibbes (1850b) reported that "in the narrow part I found two small rafts of dead timber…above the rafts it widens out again." These were apparently still present almost 30 years later, when an 1877 reclamation document recommended that obstructions be removed from the Middle River and the channel widened (Alexander 1877).

The fact that substantial rafts of woody debris were found in the same locations spanning many decades suggests that such rafts were not ephemeral features. Once the jams were established, avulsion events presumably formed secondary channels, which may have spurred the establishment of vegetation that affected subsequent channel migration and formation (O'Connor et al. 2003, Mount pers. comm.). It seems unlikely that these features can be attributed to early channel modifications or wood cutting, given their presence in 1811. The Spanish explorer accounts as well as Gibbes's 1850 survey contradict a reclamation document that seems to suggest the formation of Paradise Cut (which likely occurred in the late 1850s) caused the Old River channel to fill with drift wood (Naglee 1879).

Modern research has established the historical presence of large, persistent rafts of woody debris on large low-gradient rivers, relating them to their geomorphic effects and provision of multiple ecological functions (Triska 1984, Sedell and Froggatt 1984, Collins and Montgomery 2002, O'Connor et al. 2003). For example, on the Red River in the Midwest, the formation of backwater lakes and secondary channel systems was attributed to the obstruction of flow caused by the logjams (Triska 1984). On the Willamette River in Oregon, woody debris was related to the multi-channel river morphology, flow diversion into side channels, and the establishment of gravel bars and willow thickets (Sedell and Froggatt 1984, Benner and Sedell 1997). Ecological functions of woody debris include the provision of key habitat elements (e.g., step-pool morphology, substrate, and forage and refuge opportunities) for salmon and other aquatic species including invertebrates. Functions also include retention and cycling of organic material, which is better understood for high-energy, upper tributary

Figure 6.43. "Rafts" of woody debris are shown on Old River (just upstream of Salmon Slough) and on Middle River (A). Both of the rafts were coincident with large lakes that were fed by secondary channels leading off the river. Just downstream of the raft the river divided into the main channel (to the north, highlighted in dashed blue) and Salmon Slough. All flows were diverted to Salmon Slough by the early 1900s, with only levees marking the course of the Old River channel (B, C). Today, the old channel is barely perceptible. The change was necessary, engineers argued, because "the river was very narrow and so badly choked up with driftwood" (Tucker 1879). Note that map in (A) is not exactly aligned with other maps. (A: Gibbes 1850a, courtesy of the Map Collection of the Library of UC Davis; B: USGS 1909-1918; C: USDA 1937-1939; D: USDA 2005)

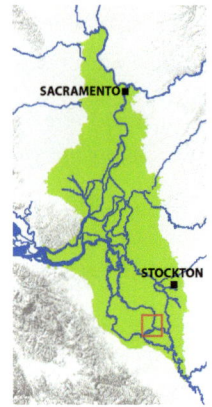

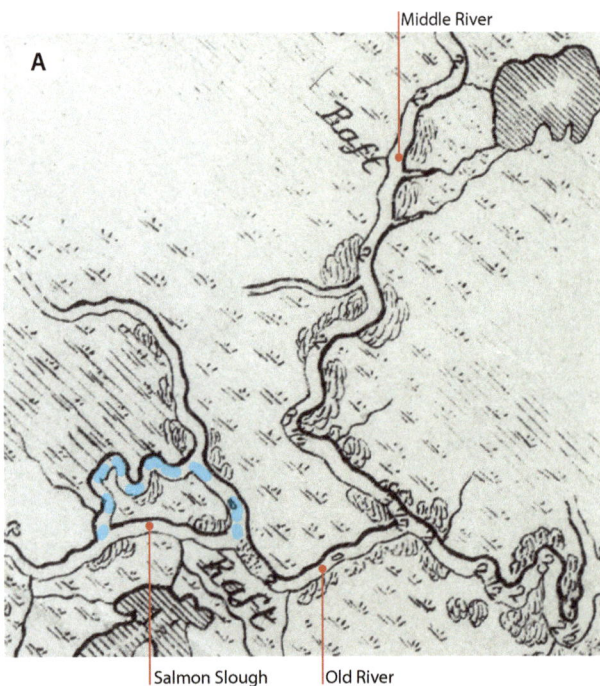

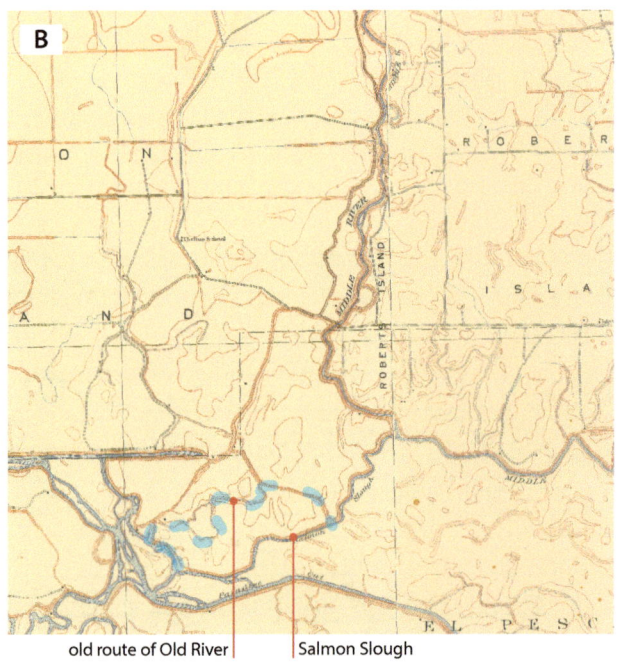

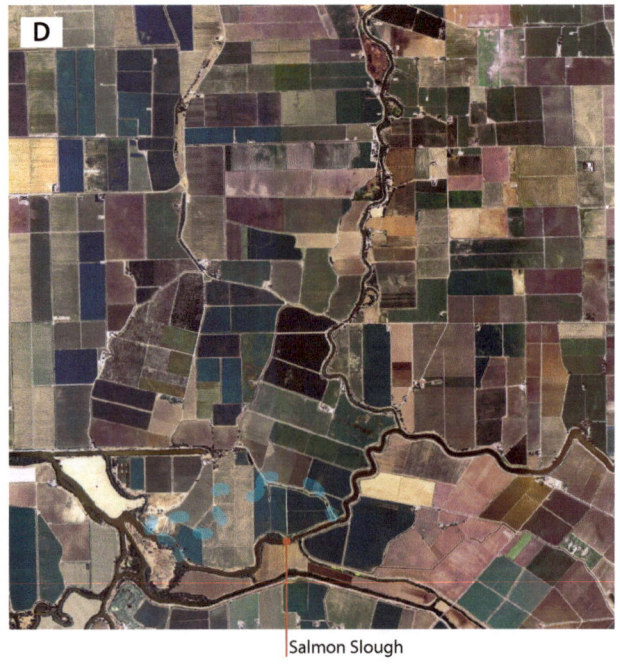

streams as opposed to low-gradient rivers (Bryant 1983, Bilby and Bisson 1998, Gregory et al. 2003).

The role of woody debris in tidal estuaries and swamps is less studied. However, a number of studies particularly from Pacific Northwest systems have demonstrated that woody debris in tidal systems influences channel morphology (e.g., step-pool, sinuosity), forms backwater habitats of ponds and side channels, stabilizes banks, reduces flow velocities, influences sedimentation processes, and affects vegetation patterns (Maser and Sedell 1994, Weinstein and Kreeger 2000, Simenstad et al. 2003, Hood 2007, Diefenderfer and Montgomery 2009, Collins B pers. comm.). These physical interactions in turn affect ecological functions, impacting the food web and nutrient dynamics and providing forage, breeding and refugia sites for fish (Maser and Sedell 1994, Hood 2007a). These linkages provide a mechanism through which many features found in the south Delta, such as side channels and lakes, may have been formed and maintained. Consequently, it is likely that the recruitment of woody debris from riparian areas was an important process affecting many features that together made the rich landscape of the south Delta. Substantial recruitment of woody debris may have also occurred elsewhere in the Delta at the head of tide (e.g., Mokelumne River above the Cosumnes confluence), though it appears the Sacramento River's flow and channel geometry prevented large rafts from forming in its channel (Gibbes 1850a, Abella and Cook 1960).

Woody debris in the south Delta likely provided important habitat for fish and other aquatic species. Fish adapted to slow-moving waters, such as Sacramento perch and Thicktail chub, would have benefited from the higher water levels maintained by the obstructed channels and from protection from predators (Moyle pers. comm.). The area may also have provided important floodplain rearing habitat for outmigrating salmon smolt. As the name implies, El Pescadero (roughly translated to "the place of fishing") was a notable fishing ground in the Delta. Spanish explorers describe pleasant meals of salmon and grapes (Abella and Cook 1960, Viader and Cook 1960). Perhaps not coincidentally, one of the most populated Indian villages in the Delta region was located in the vicinity of White House Landing.

Associated biota
Riparian forests provide important habitat for a diverse range of species yet comprise only a small proportion of the total land area in the Central Valley (Smith 1977). The south Delta forests would have provided important habitat for numerous terrestrial species, such as riparian brush rabbit (*Sylvilagus bachmani riparius*), riparian woodrat (*Neotoma fuscipes riparia*), Long-tailed weasel (*Mustela frenata xanthogenys*), coyote (*Canis latrans ochropus*), beaver (*Castor canadensis subauratus*), valley elderberry longhorn beetle (*Desmocerus californicus dimorphus*), foothill yellow-legged frog (*Rana boylii*; CNDDB 2010, MaNIS 2010) and the extirpated tule elk (*Cervus elaphus nannodes*), antelope (*Antilocapra americana*), and grizzly bear (*Ursus arctos californicus*), in addition to numerous riparian forest

These relics of the past will hold the key to relating inferences from the historical record to quantitative differences between what we now perceive as the normal condition and how the river interacted with the terrestrial ecosystem and its massive quantities of wood inputs in the past.

—SEDELL AND FROGGATT 1984

One reaches the Río del Pescadero [Old River], which has good water, depth and current, and is so called because fishing is done in it for salmon.

—SAL AND COOK 1960, IN JANUARY 1796

Some distance below the raft [vicinity of the head of Middle River] we found some very good land and plenty of timber; we also saw on the east bank several grisly [sic] bears and numerous herds of elk that resort here in the spring season from the mountains and plains and when alarmed rush into the tule, where the plunging of such herds of large animals makes a tremendous roar that can be heard for some distance.

—GIBBES 1850B

associated birds (Vahgti and Greco 2007). In his 1911 exploration northeast of Tracy, biologist Joseph Grinnell noted that "the chief feature of this levee district is the presence of timber" before listing Silky flycatcher (*Phainopepla nitens*), Slender-billed nuthatch (*Sitta carolinensis aculeata*), White breasted nuthatch (*Sitta carolinensis*), Western bluebird (*Sialia mexicana*), Oak titmouse (*Baeolophus inornatus*), Purple finch (*Carpodacus purpureus*), Ruby-crowned kinglet (*Regulus calendula*), Sacramento spurred towhee (*Pipilo maculatus falcinellus*), California woodpecker (*Melanerpes formicivora*), and Nuttall woodpecker (*Picoides nuttallii*) among the oaks (Grinnell 1911). These species were able to take advantage of the different life stages of the trees, where snags offered nesting habitat and older "badly mistletoed" valley oaks were "particularly attractive to many birds," according to Grinnell (1911). Tricolored blackbird (*Agelaius tricolor*), merlin (*Falco columbarius*), Swainson's hawk (*Buteo swainsoni*), Western yellow-billed cuckoo (*Coccyzus americanus occidentalis*), California shrike, Western meadowlark (*Sturnella neglecta*), and California horned lark (*Eremophila alpestris actia*) (CNDDB 2010, ORNIS 2010).

WILDFLOWER FIELDS AND ALKALI MEADOWS

Ecotonal environments were found throughout the south Delta floodplain (as described in previous sections), reflecting the topographic variability that influenced inundation patterns and soil characteristics and in turn affected vegetation assemblages. The ecotone at the upland margin of the floodplain provided a spatially complex edge. This zone encompassed transitions along hydrologic, topographic, and soil gradients, where lower inundation frequency and changing soil properties produced an intermixing of seasonal wetland, grassland, oak woodland and savanna, with the occasional perennial pond or wetland patch.

Between French Camp Slough and the Stanislaus River, travelers described the region of upland ecotone as an open treeless plain or as "long stretches of prairie;" a continuation of the landscape of the southern San Joaquin Valley (Fremont 1845, *Sacramento Daily Union* 1871). In the drier portions, vegetation cover was likely quite sparse. Alkali seasonal wetlands and meadows complexes were common where vernal pools were found, which were sometimes described as having a "hog-wallow" appearance of small depressions and hillocks (Unknown 1873, Hilgard 1884). Such topography was characteristic of locations throughout the San Joaquin Valley and often of the edge of the tule-dominated wetlands. The seasonal wetlands "on a large scale have a level or gently rolling surface, while on the small scale they are to a considerable extent dotted with the singular rounded hillocks, popularly known as 'hog-wallows,' from 10 to 30 feet in diameter and from 1 to 2 feet high" (Fig. 6.44; Hilgard 1884). Plant species associated with the alkali complexes include salt grass (*Distichlis spicata*), swamp grass (*Crypsis schoenoides*), button celery (*Eryngium aristulatum*), popcornflower (*Plagiobothrys leptocladus*), semaphore grass (*Pleuropogon californicus*), alkali weed (*Cressa truxillensis*), saltbush (*Atriplex* spp.), alkali heath (*Frankenia salina*), iodine bush (*Allenrolfea occidentalis*), palmate-bracted

The valley of the San Joaquin is the floweriest place of world I ever walked, one vast, level, even flower-bed, a sheet of flowers, a smooth sea, ruffled a little in the middle by the tree fringing of the river and of smaller cross-streams here and there, from the mountains.

—MUIR 1916

What then are the flowers that most attract the eye on our sandier, or lighter soils? They are the orange-colored poppy, blue and pink lupines, lovegroves, bluebells, the painted-cup, or, as it might be very suitably named, princess' plume, the flax-flower, wild chrysanthemum, star-thistle, milk-weed, dandelion, lark spurs, evening-primroses, and several others worthy of record...

—RAMBLER 1872

bird's beak (*Codylanthus palmatus*), San Joaquin spearscale (*Atriplex joaquiniana*), along with the possibly extinct caper-fruited tropidocarpum (*Tropidocarpum capparideum*; Ornduff et al. 2003, Sawyer et al. 2009, CNDDB 2010).

This alkali edge was intersected by more well drained sandy soils, which tended to define the area (Norris 1851, Handy 1864, Hilgard 1884, Lewis Publishing Co. 1890, Lapham and Mackie 1906, Nelson et al. 1918). The region was referred to by someone as the "sand plains" (Tinkham 1923). Others remarked on the "apparently unproductive" soil (Lyman and Teggart 1923) that "produces little pasturage" (Moerenhout [1849]1935). Lest the region be written off as worthless, however, a Surveyor General report stressed that the sandy loam of San Joaquin County was "by no means a barren sand, as is found in the neighborhood of San Francisco…it contains much vegetable matter" and was thus deemed adequate for agricultural production (Long in Houghton 1862). The interlacing of alkali and well drained sandy soils is seen in the soil survey maps along the eastern margin of the Delta, south of French Camp Slough (Lapham and Mackie 1905). The rolling topography and sandy soils noted by the GLO surveyors generally match the pattern from the soil survey, though alkali is not noted (Fig. 6.45; Norris 1851, Handy 1864).

The sandy soils of the upland ecotone supported annual forblands mixed with grasses that produced the "sparkling" wildflower displays celebrated by

Figure 6.44. Hog wallows describe a characteristic land surface topography consisting of small depressions and "rounded hillocks" (Hilgard 1884). They are often associated with vernal pools and alkali wetlands. This photograph was taken in 1938 at the Durham Ferry Road crossing of the the San Joaquin River. (Covello 1938, courtesy of Bank of Stockon Historical Photograph Collection)

hog wallow

Tidal channel

Fluvial channel

Tidal or Fluvial channel
(lower confidence level)

Water

Non-tidal freshwater emergent wetland

Willow riparian scrub or shrub

Valley foothill riparian

Alkali seasonal wetland complex

Grassland

Oak woodland or savanna

"occasional tree"

"land 2nd rate and sandy

"to timber"

"leave timber"

"land rolling gently"

"land sandy"

"land sandy"

"land level, sandy"

"land sandy"

"land slightly undulating, soil sandy"

"leave tule"

SACRAMENTO

STOCKTON

N   ½ mile

1 kilometer

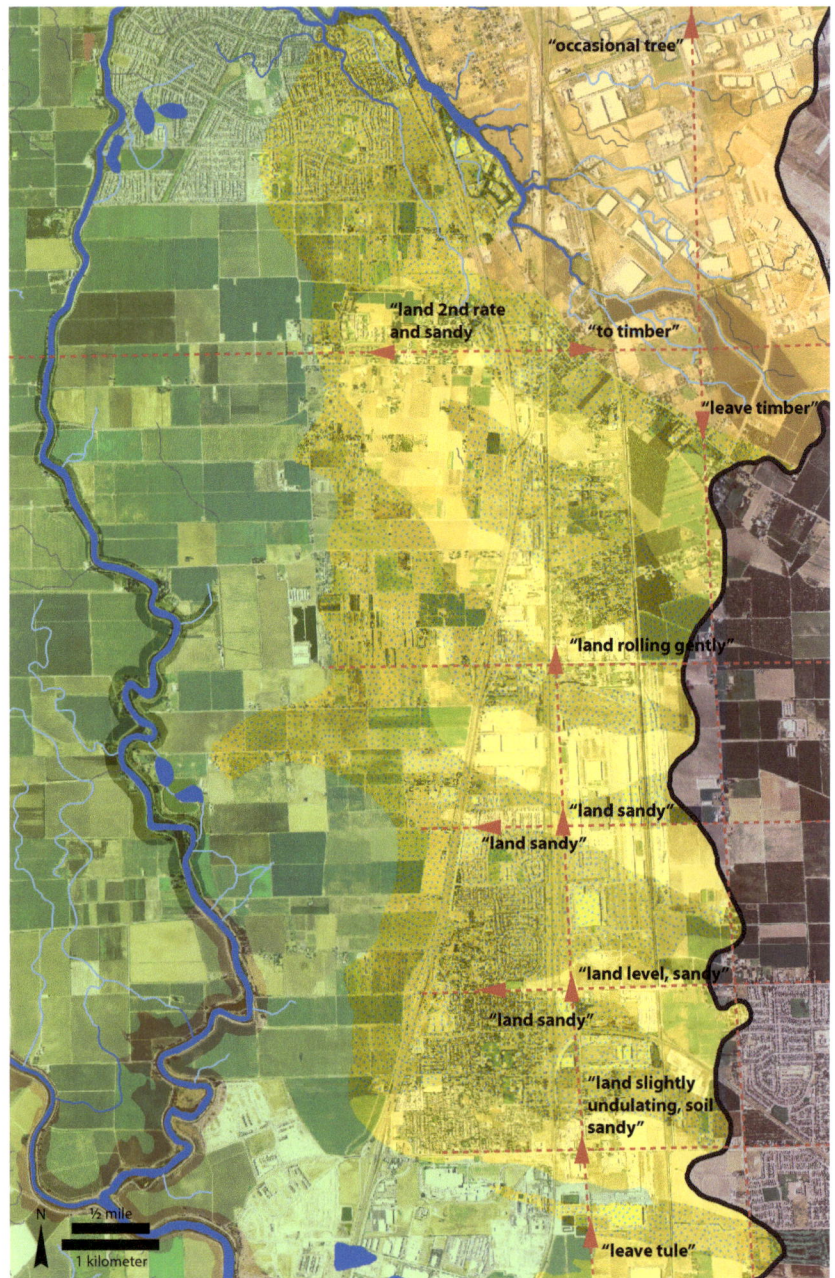

Figure 6.45. Sandy soils of the upland ecotone. The habitat map, here overlying modern imagery, illustrates how alkali seasonal wetlands along the tule edge was intersected by fingers of well drained sandy soils extending from the adjacent alluvial fan. Additional support is provided by selected notes from the GLO survey. The mapped pattern was based primarily on the 1905 Stockton soil survey. (Lapham and Mackie 1905)

many who traveled through the valley in the spring, including the renowned naturalist and conservationist John Muir (Fig. 6.46; Fremont 1845, Bryant [1848]1985, U.S. War Department 1856a, Rambler 1872, Muir 1916, Taylor 1969). Though dusty, hot, and uninviting late in the season, the landscape in the early spring months sprung forth "with a perfect carpet of flowers of every color and almost innumerable varieties" (Orr 1874). A 1905 soil survey characterized the eastern edge of the south Delta as "treeless, and unmarked by vegetation except wild grasses and a great variety of brilliantly colored wild flowers appearing during the early spring" (Lapham and Mackie 1906).

Figure 6.46. Colorful wildflower displays once covered the plains of the San Joaquin Valley in the spring. One can still observe these in places today. (photo © 1990 Dr. Oren D. Pollak)

Accounts of the character of the upland ecotone on the western edge, south of Contra Costa, are less numerous. Soil surveys and GLO notes suggest, however, that alkali seasonal wetlands were not present south of the vicinity of Bethany. Soils were mapped as Yolo adobe soils (Y) and Yolo clay loams (Ys), which were described as generally free of the effects of alkali (Nelson et al. 1918). This is likely related to the fact that slopes were much greater on this side of the valley, allowing for more effective drainage. Annual forbs probably dominated the area, as suggested by selected field notes from the GLO survey, including "level plains with weeds and flowers," "barren low land," "free from vegetation," "some vegetation," "meadow," "grassy swale," and "no timber" (Norris 1851, Hays 1853, Whitcher 1857a). The presence of wildflowers along lower plains of the Diablo range was highlighted in the report of the late 1840s U.S. Exploring Expedition survey: "a part of the surface of the plain was covered with a growth of sunflower, standing from six to ten feet high, and the blossoms very small" (U.S. War Department 1856a). It is unclear, however, exactly where the observation was made. The height of this vegetation seems rather remarkable, perhaps suggesting rich soils.

Understanding the conditions at the upland ecotone of the Delta's perennial wetlands is important for conceptualizing landscape connectivity and function. Wetland ecosystems are necessarily affect and are affected by their adjoining environments. Wetland function and process do not occur independently; rather, key ecological functions hinge upon these connections, including energy and nutrient transfer, refuge from flooding, habitat for amphibians, and access to wetlands by terrestrial species such as elk for foraging, breeding, or refuge during drought (e.g., Hulaniski 1917, Burcham 1857, Semlitsch 1998, Amexaga

In the spring there was an abundance of wild flowers, so that for great stretches one saw only the carpet of their blood, with the green of grass and foliage hidden under the riot of color. The were many pinks and whites among them, but the blue of ground lupine and larkspur and the gold of buttercups and California poppies and the many other yellow species predominated in the accepted state colors...There was no underbrush on the plains – just the iridescent green of grass interspersed with flowers, with here and there a pure golden patch where wild mustard or sunflowers had taken over.

—TAYLOR 1969, DISCUSSING CARL GRUNSKY'S CHILDHOOD NEAR STOCKTON IN THE MID-1800S

et al. 2002). These transitional areas likely supported species of concern, including the San Joaquin kit fox (*Vulpes macrotis mutica*), Swainson's hawk (*Buteo swainsonii*), and the Western burrowing owl (*Athene cunicularia hypugaea*), all of which can still be found in the area today (CALFED 2000c, CNDDB 2010). Improved understanding of ecotones is needed and involves the application of landscape ecology principles for restoration and conservation (Risser et al. 1984, Naiman et al. 1989, Naiman and Decamps 1990).

7. References

Abella R, Cook SF. 1960. *Colonial expeditions to the interior of California Central Valley, 1800-1820.* Berkeley, CA: University of California Press.

Alexander BS. 1869. Engineer's report. In *Fresh water tide lands of California*, ed. M.D. Carr & Co.

Alexander BS. 1870. *Fresh water tide lands of California.* San Francisco, CA: M.D. Carr & Co., Book and Job Printers.

Alexander BS. 1877. *Report of Gen. B.S. Alexander on the reclamation of the Rancho Pescadero, San Joaquin County.* San Jose, CA: Mercury Steam Print.

Alexander BS, Mendell GH, Davidson G. 1874. *Report of the Board of Commissioners on the irrigation of the San Joaquin, Tulare, and the Sacramento Valleys of the state of California.* Government Printing Office, Washington, DC.

Algier N. 1863. *The United States, appellants, v. John A. Sutter. No. 135. Appeal from the District Court U.S. for the Northern District of California. Supreme Court of the United States.*

Allardt GF. 1880. Map of Toland Ranch situated in the County of Solano. *Courtesy of Solano County Surveyors Office.*

Allen JRL. 1965. A review of the origin and characteristics of recent alluvial sediments. *Sedimentology* 5:89-191.

Allen-Diaz B, Bartolome JW, McClaran MP. 1999. California oak savanna. In *Savannas, barrens, and rock outcrop plant communities of North America*, ed. Roger C. Anderson, James S. Fralish, and Jerry M. Baskin, 322-339. Cambridge: Cambridge University Press.

Alley DW, Dettman DH, Hiram WL, et al. 1977. Habitats of native fishes in the Sacramento River basin. In *Riparian forests in California: Their ecology and conservation*, ed. Anne Sands, 39-46. Institute of Ecology Publication No. 15.

Amexaga JM, Santamaria L, Green AJ. 2002. Biotic wetland connectivity - supporting a new approach for wetland policy. *Acta Oecologica* 23:213-222.

Anderson K. 2005. *Tending the wild: Native American knowledge and the management of California's natural resources.* Berkeley, CA: University of California Press.

Anza JB, Bolton HE. 1930. *Anza's California expeditions.* Berkeley, CA: University of California Press.

Arguello, Cook SF. 1960. *Colonial expeditions to the interior of California Central Valley, 1800-1820* Berkeley, CA: University of California Press.

Askevold RA. 2005. Interpreting historical maps to reconstruct past landscapes in the Santa Clara Valley. Geography, San Francisco State University, San Francisco.

Atwater BF. 2011. *Loss of the Delta's historical landscape as an ecological stressor.* Delta Independent Science Board.

Atwater BF. 1979. *Ancient processes at the site of southern San Francisco Bay: movement of the crust and changes in sea level.* Fifty-eighth Annual Meeting of the Pacific Division of the American Association for the Advancement of Science, San Francisco State University, San Francisco, California, June 12-16, 1977.

Atwater BF. 1980. *Distribution of vascular-plant species in six remnants of intertidal wetland of the Sacramento-San Joaquin Delta, California.* U.S. Geological Survey.

Atwater BF. U.S. Geological Survey 1982. Geologic maps of the Sacramento-San Joaquin Delta, California. Menlo Park, CA. 1:24000.

Atwater BF, Belknap DF. 1980. Tidal-wetland deposits of the Sacramento-San Joaquin Delta, California. In *Quaternary depositional environments of the Pacific Coast: Pacific Coast*

Paleogeography Symposium 4, ed. M.E. Field, A.H. Bouma, I.P. Colburn, R.G. Douglas, and J.C. Ingle. Los Angeles, California: The Pacific Section Society of Economic Paleontologists and Mineralogists.

Atwater BF, Conard SG, Dowden JN, et al. 1979. History, landforms, and vegetation of the estuary's tidal marshes. In *San Francisco Bay, the urbanized estuary: investigations into the natural history of San Francisco Bay and Delta with reference to the influence of man*. Fifty-eighth annual meeting of the Pacific Division/American Association for the Advancement of Science held at San Francisco State University, San Francisco, California, June 12-16, 1977, ed. T. John Conomos, 493 p. San Francisco, Calif.: AAAS, Pacific Division.

Atwater BF, Hedel CW. 1976. *Distribution of seed plants with respect to tide levels and water salinity in the natural tidal marshes of the northern San Francisco Bay estuary, California*. U.S. Geological Survey Open-file report; 76-389, Menlo Park, CA. 41p. *Courtesy of U.S. Geological Survey.*

Baca JN. 1854. *U.S. v. R.L. Brown, Land Case No. 411 ND, Laguna de Santos Calle, U.S. District Court, Northern District. Courtesy of The Bancroft Library, UC Berkeley.*

Bailey EA. [1918]1927. *Historical summary of state legislative action with results accomplished in reclamation of swamp and overflowed lands of Sacramento Valley, California. Appendix D to Sacramento Flood Control Project, revised plans.* California State Printing Office, Sacramento, CA.

Bancroft HH. 1883. *The works of Hubert Howe Bancroft, Volume I. The native races. Vol. I. Wild tribes.* San Francisco, CA: A.L. Bancroft & Company, Publishers.

Barbour M, Keeler-Wolf T, Schoenherr A. 2007. *Terrestrial vegetation of California, 3rd edition.* Berkeley, Los Angeles, London: University of California Press.

Barrett SA. 1908. The geography and dialects of the Miwok Indians. *University of California Publications in American Archeology and Ethnology* 6(2):333-367.

Bartell MJ. 1912. Report showing that the waters of the McCloud river cannot be diverted to the use of San Francisco and the Bay communities owing to its present use by developed priorities in the Sacramento Valley. *Courtesy of Water Resources Collections and Archives, Riverside.*

Bates F. 1853. *The United States, appellants, v. John A. Sutter. No. 135. Appeal from the District Court U.S. for the Northern District of California. The Supreme Court of the United States.*

Baxter R, Breuer R, Brown L, et al. 2010. *2010 Pelagic Organism Decline Work Plan and Synthesis of Results.* Interagency Ecological Program.

Baye PR, Faber PM, Grewell B. 2000. Tidal marsh plants of the San Francisco Estuary. In *Baylands ecosystem species and community profiles. Life histories and environmental requirements of key plants, fish and wildlife prepared by the San Francisco Bay Area Wetlands Ecosystem Goals Project*, ed., 9-32. Oakland: USEPA, San Francisco, CA & S.F. Bay Regional Water Quality Control Board.

Beaumont D. United States District Court, Northern District. 1858. Plat of the Rancho Campo de los Franceses, finally confirmed to Charles M. Weber, California. Land Case Map E-585. *Courtesy of The Bancroft Library, UC Berkeley.*

Beaumont D. 1859a. *U.S. v. Anastasio Chaboya, Land Case No. 406 ND [Sanjon de los Mequelemnes], U.S. District Court, Northern District.* 142. *Courtesy of The Bancroft Library, UC Berkeley.*

Beaumont D. 1859b. *U.S. v. Charles Weber, Land Case No. 298 ND [Campo de Los Franceses Grant], U.S. District Court, Northern District.* 165-168. *Courtesy of The Bancroft Library, UC Berkeley.*

Beaumont D. 1861a. Map of Swamp Land District No. 17. *Courtesy of California State Lands Commission, Sacramento.*

Beaumont D. California State Land Office, 1861b. *Swamp and Overflowed Lands Reports and Correspondence.* Box 44. *Courtesy of California State Archives, Sacramento.*

Beaumont D. 1859. *U.S. v. Charles Weber, Land Case No. 298 ND [Campo de Los Franceses Grant], U.S. District Court, Northern District.* 165-168. *Courtesy of The Bancroft Library, UC Berkeley.*

Beechie TJ, Collins BD, Pess GR. 2001. Holocene and recent geomorphic processes, land use, and salmonid habitat in two North Puget Sound River Basins. In *Geomorphic processes and riverine habitat*, ed. American Geophysical Union, 37-54. Vol. 4.

Beechie TJ, Sear DA, Olden JD, et al. 2010. Process-based principles for restoring river ecosystems. *BioScience* 60(3):209-222.

Belcher E. 1843. *Narrative of a voyage round the world: performed in Her Majesty's ship Sulphur during the years 1836-1842*. London: H. Colburn.

Belcher E, Simpkinson FG, Pierce RA, et al. 1979. *H.M.S. Sulphur on the Northwest and California coasts, 1837 and 1839 : the accounts of captain Edward Belcher and midshipman Francis Guillemard Simpkinson*. Kingston, Ontario: Limestone press.

Bell SS, Fonseca MS, Motten LB. 1997. Linking Restoration and Landscape Ecology. *Resotration Ecology* 5(4):318-323.

Benner PA, Sedell JR. 1997. Upper Willamette River landscape: A historic perspective. In *River quality: Dynamics and restoration*, ed. Antonius Laenen and David A. Dunnette. Boca Raton, FL: CRC Press.

Bennyhoff JA. 1977. *Ethnogeography of the plains Miwok*. Center for Archaeological Research at Davis, CA.

Benson WF. Department of the Interior, Bureau of Land Management Rectangular Survey, California, 1877. *Field notes of the exterior lines of Township 1 North, Range 6 East, Mount Diablo Meridian, California*. Book 309-4. 151-164. *Courtesy of U.S. Bureau of Land Management, Sacramento.*

Benson WF. Department of the Interior, Bureau of Land Management Rectangular Survey, California, 1878-9. *Field notes of a portion of the exterior north, west, and south boundary lines of Township 2 North Range 5 East Mt. Diablo Meridian, California*. Book 85-1. *Courtesy of Bureau of Land Management, Sacramento.*

Benson WF. Department of the Interior, Bureau of Land Management Rectangular Survey, California, 1879. *Field notes of the meander lines of Township 2 North Range 5 East. Mt. Diablo Meridian, California*. Book 85-12. *Courtesy of U.S. Bureau of Land Management, Sacramento.*

Bettelheim MP, Thayer CH. 2006. Anniella pulchra pulchra (silvery legless lizard). Habitat. *Herpetological Review* 37(2):217-218.

Bidwell J. United States District Court, Northern District. ca. 1840. Mapa de Los Ulpinos. Land Case Map D-181. *Courtesy of The Bancroft Library, UC Berkeley.*

Bidwell J. [1842]1937. *A journey to California : with observations about the country, climate and the route to this country*. San Francisco, CA: J.H. Nash.

Bidwell J. U.S. District Court, Northern District. 1844. Rancho de Nueva Flandria, 1844, Yolo County, California. Land Case Map B-552. *Courtesy of The Bancroft Library, UC Berkeley.*

Bidwell J. U.S. District Court, Northern District. ca. 1851. Mapa del Valle del Sacramento, Land Case Map E-615. *Courtesy of The Bancroft Library, UC Berkeley.*

Bidwell J. [1884]1904. *Early California reminiscences*. Los Angeles, CA: Out West Co.

Bidwell J, Royce CC. 1907. *Addresses, reminiscences, etc. of General John Bidwell*. Chico, CA.

Bilby RE, Bisson PA. 1998. Function and distribution of large woody debris. In *River ecology and management: lessons from the Pacific Coastal ecoregion*, ed. R. J. Naiman and R. E. Bilby, 324-346. New York: Springer-Verlag.

Bingham N. 1996. Human management and development of the Sacramento-San Joaquin Delta: a historical perspective. Pages 13-18. Interagency Ecological Program Newsletter. Sacramento, CA.

Blake WP. 1858. *Report of a geological reconnaissance in California*. New York.

Blount C, Davis-King S, Milliken R. 2008. *Native American geography, history, traditional resources, and contemporary communities and concerns*. Far Western Anthropological Research Group, Inc.

Board of Swamp Land Commissioners. 1864. *Swamp and overflowed lands reports and correspondence. Annual report drafts*. R388.20, Box 44, Folder 3. *Courtesy of California State Archives, Sacramento.*

Board of Swamp Land Commissioners. 1867. *Swamp and overflowed lands reports and correspondence. Miscellaneous Correspondence and Reports. Report of Engineer.* R388.20, Box 44, Folder 5. *Courtesy of California State Archives, Sacramento.*

Board of Trustees. 1923. *Report of the Board of Trustees of Reclamation District no. 999 to the Board of Supervisors of the county of Yolo, state of California of new, supplemental and additional plans of reclamation and estimates of cost adopted by the said Trustees of Reclamation District no. 999 on December 3, 1923.* Reclamation District No. 999 *Courtesy of Water Resources Collections and Archives, Riverside.*

Boyd JC. 1895. Swamp Land Reclamation District No. 537, Yolo County, California. *Courtesy of California State Lands Commission, Sacramento.*

Boyd JC. 1903. Official map of Sacramento County, California. *Courtesy of California State Lands Commission, Sacramento.*

Brackenridge WD. [1841]1945. Journal of William Dunlop Brackenridge. October 1-28, 1841. *California Historical Society Quarterly* 24(4).

Brandegee K. 1893-4. Flora of Bouldin Island. *ZOE* IV:211-218.

Brewer WH, Watson S, Gray A. 1880. *Botany of California, Volume II.* In Geological Survey of California, ed. J. D. Whitney. Cambridge, MA: John Wilson and Son, University Press.

Brewer WH. 1974. *Up and down California in 1860-1864: the journal of William H. Brewer.* New [3rd] edition. Francis Peloubet Farquhar. Berkeley, CA: University of California Press.

Brewster J. 1856. *Annual report of the Surveyor-General of the state of California.* Surveyor General's Office. *Courtesy of California State Lands Commission online.*

Brinson MM. 1993. *A hydrogeomorphic classification for wetlands.* U.S. Army Corps of Engineers.

Brown. 1865. *U.S. v. Jonathon Stevenson et al., Land Case No. 364 ND [Los Medanos], U.S. District Court, Northern District.* 570-576. *Courtesy of The Bancroft Library, UC Berkeley.*

Brown AG. 1850. Plan of Stockton. 300 feet: 1 inch *Courtesy of Earth Sciences & Map Library, UC Berkeley.*

Brown AK. 1998. *The Anza Expedition in eastern Contra Costa and eastern Alameda Counties, California.* Anza Trail Team, Western Region, National Park Service.

Brown AK. 2005. *Reconstructing early historical landscapes in the Northern Santa Clara Valley.* Russell K. Skowronek. Santa Clara University.

Brown EC. 1901. Map showing subdivisions of the Bradford Reclamation Company Tract, Conta Costa County, California. *Courtesy of California State Lands Commission, Sacramento.*

Brown KJ, Pasternack GB. 2004. The geomorphic dynamics and environmental history of an upper deltaic floodplain tract in the Sacramento-San Joaquin Delta, California, USA. *Earth Surface Processes and Landforms* 29:1235-1258.

Brown KJ, Pasternack GB. 2005. A palaeoenvironmental reconstruction to aid in the restoration of floodplain and wetland habitat on an upper deltaic plain, California, USA. *Foundation for Environmental Conservation* 32(2):103-116.

Browning GW. 1851. *George W. Browning letter to his father: Sacramento Valley California.* California Gold Rush Papers, 1848-1859. *Courtesy of The Bancroft Library, UC Berkeley.*

Bryan K. 1923. *Geology and ground-water resources of Sacramento Valley, California.* Government Printing Office, Washington, DC.

Bryan WH. 1867. *Report of the engineer of the Sacramento Valley, irrigation and navigation canal.* Sacramento, CA: D.W. Gelwicks, State Printer.

Bryant E. [1848] 1985. *What I saw in California: being the journal or a tour; by the emigrant route and south pass of the Rocky Mountains, across the continent of North America, the Great Desert basin, and through California, in the years 1846, 1847.* Lincoln, NE & London, UK: University of Nebraska Press.

Bryant HC. 1915. Cache Slough, Solano Co., California. MVZ Archival Field Notebooks.

Bryant MD. 1983. The role and management of woody debris in West Coast salmonid nursery streams. *North American Journal of Fisheries Management* 3(3):322-330.

Bryant WC, editor. 1874. *Picturesque America, or, the land we live in. A delineation by pen and pencil of the mountains, rivers, lakes, water-falls, shores, cañons, valleys, cities, and other picturesque features of our country.* New York: D Appleton and Company.

Buchannan RB. 1853. *The United States, appellants, v. John A. Sutter. No. 135. Appeal from the District Court U.S. for the Northern District of California. Supreme Court of the United States.*

Buffum EG. 1850. *Six months in the gold mines: from a journal of three years' residence in Upper and Lower California.* Philadelphia: Lea and Blanchard.

Buordo EA. 1956. A review of the General Land Office survey and of its use in quantitative studies of former forests. *Ecology* 37:754-768.

Burcham LT. 1957. *California range land: an historico-ecological study of the range resource of California.* Sacramento, CA: Division of Forestry, Department of National Resources.

Burow KR, Shelton JL, Hevesi JA, et al. 2004. *Hydrogeologic characterization of the Modesto Area, San Joaquin Valley, California.* U.S. Geological Survey.

Buzzell W. 1859. *U.S. v. Charles Weber, Land Case No. 298 ND [Campo de Los Franceses Grant], U.S. District Court, Northern District.* 165-168. Courtesy of The Bancroft Library, UC Berkeley.

Buzzell W. 1860. *The United States, appellants, v. John A. Sutter. No. 135. Appeal from the District Court U.S. for the Northern District of California. The Supreme Court of the United States.*

Byrne R, Ingram BL, Starratt S, Malamud-Roam F, Collins JN, Conrad ME. 2001. Carbon-Isotope, diatom, and pollen evidence for late Holocene salinity change in a brackish marsh in the San Francisco Estuary. *Quaternary Research* 55:66-76.

CALFED Bay-Delta Program. 2000a. *Ecosystem Restoration Program Plan, Volume 1: Ecological Attributes of the San Francisco Bay-Delta Watershed.*

CALFED Bay-Delta Program. 2000b. *Ecosystem Restoration Program Plan, Volume II: Ecological Management Zone Visions.*

CALFED Bay-Delta Program. 2000c. *Multi-Species Conservation Strategy: Final Programmatic EIS/EIR Technical Appendix.*

California Debris Commission. 1910. *[Project for the control of the floods of the Sacramento River].* William Hammond Hall Papers. *Courtesy of California State Archives, Sacramento.*

California Debris Commission. 1914. *Mokelumne River below Galt-New Hope Bridge and sloughs interconnecting Sacramento, San Joaquin and Mokelumne rivers.* San Francisco. *Courtesy of California State Lands Commission, Sacramento.*

California Department of Fish and Game (CDFG), Yolo Basin Foundation (YBF). 2008. *Yolo Bypass Wildlife Area land management plan.* Sacramento, CA.

California Department of Fish and Game, National Marine Fisheries Service, U.S. Fish and Wildlife Service. 2010. *Ecosystem Restoration Program Conservation Strategy for Stage 2 Implementation: Sacramento-San Joaquin Delta Ecological Management Zone.*

California Department of Public Works. 1931. *Variation and control of salinity in Sacramento-San Joaquin Delta and upper San Francisco Bay.* Bulletin no. 27. California State Printing Office, Sacramento, CA.

California Department of Water Resources (CDWR). 1993. Sacramento-San Joaquin Delta atlas.

California Department of Water Resources (CDWR). 2007. California Central Valley unimpaired flow data, Fourth Edition. Sacramento.

California Department of Water Resources (CDWR). 2008. California DWR LiDAR project.

California Natural Diversity Database (CNDDB). 2010. California Department of Fish and Game, Wildlife and Habitat Data Analysis Branch, Habitat Conservation Division.

California Star. 1848. Cal. Star's Sacramento correspondence : San Pablo and Suisun Bays - the tule "Cut off" - Christmas "the Fort," and "California Battalion" on parade - New Year's Day - emigrants unheard from. January 22. *Courtesy of California Digital Newspaper Collection.*

California Swamp Land Committee. 1861. Evidence taken before the Swamp Land Committee. In *Appendix to Journals of Senate and Assembly of the 12th Session of the Legislature,* ed.

California Swampland Commissioners. 1861. *First annual report of Swamp Land Commissioners.* Appendix to Journals of Senate and Assembly of the Thirteenth Session of the Legislature.

Californian. 1847. For the Californian. August 28. *Courtesy of California Digital Newspaper Collection.*

Californian. 1848. For the Californian. April 26. *Courtesy of California Digital Newspaper Collection.*

Cappiella K, Malzone C, Smith R, et al. 1999. *Sedimentation and bathymetry changes in Suisun Bay: 1867-1990,* Open-file report 99-563.

Carpenter EJ, Cosby SW. U.S. Department of Agriculture, Bureau of Chemistry and Soils. 1930. Soil map: Suisun Area, California.

Carpenter EJ, Cosby SW. U.S. Department of Agriculture, Bureau of Chemistry and Soils. 1932. Soil map: Lodi Area, California.

Carpenter EJ, Cosby SW. U.S. Department of Agriculture, Bureau of Chemistry and Soils. 1933. Soil map: Contra Costa County, California.

Carpenter EJ, Cosby SW. 1934. *Soil survey of the Suisun area, California.* U.S. Department of Agriculture. Bureau of Chemistry and Soils. Washington, DC: Government Printing Office.

Carpenter EJ, Cosby SW. 1939. *Soil Survey of Contra Costa County, California.* U.S. Department of Agriculture. Bureau of Soils. Series 1933. Washington, DC: Government Printing Office.

Carson JH. [1852]1931. *Early recollections of the mines, and a description of the Great Tulare Valley.* Tarrytown, NY: W. Abbatt. *Courtesy of Library of Congress.*

Chamberlain GF. 1850. *George F. Chamberlain letter to father : San Francisco, Calif.* California Gold Rush Letters, 1848-1859. *Courtesy of The Bancroft Library, UC Berkeley.*

Cheney SA. California Debris Commission, Corps of Engineers, U.S. Army. 1909-12. Map of Feather River, California. San Francisco, CA. *Courtesy of California State Lands Commission, Sacramento.*

Clark CD. ca. 1905. San Joaquin River. *Gateway Magazine*:30-32.

Clark EC. 1865. *U.S. v. Jonathon Stevenson et al., Land Case No. 364 ND [Los Medanos], U.S. District Court, Northern District.* 527. *Courtesy of The Bancroft Library, UC Berkeley.*

Cleal JG. 1855. Swamp and overflowed land, Vol. 1, Survey no. 129, T5N R4E, Section 11. *Courtesy of Center for Sacramento History.*

Cleal JG. 1858. Swamp and overflowed land, Vol. 1, Survey no. 158, T8N R4E. *Courtesy of Center for Sacramento History.*

Cleal JG. 1859. Swamp and overflowed land, Vol. 1, Survey no. 134, T6N R4E, Section 26. *Courtesy of Center for Sacramento History.*

Cleal JG. 1861. *The United States, appellants, v. John A. Sutter. No. 135. Appeal from the District Court U.S. for the Northern District of California. Supreme Court of the United States.*

Clyman J, Camp CL. [1845]1928. *The adventures of a trapper and covered-wagon emigrant as told in his own reminiscences and diaries.* San Francisco, CA: California Historical Society.

Coats R, Showers MA, Pavlik B. 1988. *A management plan for the Springtown alkali sink wetlands and the endangered plant cordylanthus palmatus.* Phillip Williams & Associates, San Francisco, CA.

Colby GW. 1860. *The United States, appellants, v. John A. Sutter. No. 135. Appeal from the District Court U.S. for the Northern District of California. Supreme Court of the United States.*

Coleman JM. 1969. Brahmaputra River: channel processes and sedimentation. *Sedimentary Geology* 3:129-239.

Coleman JM. 1976. *Deltas: Processes of deposition and models for exploration.* Champaign, IL.: Continuing Education Publication Co.

Collins BD, Montgomery DR. 2001. Importance of archival and process studies to characterizing pre-settlement riverine geomorphic processes and habitat in the Puget Lowland. *Water Science and Application* 4:227-243.

Collins BD, Montgomery DR. 2002. Forest Development, Wood Jams, and Restoration of Floodplain Rivers in the Puget Lowland, Washington. *Restoration Ecology* 10(2):237-247.

Collins BD, Montgomery DR, Sheikh AJ. 2003. Reconstructing the historical riverine landscape of the Puget Lowland. In *Restoration of Puget Sound rivers*, ed. David R. Montgomery, 79-128. Seattle: Center for Water and Watershed Studies in association with University of Washington Press.

Collins BD, Sheikh AJ. 2005. *Historical reconstruction, classification, and change analysis of Puget Sound tidal marshes.* University of Washington, Seattle.

Collins JN, Grossinger RM. 2004. *Synthesis of scientific knowledge concerning estuarine landscapes and related habitats of the South Bay Ecosystem. Draft Final Technical Report of the South Bay Salt Pond Restoration Project.* San Francisco Estuary Institute, Oakland, CA. 91 pages.

Collins JN, Sutula M, Stein ED, et al. 2006. *Comparison of methods to map California riparian areas. Final report prepared for the California Riparian Habitat Joint Venture. SFEI Report No. 522.* San Francisco Estuary Institute and Southern California Coastal Water Research Project, Oakland, CA.

Colton W. 1852. *Three years in California.* New York, NY: A.S. Barnes & Co.

Compton HT. Britton and Rey. ca. 1894. Map of the County of San Joaquin.

Consortium of California Herbaria. 2009. Jepson Herbarium and University Herbarium, University of California Berkeley.

Contra Costa Board of Supervisors. 1875. Map attached to 'Viewer's report - petition of A. G. Kimbell and others'. *Courtesy of the California State Lands Commission, Sacramento.*

Contra Costa Water District. 2010. Historical fresh water and salinity conditions in the western Sacramento-San Joaquin Delta and Suisun Bay: A summary of historical reviews, reports, analyses and measurements. *Technical Memorandum WR10-001.* Water Resources Department.

Cook SF. 1955a. *Aboriginal population of the San Joaquin Valley.* Anthropological records 16.2. Berkeley and Los Angeles, CA: University of California Press. 31-80.

Cook SF. 1955b. *The epidemic of 1830-1833 in California and Oregon.* University of California publication in American archaeology and ethnology. Berkeley, CA: University of California Press.

Cook SF. 1960a. Colonial Expeditions to the Interior of California: Central Valley, 1800-1820. *University of California Anthropological Records* 16(6):239-292.

Cook SF. 1960b. *Colonial expeditions to the interior of California Central Valley, 1800-1820.* Berkeley, CA: University of California Press.

Cook SF, Elsasser AB. 1956. Burials in sand mounds of the Delta region of the Sacramento-San Joaquin River system. *University of California Archaeological Survey.* Vol. 44-46:26-46.

Cook SF, Heizer RF. 1951. *The physical analysis of nine indian mounds of the lower Sacramento Valley.* Publications in American Archaeology and Ethnology. University of California Press, Berkeley and Los Angeles, CA.

Cordell E. U.S. Coast Survey (USCS). 1867. Hydrography of part of Sacramento and San Joaquin Rivers, California. Register No. 935. 1:10,000.

Cosby SW. 1941. *Soil Survey: the Sacramento - San Joaquin Delta area, California.* U.S. Department of Agriculture. Bureau of Plant Industry. Series 1935.

Cosby SW, Carpenter EJ. 1931. *Soil survey of the Dixon area, California.* no. 7. United States Department of Agriculture. Bureau of Chemistry and Soils. Series 1931. Washington, DC: Government Printing Office.

Cosby SW, Carpenter EJ. 1932. *Soil survey of the Lodi area, California.* U.S. Department of Agriculture. Bureau of Chemistry and Soils. Washington, DC: Government Printing Office.

Cowardin LM, Carter V, Golet FC, et al. 1979. *Classification of wetlands and deepwater habitats of the United States.* Washington: Fish and Wildlife Service, Biological Services Program, U.S. Department of the Interior.

Crain PK, Moyle PB. 2011. Biology, history, status, and conservation of the Sacramento perch, *Archoplites interruptus*: a review. San Francisco Estuary and Watershed Science 9(1):1-35.

Crespí J, Bolton HE. 1927. *Fray Juan Crespí, missionary explorer on the Pacific coast, 1769-1774.* Berkeley, CA: University of California Press.

Cronise TF. 1868. *The natural wealth of California.* San Francisco, CA: H.H. Bancroft & Company.

Crow L. 2006. *High and dry: A history of the Calaveras River and its Hydrology.* Stockton: Stockton East Water District.

Dabney. 1905. *Report of the Commissioner of Public Works to the Governor of California, together with the report of the Commission of Engineers to the Commissioner of Public Works upon the rectification of the Sacramento and San Joaquin rivers and their principal tributaries, and the reclamation of the overflowed lands adjacent thereto.* Superintendent State Printing, Sacramento, CA.

Daily Alta California. 1851. Agricultural. October 2. *Courtesy of California Digital Newspaper Collection.*

Daily Alta California. 1852a. Sacramento Intelligence: the inundation nearly over. March 16. *Courtesy of California Digital Newspaper Collection.*

Daily Alta California. 1852b. The tule lands – Memorial of settlers of San Joaquin. October 9. *Courtesy of California Digital Newspaper Collection.*

Daily Alta California. 1853. Salmon fisheries on the Sacramento. February 27. *Courtesy of California Digital Newspaper Collection.*

Daily Alta California. 1862. The flood of 1862, water falling in Sacramento, the help from San Francisco. Flood at Marysville. El Dorado swept. Land slides. Flood at Tehama. Flood in Suisun Valley. Loss of life. The conditions of Sacramento. Conditions of Stockton, Knight's ferry, etc., etc., etc. General remarks on the flood. January 14. *Courtesy of California Digital Newspaper Collection.*

Daily Alta California. 1869. Reclaiming submerged islands. July 25. *Courtesy of California Digital Newspaper Collection.*

Daily Alta California. 1870. Industrial condition of the state. March 21. *Courtesy of California Digital Newspaper Collection.*

Daily Alta California. 1871. Pacific Coast dispatches. January 2. *Courtesy of California Digital Newspaper Collection.*

Daily Alta California. 1878. The San Joaquin tule lands. February 27. *Courtesy of California Digital Newspaper Collection.*

Davidson G. USCGS. 1887. Resurvey of Suisun Bay, California, Sheet no. 3, Register No. 1830, 1:10,000. *Courtesy of National Oceanic and Atmospheric Administration (NOAA).*

Davis FW, Kuhn W, Alagona P, et al. 2000. *Santa Barbara County: oak woodland inventory and monitoring program.* University of California, Santa Barbara.

Davis, Stephen. 1859. *U.S. v. Anastasio Chaboya, Land Case No. 406 ND [Sanjon de los Mequelemnes], U.S. District Court, Northern District.* 182. *Courtesy of The Bancroft Library, UC Berkeley.*

Davis WH. 1889. *Sixty years in California.* San Francisco, CA: A.J. Leary.

Dawdy DR. 1989. *Feasibility of Mapping Riparian Forests Under Natural Conditions in California.* USDA Forest Service General Technical Report PSW-110.

Dawson TE. California Department of Conservation. 2009. *Preliminary geologic map of the Lodi 30' x 60' quadrangle, California.*

Day S. 1869. Report of Sherman Day, Surveyor-General. In *Fresh water tide lands of California*, ed. M. D. Carr & Co.

de Cañizares J. 1781. Plan del gran Puerto de San Francisco [Calif.] : descubierto, y demarcado por el Alferez graduado de fragata de la Real Armada. *Courtesy of The Bancroft Library, UC Berkeley.*

de Cañizares DJ, Eldredge ZS, Molera EJ. 1909. *The march of Portolá and the discovery of the bay of San Francisco and the log of the San Carlos.* San Francisco, CA: California Promotion Committee.

de Mofras D, Wilbur ME. 1937. *Duflot de Mofras' travels on the Pacific Coast, Vol. II.* Santa Ana, CA: The Fine Arts Press.

De Pue & Company. 1879. *The illustrated atlas and history of Yolo County, California.* Oakland, CA: De Pue & Co.

Delano A. 1857. *Life on the plains and among the diggings; being scenes and adventures of an overland journey to California: with particular incidents of the route, mistakes and sufferings of the emigrants, the Indian tribes, the present and future of the great West.* New York: Miller, Orton & Co.

Demerill HL. 1890. Plan and profile of a proposed double cut-off on the San Joaquin River. Corps of Engineers.

Denton W. U.S. Department of the Interior, Bureau of Land Management Rectangular Survey, California, 1853. *Copy of field notes of surveys of section and meander lines in townships 5, 4, and 3 north of the base line of ranges 1 east of Mount Diablo Meridian, in the state of California.* Book 201-15, typed from book 50-16. *Courtesy of Bureau of Land Management, Sacramento.*

Denver F. 1860. *The United States, appellants, v. John A. Sutter. No. 135. Appeal from the District Court U.S. for the Northern District of California. Supreme Court of the United States.*

Derby. 1849. The Sacramento Valley from the American River to Butte Creek, surveyed and drawn by order of Gen Riley, commander 10th military dep. 1 in = 41 miles. *Courtesy of Earth Sciences & Map Library, UC Berkeley.*

Derby GH, Farquhar FP. 1932. The topographical reports of Lieutenant George H. Derby. *California Historical Society Quarterly* 9(2).

Dettinger MD, Cayan DR, Diaz HF, et al. 1998. North-south precipitation patterns in western North America on interannual-to-decadal timescales. *Journal of Climate* 11(12):3095-3111.

Deur D. 2009. "A caretaker responsibility": Revisiting Klamath and Modoc traditions of plant community management. *Journal of Ethnobiology* 29(2):296-322.

Diefenderfer HL, Montgomery DR. 2009. Pool spacing, channel morphology, and the restoration of tidal forested wetlands of the Columbia River, U.S.A. *Restoration Ecology* 17(1):158-168.

Dillingham WP. 1911. *Reports of the Immigration Commission: Immigrants in industries.* Government Printing Office, Washington, DC.

Drew GE. 1856-7. County Surveyor Records, San Joaquin County, Vol. 5. *Courtesy of San Joaquin County Surveyors Office, Stockton.*

Drexler JZ, de Fontaine CS, Brown TA. 2009a. Peat accretion histories during the past 6,000 years in marshes of the Sacramento-San Joaquin Delta, CA, USA. *Estuaries and Coasts* 32:871-892.

Drexler JZ, de Fontaine CS, Deverel SJ. 2009b. The legacy of wetland drainage on the remaining peat in the Sacramento-San Joaquin Delta, California, USA. *Wetlands* 29(1):372-386.

Dugin, L. D. 1859. *U.S. v. Anastasio Chaboya, Land Case No. 406 ND [Sanjon de los Mequelemnes], U.S. District Court, Northern District.* 245. *Courtesy of The Bancroft Library, UC Berkeley.*

Dumas A. [1852]1933. *A Gil Blas in California.* Los Angeles, CA: The Primavera Press. *Courtesy of Library of Congress.*

Dunn A. 1915. *Solano County, California.* San Francisco, CA: Sunset Magazine Service Bureau.

Durán FN, Cook SF. 1960. *Colonial expeditions to the interior of California Central Valley, 1800-1820.* Berkeley, CA: University of California Press.

Durán N, Chapman CE. 1911. *Expedition on the Sacramento and San Joaquin Rivers in 1817: Diary of Fray Narciso Durán.* Academy of Pacific Coast History. Berkeley, CA: University of California.

Duvall M, Rogers FB. 1957. *A navy surgeon in California, 1846-1847.* San Francisco, CA: John Howall.

Dyer EH. U.S. Department of the Interior, Bureau of Land Management Rectangular Survey, California, 1862a. *Field notes of the subdivision lines in fractional township 8 N R 3 E Mount Diablo Meridian.* Book 190-6. *Courtesy of Bureau of Land Management, Sacramento.*

Dyer EH. U.S. Department of the Interior, Bureau of Land Management Rectangular Survey, California, 1862b. *Field notes of the obsolete survey of the Rancho New Helvetia. John A. Sutter, confirmee.* Book F-6. *Courtesy of Bureau of Land Management, Sacramento.*

Eager EN. 1890. Official Map of the County of Solano, California. *Courtesy of Library of Congress.*

Eddy R. 1865. *U.S. v. Jonathon Stevenson et al., Land Case No. 364 ND [Los Medanos], U.S. District Court, Northern District.* 486. *Courtesy of The Bancroft Library, UC Berkeley.*

Edwards PL. 1890. *California in 1837: diary of Col. Philip L. Edwards containing an account of a trip to the Pacific Coast.* Sacramento, CA: A.J. Johnston & Co., Printers.

Egan D, Howell EA. 2001. *The historical ecology handbook : a restorationist's guide to reference ecosystems.* Washington, DC.: Island Press.

Eliason WA. U.S. District Court, Northern District. 1854. Map of Laguna de Santos Calle, Yolo County, California. Land Case Map E-876. *Courtesy of The Bancroft Library, UC Berkeley.*

Elmore AJ, Manning SJ, Mustard JF, et al. 2006. Decline in alkali meadow vegetation cover in California: the effects of groundwater extraction and drought. *Journal of Applied Ecology* 43:770-779.

Etcheverry BA. 1903-1954. *Effects of the 1907, 1909, and 1911 floods on the San Joaquin Delta. Courtesy of Water Resources Collections and Archives, Riverside.*

Etcheverry BA. 1924. *Report to the trustees of Reclamation District no. 551 covering adequacy of reclamation works, appraisal of lands and improvements, outstanding obligations.* Inventory of the Bernard A. Etcheverry Papers, 1903-1954. *Courtesy of Water Resources Collections and Archives.*

Evett R, Franco-Vizcaino E, Stephens S. 2007. Phytolith evidence for the absence of a prehistoric grass understory in a Jeffrey Pine – mixed conifer forest in the Sierra San Pedro Mártir, Mexico. *Canadian Journal of Forest Research* 37:306.

Fagan BM. 2003. *Before California: an archaeologist looks at our earliest inhabitants.* Lanham, MD: Rowman & Littlefield Publishers, Inc./AltaMira Press.

Fairchild MD. 1934. Reminiscences of a 'Forty-Niner. *California Historical Society Quarterly* 13(1):2-33.

Farnham JT. 1857. *Life, adventure and travels in California.* New York: Sheldon, Blakeman & Co.

Fassett HH. 1865. *U.S. v. Jonathon Stevenson et al., Land Case No. 364 ND [Los Medanos], U.S. District Court, Northern District. Courtesy of The Bancroft Library, UC Berkeley.*

Faunt CC. 2009. *Groundwater Availability of the Central Valley Aquifer, California.* U.S. Geological Survey.

Fisher G. U.S. District Court, Northern District. 1854. Campo de los Franceses, San Joaquin County, California. Land Case Map F-389. *Courtesy of The Bancroft Library, UC Berkeley.*

Flint W. 1860. Reclamation of tule lands. *The California Culturist* 3:109-112.

Florsheim JL, Mount JF. 2002. Restoration of floodplain topography by sand-splay complex formation in response to intentional levee breaches, Lower Cosumnes River, California. *Geomorphology* 44:67-94.

Florsheim JL, Mount JF, Chin A. 2008. Bank erosion as a desirable attribute of rivers. *BioScience* 58(6).

Font P, Bolton HE. 1930. *Anza's California expeditions.* Berkeley, CA: University of California Press.

Font P, Bolton HE. 1933. *Font's complete diary: a chronicle of the founding of San Francisco.* Berkeley, CA: University of California Press.

Forman RTT, Godron M. 1986. *Landscape ecology.* New York: John Wiley & Sons.

Fortier S. 1909. *Irrigation in the Sacramento Valley, California.* U.S. Department of Agriculture. Office of Experiment Stations. Bulletin 207. Washington, DC: Government Printing Office.

Foster DR. 2002. Insights from historical geography to ecology and conservation: Lessons from the New England landscape. *Journal of Biogeography* 29:1269-1275.

Foster DR, Motzkin G. 2003. Interpreting and conserving the openland habitats of coastal New England: Insights from landscape history. *Forest Ecology and Management* 185:127-150.

Fowler JS. 1853. *The United States, appellants, v. John A. Sutter. No. 135. Appeal from the District Court U.S. for the Northern District of California. Supreme Court of the United States.*

Fox P. 1987a. *Freshwater inflow to San Francisco Bay under natural conditions.* Sacramento, CA: State Water Contractors.

Fox P. 1987b. *Rebuttal to David R. Dawdy Exhibit 3 in regard to Freshwater inflow to San Francisco Bay under natural conditions.* Sacramento, CA: State Water Contractors.

Frayer WE, Peters DD, Pywell HR. U.S. Fish and Wildlife Service, 1989. *Wetlands of the California Central Valley: status and trends, 1939 to mid-1980's.*

Fremier A, Ginney EM, Merrill A, et al. 2008. *Riparian vegetation conceptual model, Delta Regional Ecosystem Restoration Implementation Plan (DRERIP).* Sacramento, CA.

Frémont JC. 1845. *Report of the exploring expedition to the Rocky Mountains, Oregon and California.* Washington, DC: Gales and Seaton.

Frémont JC, Emory WH, Abert JW. 1849. *Notes of travel in California; comprising the prominent geographical, agricultural, geological, and mineralogical features of the country; also, the route to San Diego, in California, including parts of the Arkansas, Del Norte and Gila rivers.* Dublin: J. M'Glashan.

Fugitt SC. 1859. *U.S. v. Charles Weber, Land Case No. 298 ND [Campo de Los Franceses Grant]*, U.S. District Court, Northern District. 165-168. *Courtesy of The Bancroft Library, UC Berkeley.*

Garofalo D. 1980. The Influence of Wetland Vegetation on Tidal Stream Channel Migration and Morphology. *Estuaries* 3(4):258-270.

Garone PF. 2011. *The Fall and Rise of the Wetlands of California's Great Central Valley.* Berkeley and Los Angeles: University of California Press.

Gatschet AS. 1890. *The Klamath Indians of southwestern Oregon.* Department of the Interior. Washington, DC: Government Printing Office.

Gerstäcker F. 1853. *Narrative of a journey round the world, comprising a winter-passage across the Andes to Chili, with a visit to the gold regions of California and Australia, the South Sea islands, Java, &c.* London: Hurst and Blackett.

Gibbes CD. W.B. Cooke & Co. 1850a. Map of San Joaquin River. San Francisco, CA. *Courtesy of Peter J. Shields Library Map Collection, UC Davis.*

Gibbes CD, *Stockton Times.* 1850b. Navigation on the San Joaquin. June 8. *Courtesy of The Stockton Public Library.*

Gibbes JT. 1869. Showing the lands of the Tide Land Reclamation Company. *Courtesy of Peter J. Shields Library Map Collection, UC Davis.*

Gifford EW. 1916. *Composition of California shellmounds.* Berkeley: University of California Press.

Gilbert FT. 1879. *History of San Joaquin County, California: with illustrations descriptive of its scenery, residences, public buildings, fine blocks and manufactories from original sketches by artists of the highest ability.* Oakland, CA: Thompson & West.

Gilbert GK. 1917. *Hydraulic-mining debris in the Sierra Nevada.* U.S. Geological Survey. Washington, DC: U.S. Government Printing Office.

Gillespie AH. 1860. *The United States, appellants, v. John A. Sutter. No. 135. Appeal from the District Court U.S. for the Northern District of California. The Supreme Court of the United States.*

Goals Project. 1999. *Baylands Ecosystem Habitat Goals. A report of habitat recommendations prepared by the San Francisco Bay Area Wetlands Ecosystem Goals Project.* USEPA, San Francisco, Calif./S.F. Bay Regional Water Quality Control Board, Oakland, CA.

Goman M, Wells L. 2000. Trends in river flow affecting the northeastern reach of the San Francisco Bay Estuary over the past 7000 years. *Quaternary Research* 54:206-217.

Governor's Delta Vision Blue Ribbon Task Force. 2008. *Our vision for the California Delta.*

Grant GA. 1853. *The United States, appellants, v. John A. Sutter. No. 135. Appeal from the District Court U.S. for the Northern District of California. Supreme Court of the United States.*

Gray GN. 1859. *U.S. v. Anastasio Chaboya, Land Case No. 406 ND [Sanjon de los Mequelemnes]*, U.S. District Court, Northern District. 405. *Courtesy of The Bancroft Library, UC Berkeley.*

Green, *Sacramento Daily Union.* 1881. Debris: seventeenth day in of the slickens case in court. December 8. [Transcription from People v. Gold Run court case]. *Courtesy of California Digital Newspaper Collection.*

Green JB. 1882. *People v. Gold Run Ditch and Mining Company. California Superior Court.*

Gregory SV, Boyer KL, Gurnell AM, editors, 2003. *The ecology and management of wood in world rivers.* Bethesda, Maryland: American Fisheries Society.

Gregory SV, Swanson FJ, McKee WA, et al. 1991. An ecosystem perspective of riparian zones. *BioScience* 41(8):540.

Gregory T. 1912. *History of Solano and Napa counties, California.* Los Angeles, CA: Historic Record Co.

Gregory T. 1913. *History of Yolo County, California.* Los Angeles, California: Historic Record Company.

Greiner CM. 2010. *Principles for strategic conservation and restoration*. Puget Sound Nearshore Ecosystem Restoration Project. 40.

Grinnell J. 1911. *Field Notebook*. San Joaquin Valley March 8 - May 5. *Courtesy of Museum of Vertebrate Zoology, UC Berkeley*.

Grinnell J, Dixon JS, Linsdale JM. 1937. *Fur-bearing mammals of California: their natural history, systematic status, and relations to man*. Berkeley, CA: University of California Press.

Grossinger RM. 1995. Historical evidence of freshwater effects on the plan form of tidal marshlands in the Golden Gate Estuary. Master's Marine Sciences, University of California, Santa Cruz.

Grossinger RM. 2005. Documenting local landscape change: the San Francisco Bay area historical ecology project. In *The historical ecology handbook: a restorationist's guide to reference ecosystems*, ed. Dave Egan and Evelyn A. Howell, 425-442. Washington, DC: Island Press.

Grossinger RM. 2012. *Napa Valley historical ecology atlas: Exploring a landscape of transformation and resilience*. Berkeley, CA: University of California Press.

Grossinger RM, Askevold RA. 2005. *Historical analysis of California Coastal landscapes: methods for the reliable acquisition, interpretation, and synthesis of archival data. Report to the U.S. Fish and Wildlife Service San Francisco Bay Program, the Santa Clara University Environmental Studies Institute, and the Southern California Coastal Water Research Project, SFEI Contribution 396*. San Francisco Estuary Institute, Oakland, CA. 48.

Grossinger RM, Beller EE, Salomon MN, et al. 2008. *South Santa Clara Valley Historical Ecology Study: Including Soap Lake, the Upper Pajaro River, and Llagas, Uvas-Carnadero, and Pacheco Creeks*. SFEI Publication #558, San Francisco Estuary Institute, Oakland, CA.

Grossinger RM, Striplen CJ, Askevold RA, et al. 2007. Historical landscape ecology of an urbanized California valley: wetlands and woodlands in the Santa Clara Valley. *Landscape Ecology* 22:103-120.

Grunksy CE. 1896. *Appendix no. 11: Report of C.E. Grunsky to board of consulting engineers*. Report of board of consulting engineers to the landowners of reclamation district no. 106. *Courtesy of Water Resources Collections and Archives, Riverside*.

Grunsky CE. ca. 1878. Delta. *Illustrations for Stockton Boyhood from the Grunsky family papers. Courtesy of The Bancroft Library, UC Berkeley*.

Gutierrez CI. California Department of Conservation. 2011. *Preliminary geologic map of the Sacramento 30' x 60' quadrangle, California*.

Gwynn, *Sacramento Daily Union*. 1881. Debris: twelfth day of the slickens case in court. December 2. [Transcription from People v. Gold Run court case]. *Courtesy of California Digital Newspaper Collection*.

Hall TJ. 1856. *The United States, appellants, v. John A. Sutter. No. 135. Appeal from the District Court U.S. for the Northern District of California. Supreme Court of the United States*.

Hall WH. 1879. William Hammond Hall Collection, Field Books, Box 3, Book 45, 8-32. *Courtesy of California State Archives, Sacramento*.

Hall WH. State Office, 1880. *Memorandum concerning the improvement of the Sacramento River: addressed to James B. Eads and George H. Mendell, consulting engineers. Courtesy of The Bancroft Library, UC Berkeley*.

Hall WH. California Department of Engineering. 1887. Topographical and irrigation map of the Great Central Valley of California. Sacramento, CA. *Courtesy of Peter J. Shields Library Map Collection, UC Davis*.

Hall WH, Office of State Engineer. 1886. Physical data and statistics of California. J.J. Ayers, Supt. State Printing, Sacramento.

Hall WH. ca. 1880a. Central Valley, shows land grants and major streams. *Courtesy of California State Archives, Sacramento*.

Hall WH. ca. 1880b. Grand Island and Suisun Bay to foothills and 1st Standard North. *Courtesy of California State Archives, Sacramento.*

Hall WH. ca. 1880c. Central Valley - Sacramento area. 1st and 2nd Standard North. From Grays Bend to Courtland. *Courtesy of California State Archives, Sacramento.*

Hall WH. n.d. Transit books #1-5. California State Engineering Department. *Courtesy of California State Archives, Sacramento.*

Handy HP. 1864. *Field notes of the subdivision and meanders swamp and overflowed land lines part of Township 4 North Range 5 East, Mount Diablo Meridian, California.* Book 85-4. *Courtesy of Bureau of Land Management, Sacramento.*

Handy HP. n.d. *Report of H. P. Handy acting engineer of District No. 17, to the honorable board of Swamp Land Commissioners. Courtesy of California State Archives, Sacramento.*

Hardenburgh JR. 1860. *The United States, appellants, v. John A. Sutter. No. 135. Appeal from the District Court U.S. for the Northern District of California. The Supreme Court of the United States.*

Harley JB. 1989. Historical geography and cartographic illusion. *Journal of Historical Cartography* 15:80-91.

Hart J. 2010. The once and future Delta: Mending the broken heart of California. *Bay Nature* (Apr-Jun):21-36.

Hays RB. U.S. Department of the Interior, Bureau of Land Management Rectangular Survey, California, 1852a. *Field notes of the boundary lines of townships 6-7-8-9-10 N, ranges 1-2 W and 1-2-3 E.* Book 5-4. *Courtesy of Bureau of Land Management, Sacramento.*

Hays RR. U.S. Department of the Interior, Bureau of Land Management Rectangular Survey, California, 1852b. *Field notes of the boundary lines of townships 6-7-8-9 N, ranges 1-2-3-4 E, Mt Diablo Meridian, California.* Book 5-1. *Courtesy of Bureau of Land Management, Sacramento.*

Hays RR. Department of the Interior, Bureau of Land Management Rectangular Survey, California, 1853. *Field notes of the exterior and subdivision lines of Township 1, 2, 3, 4, South Range 3, 5, 6, 7, & 8 East, Mount Diablo Meridian, California.* Book 99-1. *Courtesy of Bureau of Land Management, Sacramento.*

Henderson, G. H. P. 1865. *U.S. v. Jonathon Stevenson et al., Land Case No. 364 ND [Los Medanos], U.S. District Court, Northern District.* 496. *Courtesy of The Bancroft Library, UC Berkeley.*

Henning JS. 1871. Map of Yolo County, California. *Courtesy of Library of Congress.*

Herrmann FC. 1921. Map of Reclamation District No. 2058, San Joaquin County, California. *Courtesy of San Joaquin County Historical Society and Museum, Lodi.*

Heuer WH. 1892. *Report of Maj. W. H. Heuer, Corps of Engineers, Navigation.* House of Representatives. 52nd Congress, 1st Session. Ex. Doc. No. 108.

Heuer WH. 1900. *Surveys of San Joaquin, Sacramento.* 59 Congress, 1st Session, Doc. 262.

Heynemann L. 1869. Map of the Pierson-District and surrounding country. *Courtesy of Water Resources Collections and Archives, Riverside.*

Hickson D, Keeler-Wolf T. 2007. *Vegetation and land use classification and map of the Sacramento-San Joaquin River Delta.* California Department of Fish and Game.

Higley HA. 1859. *Annual report of the Surveyor-General for the year 1859.* State Land Office.

Higley HA. 1860. *Annual report of the Surveyor-General for the year 1860.* State Land Office.

Hilgard EW. 1884. *Report on the physical and agricultural features of the state of California, with a discussion of the present and future of cotton production in the state; also remarks on cotton culture in New Mexico, Utah, Arizona, and Mexico.* San Francisco, CA: Pacific World Press Office. 138.

History of Bacon Island. n.d. *Courtesy of The Haggin Museum.*

Hoag IN, *Sacramento Daily Union.* 1882. Early days: Personal and industrial reminiscences of California. January 2. *Courtesy of California Digital Newspaper Collection.*

Hoagland, *Sacramento Daily Union*. 1881. Debris: second day of the slickens case in court. November 17. [Transcription from People v. Gold Run court case]. *Courtesy of California Digital Newspaper Collection.*

Hobbs RJ. 1996. Towards a conceptual framework for restoration ecology. *Restoration Ecology* 4(2):93-110.

Hodgdon, Daily Alta California. 1881. Hydraulic mining litigation: trial of the case of the people against the Gold Run Mining Company at Sacramento. November 18. [Transcription from People v. Gold Run court case]. *Courtesy of California Digital Newspaper Collection.*

Holland RF. 1978. *The Geographic and Edaphic Distribution of Vernal Pools in the Great Central Valley, California.* California Native Plant Society.

Holland RF. 1986. *Preliminary descriptions of the terrestrial natural communities of California. Unpublished report.* California Department of Fish and Game, Natural Heritage Division, Sacramento, CA.

Holland RF. 1998. *Changes in Great Valley vernal pool distribution from 1989 to 1997.* Prepared for California Department of Fish and Game.

Holland RF, Griggs TF. 1976. A unique habitat - California's vernal pools. *Fremontia* 4(3):3-6.

Holmes LC, Nelson JW. 1915. *Reconnoissance soil survey of the Sacramento Valley, California.* U.S. Department of Agriculture. Bureau of Soils. Washington, DC: Government Printing Office.

Holmes LC, Watson EB, Harrington GL, et al. U.S. Department of Agriculture, Bureau of Soils. 1913. Soil Map: Reconnoissance Survey - Sacramento Valley.

Holstein G. 2000. Plant communities ecotonal to the Baylands. In *Baylands ecosystem species and community profiles. Life histories and environmental requirements of key plants, fish and wildlife. Prepared by the San Francisco Bay Area Wetlands Ecosystem Goals Project*, ed. PR Olofson, 49-68. San Francisco and Oakland, CA: U.S. Environmental Protection Agency and San Francisco Bay Regional Water Quality Control Board.

Holstein G. 2001. Pre-agricultural grassland in Central California. *Madroño* 48(4):253-264.

Hood GW. 2004. Indirect environmental effects of dikes on estuarine tidal channels: Thinking outside of the dike for habitat restoration and monitoring. *Estuaries* 27(2):273-282.

Hood W. 2007a. Scaling tidal channel geometry with marsh island area: A tool for habitat restoration, linked to channel formation process. *Water Resources Research* 43.

Hood WG. 2007b. Large woody debris influences vegetation zonation in an oligohaline tidal marsh. *Estuaries and Coasts* 30(3):441-450.

Houghton JF. 1862. *Annual report of the Surveyor-General for the year 1862.* Surveyor General's Office.

Howard AQ, Arnold RA. 1980. The Antioch Dunes - safe at last? *Fremontia* 8(3):3-12.

Howard J. 2010. *Sensitive freshwater mussel surveys in the Pacific Southwest Region: assessment of conservation status.* USDA Forest Service.

Howard JK, Cuffey KM. 2006. The functional role of native freshwater mussels in the fluvial benthic environment. *Freshwater Biology* 51:460-474.

Howland H. 1854-9. Humphrey Howland letters. *Courtesy of The Bancroft Library, UC Berkeley.*

Hulaniski FJ, editor. 1917. *The history of Contra Costa County, California.* Berkeley, CA: The Elms Publishing Co., Inc.

Hupp CR, Osterkamp WR. 1996. Riparian vegetation and fluvial geomorphic processes. *Geomorphology* 14(4):277-295.

Hutchings. 1859. Notes and sketches on the Bay and River. Hutchings' California Magazine. Vol. 37. Hutchings and Rosenfield, San Francisco, CA.

Hutchings JM. 1862. *Scenes of wonder and curiosity in California.* San Francisco, CA: Hutchings & Rosenfield.

Jackson AR. ca. 1870. Swamp Land Districts No. 4 & No. 8. *Courtesy of California State Lands Commission, Sacramento.*

Jackson JBC, Kirby MX, Berger WH, et al. 2001. Historical Overfishing and the Recent Collapse of Coastal Ecosystems. *Science* 293(5530):629-637.

Jacobs D. 1993. *California's rivers: a public trust report.* Prepared for the California State Lands Commission, Sacramento.

Jassby AD, Cloern JE. 2000. Organic matter sources and rehabilitation of the Sacramento-San Joaquin Delta (California, USA). *Aquatic Conservation: Marine and Freshwater Ecosystems* 10:323-352.

Jeffres CA, Opperman JJ, Moyle PB. 2008. Ephemeral floodplain habitats provide best growth conditions for juvenile Chinook salmon in a California river. *Environmental Biology of Fishes* 83:449-458.

Jepson WL. 1893. The riparian botany of the lower Sacramento. *Erythea* 1:238-246.

Jepson WL. 1901. *A flora of western middle California, second edition.* San Francisco: Cunningham, Curtiss & Welch.

Jepson WL. 1904. Suisun Marshes. MVZ Archival Field Notebooks. *Courtesy of Jepson Herbarium, UC Berkeley.*

Jepson WL. 1910. *The silva of California.* Berkeley, Calif.: The University Press.

Jepson WL. 1913. *Middle River, San Joaquin Delta.* Jepson field books. *Courtesy of Jepson Herbarium, UC Berkeley.*

Jessup. 1865. *U.S. v. Jonathon Stevenson et al., Land Case No. 364 ND [Los Medanos], U.S. District Court, Northern District.* 517. *Courtesy of The Bancroft Library, UC Berkeley.*

Johnson TT. 1851. *California and Oregon: or, Sights in the gold region and scenes by the way.* Philadelphia, PA: Lippincott, Grambo & Co.

Jones & Stokes Associates Inc. 1989. Results of biological resource inventories and habitat evaluations in the Kellogg Creek Watershed. Prepared for: James M. Montgomery, Consulting Engineers, Inc. Project Sponsor: Contra Costa Water District, Concord, CA.

Jones AH. U.S. Department of the Interior, Bureau of Land Management Rectangular Survey, California, 1855. *Field notes of the subdivisions and meanders of township no. 5 north, range no. 5 east, Mount Diablo Meridian.* Book 428-26. *Courtesy of Bureau of Land Management, Sacramento.*

JRP Historical. 2008. *Appendix C: Summary report Roberts Island and Union Island riparian water rights investigation, San Joaquin County, CA.*

Katibah EF, Dummer KJ, Nedeff N. 1981. Evaluation of the riparian vegetation resource in the great Central Valley of California using remote sensing techniques. *American Society of Photogrammetry and American Congress on Surveying and Mapping, Fall Technical Meeting, San Francisco, CA, September 9-11, 1981 and Honolulu, HI; United States; 14-16 Sept. 1981:234-246.*

Katibah EF. 1984. A brief history of the riparian forests in the central valley of California. In *California riparian systems: ecology conservation and productive management,* ed. R.E. Warner and K.M. Hendrix, 23-29. Berkeley, CA: University of California Press.

Keeler-Wolf T, Elam DR, Lewis K, et al. State of California Department of Fish and Game, 1998. *California vernal pool assessment preliminary report.*

Keeler-Wolf T, Evens JM, Solomeshch AI, et al. 2007. Community Classification and Nomenclature. In *California grasslands: ecology and management,* ed. Jeffrey D. Corbin Mark R. Stromberg, and Carla M. D'Antonio. Berkeley, CA: University of California Press.

Keeley JE. 2002. Native American impacts on fire regimes of the California coastal ranges. *Journal of Biogeography* 29:303-320.

Keller EA. 1977. The fluvial system: Selected observations. In *Riparian forests in California: Their ecology and conservation*, ed. Anne Sands, 39-46. Institute of Ecology Publication No. 15.

Kelley R. 1989. *Battling the Inland Sea: American Political Culture, Public Policy, and the Sacramento Valley, 1850-1986*. Berkeley: University of California Press.

Kerr T. 1850. Diary of things worth notice during my voyage to Calafornia [sic]: remarkable events there, yes and a good many things not worth notice too *Courtesy of The Bancroft Library, UC Berkeley*.

Kerr T, Camp CL. 1928. An Irishman in the Gold Rush: the journal of Thomas Kerr. *California Historical Society* 7(3):205-227.

Keseberg L. 1860. *The United States, appellants, v. John A. Sutter. No. 135. Appeal from the District Court U.S. for the Northern District of California*. Supreme Court of the United States.

Kip L. [1850]1946. *California sketches, with recollections of the gold mines*. Los Angeles, CA: N.A. Kovach.

Kip WI. 1892. *The early days of my episcopate*. New York: Thomas Whittaker.

Kirkpatrick CA. 1860. Salmon fishery on the Sacramento River. *Hutchings' California Magazine* 4(12):531-534.

Kluegul CH. 1878. S. E. D. field notes 1. California State Engineering Department. *Courtesy of California State Lands Commission, Sacramento*.

Knight W. 1980. The story of Browns Island. *The Four Seasons* 6(1).

Kroeber AL. 1932. The Patwin and their neighbors. *University of California Publications in American Archeology and Ethnology* 32(5):15-22.

Kroeber AL. [1925]1976. *Handbook of the Indians of California*. New York: Dover Publications, Inc.

Kybruz S. 1854. *The United States, appellants, v. John A. Sutter. No. 135. Appeal from the District Court U.S. for the Northern District of California*. Supreme Court of the United States.

Kyburg S. 1863. *The United States, appellants, v. John A. Sutter. No. 135. Appeal from the District Court U.S. for the Northern District of California*. Supreme Court of the United States.

Lajoie KR. 2010. *Written testimony of Kenneth R. Lajoie*. California State Water Resources Control Board.

Lambeth M. 1859. *U.S. v. Anastasio Chaboya, Land Case No. 406 ND [Sanjon de los Mequelemnes]*, U.S. District Court, Northern District. 316. *Courtesy of The Bancroft Library, UC Berkeley*.

Lapham MH, Mackie WW. U.S. Department of Agriculture, Bureau of Soils. 1905. Soil Map: Stockton.

Lapham MH, Mackie WW. 1906. *Soil survey of the Stockton area, California*. U.S. Department of Agriculture. Bureau of Soils. Washington, DC: Government Printing Office.

Larkey JL. 1969. *Davisville '68; the history and heritage of the city of Davis, Yolo County, California. In commemoration of the 100th anniversary of the founding of Davisville in 1868*. Davis, CA: Davis Historical and Landmarks Commission

Larkin TO. T Wiley Jr. 1848. Map of the valley of the Sacramento including the gold region. *Courtesy of Earth Sciences & Map Library, UC Berkeley*.

Leale. 1939. *Recollections of a Tule Sailor*. San Francisco, CA: G Fields.

Leopold LB. 1994. *A view of the river*. Cambridge, Mass.: Harvard University Press.

Leopold LB, Collins JN, Collins LM. 1993. Hydrology of some tidal channels in estuarine marshland near San Francisco. *Catena* 20(5):469-493.

Lewis HT. 1982. *A time for burning*. Occasional Publication No. 17. University of Alberta, Boreal Institute for Northern Studies, Edmonton, Alberta.

Lewis Publishing Company. 1890. *An illustrated history of San Joaquin County, California : containing a history of San Joaquin County from the earliest period of its occupancy to the present time : together with glimpses of its future prospects : with full-page portraits of some of its most eminent men and biographical mention of many of its pioneers and also prominent citizens of to-day.* Chicago, IL: Lewis Publishing Co.

Lewis Publishing Company. 1891. *A memorial and biographical history of northern California, illustrated. Containing a history of this important section of the Pacific Coast from the earliest period of its occupancy to the present time.* Chicago, IL: The Lewis Publishing Company.

Lewis WJ. U.S. Department of the Interior, Bureau of Land Management Rectangular Survey, California, 1858a. *Field notes of the subdivision lines of township 6 north range 4 east, Mt Diablo Meridian, California.* Book 66-14. *Courtesy of Bureau of Land Management, Sacramento.*

Lewis WJ. U.S. Department of the Interior, Bureau of Land Management Rectangular Survey, California, 1858b. *Field notes of the exterior lines of township 7 north range 4 east, Mount Diablo Meridian, California.* Book 66-16. *Courtesy of Bureau of Land Management, Sacramento.*

Lewis WJ. U.S. Department of the Interior, Bureau of Land Management Rectangular Survey, California, 1858c. *Field notes of the subdivision of lines of township 7 north range 4 east, Mount Diablo Meridian, California.* Book 66-17. *Courtesy of Bureau of Land Management, Sacramento.*

Lewis WJ. U.S. Department of the Interior, Bureau of Land Management Rectangular Survey, California, 1858d. *Field notes of the final survey of the Rancho Las Ulpinos, John Bidwell, confirmee.* Book A39/G1. *Courtesy of Bureau of Land Management, Sacramento.*

Lewis WJ. Department of the Interior, Bureau of Land Management Rectangular Survey, California, 1858-9. *Field notes of the boundary lines of Township 6 North Range 4 East, Mount Diablo Meridian, California.* Book 66-13. *Courtesy of Bureau of Land Management, Sacramento.*

Lewis WJ. U.S. Department of the Interior, Bureau of Land Management Rectangular Survey, California, 1859a. *Field notes of the boundary lines of Township 5 North Range 4 East, Mount Diablo Meridian, California.* Book 66-10. *Courtesy of Bureau of Land Management, Sacramento.*

Lewis WJ. U.S. Department of the Interior, Bureau of Land Management Rectangular Survey, California, 1859b. *Field notes of N.E. & W. Boundary lines of Township 6 North Range 4 East Mount Diablo Meridian, California.* 66-12. *Courtesy of Bureau of Land Management, Sacramento.*

Lewis WJ. U.S. Department of the Interior, Bureau of Land Management Rectangular Survey, California, 1859c. *Field notes of the meander lines of Township 6 North Range 4 East, Mount Diablo Meridian, California.* Book 66-15. *Courtesy of Bureau of Land Management, Sacramento.*

Lewis WJ. U.S. Department of the Interior, Bureau of Land Management Rectangular Survey, California, 1859d. *Field notes of the subdivision lines and meanders of swamp and overflowed land in township 6 north range 3 east, Mount Diablo Meridian.* Book 190-2. *Courtesy of Bureau of Land Management, Sacramento.*

Lienhard H, Wilbur ME. 1941. *A pioneer at Sutter's Fort, 1846-1850; the adventures of Heinrich Lienhard. Translated and edited by Marguerite Eyer Wilbur from the original German manuscript.* Los Angeles, CA: The Calafia Society.

Lightfoot KG, Parrish O. 2009. *California Indians and their environment: An introduction.* Berkeley, CA: University of California Press.

Logan TM. 1865. *Report on the medical topography and epidemics of California.* American Medical Association. Philadelphia: Collins.

Loring FR. U.S. Department of the Interior, Bureau of Land Management Rectangular Survey, California, 1851. *Field notes of survey of the second correction line north, of the base line extending from point of departure east and west.* Book 210-6, from Book 5-3. *Courtesy of Bureau of Land Management, Sacramento.*

Los Angeles Herald. 1890. Flooded islands. June 1. *Courtesy of California Digital Newspaper Collection.*

Lyman CS. 1848. [Journals of Chester Lyman, No. 12 and 13]. *Courtesy of California Historical Society, San Francisco.*

Lyman CS, Teggart FJ. 1923. The Gold Rush; Extracts from the diary of C. S. Lyman, 1848-1849. California Historical Society Quarterly.

Malamud-Roam F, Ingram BL. 2004. Late Holocene δ13C and pollen records of paleosalinity from tidal marshes in the San Francisco Bay estuary, California. *Quaternary Research* 62:134-145.

Malamud-Roam F, Dettinger M, Ingram B, et al. 2007. Holocene Climates and Connections between the San Francisco Bay Estuary and its Watershed: A Review. *San Francisco Estuary and Watershed Science* 5(1). Article 3.

Malamud-Roam FP, Ingram BL, Hughes M, et al. 2006. Holocene paleoclimate records from a large California estuarine system and its watershed region: Linking watershed climate and bay conditions. *Quaternary Science Reviews* 25(13-14):1570-1598.

Maloney AB. 1936. Hudson's Bay Company in California. *Oregon Historical Quarterly* 37(1).

Maloney AB, Work J. 1943. Fur brigade to the Bonaventura: John Work's California expedition of 1832-33 for the Hudson's Bay Company (Continued). *California Historical Society Quarterly* 22(4):323-348.

Mammal Networked Information System (MaNIS). 2010. Museum of Vertebrate Zoology, UC Berkeley.

Manies KL. 1997. Evaluation of General Land Office survey records for analysis of the northern Great Lakes hemlock-hardwood forests. University of Wisconsin, Madison.

Manies KL, Mladenoff DJ. 2000. Testing methods to produce landscape-scale presettlement vegetation maps from the U.S. public land survey records. *Landscape Ecology* 15:741-754.

Mann CW, Warner JF, Westover HL. U.S. Department of Agriculture, Bureau of Soils. 1909. Soil Map: Woodland.

Mann CW, Warner JF, Westover HL, et al. 1911. *Soil survey of the Woodland area, California.* Bureau of Soils. U.S. Department of Agriculture. Washington, DC: Government Printing Office.

Marlette SH. 1854. *Annual report of the Surveyor-General for the year 1854.* State Land Office.

Martin WW. 1855. Affidavit to locate swamp land. *Book A, p.28. Courtesy of Yolo County Recorder.*

Martinez D. 1998. The wisdom of fire. *Growing Native Newsletter* 21(7):5-8.

Martinez D. 2010. Fire, grass, and chaparral in prehistoric southwestern California (in prep). *Ecological Restoration.*

Maser C, Sedell JR. 1994. *From the forest to the sea: the ecology of wood in streams, rivers, estuaries, and oceans.* Delray Beach, Florida: St. Lucie Press.

Mason H. n.d. Floristics of the Sacramento-San Joaquin Delta. *Environmental Inventory, Sacramento-San Joaquin Delta.* U.S. Army Corps of Engineers.

Mason H. 1957. *A flora of the marshes of California.* Berkeley: University of California Press.

Matthewson RC. 1859. *U.S. v. Anastasio Chaboya, Land Case No. 406 ND, Sanjon de los Miquelemes, U.S. District Court, Northern District. Courtesy of The Bancroft Library, UC Berkeley.*

McAfee LC, *Pacific Rural Press.* 1874. A chapter of tule history - Staten Island. January 16. *Courtesy of California Digital Newspaper Collection.*

McClatchey J. 1860. *The United States, appellants, v. John A. Sutter. No. 135. Appeal from the District Court U.S. for the Northern District of California.* Supreme Court of the United States.

McClatchy VS. 1916. *Flood control and reclamation documents. Courtesy of Water Resources Collections and Archives, Riverside.*

McClure WF. 1927. *Sacramento flood control project : revised plans. Submitted to the Reclamation Board February 10 1925.* Sacramento, CA: California State Printing Office.

McCollum WS. [1850]1960. *California as I saw it; pencillings by the way of its gold and gold diggers, and incidents of travel by land and water.* Los Gatos, CA: Talisman Press. *Courtesy of Library of Congress.*

McConnell JI. 1887. *Yolo County, California, resources, advantages and prospects.* Woodland, CA: Yolo County Board of Trade and Immigration Association.

McCullough DR. 1969. The Tule Elk: Its History, Behavior and Ecology. *University of California Publications in Zoology* 88.

McGowan JA. 1939. San Francisco - Sacramento shipping, 1839-1854. University of California.

McGowan JA. 1961. *History of the Sacramento Valley.* New York: Lewis Historical Pub. Co.

McLeod A, Nunis DB. 1968. *The Hudson's Bay Company's first fur brigade to the Sacramento Valley: Alexander McLeod's 1829 hunt.* The Sacramento Book Collectors Club.

McQueen A. 1859. *U.S. v. Charles Weber, Land Case No. 298 ND [Campo de Los Franceses Grant], U.S. District Court, Northern District.* 200-201. *Courtesy of The Bancroft Library, UC Berkeley.*

Meko DM, Therrell MD, Baisan CH, et al. 2001. Sacramento River flow reconstructed to A.D. 869 from tree rings. *Journal of the American Water Resources Association* 37(4):1029–1040.

Mellin GF. 1918. *Prospect Island history.* Miscellaneous materials concerning flood drainage, irrigation, and land appraisal in Solano County, CA. ETCH 124. *Courtesy of Water Resources Collections and Archives, Riverside.*

Mendell. Letter from the Secretary of War, 1881. *Examination of Mokelumne River, California, from Woodbridge to its mouth.* 47th Congress, 1st session, Doc. 34.

Merriam CH. 1967. Ethnographic notes on California indian tribes, Part III: Ethnological notes on central California indian tribes. *Reports of the University of California Archaeological Survey* 68:257-448.

Milliken R. 1995a. *A time of little choice: the disintegration of tribal culture in the San Francisco Bay area, 1769-1810.* Menlo Park, CA: Ballena Press.

Milliken RT. 1995b. *Report on the 1994 archaeological excavation on the skirt of the Souza Mound, Sac-42, Sacramento County, California.* Far Western Anthropological Research Group, Inc, Davis, CA.

Mining and Scientific Press. 1869. Tule reclamation – Sherman Island and the Roberts Enterprise. July 31. *Courtesy of The Haggin Museum, Stockton.*

Minnich RA. 2008. *California's fading wildflowers: lost legacy and biological invasions.* Berkeley and Los Angeles, CA: University of California Press.

Mitsch WJ, Gosselink JG. 2007. *Wetlands.* John Wiley & Sons, Inc.

Moerenhout. [1849]1935. *The inside story of the gold rush.* San Francisco, CA: California Historical Society. *Courtesy of Library of Congress.*

Moor EN. 1878. William Hammond Hall Collection, Field Books (II), Part I, Box 4, Folder X58. *Courtesy of California State Archives, Sacramento.*

The Morning Call. 1894. Caused by 'cuts'. September 19. *Courtesy of California Digital Newspaper Collection.*

Morse F. 1888. *Descriptive report. Topographic sheet no. 1793, locality Suisun Bay (resurvey).*

Mount JF. 1995. *California rivers and streams.* Berkeley, CA: University of California Press.

Moyle PB. 2002. *Inland fishes of California. Revised and expanded.* Berkeley, CA: University of California Press.

Moyle PB, Baxter RD, Sommer T, et al. 2004. Biology and population dynamics of Sacramento splittail (*Pogonichthys macrolepidotus*) in the San Francisco Estuary: a review. San Francisco Estuary and Watershed Science 2(2):1-47.

Moyle PB, Crain PK, Whitener K. 2007. Patterns in the use of a restored California floodplain by native and alien fishes. San Francisco Estuary and Watershed Science 5(3):1-27.

Moyle PB, Lund JR, Bennett WA, et al. 2010. Habitat Variability and Complexity in the Upper San Francisco Estuary. *San Francisco Estuary and Watershed Science* 8(3):1-24.

Muir J. 1916. *A thousand-mile walk to the Gulf.* Boston and New York: Houghton Mifflin Company.

Munro-Fraser JP. 1879. *History of Solano County...and histories of its cities, towns.* San Francisco, CA: Wood, Alley, & Co.

Naglee HM. 1879. *Letter to Wm. H. Hall, and Board of Engineers of the State of California, upon the subject of the reclamation of the overflown lands of the San Joaquin Valley.* San Jose, CA.

Naiman RJ, Decamps H. 1990. *The ecology and management of aquatic-terrestrial ecotones.* J. N. R. Jeffers. Paris: UNESCO and The Parthenon Publishing Group.

Naiman RJ, Decamps H, Fournier F, editors, 1989. *The role of land/inland water ecotones in landscape management and restoration: a proposal for collaborative research:* MAB Digest 4. UNESCO, Paris.

Nelson JW. U.S. Department of Agriculture, Bureau of Soils. 1915. Soil Map: Reconnoissance Survey - Lower San Joaquin Valley Sheet.

Nelson JW, Guernsey JE, Holmes LC, et al. 1918. *Reconnoissance soil survey of the lower San Joaquin Valley, California.* U.S. Department of Agriculture. Bureau of Soils. Washington, DC: Government Printing Office.

Nelson NC. 1909. *Shellmounds of the San Francisco Bay region.* Berkeley, CA: The University Press.

Nesbit DM. 1885. *Tide marshes of the United States.* Department of Agriculture. Washington, DC: Government Printing Office.

Newell FH. 1896. *Report of progress of the Division of Hydrography for the calendar year 1895.* Geological Survey, Division of Hydrography.

Newell FH. 1907. *Fifth Annual Report of the Reclamation Service, 1906.* Washington, DC: Government Printing Office.

Norris RW. U.S. Department of the Interior, Bureau of Land Management Rectangular Survey, California, 1851a. *Field notes of the meridian and boundary lines of townships 1 N & 1 S. Ranges 3, 4, 5, 6, 7, 8, 9, 10, 11 E Mount Diablo Meridian, California.* Book 90-9. *Courtesy of Bureau of Land Management, Sacramento.*

Norris RW. U.S. Department of the Interior, Bureau of Land Management Rectangular Survey, California, 1851b. *Field notes of the survey of the bases parallel east of Monte Diablo.* 231-2. *Courtesy of Bureau of Land Management, Sacramento.*

Norris RW. U.S. Department of the Interior, Bureau of Land Management Rectangular Survey, California, 1853a. *Field Notes of the boundary lines of Township 3-4-5 North Range 3-5-6-7 East, Mount Diablo Meridian, California.* 85-11. *Courtesy of Bureau of Land Management, Sacramento.*

Norris RW. U.S. Department of the Interior, Bureau of Land Management Rectangular Survey, California, 1853b. *Field notes of the boundary lines of townships 1, 2, 3, 4, 5 North Ranges 5, 6, 7, 8 E Mount Diablo Meridian, California.* Book 88-2. *Courtesy of Bureau of Land Management, Sacramento.*

O'Connor JE, Jones MA, Haluska TL. 2003. Flood plain and channel dynamics of the Quinault and Queets Rivers, Washington, USA. *Geomorphology* 51:31-59.

Odum WE. 1988. Comparative Ecology of Tidal Freshwater and Salt Marshes. *Annual Review of Ecology and Systematics* 19:147-176.

Odum WE, Smith TJI, Hoover JK, et al. 1984. *Ecology of tidal freshwater marshes of the United States east coast: a community profile.* Virginia University, Department of Environmental Sciences, Charlottesville.

Ogden GR. 1988. *Agricultural land use and wildlife in the San Joaquin Valley, 1769-1930: an overview. Prepared for the San Joaquin Valley Drainage Program.*

Opperman J. 2008. *Floodplain conceptual model.* Delta Regional Ecosystem Restoration Implementation Plan, Sacramento, CA.

Ornduff R, Faber PM, Keeler-Wolf T. 2003. *Introduction to California plant life.* Berkeley, CA: University of California Press.

ORNIS. 2010. Museum of Vertebrate Zoology, University of California, Berkeley.

Orr NM. 1874. *The city of Stockton; its position, climate, commerce, resources, etc., together with a brief sketch of the great San Joaquin basin of California, of which Stockton is the natural business center.* Stockton, California: Stockton Board of Trade.

Pacific Rural Press. 1871. Inspection of the tule lands. November 4. *Courtesy of California Digital Newspaper Collection.*

Pacific Rural Press. 1878. Improvements on Roberts Island. April 6. *Courtesy of California Digital Newspaper Collection.*

Pacific Rural Press. 1880. Yolo. The plains. August 21. *Courtesy of California Digital Newspaper Archive.*

Pacific Rural Press. 1883. The tract of land known as the 'Honker Lake District'. October 6. *Courtesy of California Digital Newspaper Collection.*

Pacific Rural Press. 1905. Hogs getting fat on salmon. March 4. *Courtesy of California Digital Newspaper Collection.*

Palmer LL, Wallace WF, Wells HL. 1881. *History of Napa and Lake Counties, California.* San Francisco, CA: Slocum, Bowen & Co.

Parker TV, Callaway JC, Schile LM, et al. 2011. Climate change and San Francisco Bay-Delta tidal wetlands. *San Francisco Estuary and Watershed Science* (December).

Pasternack G, Brown K. ca. 2006. *McCormack-Williamson Tract Restoration Planning, Design, and Monitoring Program I: Task 1 Analysis of Historic Hydrogeomorphic Conditions.*

Paterson AM, Herbert RF, Wee SR. 1978. *Historical evaluation of the Delta waterways. Prepared for the State Lands Commission.* State Lands Commission.

Payson. 1880. Description of Bench Marks on the U.S. Government Debris Surveys. *Courtesy of the California State Archives, Sacramento.*

Payson AH. Government Printing Office, 1885. *Annual report of the Chief of Engineers, United States Army, to the Secretary of War.* House of Representatives, 49th Congress, 1st Session, Ex. Doc. 1, pt. 2, vol. II.

Peacock D. U.S. Coast and Geodetic Survey (USCGS). 1886. Sacramento and San Joaquin rivers, California. Register No. 1784. 1:10,000 *Courtesy of National Oceanic and Atmospheric Administration (NOAA).*

Pearce S, Collins JN. 2004. *Analysis of Reference Tidal Channel Plan Form For the Montezuma Wetlands Restoration Project.* Contribution No. 80, San Francisco Estuary Institute, Oakland, CA.

Phelps WD, Busch BC. 1983. *Alta California, 1840-1842 : the journal and observations of William Dane Phelps, master of the ship "Alert."* Glendale, CA: A.H. Clark Co.

Pierce P. 1988. A geoarchaeological analysis of the prehistoric Sacramento-San Joaquin Delta, California. Department of Anthropology, University of California, Davis.

Prentice J. 1856. *The United States, appellants, v. John A. Sutter. No. 135. Appeal from the District Court U.S. for the Northern District of California. Supreme Court of the United States.*

Prentice J. U.S. Department of the Interior, Bureau of Land Management Rectangular Survey, California, 1870. *Field notes of the exterior lines of township 9 north range 4 east, Mount Diablo Meridian, California, and report on public land in township 9 north, range 5 east.* Book 66-20. *Courtesy of Bureau of Land Management, Sacramento.*

Purcell MF. 1940. *History of Contra Costa County*. Berkeley, CA: The Gillick Press.

Radeloff VC, Mladenoff DJ, He HS, et al. 1999. Forest landscape change in the northwestern Wisconsin Pine Barrens from pre-European settlement to the present. *Canadian Journal of Forest Research* 29:1649-1659.

Radeloff VC, Mladenoff DJ, Manies KL, et al. 1998. Analyzing forest landscape restoration potential: presettlement and current distribution of oak in the northwest Wisconsin Pine Barrens. *Transcriptions of the Wisconsin Academy of Sciences Arts and Letters* 86:189-205.

Ralston JH, Broderick DC. California Legislature, 1852. *Reports on State lands, 1850-72. Courtesy of California State Library, Sacramento.*

Rambler R, *Pacific Rural Press*. 1872. The wild flowers of San Joaquin Valley. July 6. *Courtesy of California Digital Newspaper Collection.*

Randall, *Sacramento Daily Union*. 1882. Debris: forty-fourth day of the slickens case in court. January 18. [Transcription from People v. Gold Run court case]. *Courtesy of California Digital Newspaper Collection.*

Ray JR. 1873. Map of the City of Sacramento. *Courtesy of David Rumsey Map Collection, Cartography Associates.*

Redding BB. 1860. *The United States, appellants, v. John A. Sutter. No. 135. Appeal from the District Court U.S. for the Northern District of California. The Supreme Court of the United States.*

Reece TW. 1864. Map of the Swamp Lands in District No. 2. *Courtesy of the California State Lands Commission, Sacramento.*

Reed CF, Grunsky CE, Crawford JJ. J.D. Young, Supt. State Printing, 1890. *Report of the Examining Commission on Rivers and Harbors to the Governor of California. Courtesy of The Bancroft Library, UC Berkeley.*

Reed J. U.S. Department of the Interior, Bureau of Land Management Rectangular Survey, California, 1866. *Transcript of the field notes of a survey of the exterior boundary lines of fractional township 6 south range 1 west meridian of Mount Diablo, State of California.* Vol. R223. p. 168-195. *Courtesy of Bureau of Land Management, Sacramento.*

Rensch HE, Rensch EGH, Hoover MB, et al. 1966. *Historic spots in California.* 3d edition. Stanford, CA: Stanford University Press.

Revere JW. 1849. *A tour of duty in California.* Joseph N. Balestier. New York, NY: C. S. Francis & Co.

Rhoads JP. 1863. *The United States, appellants, v. John A. Sutter. No. 135. Appeal from the District Court U.S. for the Northern District of California. Supreme Court of the United States.*

Ringgold C. 1850a. Chart of the Sacramento River from Suisun City to the American River. *Courtesy of David Rumsey Map Collection, Cartography Associates.*

Ringgold C. 1850b. Chart of Suisun and Vallejo Bays. *Courtesy of David Rumsey Map Collection, Cartography Associates.*

Ringgold C. 1852. *A series of charts, with sailing directions.* Washington, DC: JT Towers. *Courtesy of California State Library, Sacramento.*

Risser PG, Karr JR, Forman RTT. 1984. *Landscape ecology: Directions and approaches.* Illinois Natural History Survey Special Publication Number 2.

Roberts WG, Howe GJ, Major J. 1977. A survey of riparian forest flora and fauna in California. In *Riparian forests in California: Their ecology and conservation*, ed. Anne Sands, 3-19. Institute of Ecology Publication No. 15.

Robinson ER. 1854. *The United States, appellants, v. John A. Sutter. No. 135. Appeal from the District Court U.S. for the Northern District of California. The Supreme Court of the United States.*

Robinson ER. 1860. *The United States, appellants, v. John A. Sutter. No. 135. Appeal from the District Court U.S. for the Northern District of California. Supreme Court of the United States.*

Rose AH, Manson M, Grunsky CE. State Printing Office, Sacramento, 1895. *Report of the Commissioner of Public Works to the Governor of California.*

Russell G. [ca. 1925]. [East Contra Costa County Aerial Views]. *Courtesy of California State Lands Commission, Sacramento.*

Russell WO. 1940. *History of Yolo County, California: its resources and its people...* Woodland, CA.

Ryer WM, *Pacific Rural Press.* 1884. The Sacramento River and its overflows. April 12. *Courtesy of California Digital Newspaper Collection.*

Sacramento Daily Union. 1851. Timber on the San Joaquin. November 7. *Courtesy of California Digital Newspaper Collection.*

Sacramento Daily Union. 1851b. Up-river news. October 17. *Courtesy of California Digital Newspaper Collection.*

Sacramento Daily Union. 1853. Draining the Tule Lands in Yolo. May 19. *Courtesy of California Digital Newspaper Collection.*

Sacramento Daily Union. 1854a. The Sacramento fisheries. June 6. *Courtesy of California Digital Newspaper Collection.*

Sacramento Daily Union. 1854b. Climate and diseases of California. September 14. *Courtesy of California Digital Newspaper Collection.*

Sacramento Daily Union. 1860. Delta. 2 July. *Courtesy of California Digital Newspaper Collection.*

Sacramento Daily Union. 1862a. Is the Sacramento Valley inhabitable? March 24. *Courtesy of California Digital Newspaper Collection.*

Sacramento Daily Union. 1862b. Clearing on the American. November 12. *Courtesy of California Digital Newspaper Collection.*

Sacramento Daily Union. 1862c. Swamp and Overflowed Lands. January 1. *Courtesy of California Digital Newspaper Collection.*

Sacramento Daily Union. 1864. Canal in Yolo. December 6. *Courtesy of California Digital Newspaper Collection.*

Sacramento Daily Union. 1871. Condition of the crops - the foothills and fallowed grounds. April 24. *Courtesy of California Digital Newspaper Collection.*

Sacramento Daily Union. 1873a. Our tule lands. April 12. *Courtesy of California Digital Newspaper Collection.*

Sacramento Daily Union. 1873b. Suburban sketches - No. 43, Sacramento Drainage Canal. May 31. *Courtesy of California Digital Newspaper Collection.*

Sacramento Daily Union. 1873c. Draining the tule lands in Yolo. May 19. *Courtesy of California Digital Newspaper Collection.*

Sacramento Daily Union. 1878. Firewood in the tule lands. May 4. *Courtesy of California Digital Newspaper Collection.*

Sacramento Daily Union. 1889. Abundance of Salmon. April 30. *Courtesy of California Digital Newspaper Collection.*

Sacramento Daily Union. 1892a. Out into the tules. May 30. *Courtesy of California Digital Newspaper Collection.*

Sacramento Daily Union. 1892b. Gray's Bend. Sacramento will find no relief from that source. The flood waters of the Sacramento River and the tule basin of Yolo County. December 10. *Courtesy of California Digital Newspaper Collection.*

Sacramento Transcript. 1850a. Latest from Stockton. April 5. *Courtesy of California Digital Newspaper Collection.*

Sacramento Transcript. 1850b. Tule plains on fire. June 24. *Courtesy of California Digital Newspaper Collection.*

Sacramento Transcript. 1850c. Tule plains on fire. December 30. *Courtesy of California Digital Newspaper Collection.*

Sacramento Transcript. 1851a. Cache Creek. January 24. *Courtesy of California Digital Newspaper Collection.*

Sacramento Transcript. 1851b. New Road to Putah. May 15. *Courtesy of California Digital Newspaper Collection.*

Sacramento Valley Reclamation Co. 1872. *Tule lands of Sacramento Valley in Yolo and Colusa counties, California.* Louisville: John P. Morton and Co.

Sal H, Cook SF. 1960. *Colonial expeditions to the interior of California Central Valley, 1800-1820.* Berkeley, CA: University of California Press.

San Joaquin County Surveyor. 1882. San Joaquin County plat book.

Sanderson EW. 2009. *Mannahatta: A natural history of New York City.* New York: Abrams.

Sanderson EW, Ustin SL, Foin TC. 2000. The influence of tidal channels on the distribution of salt marsh plant species in Petaluma Marsh, CA, USA. *Plant Ecology* 146:29-41.

Sands A, editor. 1977. *Riparian forests in California: Their ecology and conservation.* Institute of Ecology Publication No. 15.

Sanford JN. 1860. *The United States, appellants, v. John A. Sutter. No. 135. Appeal from the District Court U.S. for the Northern District of California.* Supreme Court of the United States.

Sarna-Wojcicki AM, Meyer CE, Bowman HR, et al. 1985. Correlation of the Rockland ash bed, a 400,000-year-old stratigraphic marker in northern California and western Nevada, and implications for middle Pleistocene paleogeography of central California. *Quaternary Research* 23(2):236-257.

Sawyer JO, Keeler-Wolf T. 1995. *A manual of California vegetation.* Sacramento, CA: California Native Plant Society.

Sawyer JO, Keeler-Wolf T, Evens J. 2009. *A manual of California vegetation, second edition.* Sacramento, CA: California Native Plant Society.

Schenck WE. 1926. Historic aboriginal groups of the California Delta region. *University of California Publications in American Archeology and Ethnology* 23(2):123-146.

Schoellhamer D, Wright S, Drexler J, et al. 2007. *Sedimentation conceptual model, Delta regional ecosystem restoration implementation plan (DRERIP).*

Schulz PD. 1979. Fish remains from a historic central California indian village. *California Fish and Game* 65(4).

Schulz PD, Simons DD. 1973. Fish species diversity in a prehistoric central California Indian midden. *California Fish and Game* 59(2):107-118.

Secretary of State, California. 1866-77. California Topographic Base Sheets. *Courtesy of California State Archives, Sacramento.*

Sedell JR, Froggatt JL. 1984. Importance of streamside forests to large rivers: The isolation of the Willamette River, Oregon, U.S.A., from its floodplain by snagging and streamside forest removal. *Verh. Internat. Verin. Limnol.* 22:1828-1834.

Semlitsch RD. 1998. Biological delineation of terrestrial buffer zones for pond-breeding salamanders. *Conservation Biology* 12(5):1113-1119.

San Francisco Estuary Institute (SFEI). 2011. Bay Area aquatic resources inventory. SFEI. http://www.californiawetlands.net/tracker/ba/map. http://www.californiawetlands.net/tracker/ba/map.

Shafer. 1882. *People v. Gold Run Ditch and Mining Company. California Superior Court.*

Shalowitz AL. 1964. *Shore and sea boundaries, with special reference to the interpretation and use of Coast and Geodetic Survey data, United States.* Coast and Geodetic Survey U.S. Department of Commerce. Washington, DC: Government Printing Office.

Sherman E. A. 1859. *U.S. v. Anastasio Chaboya, Land Case No. 406 ND [Sanjon de los Mequelemnes],* *U.S. District Court, Northern District.* 329. *Courtesy of The Bancroft Library, UC Berkeley.*

Shinn CH. 1888. The tule region. In *West of the Rocky Mountains,* ed. John Muir (1976). Philadelphia, PA: Running Press.

Sickley TA, Mladenoff DJ, Radeloff VC, et al. 2000. A Pre-European Settlement Vegetation Database for Wisconsin.

Simenstad C, Reed D, Ford M. 2006. When is restoration not? Incorporating landscape-scale processes to restore self-sustaining ecosystems in coastal wetland restoration. *Ecological Engineering* 26:27-39.

Simenstad CA, Hood WG, Thom RM, et al. 2000. Landscape structure and scale constraints on restoring estuarine wetlands for Pacific Coast juvenile fishes. In *Concepts and Controversies in Tidal Marsh Ecology,* ed. M.P. Weinstein and D.A. Kreeger. Dordrecht: Kluwer Academic Publications.

Simenstad CA, Wick A, Van De Wetering S, et al. 2003. Dynamics and ecological functions of wood in estuarine and coastal marine ecosystems. In *The ecology and management of wood in world rivers,* ed. S. V. Gregory, K. L. Boyer, and A. M. Gurnell. Bethesda, Maryland: American Fisheries Society.

Singer MB, Rolf A, Allan JL. 2008. Status of the Lower Sacramento Valley Flood-Control System within the Context of Its Natural Geomorphic Setting. *Natural Hazards Review* 9(3).

Skinner JE. 1962. *An historical review of the fish and wildlife resources of the San Francisco Bay Area.* Sacramento, CA: California. Dept. of Fish and Game. Water Projects Branch.

Smith & Elliott. [1879]1979. *Illustrations of Contra Costa Co., California with historical sketch.* Fresno, CA: Valley Publishers.

Smith CI. 1866a. Map of the Swampland District No. 57. *Courtesy of California State Lands Commission, Sacramento.*

Smith DW, Verrill WL. 1998. Vernal pool-soil-landform relationships in the Central Valley, California. In *Ecology, conservation, and management of vernal pool ecosystems: Proceedings from a 1996 conference,* ed. C. W. Witham, E. T. Bauder, D. Belk, W.R. Ferren, Jr., and R. Ornduff. Sacramento, CA: California Native Plant Society.

Smith N. 1853. *U.S. v. Charles Weber, Land Case No. 298 ND [Campo de Los Franceses Grant], U.S. District Court, Northern District.* 37. *Courtesy of The Bancroft Library, UC Berkeley.*

Smith F. 1977. A short review of the status of riparian forests in California. In *Riparian forests in California: Their ecology and conservation,* ed. Anne Sands, 1-2. Institute of Ecology Publication No. 15.

Smith ND, Perez-Arlucea M. 1994. Fine-grained splay deposition in the avulsion belt of the Lower Saskatchewan River, Canada. *Journal of Sedimentary Research, Section B: Stratigraphy and Global Studies* 64B(2):159-168.

Smith WW. 1866b. *U.S. v. Jonathon Stevenson et al., Land Case No. 364 ND [Los Medanos], U.S. District Court, Northern District.* 288. *Courtesy of The Bancroft Library, UC Berkeley.*

Sommer T, ML Nobriga, WC Harrell, W Batham, WJ Kimmerer. 2001. Floodplain rearing of juvenile chinook salmon: Evidence of enhanced growth and survival. *Canadian Journal of Fisheries and Aquatic Sciences* 58:325.

Sparks RE. 1995. Need for ecosystem management of large rivers and their floodplains. *BioScience* 45(3):168-182.

Sprague CP, Atwell HW. 1870. *Western Shore Gazetteer, Yolo County.* Woodland, CA.

Stahle D, Therrell M, Cleaveland M. 2001. Ancient blue oaks reveal human impact on San Francisco Bay salinity. *Eos* 82(12):141, 144-145.

Stahle DW, Griffin DR, Cleaveland MK, et al. 2011. A tree-ring reconstruction of the salinity gradient in the northern esturary of San Francisco Bay. *San Francisco Estuary and Watershed Science.*

Stanford B, Grossinger RM, Askevold RA, et al. Forthcoming. *Historical ecology of the Alameda Creek watershed.* San Francisco Estuary Institute, Richmond, CA.

Stanford B, Grossinger RM, Askevold RA, et al. 2011. *East Contra Costa County historical ecology study.* Prepared for Contra Costa County and the Contra Costa Watershed Forum. A report of SFEI's Historical Ecology program, SFEI Publication # 648. San Francisco Estuary Institute, Oakland, CA.

Stanger FM, Brown AK. 1969. *Who discovered the Golden Gate? The explorers' own accounts, how they discovered a hidden harbor and at last found its entrance.* San Mateo, CA: San Mateo County Historical Association.

State Agricultural Society. 1866. Annual report of the board. In *Appendix to Journals of Senate and Assembly, of the sixteenth session of the legislature of the State of California, Volume 3,* ed. Sacramento, CA: O.M. Clayes, State Printer.

State Agricultural Society. 1872. *Transactions of the State Agricultural Society during the years 1870 and 1871.*

State Journal Office. 1854. *The Sutter claim, the evidence taken in case 192 before the Board of U.S. Land Commissioners, together with the brief of the United States land agent.* Sacramento, CA.

Stephens SL, Fry DL. 2005. Fire history in Coast redwood stands in the northeastern Santa Cruz mountains, California. *Fire Ecology* 1(1).

Stewart OC, Lewis HT, Anderson K. 2002. *Forgotten fires : Native Americans and the transient wilderness.* Norman, OK: University of Oklahoma Press.

Stine S. 1996. Climate, 1650-1850. In *Sierra Nevada Ecosystem Project: Final report to Congress, Vol. II, Assessments and scientific basis for management options.* Davis: University of California, Centers for Water and Wildland Resources.

Stockton Commercial Association. 1895. *Stockton and San Joaquin County illustrated; their enterprises, attractions and buildings in part and inducements to the investment of capital and the establishment of industries.* San Francisco, CA: Trade and Commerce Publishing Co.

Stratton JT. U.S. Department of the Interior, Bureau of Land Management Rectangular Survey, California, 1861. *Field notes of the final survey of the Rancho El Pescadero, Pico and Naglee Confirmee.* Book G10/J3. *Courtesy of Bureau of Land Management, Sacramento.*

Stratton, T. 1865. *U.S. v. Jonathon Stevenson et al., Land Case No. 364 ND [Los Medanos], U.S. District Court, Northern District.* 255. *Courtesy of The Bancroft Library, UC Berkeley.*

Sullivan MS. 1934. *The travels of Jedediah Smith : a documentary outline including the journal of the great American pathfinder.* Lincoln, NE: University of Nebraska Press.

Sutter JA, Bancroft HH. 1876. *Personal reminiscences of General John A. Sutter. Courtesy of The Bancroft Library, UC Berkeley.*

Swan JA. [1848]1960. *A trip to the gold minds of California in 1848.* San Francisco, CA: Book Club of California. *Courtesy of Library of Congress.*

Sweet AT, Warner JF, Holmes LC. 1908. *Soil survey of the Modesto-Turlock area, California, with a brief report on a reconnaissance soil survey of the region east of the area.* U.S. Department of Agriculture. Bureau of Soils. Washington, DC: Government Printing Office.

Swetnam TW, Allen CD, Betancourt JL. 1999. Applied Historical Ecology: Using the Past to Manage for the Future. *Ecological Applications* 9(4):1189-1206.

Tappe DT. 1942. *The status of beavers in California.* Game Bulletin No. 3.

Taylor A. 1861. Indianology of California, May 31, 1861. *The California Farmer and Journal of Useful Sciences.*

Taylor B. 1854. *El Dorado, or, Adventures in the path of empire.* Fourth edition. New York: George P. Putnam & Co.

Taylor CG, editor. 1969. *Stockton boyhood: being the reminiscences of Carl Ewald Grunsky:* The Friends of the Bancroft Library.

Taylor KW. 1865. *U.S. v. Jonathon Stevenson et al., Land Case No. 364 ND [Los Medanos], U.S. District Court, Northern District.* 452.Courtesy of *The Bancroft Library, UC Berkeley.*

Taylor NR. 1913. *The rivers and floods of the Sacramento and San Joaquin watersheds.* U. S. Department of Agriculture. Washington, DC: Government Printing Office.

Thayer. 1859. *U.S. v. Anastasio Chaboya, Land Case No. 406 ND [Sanjon de los Mequelemnes], U.S. District Court, Northern District. Courtesy of The Bancroft Library, UC Berkeley.*

Thayer CH. 2010. Botanical priority protection areas: Delta. In *A guidebook to botanical priority protection areas of the East Bay*, ed. Heath Bartosh, Lech Naumovich, and Laura Baker: East Bay Chapter of the California Native Plants Society.

The Bay Institute (TBI). 1998. *From the sierra to the sea: the ecological history of the San Francisco Bay-Delta watershed.* The Bay Institute of San Francisco.

The PRISM Group at Oregon State University (PRISM). 2006. *United States Average Monthly or Annual Maximum Temperature, 1971-2000.* Corvallis, Oregon.

Thompson. U.S. Department of the Interior, Bureau of Land Management Rectangular Survey, California, 1862. *Field notes of the survey of Sanjon de los Moquelumnes.* Book 506-16. *Courtesy of Bureau of Land Management, Sacramento.*

Thompson. 1865. *U.S. v. Jonathon Stevenson et al., Land Case No. 364 ND [Los Medanos], U.S. District Court, Northern District.* 276-281. *Courtesy of The Bancroft Library, UC Berkeley.*

Thompson and West. 1880. *History of Sacramento County, California. With illustrations descriptive of its scenery, residences.* Oakland, CA: Thompson & West.

Thompson J. 1957. The settlement geography of the Sacramento-San Joaquin Delta, California. Geography, Stanford, CA.

Thompson J. 1996. *Flood chronologies and aftermaths affecting the lower Sacramento River, 1878-1909.* Department of Water Resources.

Thompson J. 2006. Early reclamation and abandonment of the central Sacramento-San Joaquin Delta. *Sacramento History Journal* 6(1-4):41-72.

Thompson J. in press. The legacy of fire in agriculture and subsidence in the Sacramento-San Joaquin Delta. Department of Geography, University of Illinois.

Thompson K. 1961. Riparian forests of the Sacramento Valley, California. *Annals, Association of American Geographers* 51:294-315.

Thompson K. 1977. Riparian forests of the Sacramento Valley, California. In *Riparian forests in California: their ecology and conservation*, ed. Anne Sands, 35-38. Institute of Ecology Publication no. 15.

Thornton SR. 1859. *U.S. v. Anastasio Chaboya, Land Case No. 406 ND, Sanjon de los Miquelemes, U.S. District Court, Northern District.* 200. *Courtesy of The Bancroft Library, UC Berkeley.*

Tide Land Reclamation Company. 1872. *Fresh water tide lands of California.* San Francisco, CA: M.D. Carr & Co., Book and Job Printers.

Tinkham GH. 1880. *A history of Stockton from its organization up to the present time, including a sketch of San Joaquin County* San Francisco, CA: W.M. Hinton.

Tinkham GH. 1923. *History of San Joaquin County, California: with biographical sketches of the leading men and women of the county who have been identified with its growth and development from the early days to the present.* Los Angeles, CA: Historic Record Company.

Tockner K, Stanford JA. 2002. Riverine flood plains: present state and future trends. *Foundation for Environmental Conservation* 29(3):308-330.

Tower ML. California Debris Commission, Corps of Engineers, U.S. Army. 1906. Map of American River, California. San Francisco, CA. *Courtesy of Water Resources Collections and Archives, Riverside.*

Triska FJ. 1984. Role of wood debris in modifying channel geomorphology an riparian areas of a large lowland river under pristine conditions: A historical case study. *Verhandlungen des Internationalen Verein Limnologie* 22:1876-1892.

Tucker and Smith. 1883. Map of a portion of Roberts Island.

Tucker EE. 1879a. Field notes, Book No. 89. California State Engineering Department. *Courtesy of California State Archives, Sacramento.*

Tucker EE. 1879b. Field notes, Book No. 90. California State Engineering Department. *Courtesy of California State Archives, Sacramento.*

Tucker EE. 1879c. Field notes, Book No. 91. California State Engineering Department. *Courtesy of California State Archives, Sacramento.*

Tucker EE. 1879d. Field notes, Book No. 92. California State Engineering Department. *Courtesy of California State Archives, Sacramento.*

Tucker EE. 1879e. Field notes, Book No. 93. California State Engineering Department. *Courtesy of California State Archives, Sacramento.*

Tucker EE. 1879f. Field notes, Book No. 94. California State Engineering Department. *Courtesy of California State Archives, Sacramento.*

Turner JL. 1966. *Distribution and food habits of centrarchid fishes in the Sacramento-San Joaquin Delta.* California Department of Fish and Game.

U.S. Army, California Department of Engineering, and California Debris Commission. 1913. Portion of the San Joaquin Delta, including main stream above Stockton Channel and all waterways west of main stream, Part II in 40 sheets. San Francisco, CA. *Courtesy of California State Lands Commission, Sacramento.*

U.S. Army, California Department of Engineering, and California Debris Commission. 1914-5. San Joaquin River, California, Herndon to head of Delta, Part I in 40 sheets. San Francisco, CA. *Courtesy of California State Lands Commission, Sacramento.*

U.S. Coast and Geodetic Survey (USCGS). 1881. Government Printing Office, Washington, DC.

U.S. Congress House of Representatives, Committee on Flood Control. 1916. *Control of floods on the Mississippi and Sacramento rivers.* Washington, DC.

U.S. Department of Agriculture (USDA). 1874. *Reclamation of Swamp and Overflowed Lands in California.* Washington, DC.

U.S. Department of Agriculture (USDA), Western Division Laboratory. 1937-1939. [Aerial photos of Contra Costa, Sacramento, San Joaquin, Solano, and Yolo counties]. Scale: 1:20,000. Agricultural Adjustment Administration (AAA). *Courtesy of Peter J. Shields Library, UC Davis and Earth Sciences Library, UC Berkeley.*

U. S. Department of Agriculture (USDA), Western Division Laboratory. 1939-40. [Aerial photos of Alameda County]. Scale: 1:20,000. Agricultural Adjustment Administration (AAA). *Courtesy of Earth Sciences & Map Library, UC Berkeley, and the Alameda County Resource Conservation District (ACRCD) and National Resources Conservation Service (NRCS).*

USDA (U.S. Department of Agriculture). 1942. Aerial photos of Napa County. Scale: 1:20,000. *Courtesy of Napa County Resource Conservation District and Natural Resources Conservation Service.*

U.S. Department of Agriculture (USDA). 1977. Soil Survey of Contra Costa County, California. United States Department of Agriculture, Soil Conservation Service, in cooperation with University of California Agricultural Experiment Station.

U.S. Department of Agriculture (USDA). 1992. Soil survey of San Joaquin County, California. U. S. Department of Agriculture, Soil Conservation Service, in cooperation with the Regents of the University of California and the California Department of Conservation.

U.S. Department of Agriculture (USDA). 2005. [Natural color aerial photos of Contra Costa, Sacramento, San Joaquin, Solano, Yolo counties]. Ground resolution: 1m. National Agriculture Imagery Program (NAIP). Washington, DC.

U.S. Department of Agriculture (USDA). 2009. [Natural color aerial photos of Contra Costa, Sacramento, San Joaquin, Solano, Yolo counties]. Ground resolution: 1m. National Agriculture Imagery Program (NAIP). Washington, DC.

U.S. Department of the Interior. 1896. *Bulletin of the United States Geological Survey. Issue 140.*

U.S. Department of the Interior (USDI). 1973. *Manual of Surveying Instructions: For the Survey of the Public Lands of the United States.* Bureau of Land Management. Denver, CO: Government Printing Office.

U.S. District Court, Northern District. ca. 1840a. Diseño del Rancho de los Putos. Land Case Map A-507. *Courtesy of The Bancroft Library, UC Berkeley.*

U.S. District Court, Northern District. ca. 1840b. Diseño del Rancho El Molino ó Rio Ayoska, California. Land Case Map D-493. *Courtesy of The Bancroft Library, UC Berkeley.*

U.S. District Court, Northern District. ca. 1840c. Plan del terreno denominado Campo Frances al Este del Rio Sn. Joaquin y pretende Guillermo Gulnac : [Rancho Campo de los Franceses, San Joaquin Co., Calif.]. Land Case Map D-583. *Courtesy of The Bancroft Library, UC Berkeley.*

U.S. District Court, Northern District. ca. 1840d. Desneo para mapa de Ferenol[?] sobre la orilla del Rio Cosumne. [Sacramento County.] No. 182 ND. Catherine Sheldon etc Clmt. Land Case Map D-412. *Courtesy of The Bancroft Library, UC Berkeley.*

U.S. Exploring Expedition (U.S. Ex. Ex.). Sherman and Smith. 1841. Map of Sacramento River and Bay of San Pablo with harbour of San Francisco. *Courtesy of Earth Sciences & Map Library, UC Berkeley.*

U.S. Fish and Wildlife Service (USFWS). 2009. *A system for mapping riparian areas in the western United States.* U.S. Fish and Wildlife Service, Division of Habitat and Resource Conservation, Branch of Resource and Mapping Support, Arlington, VA.

U.S. Geological Survey (USGS). 1905. *W.R. McKean in tules '05.* Rogers, Hubert F. *Courtesy of Center for Sacramento History.*

U.S. Geological Survey (USGS). 1907. *McKean's Camp at Lodi.* Rogers, Hubert F. *Courtesy of Center for Sacramento History.*

U.S. Geological Survey (USGS). 1909-1918. Topographic Quadrangles, California : 7.5-minute series 1:31,680.

U.S. Geological Survey (USGS). 1998. DRG - Digital Raster Graphic USGS 7.5' Quad Images. Washington, DC. 1:24,000.

U.S. Geological Survey (USGS), 1999. *National Hydrography Dataset (NHD).*

U.S. Surveyor General's Office. 1859. Map of Township No. 7 North, Range No. 4 East (Mount Diablo Meridian). San Francisco, CA. *Courtesy of Bureau of Land Management, Sacramento.*

U.S. War Department. 1856a. *Reports of explorations and surveys, to ascertain the most practicable and economical route for a railroad from the Mississippi River to the Pacific Ocean, Volume 5, Part 2.* Government Printing Office, Washington, DC.

U.S. War Department. 1856b. *Reports of explorations and surveys, to ascertain the most practicable and economical route for a railroad from the Mississippi River to the Pacific Ocean, Volume 5, Part 3.* Government Printing Office, Washington, DC.

U.S. War Department. 1892. *Old River, California.* House of Representatives. 52nd Congress, 2nd Session. Ex. Doc. No. 18.

U.S. War Department. 1895. *Examination of San Joaquin River, California.* House of Representatives. 54th Congress, 1st Session. Ex. Doc. No. 60.

U.S. War Department. 35th Congress, House of Representatives, 1898. *Improvement of Sacramento and Feather rivers, California. Letter from the Secretary of War, transmitting, with letter from the Chief of Engineers, report on the improvement of Sacramento and Feather rivers and their tributaries. Courtesy of Water Resources Collections and Archives, Riverside.*

U.S. War Department. 1900. *Examination of San Joaquin River, California.* House of Representatives. 56nd Congress, 2nd Session. Ex. Doc. No. 69.

Unknown. U.S. District Court, Northern District. 1854. Sketch of Rancho Campo de los Franceses : San Joaquin Co., California. Land Case Map E-691. *Courtesy of The Bancroft Library, UC Berkeley*

Unknown. U.S. District Court, Northern District. 1859. Map of C. M. Weber's grant, El Rancho del Campo de los Franceses, San Joaquin County, California. Land Case Map E-689. *Courtesy of The Bancroft Library, UC Berkeley*

Unknown. 1873. *Irrigation in California: The San Joaquin and Tulare plains.* Sacramento, CA.

Unknown. 1891. *A memorial and biographical history of northern California, illustrated. Containing a history of this important section of the Pacific Coast from the earliest period of its occupancy to the present time...* Chicago, IL: The Lewis Publishing Company.

Unknown. 1915. San Joaquin, the gateway county of California. *Courtesy of Earth Sciences & Map Library, UC Berkeley.*

Unknown. California Appellate Court. 1917a. Defendants Exhibit "A," *Strecker v. Gaul.*

Unknown. 1917b. *Herman F. Strecker v. A. Gaul.* California District Court of Appeal, third appellate district.

Unknown. California Appellate Court. 1917c. Plaintiffs Exhibit 3, *Strecker v. Gaul.*

Unknown. 1918. By-pass burning tule, Kercheval levee. *Courtesy of UC Davis Shields Library Special Collections.*

Unknown. 1919. *Report on drainage system of reclamation district no. 999 - winter of 1918-1919 as indicated by conditions existing March 1, 1919. Courtesy of RD999.*

Unknown. ca. 1870. Swamp Land District No. 72. *Courtesy of California State Lands Commission, Sacramento.*

Unknown. ca. 1894a. [Duck hunting in the Delta]. *Courtesy of The Haggin Museum, Stockton.*

Unknown. ca. 1894b. Duck hunting near 'Head Reach'. *Courtesy of The Haggin Museum, Stockton.*

Unknown. ca. 1900. *[R.D. 551].* ETCH 109. *Courtesy of Water Resources Collections and Archives, Riverside.*

Upham SC. 1878. *Notes of a voyage to California via Cape Horn, together with scenes in El Dorado, in the year 1840-'50.* Philadelphia, PA. *Courtesy of Library of Congress.*

Vaca M. 1853. *U.S. v. R.L. Brown, Land Case No. 411 ND, Laguna de Santos Calle, U.S. District Court, Northern District. Courtesy of The Bancroft Library, UC Berkeley.*

Vaghti MG, Greco SE. 2007. Riparian Vegetation of the Great Valley, Chapter 16. In *Terrestrial Vegetation of California. 3rd edition*, ed. MG Barbour, T Keeler-Wolf, and AA Schoenherr, 313-338. Berkeley, Los Angeles, London: University of California Press.

van Löben Sels PJ. n.d. *[Memoirs]*. Unpublished manuscript. *Courtesy of Pam and Russell van Löben Sels.*

van Löben Sels PJ. 1902. *[Agreement concerning Reclamation district 551]*. ETCH 109. *Courtesy of Water Resources Collections and Archives, Riverside.*

van Scoyk J. 1859. *U.S. v. Anastasio Chaboya, Land Case No. 406 ND [Sanjon de los Mequelemnes]*, U.S. District Court, Northern District. 251. *Courtesy of The Bancroft Library, UC Berkeley.*

Vaught D. 2006. A swamplander's vengeance: R.S. Carey and the failure to reclaim Putah Sink, 1855-1895. *Sacramento History Journal of the Sacramento County Historical Society* VI.

Vaught D. 2007. *After the gold rugh: tarnished dreams in the Sacramento Valley.* Baltimore, MD: The Johns Hopkins University Press.

Verix, *Californian.* 1848. Correspondence of the Californian: trip through the country on the north side of the bay of San Francisco, continued. March 22, March 29. *Courtesy of California Digital Newspaper Collection.*

Viader J, Cook SF. 1960. *Colonial expeditions to the interior of California Central Valley, 1800-1820.* Berkeley, CA: University of California Press.

Vioget J. U.S. District Court, Northern District. 1854. Mapa de los terrenos para la Colonia de Nueva Helvetia: California. Land Case Map D-624. *Courtesy of The Bancroft Library, UC Berkeley*

Vizetelly. 1849. *Four months among the gold-finders, being the diary of an expedition from San Francisco to the gold districts.* Paris: A and W Galiagnani and Co. *Courtesy of Library of Congress.*

von Schmidt AW. U.S. Department of the Interior, Bureau of Land Management Rectangular Survey, California, 1854-5. *Field notes of the subdivision and meander lines of Township 3 South Range 7 East, Mount Diablo Meridian.* 99-26. *Courtesy of Bureau of Land Management, Sacramento.*

von Schmidt AW. U.S. Department of the Interior, Bureau of Land Management Rectangular Survey, California, 1855. *Field notes of the 1st standard lines of Township 1-2-3-4-5 South Range 6-7-8, East Mount Diablo Meridian.* 99-5. *Courtesy of Bureau of Land Management, Sacramento.*

von Schmidt AW. U.S. Department of the Interior, Bureau of Land Management Rectangular Survey, California, 1858a. *Field notes of meanders and a portion of subdivisions of township 9 north, range 3 east of the Mt. Diablo base and meridian in the state of California.* Book 503-3. *Courtesy of Bureau of Land Management, Sacramento.*

von Schmidt AW. U.S. Department of the Interior, Bureau of Land Management Rectangular Survey, California, 1858b. *Field notes of the survey of Sanjon de los Moquelumnes.* Book 506-16. *Courtesy of Bureau of Land Management, Sacramento.*

von Schmidt AW. U.S. District Court, Northern District. 1859. Rancho San-Jon de los Moquelumnes, California. Land Case Map F-865. *Courtesy of The Bancroft Library, UC Berkeley.*

von Schmidt AW. 1860. *The United States, appellants, v. John A. Sutter. No. 135. Appeal from the District Court U.S. for the Northern District of California.* The Supreme Court of the United States.

Wackenreuder V. U.S. Department of the Interior, Bureau of Land Management Rectangular Survey, California, 1875. *Field notes of the subdivision lines of townships 2 N R. 2 E and T. 3 S R. 2 W, Mt. Diablo Meridian, California.* Book 291-3. *Courtesy of Bureau of Land Management, Sacramento.*

Wadsworth HH. U.S. Engineer Office. 1908a. Map of the Sacramento River: from the mouth of Feather River to Suisun Bay at Collinsville. San Francisco, CA. 1:4,800 *Courtesy of the California State Lands Commission, Sacramento.*

Wadsworth HH. U.S. Engineer Office. 1908b. Map of San Joaquin River, California: from Stockton to Suisun Bay at Collinsville. San Francisco, CA. 1:4,800 *Courtesy of California State Lands Commission, Sacramento.*

Wallace J. U.S. Department of the Interior, Bureau of Land Management Rectangular Survey, California, 1865a. *Field notes of subdivision lines of Township 3 North Range 6 East, Mount Diablo Meridian, California.* 85-13. *Courtesy of Bureau of Land Management, Sacramento.*

Wallace J. U.S. Department of the Interior, Bureau of Land Management Rectangular Survey, California, 1865b. *Field notes of the subdivisions in fractional Township 2 North, Range 6 East Mount Diablo Meridian, California.* Book 245-1. *Courtesy of Bureau of Land Management, Sacramento.*

Wallace J. U.S. Department of the Interior, Bureau of Land Management Rectangular Survey, California, 1869. *Field notes of exterior and subdivision lines of Township 5 North Range 5 East, Mount Diablo Meridian, California.* 292-1. *Courtesy of Bureau of Land Management, Sacramento.*

Wallace J. 1870. Map of the County of San Joaquin compiled from United State Surveys. *Courtesy of California State Lands Commission, Sacramento.*

Wallace J. 1876. Map of Swamp Land District No. 282, San Joaquin County. *Courtesy of California State Lands Commission, Sacramento.*

Watson WS. 1859a. Survey of the Mokelumne River, at the site of Rancho Sanjon de los Moquelumnes, California. Land Case Map E-864. *Courtesy of The Bancroft Library, UC Berkeley.*

Watson WS. 1859b. *U.S. v. Anastasio Chaboya, Land Case No. 406 ND [Sanjon de los Mequelemnes], U.S. District Court, Northern District.* 269. *Courtesy of The Bancroft Library, UC Berkeley.*

Weinstein MP, Kreeger DA, editors, 2000. *Concepts and cotnroversies in tidal marsh ecology.* Dordrecht: Kluwer Academic Publishers.

Wells AJ. 1909. *The Sacramento Valley of California.* San Francisco, CA: Southern Pacific Co.

West JG. 1977. *Late Holocene Vegetation History of the Sacramento-San Joaquin Delta, California.* Cultural Heritage Section, California Department of Parks and Recreation.

Wheeler AS. 1920. Map of the Liberty Reclamation District. *Courtesy of Solano County Surveyors Office.*

Whipple AA, Grossinger RM, Davis FW. 2011. Shifting baselines in a California oak savanna: nineteenth century data to inform restoration scenarios. *Restoration Ecology* 19(101):88-101.

Whitcher JE. United States District Court, Northern District. 1853a. [Map of the Rancho Los Meganos: Calif.] Land Case Map F-250. [ca. 1:31,680] *Courtesy of The Bancroft Library, UC Berkeley.*

Whitcher JE. 1853b. *U.S. v. John Marsh, Land Case No. 107 ND [Los Meganos], U.S. District Court, Northern District.* 17. *Courtesy of The Bancroft Library, UC Berkeley.*

Whitcher JE. 1857a. *Field notes of the final survey of the Rancho El Pescadero, Hiram Grimes, et al. confirmee.* Book C27/G3. *Courtesy of Bureau of Land Management, Sacramento.*

Whitcher JE. U.S. Surveyor General. 1857b. Plat of the Rancho El Pescadero. *Courtesy of Bureau of Land Management, Sacramento.*

White CA. 1983. *A history of the rectangular survey system.* Washington, DC: U.S. Government Printing Office.

Whiting JS. 1854. *U.S. v. Charles Weber, Land Case No. 298 ND [Campo de Los Franceses Grant], U.S. District Court, Northern District.* 71. *Courtesy of The Bancroft Library, UC Berkeley.*

Whitlow TH, Harris RW, Lieser AT. 1984. Experimenting with levee vegetation: Some unexpected findings. In *California riparian systems: ecology conservation and productive management*, ed. RE Warner and KM Hendrix, 558-565. Berkeley, CA: University of California Press.

Wiens J, Moss M, editors, 2005. *Issues and perspectives in landscape ecology.* Cambridge, UK: Cambridge University Press.

Wilkes C. 1845. *Narrative of the United States Exploring Expedition during the years 1838,1839, 1840, 1841, 1842.* London: Wiley and Putnam.

Wilkes C. 1849. Harbours - California. In *Western America, including California and Oregon, with maps of those regions, and of "The Sacramento Valley" from actual surveys*, ed.: Lea and Blanchard.

Williams EE. 1973. Tales of Old San Joaquin City. *San Joaquin Historian* 9(3).

Williamson RS. 1856. *Reports of explorations and surveys, to ascertain the most practicable and economical route for a railroad from the Mississippi River to the Pacific Ocean: Geological Report.* House of Rep, 33d Congress, 2d Session, Ex. Doc. No. 91.

Williamson RS. 1857. *Reports of explorations and surveys, to ascertain the most practicable and economical route for a railroad from the Mississippi River to the Pacific Ocean: Botanical Report.* House of Rep, 33d Congress, 2d Session, Ex. Doc. No. 91.

Wood EL. 1941. *George Yount: the kindly host of Caymus Rancho.* Grabhorn Press.

Woodruff DS. 1865. *U.S. v. Jonathon Stevenson et al., Land Case No. 364 ND [Los Medanos], U.S. District Court, Northern District.* 519. Courtesy of The Bancroft Library, UC Berkeley.

Wright SA, Schoellhamer DH. 2004. Trends in the sediment yield of the Sacramento River, California, 1957-2001. *San Francisco Estuary and Watershed Science* 2(2).

Wright SA, Schoellhamer DH. 2005. Estimating sediment budgets at the interface between rivers and estuaries with application to the Sacramento-San Joaquin River Delta. *Water Resources Research* 41:1-17.

Wright W. ca. 1850a. *Lost in the tule marshes. Courtesy of California Historical Society, San Francisco.*

Wright W. ca. 1850b. *Hunting for market. Courtesy of California Historical Society, San Francisco.*

Yoshiyama RM, Fisher FW, Moyle PB. 1998. Historical abundance and decline of chinook salmon in the Central Valley region of California. *North American Journal of Fisheries Management* 18:487-521.

Young JD. California Department of Engineering, 1880. *Report of the state engineer to the Legislature of the State of California.* 2. 15-17.

Personal Communications

Anderson, Kat. September 2010.

Atwater, Brian. 2010-2012.

Burmester, Daniel. May 2012.

Collins, Brian. September 2011.

Collins, Josh. 2009-2012.

Enright, Christopher. November 2009.

Fleenor, William. June 2011, July 2012.

Hoppe, Walter. July 2010.

Keeler-Wolf, Todd, 2011.

Mount, Jeffrey. June 2011.

Moyle, Peter B. June 2011.

Soares, John. May 1995.

Thayer, Christopher H. October 2011.

Thompson, John. 2010.

van Löben Sels, Pam. July 2011.

van Löben Sels, Russell. July 2011.

Windham-Myers, Lisamarie. September 2011.

www.ingramcontent.com/pod-product-compliance
Lightning Source LLC
Chambersburg PA
CBHW041724210326
41598CB00008B/772